ADVANCED MODELING IN COMPUTATIONAL ELECTROMAGNETIC COMPATIBILITY

BICENTENNIAL 1807 WILEY 2007 BICENTENNIAL

THE WILEY BICENTENNIAL–KNOWLEDGE FOR GENERATIONS

Each generation has its unique needs and aspirations. When Charles Wiley first opened his small printing shop in lower Manhattan in 1807, it was a generation of boundless potential searching for an identity. And we were there, helping to define a new American literary tradition. Over half a century later, in the midst of the Second Industrial Revolution, it was a generation focused on building the future. Once again, we were there, supplying the critical scientific, technical, and engineering knowledge that helped frame the world. Throughout the 20th Century, and into the new millennium, nations began to reach out beyond their own borders and a new international community was born. Wiley was there, expanding its operations around the world to enable a global exchange of ideas, opinions, and know-how.

For 200 years, Wiley has been an integral part of each generation's journey, enabling the flow of information and understanding necessary to meet their needs and fulfill their aspirations. Today, bold new technologies are changing the way we live and learn. Wiley will be there, providing you the must-have knowledge you need to imagine new worlds, new possibilities, and new opportunities.

Generations come and go, but you can always count on Wiley to provide you the knowledge you need, when and where you need it!

WILLIAM J. PESCE
PRESIDENT AND CHIEF EXECUTIVE OFFICER

PETER BOOTH WILEY
CHAIRMAN OF THE BOARD

ADVANCED MODELING IN COMPUTATIONAL ELECTROMAGNETIC COMPATIBILITY

DRAGAN POLJAK, PhD
Department of Electronics
University of Split, Croatia

BICENTENNIAL 1807 WILEY 2007 BICENTENNIAL

WILEY-INTERSCIENCE
A JOHN WILEY & SONS, INC., PUBLICATION

Published by John Wiley & Sons, Inc., Hoboken, New Jersey.
Published simultaneously in Canada.

Wiley Bicentennial Logo: Richard J. Pacifico

Library of Congress Cataloging-in-Publication Data:

Poljak, D. (Dragan)
Advanced modeling in computational electromagnetic compatibility / by Dragan Poljak.
p. cm.
Includes bibliographical references.
ISBN: 978-0-470-03665-5
1. Electromagnetic compatibility–Mathematical models. 2. Electromagnetic compatibility–Data processing. I. Title.
TK7867.2.P65 2007
621.382'24–dc22 2006027628

Printed in the United States of America.
10 9 8 7 6 5 4 3 2 1

To my beloved wife, my daughter and my sister

CONTENTS

PREFACE

Electromagnetic compatibility (EMC) is the applied discipline within the science of electromagnetism including almost all relevant areas of theoretical (computational) and experimental electromagnetics. Theoretical methods in electromagnetics can be classified as analytical or numerical, and this book is strictly related to the numerical methods in EMC.

Computational models in EMC provide the foundation for numerous applications in both research and industry purposes. A powerful computer is an everyday tool of engineers and researchers and it is expected to become more powerful, allowing for widespread modeling of EMC problems. Computational models have become more and more important, especially when applied to problems that are not easily handled with experimental methods, like human exposure to electromagnetic fields. This book aims to provide researchers, postgraduate students, and professionals to approach computational models in EMC.

Though many significant strides have been made in EMC modeling in the past few decades, progress in this research topic is expected to continue at a rapid pace. There are a number of books on EMC already published which can be selected in several categories. Some books written on the subject are handbooks or general introductions to the subject, while others are focused to a particular topic where authors are experts in the field, or are practical and knowledgeable authors. Nevertheless, it is hard to find a book which gives a good overview of the subject and, at the same time, efficiently covers some advanced EMC topics.

My impression is that the book by Tesche et al. entitled *EMC Analysis Methods and Computational Models* published in 1997 maybe did the best job in describing and illustrating various modeling techniques applicable to a wide area of EMC. This excellent-organized book built analysis models from the very first principles of electromagnetic theory and describes its use for practical problems in EMC, thus being a very useful and instructive textbook for advanced undergraduate and graduate courses not only in EMC, but also in applied electromagnetic theory.

Anyway, my point is that most of the general books on EMC did a good job with introductory topics, but then do not cover some advanced topics on EMC modeling or important applications, while more advanced books covering specific topics are usually too specialized on a particular subject and often omit important topics that are outside of the scope of the problem of interest.

So, a decade after Tesche et al. book, as an academic person who has been tackling both research and practical engineering problems, I tried to write a book with the intention to do both, that is, to cover a variety of EMC problems, from introductory to specialized topics, and to address researchers and engineers who are dealing with modeling and applications.

The book starts providing a crash course in fundamentals of electromagnetic theory and numerical modeling, then covers a frequency and time domain wire antenna analysis, and finally deals with modeling of a broad range of EMC problems of interest.

As a researcher who came from *Antennas and Propagation Society* into the EMC community, I had a strong background in antenna theory and numerical methods which both have provided me with the most powerful modeling tools in the electromagnetic theory to attack a variety of typical EMC problems like lightning, transmission lines, grounding systems, and even human exposure to electromagnetic fields. It is worth mentioning that many of these topics have been traditionally treated by quasistatic solution methods.

The genesis of this book has been based on valuable experience I got from working on my previous books on wire antenna analysis, numerical modeling, and human exposure to electromagnetic radiation, all published by WIT Press. Many concepts used in this book have been clarified through presenting many postgraduate courses on different topics in electromagnetics at the Department of Electronics, at the University of Split, Croatia and Wessex Institute of Technology, UK.

So to say, the basic viewpoint in writing strategy of the book is concerned with figuring out how to approach and solve EMC problems numerically by using originally developed antenna models.

I hope that the present material will help readers to handle various EMC problems, to develop their own EMC computational models in research and in industry applications, and also to better understand numerical methods developed and used by other researchers and engineers not only in EMC but also in other areas of electrical engineering.

The book is divided in three parts. The first part deals with introductory topics in EMC, namely, it is concerned with the fundamentals of electromagnetic theory, basics in numerical modeling, and simple computational models in the analysis of static, quasistatic, and scattering problems. All these models are based on integral equation formulation and related boundary element analysis.

As in many cases electromagnetic interferences are radiative phenomena in nature. The second part of the book deals with an analysis of the wire antennas using the frequency domain (FD) and the time domain (TD) integral equation formulation, respectively.

The analysis of radiation and scattering from thin wires is an important issue in antenna theory and applications but is also very convenient for testing the newly developed numerical techniques. Moreover, the wire antenna models may have a number of applications in the area of EMC. The analysis of wire antennas in frequency and time domain using numerical methods is presented through Chapters 7 to 9.

The art of computational electromagnetics has significantly grown in the last decade, particularly due to the availability of highly powerful computer resources having released (yielded) some widely used numerical methods such as finite difference method (FDM), finite element method (FEM), method of moments (MoM), and the boundary element method (BEM).

What I mostly used in this book is the originally developed specific scheme of the Galerkin–Bubnov boundary element method (GB-BEM), also entitled as the finite element integral equation method (FEIEM), which is presented in details through the both frequency and time domain procedures. Why BEM? To the best of my knowledge BEM represents the most suitable numerical method for handling many EMC problems involving analysis of wire configuration of arbitrary shape. There are many reasons for that, but the most important ones can be summarized in the text to follow.

In the last decade BEM has become a well-established technique widely used for solving a variety of problems in electromagnetics. BEM can be referred to as a combination of classical integral equation methods and finite element discretization concepts. This method provides a high degree of accuracy associated with integral equations in conjunction with the versatility of finite element type meshes, with the added advantage that the discretization is only required over the surface of the region or boundary under consideration. Therefore, BEM is a sophisticated tool that can be efficiently applied to solve a variety of problems for which finite elements become inefficient. Such is the case in many applications in electromagnetics where the functions present high gradients or they may be singular. Furthermore, many of these problems extend to infinity, a case for which boundary elements is particularly well adapted.

Finally, the third part of the book deals with the solution of some specific EMC problems by means of the wire antenna theory presented in Part II. The applications of antenna models are related to aboveground and belowground transmission lines, respectively, and grounding systems and the interaction of the human body with the electromagnetic radiation. Chapters 10 to 12 deals with aboveground and belowground cables, respectively, while the analysis of vertical and horizontal electrodes, respectively, is undertaken in Chapter 13.

As the presence of electromagnetic fields in the environment has recently been associated with the controversy on possible adverse health effects on humans which generally depend on their strength and frequency, the last chapter of the third part of the book also contains the study on human exposure to low frequency and transient electromagnetic fields, respectively.

The book contains several numerical examples pertaining to academic and real world problems.

What was borne in my mind while writing this book was that, in spite of the fact that the computational EMC has been growing rapidly in the last two decades, there has still been a certain lack of suitable texts on the manifold aspects of EMC.

I would like to believe that this work represents a sort of modest attempt to fill this gap and could be useful to academic researchers and engineers from industry working in this area.

To summarize, I am really pleased to state that working on this particular material, as always when I write a book on electromagnetics, has been a kind of exciting adventure and a pleasure for me. As a matter of fact, this book results from more than 15 years of continuing work in the area of wire antennas and related EMC applications, not only by myself but also by some of my colleagues from Department of Electronics, University of Split, Croatia and Wessex Institute of Technology, UK. Still, there is a number of challenging problems to work out in future.

DRAGAN POLJAK

Split, Croatia-Southampton, UK
Summer 2006

The author would like to thank his PhD students and friends Clapton and Damir for great help in preparing the camera ready copy of the manuscript.

PART I

FUNDAMENTAL CONCEPTS IN COMPUTATIONAL ELECTROMAGNETIC COMPATIBILITY

The scientific man does not aim at an immediate result. He does not expect that his advanced ideas will be readily taken up. His work is like that of a planner for the future. His duty is to lay foundation of those who are to come and point the way.

Nikola Tesla

1

INTRODUCTION TO COMPUTATIONAL ELECTROMAGNETICS AND ELECTROMAGNETIC COMPATIBILITY

This chapter deals with some introductory considerations related to the two basic topics of this book: computational electromagnetics (CEM) and electromagnetic compatibility (EMC). Specific topics on CEM and EMC topics have been dealt throughout this book.

1.1 HISTORICAL NOTE ON MODELING IN ELECTROMAGNETICS

Electromagnetics as a rigorous theory started when James Clerk Maxwell derived his celebrated four equations and published this work in the famous treatise in 1865 [1]. In addition to Maxwell's equations themselves, relating the behavior of electromagnetic fields and sources, several other physical relationships are necessary for their solution. The most important are Ohm's law, the equation of continuity, and the constitutive relations of the medium and the imposed boundary conditions of the physical problem of interest. Before Maxwell, the science of electromagnetism had existed mostly as an experimental discipline for several centuries through the works of scientists such as Benjamin Franklin, Charles Augustin de Coulomb, André Marie Ampère, Hans Christian Oersted, and Michael Faraday. The early doubt about Maxwell's theory vanished in 1888 when Heinrich Hertz transmitted and received radio waves, thus having demonstrated the validity of the Maxwell theory.

The early works on analytical solution methods in electromagnetics, based on Maxwell's equations, were mainly focused on the area of radio science. Some of such applications of the electromagnetic theory started to appear not long after Maxwell's treatise had appeared. Among the analyzed simple geometries were the fields radiated from the Hertzian dipole, an infinitely long straight circular wire and two coaxial cones [2]. In most of the cases, the equations were solved as boundary value problems having yielded to the solution in terms of infinite series expansions.

Analytical methods in electromagnetics were unable to handle the practical engineering problems until computational electromagnetics came along. This was brought on by the advent of the digital computer in the 1960s. While much effort of the early research based on analytical methods was focused on the study of antennas in radio science, the emergence of computational electromagnetics opened up a number of new areas of applications. At the beginning, the emphasis was on high frequency radar systems related to defense. This was mainly due to the concentration of available research funds for such a work during that time. Indeed, in what is reported as one of the earliest cases of solving electromagnetic problems on a digital computer in the 1950s, the machine was built prompted by the need for calculating artillery ballistic tables [3].

Since that time, a continuing advancement in computational methods has been prompted by the rapid progress in computer hardware. Frequency domain integral equation techniques having simpler mathematical framework than their time domain counterpart started to appear in the mid 1960s [4–6]. One of the first digital computer solution of the Pocklington's equation was reported in 1965 [7]. This was followed by one of the first implementations of the finite difference method (FDM) to the solution of partial differential equations in 1966 [8] and time domain integral equation formulations in 1968 [9,10] and 1973 [11,12]. Through 1970s, the finite element method (FEM) became widely used in almost all areas of applied electromagnetism encompassing power engineering and electronics applications, microwaves, antennas and propagation, and electromagnetic compatibility. Good review of many important FEM applications in electrical engineering and electronics till date has been presented in Ref. [13]. The boundary element method (BEM) developed in the late seventies for the purposes of civil and mechanical engineering [14] started to be used in electromagnetics in 1980s [15,16]. Nevertheless, there have been many applications of BEM in electromagnetics; the primer of BEM for electrical engineers appeared quite recently [17].

With up to a dozen or more computational methods are nowadays commonly in use for electromagnetic modeling purposes, there is no particular method which can claim superiority over the whole range of applications [18]. The physical and mathematical base on which specific method has been built often gives it some advantage when dealing with a particular class of problem. Nevertheless, the integral equation formulation handled via the *method of moments* (MoM) with its wide application and versatility is accepted by many researchers to be a sort of "workhorse" in computational electromagnetics [19]. An excellent review of the numerical methods used in computational electromagnetics has been given in paper by Miller in 1998 [20]. Among many others, a rather comprehensive textbook on numerical methods in electromagnetics is the one by Sadiku [21], whereas a relevant review of theoretical models and computational methods used in electromagnetic compatibility is available in Ref. [22].

Analysis of wire antennas and scatterers and related EMC applications by using the integral equation approach has been presented in Ref. [23]. Direct time domain techniques for the solution of certain classes of EMC problems have been documented in Refs. [24, 25].

The present book features both frequency and time domain integral equation approach to the analysis of various problems arising in electromagnetic compatibility by using some direct and indirect schemes of the BEM.

Topics of particular interest covered in this book are related to wire configurations, that is, antennas, transmission lines, and grounding systems. The last chapter is devoted to human exposure to extremely low frequency (ELF) and transient exposures.

1.2 ELECTROMAGNETIC COMPATIBILITY AND ELECTROMAGNETIC INTERFERENCE

EMC is usually regarded as the ability of a device to function satisfactorily within its electromagnetic environment, that is a device, system, or equipment is assumed not only to be unaffected by external fields but also not to cause interference in sense of intolerable electromagnetic disturbances to a nearby system or anything in that environment. Satisfactory operation of a device, equipment, or system implies their functional work and immunity to certain interference levels which can be regarded as normal in the environment even under these circumstances.

Therefore, the principal task of EMC is to suppress any kind of electromagnetic interference (EMI). The first request is often regarded as *immunity testing*, that is, once the device is constructed it is necessary to check if it can be a potential victim of EMI or if it satisfies the EMC request of being unaffected by an external source produced by its electromagnetic environment.

The second request raised in the design process that device is not a potential EMI source that is its normal operation does not interfere with other electrical systems, is referred to as the *emission testing*.

In a theoretical sense, the aspects of immunity and emission testing, respectively, are related through the reciprocity theorem in electromagnetics [22, 23].

Generally, the methods used in EMC are not only to visualize electromagnetic phenomena but also to predict and suppress interferences can be regarded as either theoretical or experimental. This book is strictly oriented to the consideration of theoretical approach and related computational models in the analysis of EMC problems.

1.2.1 EMC Computational Models and Solution Methods

It is the rapid progress in the development of digital computers that has provided advances in EMC computational models in last few decades.

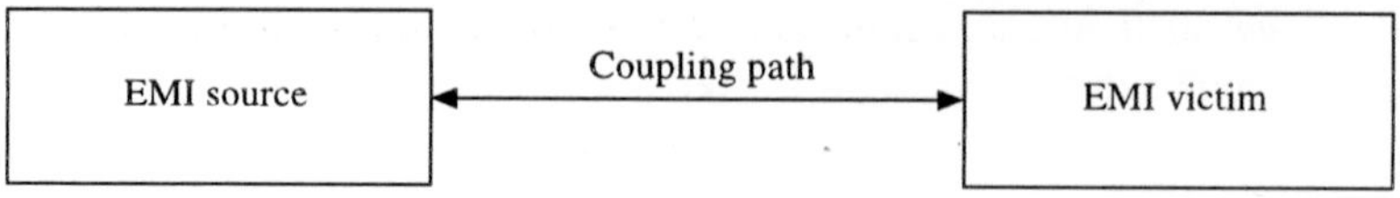

Figure 1.1 A basic EMC model.

Electromagnetic modeling provides the simulation of an electrical system electromagnetic behavior for a rather wide variety of parameters including different initial and boundary conditions, excitation types, and different configuration of the system itself. The important fact is that modeling can be undertaken within a significantly shorter time than it would be necessary for building and testing the appropriate prototype via experimental procedures. The basic purpose of an EMC computational model is to predict a victim response to the external excitation generated by a certain EMI source.

A basic EMC model, Figure 1.1, includes EMI source (radio transmitter, mobile phone, lightning strike, or any kind of undesired electromagnetic pulse (EMP), coupling path which is related to electromagnetic fields propagating in free space, material medium, or conductors, and finally, EMI victim that represents any kind of electrical equipment (e.g., radio-receiver), medical electronic equipment (e.g., pacemaker), or even the human body itself.

It is worth mentioning that the cost of EMC analysis at the design level represents up to 7% of the total product cost [22]. On the contrary, if a prototype is already produced, the subsequent incorporation of EMC measures can increase the cost of the product up to 50% of the total cost [22].

In principle, all EMC models arise from the rigorous electromagnetic theory concepts and foundations are based on Maxwell equations. The governing equations of a particular problem in differential, integral, or integro-differential form can be readily derived from the four Maxwell equations.

EMC models are analyzed using either analytical or numerical methods. Though both approaches can be used in the design of the electrical systems, analytical models are not useful for accurate simulation of electric systems or their use is restricted to the solution of rather simplified geometries with a high degree of symmetry (canonical problems).

On the contrary, a more accurate simulation of various practical engineering problems is possible by the use of numerical methods. In this case errors are primarily related to the limitations of mathematical model itself and the applied numerical method of solution, respectively.

In general, analytical and numerical techniques are used for a wide variety of purposes such as

- predicting system-level responses to EMI sources,
- evaluating the behavior of EMC protection measures,
- processing measures system test data.

The development of analytical and numerical modeling techniques has had a marked impact on the area of EMC. These techniques are used in the design, construction, test, and evaluation phases of Ref. [22]:

- defense electronic systems,
- communications and data transmission systems,
- power utilities,
- consumer electronics.

EMC computational models can be validated via experimental measurements or theoretically comparing the results to already well-established numerical models.

It is also possible to test a new model on some standard benchmark problems or on some canonical problems for which the closed-form solution is available.

1.2.2 Classification of EMC Models

There are many possible classifications of EMC computational models used in research and practical purposes [22–26].

Regarding underlying theoretical background EMC models can be classified as

- circuit theory models featuring the concentrated electrical parameters,
- transmission line models using distributed parameters in which low frequency electromagnetic field coupling are taken into account,
- models based on the full-wave approach taking into account radiation effects for the treatment of electromagnetic wave propagation problems

It is worth emphasizing that this book deals mostly with the full-wave EMC models based on the thin-wire antenna (scattering) theory.

Furthermore, taking into account different character of EMI sources, EMC problems [22–26]can be classified as

- continuous wave (CW) problems,
- transient phenomena.

The most general classification of EMI sources is the one related to natural and artificial (man-made) sources, respectively. The natural EMI source most commonly being analyzed is lightning. Regarding coupling path, EMI can be divided in two groups:

- conducted disturbances (induced overvoltages, voltage dips, switching, harmonics),
- radiated disturbances (lightning induced voltages, antenna radiation, crosstalk).

1.2.3 Summary Remarks on EMC Modeling

To reliably separate EMI source from its victim, regulations and standards become necessary. Standards set limits defining the acceptable and plausible (reasonable) level of susceptibility and provide individual testing of equipment.

EMI measurements and calculations carried out are related to

- radiated and conducted emission,
- radiated and conducted susceptibility.

Physical phenomena that represent EMI source, EMI victim, and coupling path between an EMI source and susceptible device can be modeled to a certain degree. The most important question is the level of the accuracy achieved within a given model. The main limits to EMC modeling arise from the physical complexity of the considered electric system. Sometimes even the electrical properties of the system are too difficult to determine or the number of independent parameters necessary for building a valid EMC model is too large for a practical computer code to handle.

The EMC modeling approach presented in this book is based on integral equation formulations in the frequency and time domain and related boundary element method of solution featuring the direct and indirect approach, respectively.

This approach is preferred over a partial differential equation formulations and related numerical methods of solution, as the integral equation approach is based on the corresponding fundamental solution of the linear operator and, therefore, provides more accurate results. This higher accuracy level is paid with more complex formulation than is required within the framework of the partial differential equation approach and related computational cost.

1.3 REFERENCES

[1] J. C. Maxwell: A Treatise on Electricity and Magnetism, *Oxford University Press*, 1865, 1904 (also, *Dover, New York*, 1954, republication of the 3rd ed., *Clarendon Press*, 1891).

[2] J. A. Aharoni: An Introduction to Their Theory, *Oxford University Press*, 1946.

[3] R. C. Hansen: Early computational electromagnetics, *IEEE Antennas Propagat. Mag., Vol. 38, No. 3*, June 1996, pp 60–61.

[4] M. G. Andreasen: Scattering from parallel metallic cylinders with arbitrary cross-section, *IEEE Trans. Antennas Propagat., Vol. 12*, November 1964, pp 746–754.

[5] M. G. Andreasen: Scattering from bodies of revolution, *IEEE Trans. Antennas Propagat., Vol. 13*, March 1965, p 303.

[6] K. K. Mei: On the integral equation of thin wire antennas, *IEEE Trans. Antennas Propagat. Vol. 13*, 1965, pp 59–62.

[7] J. H. Richmond: Digital computer solutions of the rigorous equations for scattering problems, *Proc. IEEE, Vol. 53*, August 1965, pp 796–804.

[8] K. S. Yee: Numerical solution of initial boundary value problems involving Maxwell's equations in isotropic media, *IEEE Trans. Antennas Propagat., Vol. 14*, May 1966, pp 302–307.

[9] C. L. Bennett, W. L. Weeks: Electromagnetic pulse response of cylindrical scatterers, *Int. Antennas and Propogat. Symp.*, Boston, MA, September 1968.

[10] E. P. Sayre, R. F. Harrington: Transient response of straight wire scatterers and antennas, *Int. Antennas and Propogat. Symp.*, Boston, MA, September 1968.

[11] E. K. Miller, A. J. Poggio, G. J. Burke: An integro-differential equation technique for the time-domain analysis of thin wire structures—I. The numerical method, *J. Comput. Phys., Vol. 12, No. 1*, May 1973, pp 24–48.

[12] E. K. Miller, A. J. Poggio, G. J. Burke: An integro-differential equation technique for the time-domain analysis of thin wire structures—II. Numerical results, *J. Comput. Phys., Vol. 12, No. 2*, June 1973, pp 210–233.

[13] P. P. Silvester, R. L. Ferrari: Finite Elements for Electrical Engineers, 3rd ed., *Cambridge University Press, Cambridge*, 1996.

[14] C. A. Brebbia: Boundary Element Methods for Engineers, *Pentech Press, London*, 1978.

[15] K. I. Yashiro, S. Ohkama: Application of Boundary Element Method to Electromagnetic Field Problems, *IEEE Trans. Microwave Theory and Tec., Vol. 32, No. 4*, April 1984, pp 455–461.

[16] K. I. Yashiro, S. Ohkama: Boundary Element Method for Electromagnetic Scattering from Cylinders, *IEEE Trans. Antennas Propagat., Vol. 33, No. 4*, April 1985, pp 383–389.

[17] D. Poljak, C. A. Brebbia: Boundary Element Methods for Electrical Engineers, *WIT Press, Southampton-Boston*, 2005.

[18] P. R. Foster: CAD for antenna systems, *IEE Electronics & Commun. Eng. J., Vol. 12, No. 1*, February 2000, pp 3–14.

[19] J. L. Volakis, L. C. Kempel: Electromagnetics: computational methods and considerations, *IEEE Computational Sci. & Eng.*, 1995, pp 42–57.

[20] E. K. Miller: A selective survey of computational electromagnetics, *IEEE Trans. Antennas Propagat., Vol. 36, No. 9*, September 1988, pp 1281–1305.

[21] M. N. O. Sadiku: Numerical Techniques in Electromagnetics, 2nd ed., *CRC Press, Boca Raton*, 2001.

[22] F. Tesche, M. Ianoz, T. Karlsson: EMC Analysis Methods and Computational Models, *John Wiley & Sons, New York*, 1997.

[23] D. Poljak: Electromagnetic Modeling of Wire Antenna Structures, *WIT Press, Southampton-Boston*, 2002.

[24] D. Poljak, C. Y. Tham: Integral Equation Techniques in Transient Electromagnetics, *WIT Press, Southampton-Boston*, 2003.

[25] D. Poljak (Ed.): Time Domain Techniques in Computational Electromagnetics, *WIT Press, Southampton-Boston*, 2004.

[26] F. J. K. Buesink: *Engineering Electromagnetic Compatibility, EMV 99, Workshop Notes, Dusseldorf, Germany*, March 1999.

2

FUNDAMENTALS OF ELECTROMAGNETIC THEORY

The electromagnetic field laws can be expressed very concisely by a single set of four differential equations. These were evolved through the efforts of many prominent scientists, mainly during the nineteenth century, and were finally cast into their now well-known form as Maxwell equations. There is also an equivalent integral form of these equations. The differential form of Maxwell equations is used most frequently in solving problems. However, the integral form of these equations is important as it better explains the underlying physical law.

2.1 DIFFERENTIAL FORM OF MAXWELL EQUATIONS

Time-varying electromagnetic fields can be represented by the following variables:

- E is the electric field intensity,
- H is the magnetic field intensity,
- B is the magnetic flux density,
- D is the electric flux density,
- J is the current density,
- ρ is the charge density,
- φ is the electric scalar potential,
- A is the magnetic vector potential;

and these fields are governed by physical laws expressed mathematically by four Maxwell equations.

The *first Maxwell equation* is the differential form of Faraday's law (The time-changing magnetic flux density $\vec{B}$ causes the curl of electric field $\vec{E}$)

$$\nabla x\vec{E} = -\frac{\partial \vec{B}}{\partial t} \tag{2.1}$$

An important consequence of Faraday's law is that no magnetic poles can exist, as the divergence of the above equation is identically zero. Hence, the time-varying magnetic fields are vortex sources of electric fields.

The *second Maxwell equation* is the differential form of Ampere's law stating that either an electric current density $\vec{J}$ or a time-varying electric flux density $\vec{D}$ gives rise to a magnetic field $\vec{H}$. This can be expressed as

$$\nabla x\vec{H} = \vec{J} + \frac{\partial \vec{D}}{\partial t} \tag{2.2}$$

It is worth noting that the term $\partial\vec{D}/\partial t$ was added by Maxwell to the original expression of Ampere's law to make the law consistent with the conservation of electric charge.

The *third Maxwell equation* states that electric monopoles exist, so that

$$\nabla\vec{D} = \rho \tag{2.3}$$

That is, charge densities ρ are the sources of the electric field.

Finally, the *fourth Maxwell equation* states that all magnetic poles occur in pairs and are due to electric currents; no free poles can exist in the electromagnetic field theory. This is expressed by the divergence Maxwell equation

$$\nabla\vec{B} = 0 \tag{2.4}$$

which implies that the magnetic field is always solenoidal.

2.2 INTEGRAL FORM OF MAXWELL EQUATIONS

Integral formulation of the Faraday's law states that any change of flux of magnetic induction B through any closed loop, as shown in Figure 2.1, induces an electromotive force around the loop. The surface integration over the first Maxwell eqn (2.1) results in a following expression:

$$\int_S \nabla x\vec{E}d\vec{S} = -\int_S \frac{\partial \vec{B}}{\partial t} d\vec{S} \tag{2.5}$$

where $d\vec{S} = \vec{n}dS$.

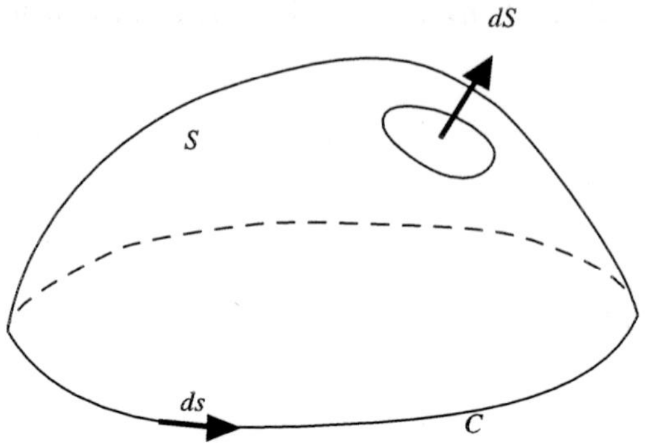

Figure 2.1 The surface S with related contour C.

The surface integral is taken over any surface S bounded by the loop c. Applying the Stokes theorem to the left-hand side of eqn (2.5), one obtains

$$\int_S \nabla x\vec{E}d\vec{S} = \oint_c \vec{E}d\vec{s} \tag{2.6}$$

where the line integral is taken around the loop and the integral form of the first Maxwell equation is given by

$$e = -\frac{\partial \phi}{\partial t} \tag{2.7}$$

where e is an electromotive force (EMF):

$$e = \oint_c \vec{E}d\vec{s} \tag{2.8}$$

which is effectively a voltage around the closed path and ϕ is the magnetic flux defined as

$$\phi = \int_S \vec{B}d\vec{S} \tag{2.9}$$

The voltage induced by a varying flux has a polarity such that the induced current in a closed path gives rise to a secondary magnetic flux which opposes the change in time-varying source magnetic flux.

The integral form of the Ampere's law can be obtained by integrating the second Maxwell equation

$$\int_S \nabla x\vec{H}d\vec{S} = \int_S \left[\vec{J} + \frac{\partial \vec{D}}{\partial t}\right] d\vec{S} \tag{2.10}$$

The term $\partial\vec{D}/\partial t$ was added to the original Ampere's law to make it consistent with the electric charge conservation. This term is usually referred to as a displacement current.

Again, applying the Stokes theorem to the left-hand side integral in eqn (2.10) one obtains

$$\int_S \nabla x\vec{H}\mathrm{d}\vec{S} = \oint_c \vec{H}\mathrm{d}\vec{s} \tag{2.11}$$

The integral form of the second Maxwell equation is then given as follows:

$$\oint_c \vec{H}\mathrm{d}\vec{s} = I + \frac{\partial\psi}{\partial t} \tag{2.12}$$

where I is the electric current given by

$$I = \int_S \vec{J}\mathrm{d}\vec{S} \tag{2.13}$$

and ψ is the electric flux defined as

$$\psi = \int_S \vec{D}\mathrm{d}\vec{S} \tag{2.14}$$

Notice that eqn (2.12) is Ampere circular rule with Maxwell addition of the term $\partial\psi/\partial t$, which is called displacement current. The generalized Ampere's law states that either an electric current or a time-varying electric flux gives rise to a magnetic field.

Taking the volume integral over the third Maxwell eqn (2.3) yields

$$\int_V \nabla\vec{D}\mathrm{d}V = \int_V \rho\mathrm{d}V \tag{2.15}$$

and applying the Gauss's divergence theorem results in

$$\int_V \nabla\vec{D}\mathrm{d}V = \oint_S \vec{D}\mathrm{d}\vec{S} \tag{2.16}$$

it follows

$$\oint_S \vec{D}\mathrm{d}\vec{S} = Q \tag{2.17}$$

where Q is the total electric charge within the volume V, that is,

$$Q = \int_V \rho \mathrm{d}V \tag{2.18}$$

Equation (2.17) is the Gauss's flux law for the electric field stating that the flux of D is equal to all the sources in the domain in the form of an electric charge.

The Gauss's flux law for the magnetic field can be derived by taking the volume integral of the fourth Maxwell eqn (2.4), that is,

$$\int_V \nabla\vec{B}\mathrm{d}V = 0 \tag{2.19}$$

After applying the Gauss's divergence theorem, that is,

$$\int_V \nabla\vec{B}\mathrm{d}V = \oint_S \vec{B}\mathrm{d}\vec{S} \tag{2.20}$$

it follows

$$\oint_S \vec{B}\mathrm{d}\vec{S} = 0 \tag{2.21}$$

stating that the flux of B over any closed surface S is identically zero.

The solution of the four Maxwell equations in either differential or integral form is possible only if additional constitutive equations are available connecting D to E, J to E, and H to B such as

$$\vec{D} = \varepsilon\vec{E} \tag{2.22}$$

$$\vec{J} = \sigma\vec{E} \tag{2.23}$$

$$\vec{B} = \mu\vec{H} \tag{2.24}$$

for a linear medium, where ε is the permittivity, σ is the conductivity, and μ is the permeability of a medium, or whatever forms apply for a nonlinear medium.

2.3 MAXWELL EQUATIONS FOR MOVING MEDIA

The Maxwell equations in their differential and integral form, respectively, considered so far are valid for any stationary media. The third and fourth Maxwell

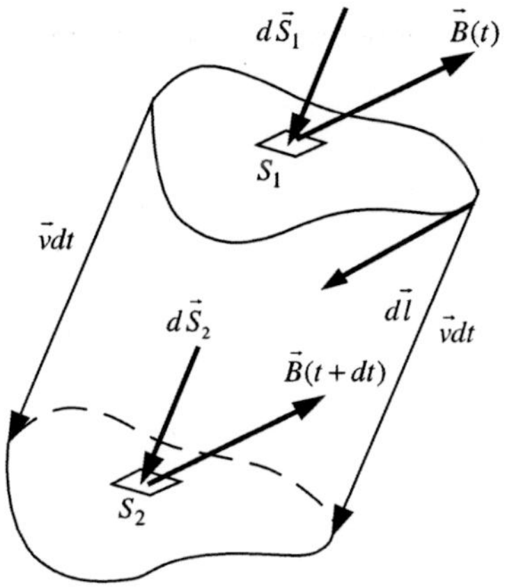

Figure 2.2 Geometry of a moving medium.

equations are not affected by the motion of the medium. On the contrary, the extension of the law of induction expressed by the first Maxwell equation, to take into account the motion of the medium, requires considerable care.

In other words, the first Maxwell equation must contain the total time rate of magnetic flux density change

$$\nabla x\vec{E} = -\frac{\mathrm{d}\vec{B}}{\mathrm{d}t} \tag{2.25}$$

The total time rate of flux changes across a given surface, when the surface itself across which the flux is evaluated is in motion, as shown in Figure 2.2 (The velocity of a moving medium is $\vec{v}$) is given by

$$\frac{\mathrm{d}\phi}{\mathrm{d}t} = \frac{\mathrm{d}}{\mathrm{d}t}\int\limits_S \vec{B}\mathrm{d}\vec{S} \tag{2.26}$$

where the time rate of magnetic flux density variation in Cartesian coordinates can be expressed as

$$\frac{\mathrm{d}\vec{B}}{\mathrm{d}t} = \frac{\partial\vec{B}}{\partial x}\frac{\mathrm{d}x}{\mathrm{d}t} + \frac{\partial\vec{B}}{\partial y}\frac{\mathrm{d}y}{\mathrm{d}t} + \frac{\partial\vec{B}}{\partial z}\frac{\mathrm{d}z}{\mathrm{d}t} + \frac{\partial\vec{B}}{\partial t} \tag{2.27}$$

which can also be written in the form

$$\frac{\mathrm{d}\vec{B}}{\mathrm{d}t} = (\vec{v}\cdot\nabla)\vec{B} + \frac{\partial\vec{B}}{\partial t} \tag{2.28}$$

As the first term on the right-hand side of eqn (2.28) can be written as

$$(\vec{v}\cdot\nabla)\cdot\vec{B} = -\nabla x(\vec{v}x\vec{B}) \tag{2.29}$$

it follows

$$\nabla x\vec{E} = -\frac{\partial\vec{B}}{\partial t} + \nabla x(\vec{v}x\vec{B}) \tag{2.30}$$

and the first Maxwell equation is then given by

$$\nabla x(\vec{E} - \vec{v}x\vec{B}) = -\frac{\partial\vec{B}}{\partial t} \tag{2.31}$$

and represents the differential form of the Faraday's law in a moving medium.

The expression $\vec{E} - \nabla x\vec{B}$ represents the total field measured by a stationary observer.

The integral form of the Faraday's law in a moving medium can be obtained by integrating eqn (2.31),

$$\int_S \nabla x\vec{E}\mathrm{d}\vec{S} = -\int_S \frac{\partial\vec{B}}{\partial t}\mathrm{d}\vec{S} + \int_S \nabla x(\vec{v}x\vec{B})\mathrm{d}\vec{S} \tag{2.32}$$

Applying the Stokes theorem to eqn (2.32) yields

$$\oint_c \vec{E}\mathrm{d}\vec{s} = -\int_S \frac{\partial\vec{B}}{\partial t}\mathrm{d}\vec{S} + \oint_c (\vec{v}x\vec{B})\mathrm{d}\vec{s} \tag{2.33}$$

The second Maxwell equation for a moving medium must contain the total time rate of electric flux density change, that is,

$$\nabla x\vec{H} = \vec{J} + \frac{\mathrm{d}\vec{D}}{\mathrm{d}t} \tag{2.34}$$

where the time rate of vector $\vec{D}$ change in Cartesian coordinates can be expressed as

$$\frac{\mathrm{d}\vec{D}}{\mathrm{d}t} = \frac{\partial\vec{D}}{\partial x}\frac{\mathrm{d}x}{\mathrm{d}t} + \frac{\partial\vec{D}}{\partial y}\frac{\mathrm{d}y}{\mathrm{d}t} + \frac{\partial\vec{D}}{\partial z}\frac{\mathrm{d}z}{\mathrm{d}t} + \frac{\partial\vec{D}}{\partial t} \tag{2.35}$$

or

$$\frac{\mathrm{d}\vec{D}}{\mathrm{d}t} = (\vec{v}\cdot\nabla)\cdot\vec{D} + \frac{\partial\vec{D}}{\partial t} \tag{2.36}$$

As the first term on the right-hand side of eqn (2.36) can be written as

$$(\vec{v} \cdot \nabla) \cdot \vec{D} = -\nabla x(\vec{v} x \vec{D}) + \rho \vec{v} \tag{2.37}$$

it follows

$$\nabla x \vec{H} = \vec{J} - \nabla x(\vec{v} x \vec{D}) + \rho \vec{v} + \frac{\partial \vec{D}}{\partial t} \tag{2.38}$$

Equation (2.38) contains additional terms as correction to the current density in second Maxwell equation for stationary medium (2.2), the convection current term $\rho \vec{v}$ due to the motion of charge density and the current term due to the motion of polarized medium $\vec{v} x \vec{D}$.

The second Maxwell equation for a moving medium can then be written in the form

$$\nabla x(\vec{H} + \vec{v} x \vec{D}) = \vec{J} + \rho \vec{v} + \frac{\partial \vec{D}}{\partial t} \tag{2.39}$$

The integral form of the second Maxwell equation for a moving polarized dielectric medium can be obtained by integrating eqn (2.39),

$$\int_S \nabla x(\vec{H} + \vec{v} x \vec{D}) \mathrm{d}\vec{S} = -\int_S \vec{J} \mathrm{d}\vec{S} + \int_S \rho \vec{v} \mathrm{d}\vec{S} + \int_S \frac{\partial \vec{D}}{\partial t} \mathrm{d}\vec{S} \tag{2.40}$$

Applying the Stokes theorem to eqn (2.40) yields

$$\oint_c \vec{H} \mathrm{d}\vec{s} = -\oint_c (\vec{v} x \vec{D}) \mathrm{d}\vec{s} + \int_S \vec{J} \mathrm{d}\vec{S} + \int_S \rho \vec{v} \mathrm{d}\vec{S} + \int_S \frac{\partial \vec{D}}{\partial t} \mathrm{d}\vec{S} \tag{2.41}$$

The terms at the right-hand side of eqn (2.41) represent the total current which gives rise to the magnetic field composed from conductive currents, convective currents, and the currents generated by the polarization change rate and the motion of polarized media.

2.4 THE CONTINUITY EQUATION

The equation of continuity is a statement of charge conservation coupling the charges and current densities and can be readily derived from Maxwell eqn (2.4).

Taking the divergence of the Maxwell eqn (2.4) that results in

$$\nabla(\nabla x\vec{H}) = \nabla\vec{J} + \nabla\left(\frac{\partial\vec{D}}{\partial t}\right) \tag{2.42}$$

As the left-hand side of eqn (2.42) vanishes identically, the rest can be written as

$$\nabla\vec{J} + \frac{\partial}{\partial t}(\nabla\vec{D}) = 0 \tag{2.43}$$

Equation (2.43), combined with eqn (2.3), gives the equation of continuity, that is,

$$\nabla\vec{J} = -\frac{\partial\rho}{\partial t} \tag{2.44}$$

The rate of charge moving out of a region is equal to the time rate of charge density decrease.

The integral form of the continuity equation is obtained by performing the volume integration

$$\int_V \nabla\vec{J}\mathrm{d}V = -\frac{\partial}{\partial t}\int_V \rho\,\mathrm{d}V \tag{2.45}$$

and applying the Gauss's divergence theorem

$$\int_V \nabla\vec{J}\mathrm{d}V = \oint_S \vec{J}\mathrm{d}\vec{S} \tag{2.46}$$

The integral form of the continuity equation is then given by

$$\oint_S \vec{J}\mathrm{d}\vec{S} = -\frac{\partial Q}{\partial t} \tag{2.47}$$

where the unit normal in $\mathrm{d}\vec{S}$ is the outward-directed normal and Q is the total charge within the volume

$$Q = \int_V \rho\,\mathrm{d}V \tag{2.48}$$

Equation (2.47) represents the Kirchhoff's conservation law widely used in the electric circuit theory. Namely, if the surface S is closed, in order for net current to come out, there must be a decrease of positive charge within.

In the case of a moving medium, eqn (2.44) must contain the total time rate of charge density ρ change, that is,

$$\nabla\vec{J} + \frac{\mathrm{d}\rho}{\mathrm{d}t} = 0 \tag{2.49}$$

where the time rate of charge density ρ change in Cartesian coordinates can be written in the form

$$\frac{\mathrm{d}\rho}{\mathrm{d}t} = \frac{\partial\rho}{\partial x}\frac{\mathrm{d}x}{\mathrm{d}t} + \frac{\partial\rho}{\partial y}\frac{\mathrm{d}y}{\mathrm{d}t} + \frac{\partial\rho}{\partial z}\frac{\mathrm{d}z}{\mathrm{d}t} + \frac{\partial\rho}{\partial t} \tag{2.50}$$

which can also be expressed as

$$\frac{\mathrm{d}\rho}{\mathrm{d}t} = \vec{v}\cdot\nabla\rho + \frac{\partial\rho}{\partial t} \tag{2.51}$$

For the constant velocity of a moving medium it follows

$$\vec{v}\cdot\nabla\rho = \nabla(\rho\vec{v}) \tag{2.52}$$

and the continuity equation in the case of a moving medium is then

$$\nabla(\vec{J} + \rho\vec{v}) + \frac{\partial\rho}{\partial t} = 0 \tag{2.53}$$

where the second divergence term with the argument $\rho\vec{v}$ is the correction term to the conductive current due to the relative motion of free charges, where v is the charge velocity.

Equation (2.53) is often used in the analysis of solid state devices.

2.5 OHM'S LAW

The conservation of charge in a continuous medium is expressed by the equation of continuity

$$\nabla\vec{J} + \frac{\partial\rho}{\partial t} = 0 \tag{2.54}$$

The current density $\vec{J}$ is considered to be stationary if there is no accumulation of charge density at any point. As the term $\partial\rho/\partial t$ can be nonzero within a conducting medium only transiently, the stationary flow is then given by

$$\nabla\vec{J} = 0 \tag{2.55}$$

The continuity eqn (2.55) represents the field equivalent of Kirchhoff's current law stating that the net current leaving a junction of several conductors is zero.

The expression which relates the current density and electric field at any point within the conducting material is

$$\vec{J} = \sigma\vec{E} \tag{2.56}$$

where σ is the electrical conductivity of the particular material.

Equation (2.56) is considered to be a differential form of the Ohm's law. It is worth mentioning that this relation expresses a phenomenological characteristic and is by no means universally valid. The range of current values which eqn (2.56) holds is called the linear range of the particular material and can be very large as in metals or very small as in a semiconductor. Thus, eqn (2.56) implies that continuum is isotropic.

Stationary current density is impossible in a purely irrotational electric field, since, in a stationary current density related energy is expended at a rate $\vec{J} \cdot \vec{E}$ per unit volume and this energy cannot be ensured by an irrotational field. Stationary currents are possible only if there are electric field sources referred to as electromotive force (EMF) which produce nonirrotational fields. Assuming that such electromagnetic fields E' exist, the conduction equation becomes

$$\vec{J} = \sigma(\vec{E} + \vec{E}') \tag{2.57}$$

The electromotive force e can be defined as

$$e = \oint_c (\vec{E} + \vec{E}')\mathrm{d}\vec{s} \tag{2.58}$$

The conservative part of eqn (2.57) drops out of the closed line integration, that is,

$$\oint_c \vec{E}\mathrm{d}\vec{s} = 0 \tag{2.59}$$

which means that the current density is entirely due to the nonconservative forces

$$e = \oint_c \vec{E}'\mathrm{d}\vec{s} = \oint_c \frac{\vec{J}}{\sigma}\mathrm{d}\vec{s} \tag{2.60}$$

In the case where current density is assumed to be nearly constant over integration path, as quite frequently is the case, eqn (2.60) can be written as

$$e = \oint_c \vec{E}'\mathrm{d}\vec{s} = J \cdot S \cdot \oint_c \frac{\mathrm{d}s}{\sigma S} = I \cdot R \tag{2.61}$$

where R is the conductor resistance.

Equation (2.61) is the well-known integral and widely used integral form of the Ohm's law.

2.6 CONSERVATION LAW IN THE ELECTROMAGNETIC FIELD

A general relationship giving power and energy expressed in terms of electric and magnetic fields is written in the form of Poyinting theorem which is one of the most fundamental laws of electromagnetic theory.

The rate of increase of electromagnetic energy in the domain equals the rate of flow of energy in through the domain surface less the Joule heat production in the domain.

The conservation law of electromagnetic energy can be obtained from curl Maxwell equations. An equivalence of vector operators is given as follows:

$$\nabla \cdot (\vec{E}x\vec{H}) = \vec{H} \cdot \nabla x\vec{E} + \vec{E} \cdot \nabla x\vec{H} \tag{2.62}$$

Combining eqns (2.3), (2.4), and (2.62) yields

$$\vec{E} \cdot \frac{\partial \vec{D}}{\partial t} + \vec{H} \cdot \frac{\partial \vec{B}}{\partial t} = -\vec{E} \cdot \vec{J} + \vec{E} \cdot (\nabla x\vec{H}) - \vec{H} \cdot (\nabla x\vec{E}) \tag{2.63}$$

or in the alternative form

$$\frac{\partial w}{\partial t} = -\vec{E} \cdot \vec{J} - \nabla \cdot (\vec{E} \times \vec{H}) \tag{2.64}$$

where w term represents the energy storage per unit volume for an electromagnetic field

$$w = \frac{1}{2}(\vec{E}\vec{D} + \vec{H}\vec{B}) \tag{2.65}$$

Integrating eqn (2.64) over some finite region in the space, one obtains the integral form of the electromagnetic energy conservation law,

$$\int_V w dV = -\int_V \vec{E} \cdot \vec{J} dV - \int_V \nabla \cdot (\vec{E}x\vec{H}) dV \tag{2.66}$$

The left-hand side term represents the time rate of the stored energy in the electric and magnetic fields of the region.

The first term on the right-hand side is related to the Joule heat. Namely, this term represents either the ohmic power loss if J is a conduction current density or the power required to accelerate charges if J is a convection current arising from moving charges. In particular, if there is an energy source, then the product EJ is negative for that source and represents energy flow out of the region.

The other term on the right-hand side gives the flow in through the domain boundary.

Finally, applying the Gauss's integral theorem to the last term of eqn (2.66)

$$\int_V \nabla \cdot (\vec{E}x\vec{H})\mathrm{d}V = \oint_S (\vec{E}x\vec{H}) \cdot \mathrm{d}\vec{S} \tag{2.67}$$

the volume integral transforms to the surface integral over the boundary, where $\mathrm{d}\vec{S}$ is the outward drawn normal vector surface element.

Since all the energy changes must be supplied externally, this term represents the energy flow into the volume per unit time due to a minus sign of the surface integral. Changing sign, the rate of energy flow, or power flow, out through the enclosing surface is

$$P = \oint_S (\vec{E}x\vec{H})\mathrm{d}\vec{S} \tag{2.68}$$

where the vector $\vec{E}x\vec{H}$ is the flow of energy per unit area per unit time at the surface (power density flow), known as Poyinting vector

$$\vec{P}_d = \vec{E}x\vec{H} \tag{2.69}$$

The Poyinting vector (2.69) gives direction and magnitude of energy flow density at any point in space.

Power flow does not exist in the vicinity of a system of static charges having electric but no magnetic field. Also, in the vicinity of a perfect conductor, there is a zero tangential component normal to the conductor and there cannot be power flow into the perfect conductor.

Therefore, the final integral form of the conservation law in the electromagnetic field is then given by

$$\frac{\partial}{\partial t}\int_V \frac{1}{2}(\vec{E} \cdot \vec{D} + \vec{H} \cdot \vec{B})\mathrm{d}V = -\int_V \vec{E} \cdot \vec{J}\mathrm{d}V + \oint_S (\vec{E}x\vec{H}) \cdot \mathrm{d}\vec{S} \tag{2.70}$$

In other words, the rate of increase of electromagnetic energy in the domain equals the rate of flow of energy in through the domain surface less the Joule heat production in the domain.

Equation (2.70) is the Poyinting theorem valid for general media since there are no specializations made with respect to the medium.

For a battery with a nonelectrostatic field E' pumping energy both into heat losses and into a magnetic field is considered the corresponding current density and can be written as

$$\vec{J} = \sigma(\vec{E} + \vec{E}') \tag{2.71}$$

The electric field under the integral sign from eqn (2.70) is then

$$\vec{E} = \frac{\vec{J}}{\sigma} - \vec{E}' \tag{2.72}$$

and the first term on the right-hand side of eqn (2.70) can be written in the following form:

$$\int_V \vec{E}\vec{J}\mathrm{d}V = \int_V \left. \frac{|\vec{J}|}{\sigma} - \vec{E}'\vec{J} \right)\mathrm{d}V \tag{2.73}$$

Equation (2.73), therefore, represents the sum of the Joule heat losses and the negative rate at which the electromotive forces are doing work.

The last term on the right-hand side requires deeper consideration. Namely, the term

$$\int_V \vec{E}' \cdot \vec{J}\mathrm{d}V \tag{2.74}$$

represents an energy which has been hitherto neglected, since for static and quasistatic fields, the integral tends to vanish by taking an arbitrarily large enclosing surface.

The first volume integral from the left-hand side of eqn (2.70)

$$W_{el} = \int_V \frac{1}{2}(\vec{E} \cdot \vec{D})\mathrm{d}V \tag{2.75}$$

represents a total energy stored in the electric field system.

For linear materials, electric field strength $\vec{E}$ is proportional to electric flux density $\vec{D}$, so eqn (2.75) is simplified into

$$W_{el} = \int_V \frac{\varepsilon}{2}E^2\mathrm{d}V \tag{2.76}$$

The second volume integral from the left-hand side of eqn (2.70)

$$W_{mag} = \int_V \frac{1}{2}(\vec{H} \cdot \vec{B})\mathrm{d}V \tag{2.77}$$

represents a total energy stored in the magnetic field system.

For linear materials, magnetic field strength $\vec{H}$ is proportional to magnetic flux density $\vec{B}$, so eqn (2.77) is simplified into

$$W_{mag} = \int_V \frac{\mu}{2}H^2\mathrm{d}V \tag{2.78}$$

The analogy of relations (2.77) and (2.78) to eqns (2.75) and (2.76) is apparent. Equations (2.75)–(2.78) interpret the energy of a system of sources as actually stored in the fields produced by these sources.

2.7 THE ELECTROMAGNETIC WAVE EQUATIONS

Maxwell equations are coupled first-order space-time partial differential equations which are very difficult to apply when solving boundary-value problems. A way to overcome the difficulty of solving coupling equations is to decouple these first order equations, thereby, obtaining the second-order electromagnetic wave equations.

The wave equations can be easily derived from the Maxwell curl equations by differentiation and substitution. Taking curl on both sides of eqn (2.4) leads to

$$\nabla x \nabla x \vec{H} = \nabla x \vec{J} + \frac{\partial}{\partial t}(\nabla x \vec{D}) \tag{2.79}$$

In order to eliminate vectors J and D in favor of E, the constitutive equations

$$\vec{J} = \sigma \vec{E} \tag{2.80}$$

and

$$\vec{D} = \varepsilon \vec{E} \tag{2.81}$$

can be used.

Assuming uniform scalar material properties, it follows

$$\nabla x \nabla x \vec{H} = \sigma \nabla x \vec{E} + \varepsilon \frac{\partial}{\partial t}(\nabla x \vec{E}) \tag{2.82}$$

According to the Maxwell eqn (2.1), curl of E is replaced by the rate of change of magnetic flux density resulting in the following relation:

$$\nabla x \nabla x \vec{H} = -\sigma \frac{\partial \vec{B}}{\partial t} - \varepsilon \frac{\partial^2 \vec{B}}{\partial t^2} \tag{2.83}$$

If the magnetic material is simple enough to be described by the following constitutive equation:

$$\vec{B} = \mu \vec{H} \tag{2.84}$$

then the wave equation is given by

$$\nabla x \nabla x \vec{H} = -\mu\sigma \frac{\partial \vec{H}}{\partial t} - \mu\varepsilon \frac{\partial^2 \vec{H}}{\partial t^2} \tag{2.85}$$

Starting from the Maxwell electric curl eqn (2.1) and performing the same mathematical manipulation, a similar equation in the electric field E is obtained,

$$\nabla x \nabla x \vec{E} = -\mu\sigma \frac{\partial \vec{E}}{\partial t} - \mu\varepsilon \frac{\partial^2 \vec{E}}{\partial t^2} \tag{2.86}$$

Using the standard vector identity, valid for any vector E,

$$\nabla x \nabla x \vec{E} = \nabla \cdot (\nabla \vec{E}) - \nabla^2 \vec{E} \tag{2.87}$$

yields the final form of the wave equations as

$$\nabla^2 \vec{H} - \mu\sigma \frac{\partial \vec{H}}{\partial t} - \mu\varepsilon \frac{\partial^2 \vec{H}}{\partial t^2} = 0 \tag{2.88}$$

$$\nabla^2 \vec{E} - \mu\sigma \frac{\partial \vec{E}}{\partial t} - \mu\varepsilon \frac{\partial^2 \vec{E}}{\partial t^2} = -\frac{1}{\varepsilon} \nabla \rho \tag{2.89}$$

It is visible that the magnetic wave eqn (2.88) is homogeneous, while its electric counterpart implies the important fact that all electromagnetic phenomena are due to electric charges.

The wave equations (2.88) and (2.89) are valid for uniform regions and represent a significant degree of mathematical simplification as compared to the Maxwell equations. Many practical problems can be handled by solving one wave equation without reference to the other, so that only a single boundary-value problem in a single vector variable remains.

If a linear, isotropic, homogeneous, source-free medium is considered, then the set of eqns (2.88) and (2.89) are simplified into

$$\nabla^2 \vec{H} - \mu\varepsilon \frac{\partial^2 \vec{H}}{\partial t^2} = 0 \tag{2.90}$$

$$\nabla^2 \vec{E} - \mu\varepsilon \frac{\partial^2 \vec{E}}{\partial t^2} = 0 \tag{2.91}$$

Equations (2.90) and (2.91) are the equation of motion of electromagnetic waves in free space. The velocity of wave propagation is the velocity of light,

$$c = \frac{1}{\sqrt{\mu\varepsilon}} \tag{2.92}$$

where $c = 3 \times 10^8$ m/s, approximately.

2.8 BOUNDARY RELATIONSHIPS FOR DISCONTINUITIES IN MATERIAL PROPERTIES

An electromagnetic field may occur in a material medium usually characterized by its constitutive parameters conductivity σ, permeability μ, and permittivity ε. The material is linear if σ, μ, and ε are independent of E and H, and nonlinear if otherwise. Similarly, the medium is homogeneous if σ, μ, and ε are not space dependent and inhomogeneous otherwise. Finally, the medium is isotropic if σ, μ, and ε are independent of direction or anisotropic otherwise.

The boundary conditions at the interface separating two different media 1 and 2 with parameters σ_1, μ_1, ε_1 and σ_2, μ_2, and ε_2, respectively, as shown in Figure 2.3, can be easily derived from the Maxwell's equations in their integral form.

A relation for the tangential components of the electric field may be found by taking a line integral along a closed path of length Δl on one side of the boundary and returning on the other side as it is indicated in Figure 2.2.

The general conservative property of the electric field implies that any closed line of electrostatic field must be zero, that is,

$$\oint_c \vec{E}d\vec{s} = 0 \tag{2.93}$$

The sides normal to the boundary are assumed to be small enough that their contributions to the line integral vanish when compared with those of the sides parallel to the surface. Therefore, one has

$$\oint_c \vec{E}d\vec{s} = E_{t1}\Delta l - E_{t2}\Delta l \tag{2.94}$$

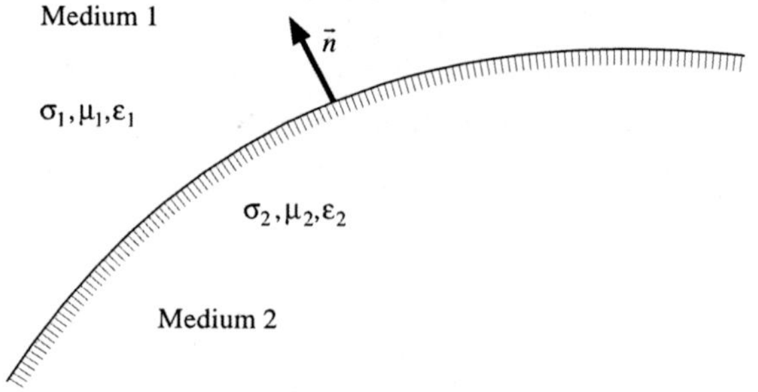

Figure 2.3 Two media configuration.

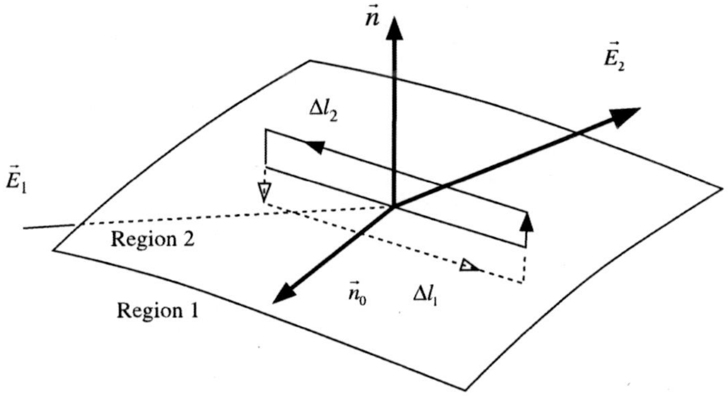

Figure 2.4 Electric field at the boundary between two different media.

The subscript t denotes component tangential to the interface. The length of the tangential loop sides is small enough to take E_t as constant over the length.

From eqns (2.93) and (2.94), it simply follows

$$E_{t1}\Delta l - E_{t2}\Delta l = 0 \tag{2.95}$$

or

$$E_{t1} = E_{t2} \tag{2.96}$$

The tangential components of electric field are the same on both sides of the boundary.

With the direction information included, it can be written as

$$\vec{n}x(\vec{E}_1 - \vec{E}_2) = 0 \tag{2.97}$$

where $\vec{n}$ is a unit normal vector directed from medium 1 to medium 2, subscripts 1 and 2 denote fields in regions 1 and 2, respectively.

In order to determine the behavior of the electric flux densities at a boundary, a small disk shape volume is considered as shown in Figure 2.5.

The thickness of the disk is assumed to be small enough that the net flux out of the sides vanishes in comparison with that out of the flat faces.

Assuming the existence of net surface charge density ρ_s on the boundary, the total flux out of the disk has to equal ρ_s. Applying the Gauss's law

$$\oint_S \vec{D}d\vec{S} = Q \tag{2.98}$$

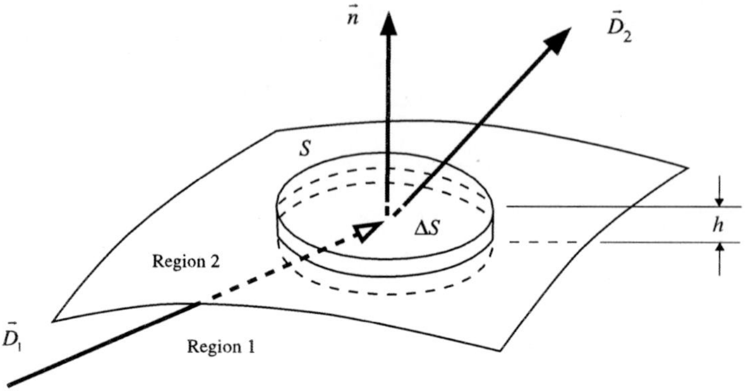

Figure 2.5 Electric flux density at the boundary between two different media.

where,

$$Q = \int_S \rho_s \mathrm{d}S \tag{2.99}$$

The side integral approaches zero if the disk thickness goes to zero. Therefore, it follows

$$\oint_S \vec{D}\mathrm{d}\vec{S} = D_{1n}\Delta S - D_{2n}\Delta S \tag{2.100}$$

From eqns (2.98) to (2.100) one obtains

$$D_{1n}\Delta S - D_{2n}\Delta S = \rho_s \Delta S \tag{2.101}$$

or

$$D_{1n} - D_{2n} = \rho_s \tag{2.102}$$

With the direction information included, where n is the unit vector normal to the surface,

$$\vec{n}(\vec{D}_1 - \vec{D}_2) = \rho_s \tag{2.103}$$

The boundary conditions at an interface between two regions with different permeabilities can be found in the same way as was done for the case of different permittivities.

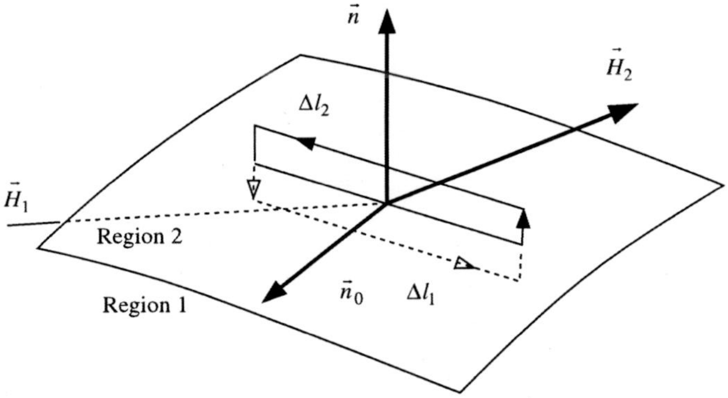

Figure 2.6 Magnetic field at the boundary between two different media.

The relation between transverse magnetic fields may be found by integrating the magnetic field H along a line enclosing the interface plane as shown in Figure 2.6.

The variation in magnetic field strength across an interface is obtained by applying the Ampere's law around a closed rectangular path.

The Ampere's law states that the line integral of the tangential component of the magnetic field strength around a closed path is equal to the current encircled by the path,

$$\oint_c \vec{H} d\vec{s} = I \tag{2.104}$$

The sides normal to the boundary are assumed to be small enough that their contributions to the line integral vanish when compared with those of the sides parallel to the surface. Therefore, it follows

$$\oint_c \vec{H} d\vec{s} = H_{t1}\Delta l - H_{t2}\Delta l \tag{2.105}$$

From eqns (2.104) and (2.105), one has

$$H_{t1}\Delta l - H_{t2}\Delta l = J_s \Delta l \tag{2.106}$$

The length Δl of the sides are arbitrarily small so that H_t may be uniformly assumed. The rest of the integration path is effectively reduced to zero length, that is,

$$H_{t1} - H_{t2} = J_s \tag{2.107}$$

where J_s is a surface current density in amperes per meter width flowing along the boundary.

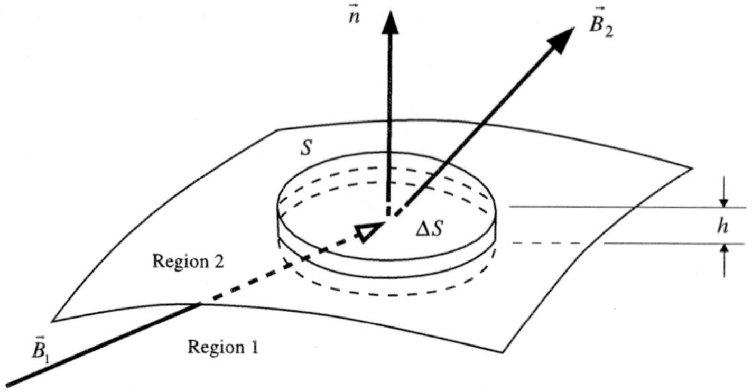

Figure 2.7 Magnetic flux density at the boundary between two different media.

With the direction information included, where $\vec{n}$ is the unit vector normal to the surface,

$$\vec{n}x(\vec{H}_1 - \vec{H}_2) = \vec{J}_s \tag{2.108}$$

There is a discontinuity of the tangential field at the boundary between two regions equal to any surface current density that may exist on the boundary.

A disk shaped volume enclosing the boundary between the two media is considered as shown in Figure 2.7.

The surfaces ΔS of the volume are assumed to be arbitrarily small so that the normal flux density B_n does not vary across the surface. Also, the thickness of the disk is vanishingly small so that there is negligible flux through the side wall.

Starting from the fourth Maxwell equation in the integral form,

$$\oint_S \vec{B}d\vec{S} = 0 \tag{2.109}$$

stating that magnetic flux is always zero for closed surfaces it follows:

$$\oint_S \vec{B}d\vec{S} = B_{1n}\Delta S - B_{2n}\Delta S \tag{2.110}$$

Now, if the disk surfaces approach one another, keeping the interface between them, the net outward flux from the disk is

$$B_{1n}\Delta S - B_{2n}\Delta S = 0 \tag{2.111}$$

or

$$B_{1n} = B_{2n} \tag{2.112}$$

In other words, the normal component of magnetic flux density is continuous across an interface. With the direction information included, where n is the unit vector normal to the surface,

$$\vec{n}(\vec{B}_1 - \vec{B}_2) = 0 \tag{2.113}$$

where $\vec{n}$ is a unit normal vector directed from medium 1 to medium 2, subscripts 1 and 2 denote fields in regions 1 and 2, respectively.

To summarize, the interface conditions are given by the following set of equations:

$$\vec{n}x(\vec{E}_1 - \vec{E}_2) = 0 \tag{2.114}$$

$$\vec{n}x(\vec{H}_1 - \vec{H}_2) = \vec{J}_s \tag{2.115}$$

$$\vec{n}(\vec{D}_1 - \vec{D}_2) = \rho_s \tag{2.116}$$

$$\vec{n}(\vec{B}_1 - \vec{B}_2) = 0 \tag{2.117}$$

where $\vec{n}$ is a unit normal vector directed from medium 1 to medium 2, subscripts 1 and 2 denote fields in regions 1 and 2, respectively. Equations (2.114) and (2.117) state that the tangential components of E and the normal components of B are continuous across the boundary. Equation (2.115) represents that the tangential component of H is discontinuous by the surface current density J_s on the boundary. Equation (2.116) means that the discontinuity in the normal component of D is the same as the surface charge density ρ_s on the boundary.

In engineering applications, it is sufficient to make the tangential components of the fields satisfy the necessary interface conditions as the normal components implicitly satisfy the associated boundary conditions.

In addition, the surface current J_s and surface charge ρ_s are very often encountered when one of the materials is a perfect or good conductor. In the case of a perfect conductor, the electric field E and magnetic field H vanish within the perfectly conducting medium. These fields are replaced by the surface charge density ρ_s and surface current density J_s. At higher frequencies, there is a well-known effect which confines current largely to surface regions. The so-called skin depth in common situations is often sufficiently small for the surface phenomenon to be an accurate representation. Therefore, the familiar rules for the behavior of time-varying fields at a boundary defined by good conductors follow directly from consideration of the limit condition, that is, when the conductor is perfect.

No time-varying field can exist in a perfect conductor, so the related electric field is entirely normal to the conductor and it is supported by a surface charge density, that is,

$$D_n = \rho_s \tag{2.118}$$

The magnetic field is entirely tangential to the conductor and is equilibrated by a surface current density term, that is,

$$H_s = J_s \tag{2.119}$$

True boundary conditions, that is, conditions at the extremes of the boundary value problem, are obtained by extending the interface conditions.

2.9 THE ELECTROMAGNETIC POTENTIALS

The mathematical treatment of electric and magnetic fields can be simplified by using auxiliary potential functions instead of fields. These auxiliary functions are the electric scalar potential φ, the magnetic vector potential A, and the magnetic scalar potential φ_m. These potentials are derived directly from the Maxwell equations.

Taking into account the well-known vector identity

$$\nabla x \nabla x \vec{f} = 0 \tag{2.120}$$

which is valid for any twice differentiable vector f, the Maxwell magnetic divergence eqn (2.4) is always satisfied if the flux density B is expressed by means of an auxiliary vector A, that is,

$$\vec{B} = \nabla x \vec{A} \tag{2.121}$$

The first Maxwell curl eqn (2.3) then becomes

$$\nabla x \vec{E} = -\frac{\partial}{\partial t}(\nabla x \vec{A}) \tag{2.122}$$

and by rearranging eqn (2.122), it follows that

$$\nabla x \left(\vec{E} + \frac{\partial \vec{A}}{\partial t} \right) = 0 \tag{2.123}$$

Since the curl of the gradient of any differentiable scalar function U always vanishes, that is,

$$\nabla x \nabla U = 0 \tag{2.124}$$

the bracket quantity in eqn (2.123) can be written as the gradient of the scalar potential function φ,

$$\vec{E} + \frac{\partial \vec{A}}{\partial t} = -\nabla \varphi \tag{2.125}$$

or

$$\vec{E} = -\frac{\partial \vec{A}}{\partial t} - \nabla\varphi \tag{2.126}$$

Here, A and φ are usually referred as the magnetic vector potential and electric scalar potential, respectively.

Thus, knowing the potential functions A and φ, the magnetic and electric fields can be determined from eqns (2.121) and (2.126).

There are also many magnetostatic problems concerned with finding magnetic fields, in which at least a part of the domain of interest is free of electric currents. For such source-free domains, the curl of the magnetic field H is equal to zero,

$$\nabla x\vec{H} = 0 \tag{2.127}$$

Since any zero-curl vector can be represented in terms of the gradient of a scalar function, the magnetic field intensity for such cases may be written as

$$\vec{H} = -\nabla\varphi_m \tag{2.128}$$

where the minus sign is taken to provide a convenient analogy with the case of electrostatic potential.

2.10 BOUNDARY RELATIONSHIPS FOR POTENTIAL FUNCTIONS

The potential equations must be complemented by appropriate interface and boundary conditions [1]. Figure 2.8 shows an interface between materials 1 and 2. A closed contour parallels the interface within material 1 and returns in material 2, very nearly paralleling its path in material 1, as shown in Figure 2.9.

The magnetic flux flowing through the contour is given by the contour integral of the vector potential A. Namely, the flux ϕ can be calculated by integrating the local flux density B over the surface bounded by the contour, that is,

$$\phi = \int_S \vec{B}\mathrm{d}\vec{S} \tag{2.129}$$

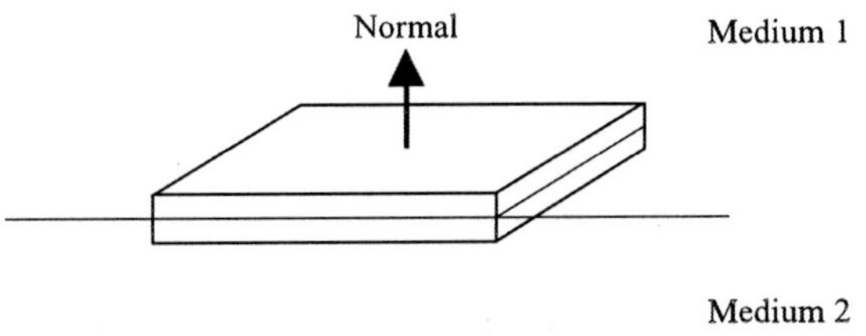

Figure 2.8 The interface between two media.

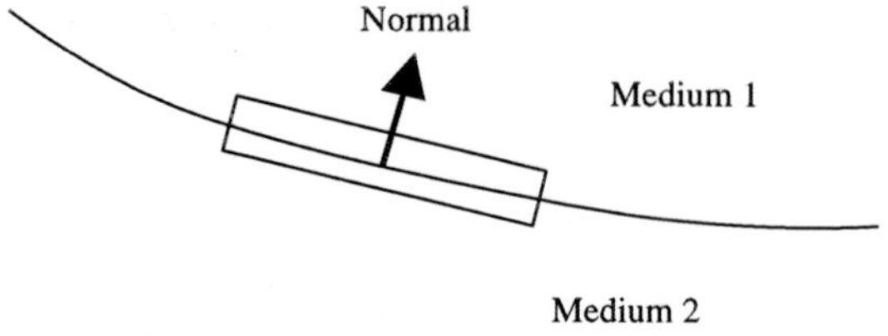

Figure 2.9 Interface at a surface between two different materials.

Substituting B in terms of the vector potential A, it follows

$$\phi = \int_S \nabla x \vec{A} d\vec{S} \tag{2.130}$$

Applying the Stokes theorem, the integration can be performed over the close contour c as follows:

$$\phi = \oint_c \vec{A} d\vec{s} \tag{2.131}$$

The surface spanned by the contour can be made arbitrarily small by keeping the two long sides of the contour arbitrarily close to each other while remaining at opposite sides of the material interface. The flux enclosed by the contour can, therefore, become arbitrarily small as well, so that it vanishes in the limit of the vanishing area. This can happen in the case when the tangential component of A has the same value on both sides of the interface. Consequently, the appropriate boundary condition in the limit becomes

$$\vec{n} x(\vec{A}_1 - \vec{A}_2) = 0 \tag{2.132}$$

A corresponding condition for the electric scalar potential can be obtained from Faraday's law relating the line integral of electric field to the time derivative of magnetic flux. Since the flux vanishes for the contour shown in Figure 2.9, the contour integral of the electric field must also vanish, that is,

$$\oint_c \vec{E} d\vec{s} = -\frac{\partial \phi}{\partial t} = -\frac{\partial}{\partial t} \oint_c \vec{A} d\vec{s} = 0 \tag{2.133}$$

Expressing the electric field in terms of its potentials results in the following integral relationship:

$$\oint_c \vec{E} d\vec{s} = -\oint_c \nabla \varphi d\vec{s} - \frac{\partial}{\partial t} \oint_c \vec{A} d\vec{s} \tag{2.134}$$

Since the contour integrals of both E and A vanish, eqn (2.134) can only be satisfied if

$$\oint_c \nabla\varphi d\vec{s} = 0 \tag{2.135}$$

Equation (2.135) implies the continuity of the scalar potential across the material interface as

$$\varphi_1 = \varphi_2 \tag{2.136}$$

This interface condition is particularly useful in solving electrostatic problems.

2.11 POTENTIAL WAVE EQUATIONS

The electromagnetic potentials must satisfy the wave equations. Assuming that all materials have single-valued scalar magnetic properties, it can be written as

$$\frac{1}{\mu}\nabla x\vec{A} = \vec{H} \tag{2.137}$$

Now taking the curls on both sides, it follows

$$\nabla x\left(\frac{1}{\mu}\nabla x\vec{A}\right) = \nabla x\vec{H} \tag{2.138}$$

and combining eqn (2.138) with eqn (2.2) one obtains

$$\nabla x\left(\frac{1}{\mu}\nabla x\vec{A}\right) = \vec{J} + \frac{\partial\vec{D}}{\partial t} \tag{2.139}$$

Substituting the constitutive eqns (2.22) and (2.23) results in

$$\nabla x\left(\frac{1}{\mu}\nabla x\vec{A}\right) = \sigma\vec{E} + \varepsilon\frac{\partial\vec{E}}{\partial t} \tag{2.140}$$

Now, expressing the electric field in terms of scalar and vector potential yields

$$\nabla x\left(\frac{1}{\mu}\nabla x\vec{A}\right) + \sigma\frac{\partial\vec{A}}{\partial t} + \varepsilon\frac{\partial^2\vec{A}}{\partial t^2} = -\sigma\nabla\varphi - \varepsilon\nabla\left(\frac{\partial\varphi}{\partial t}\right) \tag{2.141}$$

The double curl operator is changed to the equivalent laplace operator using the following identity, valid for any sufficiently differentiable vector A and scalar g,

$$\nabla x g \nabla x \vec{f} = -g\nabla^2\vec{f} + g\nabla \cdot (\nabla\vec{f}) + (\nabla g)gx(\nabla x\vec{f}) \tag{2.142}$$

The result is as follows:

$$\frac{1}{\mu}\nabla^2\vec{A} - \nabla\frac{1}{\mu}x\nabla x\vec{A} - \sigma\frac{\partial\vec{A}}{\partial t} - \varepsilon\frac{\partial^2\vec{A}}{\partial t^2} = \nabla\left(\nabla\vec{A} + \sigma\varphi + \varepsilon\frac{\partial\varphi}{\partial t}\right) \tag{2.143}$$

The left-hand side of eqn (2.143) includes an extra term, as compared to the electromagnetic wave equations, it permits material inhomogenity. For regions with homogeneous and uniform magnetic material properties, eqn (2.143) simplifies into the form of a standard wave equation on the left-hand side as follows:

$$\frac{1}{\mu}\nabla^2\vec{A} - \sigma\frac{\partial\vec{A}}{\partial t} - \varepsilon\frac{\partial^2\vec{A}}{\partial t^2} = \frac{1}{\mu}\nabla\left(\nabla\vec{A} + \mu\sigma\varphi + \mu\varepsilon\frac{\partial\varphi}{\partial t}\right) \tag{2.144}$$

However, the right-hand side of eqn (2.144) involves both electric scalar and magnetic vector potentials.

2.11.1 Coulomb Gauge

The potentials A and φ are not completely determined by the wave equations (2.143) or (2.144) and boundary rules (2.132) and (2.136), since any vector field can be uniquely determined by specifying both its curl and its divergence. The curl of A is equal to the magnetic flux density B and the divergence of A is not yet specified. The value of the divergence can be chosen at will without affecting the physical problem, namely, for all possible values always yield the same magnetic flux density B.

Various choices for the divergence of A are referred to as choices of gauge. In other words, a single physical problem can be described by different potential fields (one for each choice of gauge) and each particular choice of divergence of A is a gauge condition.

The best-known gauge condition in electromagnetics is the so-called Coulomb gauge that is given by

$$\nabla\vec{A} = 0 \tag{2.145}$$

or in the equivalent integral form

$$\oint_S \vec{A}d\vec{S} = 0 \tag{2.146}$$

and the appropriate boundary rule is given by

$$\vec{n}(\vec{A}_1 - \vec{A}_2) = 0 \tag{2.147}$$

Thus, the normal components of A must now be continuous under this choice of gauge, as well as the tangential components which have to be continuous under all circumstances.

The Coulomb gauge is mostly employed in magnetostatic problems. If there is no time variation, eqn (2.144) becomes

$$\frac{1}{\mu}\nabla^2\vec{A} = \nabla(\sigma\varphi) \tag{2.148}$$

If the conductivity σ is uniform, the current density can be expressed by the relation

$$\vec{J} = -\sigma\nabla\varphi \tag{2.149}$$

Equation (2.148) can be written as follows:

$$\frac{1}{\mu}\nabla^2\vec{A} = -\vec{J} \tag{2.150}$$

It is worth noting that the major part of computational magnetostatics problems is based on this equation.

2.11.2 Diffusion Gauge

An additional possible choice of gauge, also widely used, is the so called diffusion gauge that requires the divergence of A to be

$$\nabla\vec{A} + \mu\varepsilon\frac{\partial\varphi}{\partial t} = 0 \tag{2.151}$$

This choice removes the time-changing term on the right-hand side of eqn (2.144) leading to the relation

$$\frac{1}{\mu}\nabla^2\vec{A} - \sigma\frac{\partial\vec{A}}{\partial t} - \varepsilon\frac{\partial^2\vec{A}}{\partial t^2} = \frac{1}{\mu}\nabla(\mu\sigma\varphi) \tag{2.152}$$

Since the divergence of A no longer vanishes, the related boundary condition is now

$$\oint_S \vec{A}\mathrm{d}\vec{S} = \int_V \nabla\vec{A}\mathrm{d}V = \int_V \mu(\varepsilon_1 - \varepsilon_2)\frac{\partial\varphi}{\partial t}\mathrm{d}V \tag{2.153}$$

Therefore, only a single value of scalar potential appears in eqn (2.153), since the scalar potential must be continuous regardless of the choice of gauge. As the volume of integration approaches zero, the appropriate boundary conditions for A are again given by eqn (2.132). Namely, the normal components of A must still be continuous, just as it is the case under the Coulomb gauge.

The resulting quasiwave equation is then

$$\frac{1}{\mu}\nabla^2\vec{A} - \sigma\frac{\partial\vec{A}}{\partial t} - \varepsilon\frac{\partial^2\vec{A}}{\partial t^2} = \nabla(\sigma\varphi) \tag{2.154}$$

and the right-hand side can be written as

$$\nabla(\sigma\varphi) = -\vec{J} \tag{2.155}$$

where the actual current density physically signifies the current density that would flow if all time variations were slow. A major part of eddy-current and skin-effect problems relies on this formulation. Also, it is convenient in low-frequency problems to neglect displacement currents. This results in the following potential equation:

$$\nabla^2\vec{A} - \mu\sigma\frac{\partial\vec{A}}{\partial t} = -\mu\vec{J} \tag{2.156}$$

This interpretation is widely used in a large class of problems, both static and time-varying.

2.11.3 Lorentz Gauge

The most commonly used gauge in wave propagation problems is the so-called Lorentz gauge. The divergence of A in this case is defined as

$$\nabla\vec{A} + \mu\sigma\varphi + \mu\varepsilon\frac{\partial\varphi}{\partial t} = 0 \tag{2.157}$$

The resulting wave equation is homogeneous and the corresponding solution is entirely driven by the given boundary conditions. In addition, the scalar potential φ is governed by a similar but separate wave equation.

The potential wave equations for A and φ under the Lorentz gauge follow directly from the Maxwell electric divergence eqn (2.3) written in the form

$$\nabla\vec{E} = \frac{\rho}{\varepsilon} \tag{2.158}$$

and from eqn (2.126) when performed as a divergence operator by both sides,

$$\nabla\vec{E} = -\frac{\partial}{\partial t}\nabla\vec{A} - \nabla^2\varphi \tag{2.159}$$

Equating the preceding two expressions for divergence of E and using eqn (2.157) yields an inhomogeneous wave equation for scalar potential as follows:

$$\nabla^2\varphi - \mu\sigma\frac{\partial\varphi}{\partial t} - \mu\varepsilon\frac{\partial^2\varphi}{\partial t^2} = -\frac{\rho}{\varepsilon} \tag{2.160}$$

The wave equation for A follows from eqn (2.144) using the Lorentz gauge and it is given by

$$\nabla^2\vec{A} - \mu\sigma\frac{\partial\vec{A}}{\partial t} - \mu\varepsilon\frac{\partial^2\vec{A}}{\partial t^2} = 0 \tag{2.161}$$

The boundary conditions imposed by this choice of gauge are the same as in the diffusion gauge.

If a linear, isotropic, homogeneous, lossless medium is considered, then the set of eqns (2.160) and (2.161) are simplified into

$$\nabla^2\varphi - \frac{1}{c^2}\frac{\partial^2\varphi}{\partial t^2} = -\frac{\rho}{\varepsilon} \tag{2.162}$$

and

$$\nabla^2\vec{A} - \frac{1}{c^2}\frac{\partial^2\vec{A}}{\partial t^2} = 0 \tag{2.163}$$

The velocity of wave propagation is the velocity of light given by eqn (2.92).

In addition, if there is no time variation, eqn (2.162) results in the well-known Poisson potential equation

$$\nabla^2\varphi = -\frac{\rho}{\varepsilon} \tag{2.164}$$

It is worth noting that the major part of computational electrostatics problems is based on this equation. If the source-free region is of interest, eqn (2.164) simplifies into the Laplace equation

$$\nabla^2\varphi = 0 \tag{2.165}$$

which is also widely used in a major part of electrostatic problems.

2.12 RETARDED POTENTIALS

The potential wave equations can be solved completely in a relatively small number of special cases. Integral solutions to potential wave equations (2.160) and (2.161) are given in the form of so-called retarded potentials.

Assuming that the solution of the potential wave equation

$$\nabla^2 \varphi - \mu\sigma \frac{\partial \varphi}{\partial t} - \mu\varepsilon \frac{\partial^2 \varphi}{\partial t^2} = -\frac{\rho(t)}{\varepsilon} \tag{2.166}$$

is desired in the unbounded space, the charge density $\rho(t)$ differs from zero only in an infinitesimal space dV surrounding the origin. Namely, the right-hand side represents a time-varying point charge.

The solution of eqn (2.166) is given [2] in the form

$$\varphi(r, t) = \frac{\rho(t - r/c)}{4\pi\varepsilon r} \tag{2.167}$$

The physical meaning of eqn (2.167) is that the potential still corresponds to the charge causing it, but with a time retardation which equals the time taken for light to propagate from the charge to the point of potential observation. The electrostatic potential, on the contrary may be viewed simply as the special case of very small retardation, that is, a simplification valid at close distances.

Nevertheless, the point charge is a very special and quite artificial case; (2.167) can be applied to the more general case of time-varying charge distributed over some finite volume of space V by dividing the volume into small portions and treating the charge in each as a point charge at the given point.

Summing over all the elementary volumes, the potential at a given [1,2] observation point is then

$$\varphi(\vec{r}, t) = \int_{V'} \frac{\rho(\vec{r}', t - R/c)}{4\pi\varepsilon R} \mathrm{d}V' \tag{2.168}$$

where R is the distance from the source point to the observation point.

Integral (2.168) is often called a retarded potential.

The magnetic vector potential may be derived following a similar treatment. For the purpose of the antenna theory, the conduction current is taken to be prescribed by the sources feeding the antenna, that is,

$$-\mu\sigma \frac{\partial \vec{A}}{\partial t} = \mu \vec{J} \tag{2.169}$$

so that the vector potential wave equation is given in the form

$$\nabla^2 \vec{A} - \mu\varepsilon \frac{\partial^2 \vec{A}}{\partial t^2} = \mu \vec{J} \tag{2.170}$$

The solution [1,2] is given by

$$\vec{A}(r,t) = \int_{V'} \frac{\vec{J}(\vec{r}',t-R/c)}{4\pi\varepsilon R} \mathrm{d}V' \tag{2.171}$$

The solution of eqn (2.171) is carried out by separating this vector equation into its Cartesian components. The result is a set of equations identical in form to the scalar potential equation. The recombination of these components results in solution (2.171).

2.13 GENERAL BOUNDARY CONDITIONS AND UNIQUENESS THEOREM

For the realistic problem under consideration, the boundary ∂V of a problem domain V is generally composed of three distinct boundary segments ∂D, ∂N, and ∂C. Now, let the solution u (representing electric scalar potential, magnetic vector potential, electric field, or magnetic field) satisfy the following conditions on the three segments:

$$u = u_0 \quad \text{on } \partial D \text{ (Dirichlet)} \tag{2.172}$$

$$\frac{\partial u}{\partial n} = q_0 \quad \text{on } \partial N \text{ (Neumann)} \tag{2.173}$$

$$\frac{\partial u}{\partial n} + au = p_0 \quad \text{on } \partial C\text{(Cauchy)} \tag{2.174}$$

The Dirichlet conditions are called essential or principal boundary conditions, whereas the Neumann boundary conditions are called natural boundary conditions. The Cauchy boundary conditions are often called mixed boundary conditions.

Uniqueness theorem generally ensures that the solution of a partial differential equation with some associated boundary conditions is unique. In electromagnetics, the theorem may be stated as follows: If in any way a set of electric and magnetic fields is found satisfying Maxwell equations and the prescribed boundary conditions simultaneously, the set is considered to be unique. Consequently, a field is uniquely specified by the sources in terms of volume charge and current density, respectively, within the medium plus the tangential components of the electric or magnetic field over the boundary.

2.14 ELECTRIC AND MAGNETIC WALLS

The homogeneous Neumann boundary conditions for the formulations when $u = H$ and $u = E$ have important physical interpretations in engineering.

If a field problem is formulated in terms of electric field E, the natural boundary condition is given by

$$\vec{n}\nabla x\vec{E} = 0 \tag{2.175}$$

which can also be written in accordance with the Maxwell's electric curl eqn (2.1) as

$$\vec{n}\frac{\partial\vec{B}}{\partial t} = \frac{\partial}{\partial t}(\vec{n}\vec{B}) = 0 \tag{2.176}$$

which states that the magnetic flux density B must lie parallel to the bounding surface.

Such boundaries corresponding to perfect conductors are often referred to as *electric walls* (the surfaces through which no magnetic flux can penetrate, a condition satisfied by a perfect electric conductor).

On the contrary, if a field problem is formulated in terms of the magnetic field H, the natural boundary condition is then given by

$$\vec{n}\nabla x\vec{H} = 0 \tag{2.177}$$

Using the Maxwell's magnetic curl eqn (2.2), eqn (2.176) can be rewritten in the form

$$\vec{n}\left[\vec{J} + \frac{\partial\vec{D}}{\partial t}\right] = 0 \tag{2.178}$$

Inside any dielectric medium there cannot exist any conduction current density J, consequently it follows:

$$\frac{\partial}{\partial t}[\vec{n}\vec{D}] = 0 \tag{2.179}$$

The electric flux density vector D and, therefore, the electric field vector E must lie parallel to the bounding surface. In other words, there cannot be any components of vector D normal to the boundary. Such boundaries are then referred to as *magnetic walls* (no electric flux can penetrate them).

2.15 THE LAGRANGIAN FORM OF ELECTROMAGNETIC FIELD LAWS

The Hamilton variation principle provides a single equation from which the Maxwell's equation can be deduced. A variational principle is applicable to continuous media, thus, ensuring its applicability to the behavior of fields. An electromagnetic energy density can be defined as a function of position in space in terms of the electromagnetic field variables. It can be stated that in every essential way, the electromagnetic field can be treated as a mechanical system and it is, therefore, expected that the Hamilton variation principle is applicable in electromagnetic field theory. Basically, this means that one should be able to set up a suitable Lagrangian function of the field variables, such that Hamilton variation principle shall result in the equations of motion of the field, that is, in Maxwell's equations. As it is well-known, there is no general rule for the derivation of the Lagrangian function. One might be tempted to guess that it is the difference between the kinetic and potential energy of the field, but the difficulty arises as the meaning of kinetic and potentials energy in the electromagnetic field is not quite clear [3].

2.15.1 Lagrangian Formulation and Hamilton Variational Principle

Hamilton variational principle has become the basis of modern analytical dynamics, but it has also been considered as a model for universal physical laws.

For simplicity, a system with one degree of freedom represented by a coordinate q is considered. It is also assumed that some function of position, velocity, and time $L(q, \dot{q}, t)$ is given where $\dot{q}$ denotes the time derivation of t.

What has to be determined is how a point should move in this one-dimensional space so that the time integral of L is minimized compared with the integral over the conceivable paths between the same starting and end points as shown in Figure 2.10. The solution is given by stating q as a function of time $q = q(t)$.

In order to compare all paths having the same starting and end points, the variation of function q is zero at both ends,

$$\delta q = 0 \tag{2.180}$$

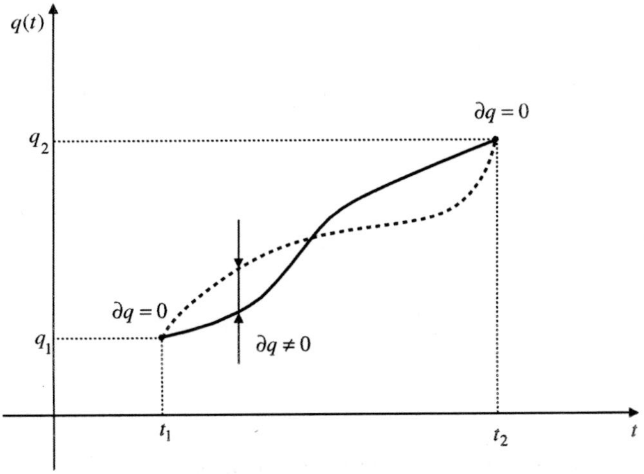

Figure 2.10 The varied function $q(t)$.

In other words, all alternatives start at a given instant t_1 and arrive together at another instant t_2.

The minimum condition is then given by a functional F expressed [3] by the integral

$$F = \int_{t_1}^{t_2} L(q, \dot{q}, t)\mathrm{d}t = \min \qquad (2.181)$$

In classical mechanics function L is expressed as

$$L = W_{kin} - W_{pot} \qquad (2.182)$$

where W_{kin} and W_{pot} are the kinetic and potential energies, respectively.

According to the calculus of variation, the functional

$$F = \int_{t_1}^{t_2} (W_{kin} - W_{pot})\mathrm{d}t = \min \qquad (2.183)$$

approaches minimum value when its variation vanishes, that is,

$$\delta F = 0 \qquad (2.184)$$

or

$$\delta \int_{t_1}^{t_2} L\mathrm{d}t = 0 \qquad (2.185)$$

From the mathematical point of view, the function $q(t)$ minimizes the functional (2.181) or (2.183) respectively. Hence, it follows:

$$\delta F = \int_{t_1}^{t_2} \delta L\mathrm{d}t \qquad (2.186)$$

For the simplest case given by $L(q, \dot{q}, t)$, the variation of function L is given by

$$\delta L = \frac{\partial L}{\partial q}\delta q + \frac{\partial L}{\partial \dot{q}}\delta \dot{q} \qquad (2.187)$$

Performing some further mathematical manipulation one obtains

$$\delta F = \int_{t_1}^{t_2} \left[\frac{\partial L}{\partial q} - \frac{\mathrm{d}}{\mathrm{d}t}\left(\frac{\partial L}{\partial \dot{q}}\right) \right] \delta q dt + \frac{\partial L}{\partial \dot{q}} \delta q \Big|_{t_1}^{t_2} = 0 \tag{2.188}$$

As $\delta q = 0$ at ends of the path, the second term at the right-hand side automatically vanishes.

Furthermore, according to the *fundamental lemma* of variational calculus [3], the first integral term at the right-hand side of eqn (2.188) vanishes if the following condition is satisfied:

$$\frac{\partial L}{\partial q} - \frac{\mathrm{d}}{\mathrm{d}t}\left(\frac{\partial L}{\partial \dot{q}}\right) = 0 \tag{2.189}$$

This second-order differential equation relates the position q for time t and also determines the true path of the system when two end positions and times are given. This equation is known as Lagrange equation of motion.

Hamilton variational principle can be considered as a general law not only for particle dynamics but also for the dynamics of continuous materials.

As an extension to three-dimensional problems in continuous materials, the variational principle corresponding to eqn (2.186) is given by

$$\delta \int_{t_1}^{t_2} \int_V \bar{L}_d \mathrm{d}V \mathrm{d}t = 0 \tag{2.190a}$$

where $\bar{L}_d$ is so-called the Lagrange density defined by

$$L = \int_V \bar{L}_d \mathrm{d}V \tag{2.190b}$$

and has a unit of energy per volume.

It is worth noting that the variational principle is an invariant scalar equation for coordinate transformations [3].

2.15.2 Lagrangian Formulation and Hamilton Variational Principle in Electromagnetics

In an electromagnetic oscillation, energy oscillates between electric and magnetic energy just as in a mechanical oscillation energy oscillates between kinetic and potential energy. The minimum condition is again defined by the integral,

$$F = \int_{t_1}^{t_2} L\mathrm{d}t = \min \tag{2.191}$$

The electromagnetic energy functional F corresponding to eqn (2.183) is given by

$$F = \int_{t_1}^{t_2} (W_{mag} - W_{el})\mathrm{d}t = \min \tag{2.192}$$

The related Lagrange equation of motion can be derived in two steps, first for the case of the source free region and then for the region containing sources.

If the derivation is, for the sake of simplicity, first restricted to the vacuum fields only, functional (2.192) is given in the form

$$F = \int_{t_1}^{t_2}\int_V \frac{1}{2}(\vec{B} \cdot \vec{H} - \cdot\vec{D} \cdot \vec{E})\mathrm{d}V\mathrm{d}t = \min \tag{2.193}$$

According to the analogy with the discussion presented in Section 2.15.1, the correct Lagrangian for the vacuum electromagnetic field has been considered to be the following expression:

$$L = \frac{1}{2}[\vec{B} \cdot \vec{H} - \vec{E} \cdot \vec{D}] \tag{2.194}$$

where the term $1/2[\vec{E} \cdot \vec{D}]$ can be interpreted as the potential energy density, and $\frac{1}{2}[\vec{B} \cdot \vec{H}]$ as the kinetic energy density.

As the Hamilton variation principle demands that variations in the field variables have to be made in a way to minimize the Lagrangian integral, it follows:

$$\delta \int_{t_1}^{t_2}\int_V L\mathrm{d}V\mathrm{d}t = 0 \tag{2.195}$$

and in Cartesian coordinates it can be written as

$$\delta \iiint_V \int_{t_1}^{t_2} L\mathrm{d}x\mathrm{d}y\mathrm{d}z\mathrm{d}t = 0 \tag{2.196}$$

However, six electromagnetic field variables E_x, E_y, E_z, B_x, B_y, and B_z are in a sense redundant because these four variables φ, A_x, A_y, A_z, representing the four potential components of the electromagnetic field, are in fact sufficient to define the field completely.

The vectors $\vec{E}$ and $\vec{B}$ are defined in terms of φ and $\vec{A}$ by the following equations:

$$\vec{E} = -\frac{\partial \vec{A}}{\partial t} - \nabla\varphi \tag{2.197}$$

$$\vec{B} = \nabla x \vec{A} \tag{2.198}$$

Two sets of Maxwell curl equations are simple identities stemming from these identities.

The Lagrangian in terms of the electric and magnetic potential is then given by

$$L = \frac{1}{2\mu}(\nabla x \vec{A})^2 - \frac{\varepsilon}{2}\ \nabla\varphi + \frac{\partial \vec{A}}{\partial t}\Big)^2 \tag{2.199}$$

Thus, the Lagrange equations of motion, in Cartesian coordinates, corresponding to eqn (2.188) are given by the following set of partial differential equations:

$$\frac{\partial L}{\partial q_m} = \frac{\partial}{\partial t}\frac{\partial L}{\partial\left(\frac{\partial q_m}{\partial t}\right)} + \frac{\partial}{\partial x}\frac{\partial L}{\partial\left(\frac{\partial q_m}{\partial x}\right)} + \frac{\partial}{\partial y}\frac{\partial L}{\partial\left(\frac{\partial q_m}{\partial y}\right)} + \frac{\partial}{\partial z}\frac{\partial L}{\partial\left(\frac{\partial q_m}{\partial z}\right)} \tag{2.200}$$

where $m = 1,2,3,4$ and $q_1 = \varphi$, $q_2 = A_x$, $q_3 = A_y$, $q_4 = A_z$.

Furthermore, the complete set of equations for vector and scalar potential is then

$$\frac{\partial L}{\partial \varphi} = \frac{\partial}{\partial t}\frac{\partial L}{\partial\left(\frac{\partial \varphi}{\partial t}\right)} + \frac{\partial}{\partial x}\frac{\partial L}{\partial\left(\frac{\partial \varphi}{\partial x}\right)} + \frac{\partial}{\partial y}\frac{\partial L}{\partial\left(\frac{\partial \varphi}{\partial y}\right)} + \frac{\partial}{\partial z}\frac{\partial L}{\partial\left(\frac{\partial \varphi}{\partial z}\right)} \tag{2.201}$$

$$\frac{\partial L}{\partial A_x} = \frac{\partial}{\partial t}\frac{\partial L}{\partial\left(\frac{\partial A_x}{\partial t}\right)} + \frac{\partial}{\partial x}\frac{\partial L}{\partial\left(\frac{\partial A_x}{\partial x}\right)} + \frac{\partial}{\partial y}\frac{\partial L}{\partial\left(\frac{\partial A_x}{\partial y}\right)} + \frac{\partial}{\partial z}\frac{\partial L}{\partial\left(\frac{\partial A_x}{\partial z}\right)} \tag{2.202}$$

$$\frac{\partial L}{\partial A_y} = \frac{\partial}{\partial t}\frac{\partial L}{\partial\left(\frac{\partial A_y}{\partial t}\right)} + \frac{\partial}{\partial x}\frac{\partial L}{\partial\left(\frac{\partial A_y}{\partial x}\right)} + \frac{\partial}{\partial y}\frac{\partial L}{\partial\left(\frac{\partial A_y}{\partial y}\right)} + \frac{\partial}{\partial z}\frac{\partial L}{\partial\left(\frac{\partial A_y}{\partial z}\right)} \tag{2.203}$$

$$\frac{\partial L}{\partial A_z} = \frac{\partial}{\partial t}\frac{\partial L}{\partial\left(\frac{\partial A_z}{\partial t}\right)} + \frac{\partial}{\partial x}\frac{\partial L}{\partial\left(\frac{\partial A_z}{\partial x}\right)} + \frac{\partial}{\partial y}\frac{\partial L}{\partial\left(\frac{\partial A_z}{\partial y}\right)} + \frac{\partial}{\partial z}\frac{\partial L}{\partial\left(\frac{\partial A_z}{\partial z}\right)} \tag{2.204}$$

Inserting Lagrangian (2.199) into eqn (2.201) yields

$$\frac{\partial L}{\partial \varphi} = 0 \tag{2.205}$$

$$\frac{\partial L}{\partial\left(\dfrac{\partial \varphi}{\partial x}\right)} = -\varepsilon\frac{\partial \varphi}{\partial x} = -\varepsilon E_x \tag{2.206}$$

$$\frac{\partial L}{\partial\left(\dfrac{\partial \varphi}{\partial y}\right)} = -\varepsilon\frac{\partial \varphi}{\partial y} = -\varepsilon E_y \tag{2.207}$$

$$\frac{\partial L}{\partial\left(\dfrac{\partial \varphi}{\partial z}\right)} = -\varepsilon\frac{\partial \varphi}{\partial z} = -\varepsilon E_z \tag{2.208}$$

The resulting expression obtained from eqn (2.205) to eqn (2.208) is then

$$\frac{\partial}{\partial x}(\varepsilon E_x) + \frac{\partial}{\partial y}(\varepsilon E_y) + \frac{\partial}{\partial z}(\varepsilon E_z) = 0 \tag{2.209}$$

which is the third Maxwell equation for a source free region, that is,

$$\nabla \vec{D} = 0 \tag{2.210}$$

Furthermore, inserting Lagrangian (2.199) into eqn (2.202) results in

$$\frac{\partial L}{\partial A_x} = 0 \tag{2.211}$$

$$\frac{\partial L}{\partial\left(\dfrac{\partial A_x}{\partial t}\right)} = -\varepsilon\frac{\partial \varphi}{\partial x} = -D_x \tag{2.212}$$

$$\frac{\partial L}{\partial\left(\dfrac{\partial A_x}{\partial y}\right)} = H_z \tag{2.213}$$

$$\frac{\partial L}{\partial\left(\dfrac{\partial A_x}{\partial z}\right)} = -H_y \tag{2.214}$$

which gives

$$\frac{\partial D_x}{\partial t} - \frac{\partial H_z}{\partial y} + \frac{\partial H_y}{\partial z} = 0 \tag{2.215}$$

Furthermore, inserting Lagrangian (2.199) into eqn (2.203) results in

$$\frac{\partial L}{\partial A_y} = 0 \tag{2.216}$$

$$\frac{\partial L}{\partial\left(\frac{\partial A_y}{\partial t}\right)} = -\varepsilon\frac{\partial \varphi}{\partial y} = -D_y \tag{2.217}$$

$$\frac{\partial L}{\partial\left(\frac{\partial A_y}{\partial y}\right)} = H_z \tag{2.218}$$

$$\frac{\partial L}{\partial\left(\frac{\partial A_y}{\partial z}\right)} = -H_x \tag{2.219}$$

which can be written as

$$\frac{\partial D_y}{\partial t} - \frac{\partial H_x}{\partial z} + \frac{\partial H_z}{\partial x} = 0 \tag{2.220}$$

Finally, inserting Lagrangian (2.199) into eqn (2.204) results in

$$\frac{\partial L}{\partial A_z} = 0 \tag{2.221}$$

$$\frac{\partial L}{\partial\left(\frac{\partial A_z}{\partial t}\right)} = -\varepsilon\frac{\partial \varphi}{\partial z} = -D_z \tag{2.222}$$

$$\frac{\partial L}{\partial\left(\frac{\partial A_z}{\partial x}\right)} = -H_y \tag{2.223}$$

$$\frac{\partial L}{\partial\left(\frac{\partial A_y}{\partial z}\right)} = H_x \tag{2.224}$$

and it follows

$$\frac{\partial D_z}{\partial t} + \frac{\partial H_x}{\partial y} - \frac{\partial H_y}{\partial x} = 0 \tag{2.225}$$

Now, rearranging the relations (2.215), (2.220), and (2.225) the following set of scalar equations is obtained:

$$\frac{\partial H_z}{\partial y} - \frac{\partial H_y}{\partial z} = \frac{\partial D_x}{\partial t} \tag{2.226}$$

$$\frac{\partial H_x}{\partial z} - \frac{\partial H_z}{\partial x} = \frac{\partial D_y}{\partial t} \tag{2.227}$$

$$\frac{\partial H_y}{\partial x} - \frac{\partial H_x}{\partial y} = \frac{\partial D_z}{\partial t} \tag{2.228}$$

Set of equations (2.226–2.228) represents the second curl Maxwell equation for free-space region,

$$\nabla x\vec{H} = \frac{\partial \vec{D}}{\partial t} \tag{2.229}$$

The Lagrangian formulation presented so far leads to the Maxwell equations of motion of the electromagnetic field. This derivation implies that the Maxwell equations are invariant for coordinate transformations. The present discussion is related only to the Lagrangian for vacuum fields.

It is also possible to set up a Lagrangian for the field in materials carrying free charge and current densities. The resulting motion equations then turn out to be Maxwell equations with the proper modifications due to the material, the charge density, and current density terms.

The energy functional for a complete description of an electromagnetic field in the source region is

$$F = \int_{t_1}^{t_2}\int_V \left(\frac{1}{2}\cdot\vec{B}\cdot\vec{H} - \frac{1}{2}\cdot\vec{D}\cdot\vec{E} + \vec{J}\cdot\vec{A} - \varphi\cdot\rho\right)\mathrm{d}V\mathrm{d}t = \min \tag{2.230}$$

while the corresponding Lagrangian density is then found to be

$$L = \frac{1}{2}[\vec{B}\cdot\vec{H} - \vec{E}\cdot\vec{D}] + [\vec{J}\cdot\vec{A} - \rho\cdot\varphi] \tag{2.231}$$

The part of the Lagrangian depending on the charge and current sources may be considered as the difference between a kinetic energy density $\vec{J}\cdot\vec{A}$ and a potential energy density $\rho\cdot\varphi$. Hence, the full Lagrangian density is then the sum of the field Lagrangian having the same form as eqn (2.194) and this new part depends on the charge and current density.

Maxwell equations (2.1) and (2.4) can be derived by using functional (2.230) with Lagrangian (2.231).

2.16 COMPLEX PHASOR NOTATION OF TIME-HARMONIC ELECTROMAGNETIC FIELDS

In many practical circumstances, particularly at low frequencies, systems are excited sinusoidally and a time-harmonic variation of electromagnetic fields can be assumed. In such cases, it is convenient to represent the variables of interest in a complex phasor form. Thus, an arbitrary time-dependent vector field $F(r, t)$ can be expressed as follows [2]:

$$\vec{F}(\vec{r}, t) = \mathrm{Re}[\vec{F}_s(\vec{r})e^{j\omega t}] \tag{2.232}$$

where $F_s(r)$ is the phasor form of $F(r, t)$, and $F_s(r)$ is in general complex with an amplitude and a phase changing with position. Then, Re[ ] implies taking the real part of quantity in brackets and ω is the angular frequency of the sinusoidal excitation.

In addition, using the phasor representation, the derivatives with respect to time results in

$$\frac{\partial}{\partial t}[\vec{F}_s(\vec{r})e^{j\omega t}] = j\omega\vec{F}_s(\vec{r})e^{j\omega t} \tag{2.233}$$

Consequently, the set of differential Maxwell's equations for the sinusoidal steady state becomes

$$\nabla x\vec{E} = -j\omega\vec{B} \tag{2.234}$$

$$\nabla x\vec{H} = \vec{J} + j\omega\vec{D} \tag{2.235}$$

$$\nabla\vec{D} = \rho \tag{2.236}$$

$$\nabla\vec{B} = 0 \tag{2.237}$$

It should be observed that the assumption of the time-harmonic variation of fields eliminates the time dependence from Maxwell's equations, thereby reducing the space-time dependence to space dependence only.

However, this simplification does not exclude the possibility of describing more general time-changing fields if ω is considered to be one element of the entire frequency spectrum, with all the Fourier components superposed.

Namely, any nonsinusoidal field can be expressed as

$$\vec{F}(\vec{r}, t) = \mathrm{Re}\left[\int_{-\infty}^{+\infty} \vec{F}_s(\vec{r}, \omega)e^{j\omega t}\mathrm{d}\omega\right] \tag{2.238}$$

Therefore, the solutions to Maxwell's equations for a nonsinusoidal field can be obtained by summing all the Fourier components $F_s(r, \omega)$ over ω.

2.16.1 Poyinting Theorem for Complex Phasors

Starting with Maxwell's equation in the complex form (2.234)–(2.237) applicable to time-harmonic electromagnetic fields, the complex Poyinting theorem can be derived by steps parallel to those used for the theorem associated with general time varying quantities.

Using the vector identity,

$$\nabla(\vec{E}x\vec{H}^*) = \vec{H}^*\nabla x\vec{E} + \vec{E}\nabla x\vec{H}^* \tag{2.239}$$

where the asterisk stands for the complex conjugate and taking into account the complex phasor form of curl Maxwell's equations (2.234) and (2.235), it follows:

$$\nabla(\vec{E}x\vec{H}^*) = \vec{H}^*(-j\omega\vec{B}) - \vec{E}^*(\vec{J}^* - j\omega\vec{D}^*) \tag{2.240}$$

Integrating eqn (2.240) through volume V and applying the divergence theorem, one has

$$\int_V \nabla\left(\vec{E}x\vec{H}^*\right)\mathrm{d}V = \oint_S \left(\vec{E}x\vec{H}^*\right)\mathrm{d}\vec{S} = -\int_V \left[\vec{E}\vec{J}^* + j\omega\left(\vec{H}^*\vec{B} - \vec{E}\vec{D}^*\right)\right]\mathrm{d}V \tag{2.241}$$

Equation (2.241) is the general Poyinting theorem as it applies to complex phasors.

In addition, if an isotropic medium (in which all losses occur through conduction currents $J = \sigma E$) then eqn (2.241) becomes

$$\oint_S \left(\vec{E}x\vec{H}^*\right)\mathrm{d}\vec{S} = -\int_V \sigma\vec{E}\vec{E}^*\mathrm{d}V - j\omega\int_V \left(\mu\vec{H}\vec{H}^* - \varepsilon\vec{E}\vec{E}^*\right)\mathrm{d}V \tag{2.242}$$

In relations (2.234)–(2.242), all quantities are space-time dependent and assumed to be time-harmonic values ($e^{j\omega t}$ dependence is assumed).

Using definition (2.232) and the identity

$$\mathrm{Re}\left[\vec{F}_s(\vec{r})e^{j\omega t}\right] = \frac{1}{2}\left[\vec{F}_s(\vec{r})e^{j\omega t} + \vec{F}_s^*(\vec{r})e^{-j\omega t}\right] \tag{2.243}$$

it follows

$$\begin{aligned}\vec{P}_d(\vec{r},t) &= \mathrm{Re}\left[\vec{E}(\vec{r})e^{j\omega t}\right]x\mathrm{Re}\left[\vec{H}(\vec{r})e^{j\omega t}\right] \\ &= \frac{1}{2}\left[\vec{E}(\vec{r})x\vec{H}^*(\vec{r})\right] + \frac{1}{2}\left[\vec{E}(\vec{r})x\vec{H}^*(\vec{r})e^{j2\omega t}\right]\end{aligned} \tag{2.244}$$

The first term of eqn (2.244) is not a time function and the time variations of the second term are twice the given frequency.

The time average Poyinting vector or in other words the average power density is then

$$\vec{P}_{d,av} = \frac{1}{2\pi} \int_0^{2\pi} \vec{P}_d(\vec{r}, t) d(\omega t) = \frac{1}{2} \text{Re}\left[\vec{E} x \vec{H}^*\right] \tag{2.245}$$

The $\frac{1}{2}$ factor appears because E and H fields represent peak values and it should be omitted for root-mean-square (rms) values.

The total average power is then given by

$$P_{av} = \oint_S \frac{1}{2} \text{Re}[\vec{E} x \vec{H}^*] d\vec{S} \tag{2.246}$$

and it can, for example, represent the radiated power by an antenna.

In addition, the first volume integral in the right-hand side of eqn (2.242) represents power loss in the conduction currents and it is just twice the average power loss, which is given by

$$P_L = \frac{1}{2} \int_V \sigma |\vec{E}|^2 dV \tag{2.247}$$

Finally, the second volume integral in the right-hand side of eqn (2.242) is proportional to the difference between the average stored magnetic energy in the volume and the average stored electric energy.

2.16.2 Complex Phasor Form of Electromagnetic Wave Equations

The complex phasor representation applied to wave equations (2.88) and (2.89) results in the following set of the Helmholtz type equations:

$$\nabla^2 \vec{H} - j\omega\mu\sigma\vec{H} + \omega^2 \mu\varepsilon\vec{H} = 0 \tag{2.248}$$

$$\nabla^2 \vec{E} - j\omega\mu\sigma\vec{E} + \omega^2 \mu\varepsilon\vec{E} = -\frac{1}{\varepsilon} \nabla\rho \tag{2.249}$$

which is usually written in the form

$$\nabla^2 \vec{H} - \gamma^2 \vec{H} = 0 \tag{2.250}$$

$$\nabla^2 \vec{E} - \gamma^2 \vec{E} = -\frac{1}{\varepsilon} \nabla\rho \tag{2.251}$$

where γ is the complex propagation constant,

$$\gamma = \sqrt{j\omega\mu\sigma - \omega^2 \mu\varepsilon} \tag{2.252}$$

For a linear, isotropic, homogeneous, source-free medium the set of eqns (2.248) and (2.249) are simplified into

$$\nabla^2 \vec{H} + k^2 \vec{H} = 0 \tag{2.253}$$

$$\nabla^2 \vec{E} + k^2 \vec{E} = -\frac{1}{\varepsilon} \nabla \rho \tag{2.254}$$

where k is a wave number of a lossless medium that is given by

$$k = \omega\sqrt{\mu\varepsilon} \tag{2.255}$$

For the case of a source-free medium eqn (2.254) becomes homogeneous,

$$\nabla^2 \vec{E} + k^2 \vec{E} = 0 \tag{2.256}$$

The complex form of potential wave equations could be derived similarly.

2.16.3 The Retarded Potentials for the Time-Harmonic Fields

If all electromagnetic quantities of interest are varying harmonically in time, the particular integrals for the retarded potentials are given by

$$\varphi(\vec{r}) = \int_{V'} \frac{\rho(\vec{r}')e^{-jkR}}{4\pi\varepsilon R} \mathrm{d}V' \tag{2.257}$$

$$\vec{A}(r) = \int_{V'} \frac{\vec{J}(\vec{r}')e^{-jkR}}{4\pi\varepsilon R} \mathrm{d}V' \tag{2.258}$$

while the complex notation $e^{j\omega t}$ is understood and omitted.

2.17 TRANSMISSION LINE THEORY

At low frequencies where the wavelength is much larger than the system size, electromagnetic interferences can be modeled using the circuit theory [4], whereas at higher frequencies, due to wave-like behavior of the signals the simple circuit models are not valid. In this case, however, conducted interferences can be treated by means of transmission line (TL) theory.

The request that at high frequencies the propagation effects must be taken into account requires the conductor modeling via distributed parameters as illustrated in Figure 2.11, where a differential section of a two conductor transmission line is shown.

The distributed parameters R', G', L', and C' can be determined by calculation or measurement. Excited transmission line can produce two different kinds of

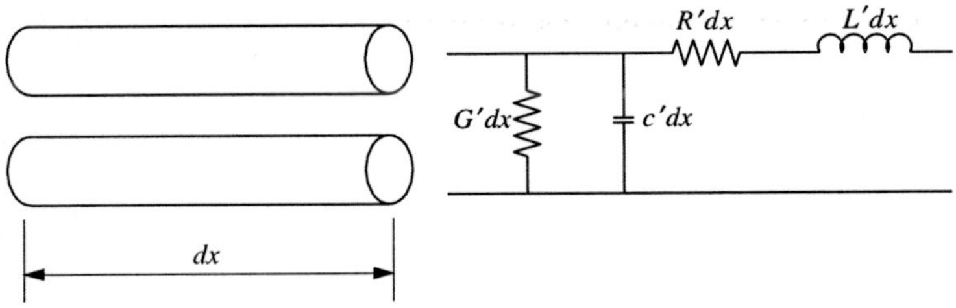

Figure 2.11 Distributed parameters of a two-conductor system.

responses: transmission line mode currents and antenna mode currents. The transmission line mode currents consist of equal and opposite currents flowing in each conductor and radiate negligible amounts of energy because the currents flow in the opposite direction. The antenna mode currents flow in the same direction on the line vanishing at each end of the line. Consequently, this set of currents radiates energy away from the line very well.

For a rigorous analysis of the current at a general point on the line it is necessary to take into account both the modes. However, in many applications transmission line mode is the dominant one and the antenna mode currents can be neglected, especially if only the responses at the loads of the line are considered. Also, under some circumstances, the antenna mode can vanish due to the symmetries in the transmission line configuration.

2.17.1 Field Coupling Using Transmission Line Models

Instead of lumped voltage or current sources, the transmission line can be excited by electromagnetic fields generated by distant electromagnetic interference (EMI) sources which induce currents and voltages on the line and load impedance at the ends. These responses can be rigorously analyzed using the antenna (scattering) theory or approximately using the transmission line models. It is worth emphasizing that transmission line systems support two types of wave propagation: free space propagation of an incident field illuminating the line and the propagation of the induced currents and voltages along the line.

There are three different approaches to analyze the coupling of an electromagnetic field to a line via transmission line theory [4]

- Line is excited by the incident magnetic flux linking the two conductors and incident electric flux terminating on the two conductors, giving rise to distributed voltage and current sources on the line. This is known as the Taylor approach.
- The excitation function is a tangential electric field along the conductors that can be represented in terms of distributed voltage sources existing in the transmission line. This approach is referred to as the Agrawal method.

- The line is excited only by an incident magnetic field that gives rise only to distributed current sources on the line. This approach has been derived by Rachidi.

The use of transmission line models requires that the line length be significantly larger than the separation of the conductors. If this is not the case the line acts as a loop antenna and TL models are not appropriate. Therefore, by using these models it is possible to compute only the transmission components of the current, and if the antenna mode current along the line is of interest the full field analysis is required.

2.17.2 Derivation of Telegrapher's Equation for the Two-Wire Transmission Line

This section deals with the derivation of Telegrapher's equations for the two-wire transmission line for the analysis of the behavior of currents and voltages induced along the lines [4]. Differential cross-section of a two-wire transmission line is shown in Figure 2.12. A vertically polarized field is considered but the derivation is also valid for the case of horizontally polarized fields. The incident field illuminating the conductors induces a current and charge on the wires. The charge also causes the voltage between the conductors.

Two identical wires are assumed to be perfectly conducting with radius a and insulated in free space.

The separation of the conductors d is large compared with radius $a(d \gg a)$ and is small compared with incident field wavelength $(d \ll \lambda)$. Therefore, only the transmission line mode currents are taken into account.

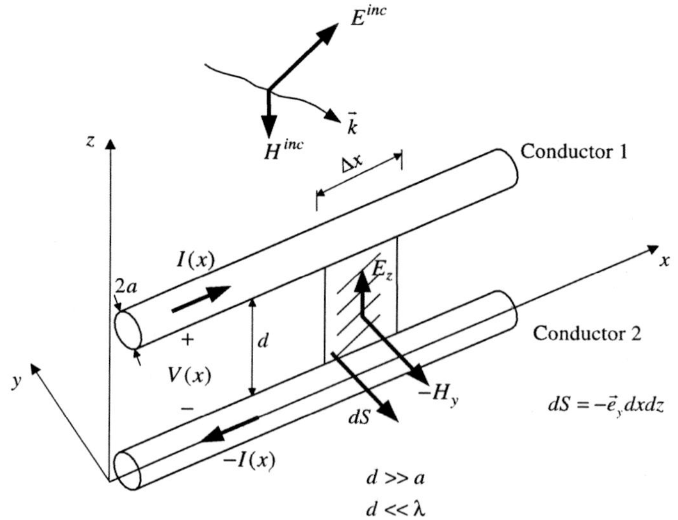

Figure 2.12 Differential section of a two-wire transmission line.

2.17.2.1 First Telegrapher's Equation First Telegrapher's equation can be derived starting from the Maxwell equation,

$$\nabla x\vec{E} = -j\omega\mu_0\vec{H} \tag{2.259}$$

whose integral form is

$$\int_c \vec{E}d\vec{s} = -j\omega\mu_0 \iint_S \vec{H}d\vec{S} \tag{2.260}$$

Total electric and magnetic field, respectively, may be decomposed into [4]

- incident electric and magnetic fields E^{inc} and H^{inc}, respectively, existing in the absence of the transmission line;
- scattered electric and magnetic fields E^{sct} and H^{sct}, respectively, generated by the currents and charges induced along the wires;
- quasistatic fields produced by the separation of charge and current around the circumference of the wire.

Equation (2.260) can be integrated over the differential area dS as follows:

$$\int_0^d [E_z(x+\Delta x,z)-E_z(x,z)]dz - \int_x^{x+\Delta x} [E_x(x,d)-E_x(x,0)]dx = -j\omega\mu_0 \int_0^d \int_0^{x+\Delta x} (-H_y)dxdz \tag{2.261}$$

where all quantities in eqn (2.261) are related to the total fields.

Since $d \ll \lambda$, the total voltage between the conductors can be obtained from the expression

$$V(x) = -\int_0^d E_z(x,z)dz \tag{2.262}$$

which is in accordance to the quasistatic concept.

If the wires are perfectly conducting, the total fields tangential to the wire surface vanish.

Dividing eqn (2.261) by Δx and by taking the limit $\Delta x \to 0$ yields:

$$\frac{dV(x)}{dx} = -j\omega\mu_0 \int_0^d H_y(x,z)dz = -j\omega\mu_0 \int_0^d H_y^{inc}(x,z)dz - j\omega\mu_0 \int_0^d H_y^{sct}(x,z)dz \tag{2.263}$$

Note that the total magnetic field consists of incident and scattered component, respectively.

The second term at the right-hand side of eqn (2.263) is the magnetic flux flowing between the two conductors generated by the currents flowing along the both conductors. Assuming the current $I(x)$ to be uniformly distributed around the circumference of the wire ($d \gg a$) and also assuming the separation between the conductors to be electrically small ($d \ll \lambda$), the flux density can be evaluated by analytically integrating the second term at the right-hand side of eqn (2.263), that is as follows:

$$\Phi(x) = \mu_0 \int_0^d H_y^{sct}(x, z)\mathrm{d}z = L'I(x) \tag{2.264}$$

where L' is the inductance per unit length,

$$L' = \frac{\mu_0}{\pi} \ln \frac{d}{a} \tag{2.265}$$

Substituting eqns (2.264) and (2.265) into eqn (2.263) yields

$$\frac{\mathrm{d}V(x)}{\mathrm{d}x} + j\omega L'I(x) = V'_{S1}(x) \tag{2.266}$$

where the distributed voltage source along the line is given by

$$V'_{S1}(x) = -j\omega\mu_0 \int_0^d H_y^{inc}(x, z)\mathrm{d}z \tag{2.267}$$

Equation (2.266) is the *first telegrapher's equation* for the case of a distributed field excitation along the transmission line.

2.17.2.2 Second Telegrapher's Equation Second Telegrapher's equation can be derived starting from the Maxwell equation,

$$\nabla x\vec{H} = \vec{J} + j\omega\varepsilon\vec{E} \tag{2.268}$$

whose integral form is:

$$\int_S \nabla x\vec{H}\mathrm{d}\vec{S} = \int_S [\vec{J} + j\omega\vec{E}]\mathrm{d}\vec{S} \tag{2.269}$$

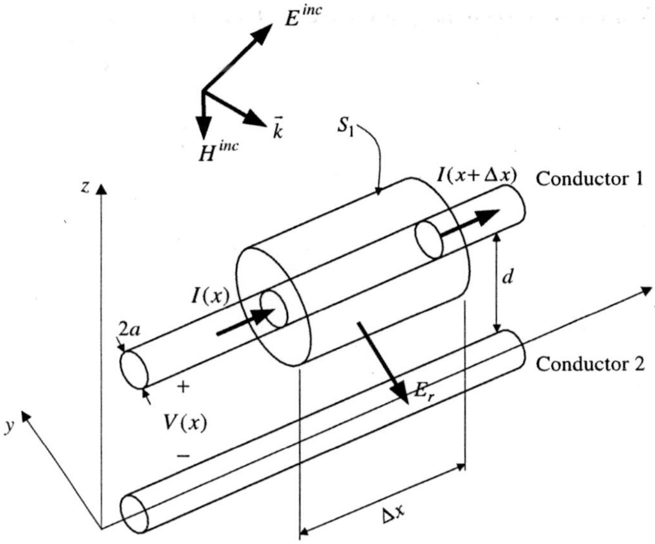

Figure 2.13 Closed surface surrounding the wire.

For the case of a closed surface S, surrounding one of the conductors as shown in Figure 2.13 left-hand side of eqn (2.269) vanishes, that,

$$\int_S \nabla x \vec{H} d\vec{S} = 0 \tag{2.270}$$

Rearranging eqn (2.269) yields

$$I(x + \Delta x) - I(x) + j\omega\varepsilon \iint_{S_1} E_r r d\Phi dx = 0 \tag{2.271}$$

where E_r is the total radial field in the vicinity of the wire.

The total electric field can be represented as a sum of the incident and scattered component. Dividing eqn (2.271) with Δx and taking the limits $r \to a$ and $\Delta x \to 0$ results in the following equation:

$$\frac{dI(x)}{dx} + j\omega\varepsilon \int_0^{2\pi} E_r^{sct}(x) a d\Phi + j\omega\varepsilon \int_0^{2\pi} E_r^{inc}(x) a d\Phi = 0 \tag{2.272}$$

Assuming $a \ll d$, the first integral at the left-hand side in eqn (2.272) becomes

$$j\omega\varepsilon \int_0^{2\pi} E_r^{sct}(x) a d\Phi = j\omega 2\pi a \varepsilon E_r^{sct}(x) = j\omega q'(x) \tag{2.273}$$

where $q'(x)$ represents the linear charge density along the conductor.

The second integral at the left-hand side of eqn (2.272) vanishes because the free charge in the vicinity of the wire does not exist if the wire is removed. Therefore, it follows

$$\frac{dI(x)}{dx} + j\omega q'(x) = 0 \tag{2.274}$$

Equation (2.274) is the second telegrapher's equation given in the form of one-dimensional equation of continuity relating current and charge along the wire.

The induced charge along the wire charge can be expressed in terms of capacitance per unit length of the wire C' and scattered voltage,

$$q'(x) = C' V^{sct}(x) \tag{2.275}$$

where C' is given by

$$C' = \frac{\pi\varepsilon}{\ln\frac{d}{a}} (a \ll d) \tag{2.276}$$

Furthermore, the total voltage between the wires is given by

$$V(x) = V^{inc}(x) + V^{sct}(x) = -\int_0^d E_z^{inc}(x, z)dz + V^{sct}(x) \tag{2.277}$$

Now the second telegrapher's equation can be written as

$$\frac{dI(x)}{dx} + j\omega C' V(x) = I'_{S_1}(x) \tag{2.278}$$

where $I'(x)$ is defined by relation

$$I'_{S_1}(x) = -j\omega C' \int_0^d E_z^{inc}(x, z)dz \tag{2.279}$$

Telegrapher's equations (2.266) and (2.278) require a knowledge of the termination conditions at the line ends.

2.17.2.3 Telegrapher's Equations for a Finitely Conducting Wire If the transmission line is not perfectly conducting, the total tangential electric field along the wires is not zero. The telegrapher's equations can be modified to account for the presence of finitely conducting wire by using the concept of surface impedance that relates the total field on the surface of the wire to the current flowing on the wire [4],

$$E_x(x) = \frac{Z_{cw}}{2\pi a}\frac{I_0(\gamma_w a)}{I_1(\gamma_w a)}I(x) = Z'_w(x)I(x) \tag{2.280}$$

where $Z'_w(x)$ is the impedance per unit length of the wire that is given by

$$Z_{cw} = \sqrt{\frac{j\omega\mu_w}{\sigma_w + j\omega\varepsilon_w}} \tag{2.281}$$

and γ is the propagation constant in the wire material,

$$\gamma_w = \sqrt{j\omega\mu_w(\sigma_w + j\omega\varepsilon_w)} \tag{2.282}$$

At low frequencies ($|\gamma_w a| \ll 1$) this impedance can be approximated by expression

$$Z'_w = \frac{1}{a^2\pi\sigma_w} + j\omega\frac{\mu\omega}{8\pi} \tag{2.283}$$

while at high frequencies ($|\gamma_w a| \gg 1$) the surface impedance becomes

$$Z'_w = \frac{1+j}{2\pi a}\sqrt{\frac{\omega\mu_w}{2\sigma\omega}} \tag{2.284}$$

Performing some mathematical manipulations, the following telegrapher's equation is obtained:

$$\frac{\mathrm{d}V(x)}{\mathrm{d}x} + Z'I(x) = V'_{S_1}(x) \tag{2.285}$$

where the per-unit-length impedance term Z' is given by [4]

$$Z' = j\omega L' + 2Z'_w \tag{2.286}$$

Note that the distributed voltage source $V'(x)$ remains unmodified.

2.17.2.4 Telegrapher's Equations for Lossy Media If the transmission line is immersed into the medium having finite conductivity σ_m and dielectric constant ε_r a conductive current flowing between the lines may appear. This phenomenon can be taken into account via a shunt per-unit-length conductance G' that is defined as

$$G' = \frac{\sigma_m}{\varepsilon_m}C' \tag{2.287}$$

The corresponding second telegrapher's equation now becomes

$$\frac{\mathrm{d}I(x)}{\mathrm{d}x} + Y'V(x) = I'_{S_1}(x) \tag{2.288}$$

where the per-unit-length admittance of the line is given by

$$Y' = j\omega C' + G' = \left(j\omega + \frac{\sigma_m}{\varepsilon_m}\right)C' \tag{2.289}$$

while the distributed current source is given by eqn (2.279).

2.17.2.5 ***Single Line Over a Perfectly Conducting Ground Plane*** A transmission line structure may consist of a single conductor above a conducting ground plane that serves as the return conductor. The current excited by an incident field on the line consists primarily of the transmission line mode with the antenna mode being negligibly small due to the symmetry of the excitation.

The induced currents and voltages along the line can also be analyzed using somewhat modified telegrapher's equations.

Namely, the incident electric and magnetic fields exciting the line that exist near the line but with the line removed, as shown in Figure 2.14, are now composed from incident- and ground-reflecting components, respectively. On the PEC ground the

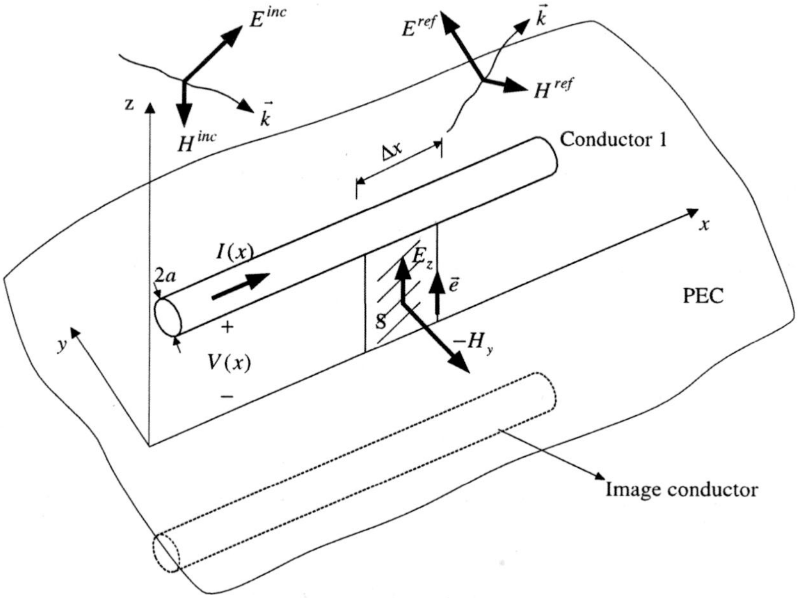

Figure 2.14 Differential section of a straight wire above a perfect ground.

boundary condition for electric field requires $E_x = 0$, $E_y = 0$, and $E_z \neq 0$. Similarly, for magnetic field it follows $H_z = 0$, H_x, and $H_y \neq 0$. Further to these modifications, the per-unit-length parameters of the line are different from those related to the two-conductor line in free space.

The derivation of the corresponding telegrapher's equations can be carried out using the total voltage formulation as in the case of two wires in free space. The Maxwell eqn (2.259) is integrated over the shaded area shown in Figure 2.14. Instead of decomposing the magnetic field into incident and scattered components, the magnetic field is divided into excitation and scattered components, respectively, that is,

$$\vec{H}^{exc} = \vec{H}^{inc} + \vec{H}^{ref} \tag{2.290}$$

Following the same steps in the derivation as in section 2.17.2.1, the first telegrapher's equation is obtained,

$$\frac{dV(x)}{dx} = j\omega L' I(x) = V'_{S_1}(x) \tag{2.291}$$

where per-unit-length inductance is

$$L' = \frac{\mu_0}{2\pi} \ln \frac{2h}{a} \tag{2.292}$$

The voltage source $V'_{S_1}(x)$ distributed along the line is expressed in terms of the incident- and the ground-reflecting magnetic fields:

$$\begin{aligned} V'_{S_1}(x) &= -j\omega\mu_0 \int_0^h H_y^{exc}(x,z)\,dz \\ &= -j\omega\mu_0 \int_0^h \left[H_y^{inc}(x,z) + H_y^{ref}(x,z) \right] dz \end{aligned} \tag{2.293}$$

The second telegrapher's equation can be derived by decomposing the electric field into excitation components and line scattered components, that is,

$$\vec{E}^{exc} = \vec{E}^{inc} + \vec{E}^{ref} \tag{2.294}$$

Incident, reflected, and scattered electric and magnetic fields are shown in Figure 2.15.

Performing some mathematical manipulations the second telegrapher's equation is obtained,

$$\frac{dI(x)}{dx} = j\omega C' V(x) = I'_{S_1}(x) \tag{2.295}$$

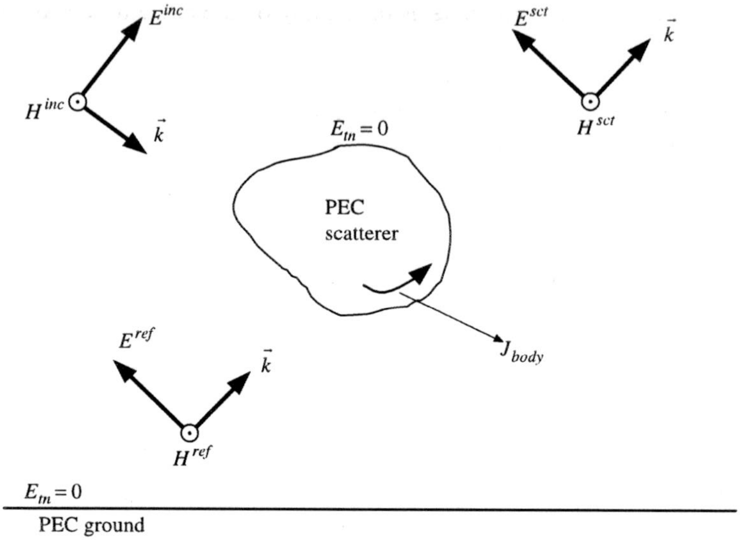

Figure 2.15 Scattering from the PEC object above a PEC ground.

where per-unit-length capacitance is

$$C' = \frac{2\pi\varepsilon}{\ln\dfrac{2h}{a}} \tag{2.296}$$

The current source $I'_{S_1}(x)$ distributed along the line is in terms of the incident- and ground-reflecting electric fields,

$$\begin{aligned} I'_{S_1}(x) &= -j\omega C' \int_0^h E_z^{exc}(x,z)\mathrm{d}z \\ &= -j\omega C' \int_0^h \left[E_z^{inc}(x,z) + E_z^{ref}(x,z)\right]\mathrm{d}z \end{aligned} \tag{2.297}$$

Telegrapher's equations (2.291) and (2.295) can be solved if the associated termination conditions at the line ends are known.

2.17.2.6 Effects of a Lossy Ground on Transmission Lines If the ground plane is not perfectly conducting, the telegrapher's equations have to be modified to take into account the effects of a lossy ground. In the case of a lossy ground, both the impedance and propagation constants become complex, while the induced currents and voltages propagating along the line are attenuated. Furthermore, the reflected and transmitted plane wave fields can be represented by the corresponding reflection and transmission coefficients depending on the angle of incidence and on the electrical properties of the ground. Contrary to the case of the perfect ground, where the amplitude and waveform of the reflected field is identical to the incident field, field reflected from a lossy ground is attenuated exponentially with depth.

In this case the first Telegrapher's equation [4] is given by

$$\frac{dV(x)}{dx} + Z'(\omega)I(x) = V'_{S_1}(x) \tag{2.298}$$

and the per-unit-length impedance Z' is defined as

$$Z'(\omega) = j\omega L' + Z'_g + Z_w \tag{2.299}$$

where Z'_g denotes the ground impedance and Z_w accounts for the wire internal impedance.

The source term in eqn (2.298) is given by

$$V'_{S1}(x) = -j\omega\mu_0 \int_0^h H_y^{exc}(x,z)dz + E_x^{exc}(x,0) \tag{2.300}$$

The second Telegrapher's equation can be written as [4]

$$\frac{dI(x)}{dx} + Y'V(x) = I'_{S_1}(x) \tag{2.301}$$

where the line admittance Y' is defined as

$$Y' = \left(\frac{1}{j\omega C'} + \frac{1}{Y'_g}\right)^{-1} \tag{2.302}$$

It is worth noting that Y' is related to the capacitance of the line over a perfect ground, while the term Y'_g stands for the ground admittance.

Ground impedance Z'_g and admittance Y'_g are related as follows:

$$Z'_g Y'_g \doteq \gamma_g^2 = j\omega\mu_0(\sigma_g + j\omega\varepsilon_g) \tag{2.303}$$

The distributed current source is defined by the expression

$$I'_{S_1}(x) \doteq -Y'\left[\int_0^h \left[E_z^{inc}(x,z) + E_z^{ref}(x,z)\right]dz\right] \tag{2.304}$$

There are several different expressions developed in relevant literature for the ground impedance Z'_g. Many of these expressions have been reviewed in Ref. [4].

One of the usually used formulas is given by Sunde,

$$Z'_g \doteq \frac{j\omega\mu_0}{\pi} \int_0^\infty \frac{e^{-2hu}}{u + \sqrt{u^2 + \gamma_g^2}} du \tag{2.305}$$

Good approximation of eqn (2.305) for low frequencies $\sigma_g \gg \omega\varepsilon_r\varepsilon_0$ has been derived by Carson,

$$Z'_g \doteq \frac{\omega\mu_0}{\pi} \int_0^\infty (\sqrt{j + v^2} - v)e^{-2h'v} dv \tag{2.306}$$

where

$$v = \frac{u}{\sqrt{\omega\mu_0\sigma_g}} \quad \text{and} \quad h' = h\sqrt{\omega\mu_0\sigma_g}; \quad u = (\xi^2 - k^2)^{1/2} \tag{2.307}$$

Approximation proposed by Vance is given by

$$Z'_g = -\frac{j\gamma_g}{2\pi h\sigma_g} \frac{H_0^{(1)}(j\gamma_g h)}{H_1^{(1)}(j\gamma_g h)} \tag{2.308}$$

where $H^{(1)}$ are cylindrical Hankel functions.

2.18 PLANE WAVE PROPAGATION

The existence of electromagnetic wave propagation is proved mathematically by the solution of Maxwell equations in conjunction with the appropriate material properties.

Plane waves are good approximations to real waves in various practical situations. For example, radio waves at very large distances from the transmitter, or from diffracting objects, have negligible curvature and are well represented by plane waves. Moreover, complicated wave patterns can be considered as a superposition of plane waves. Even in the situation when the plane wave concept is not followed, the basic ideas of propagation, reflection, and refraction met in the plane wave approach help to understand more complex wave problems.

If one starts by finding the simplest possible solution to eqn (2.256) for the case of source-free homogeneous medium and x-directed component of the vector field intensity,

$$\frac{\partial^2 E_y}{\partial x^2} + \frac{\partial^2 E_y}{\partial y^2} + \frac{\partial^2 E_y}{\partial z^2} = -k^2 E_y \tag{2.309}$$

If there is no variation of the fields in both x and y directions then eqn (2.309) is simplified into

$$\frac{\partial^2 E_y}{\partial x^2} = -k^2 E_y \tag{2.310}$$

and the solution is given by an expression

$$E_y = Ae^{-jkx} + Be^{+jkx} \tag{2.311}$$

where k is the wave number of the corresponding lossless medium defined by relation (2.255), A and B are magnitudes of forward and backward wave, respectively.

It can be observed that eqn (2.311) represents a transmission-line type of wave that propagates in the x-direction. E_y component is one of the field component that acts in a direction perpendicular to the direction of propagation of the wave.

There is no variation of the field quantities in the plane perpendicular to the propagation direction, therefore, the wave is simply called a *plane wave*.

The expression for the forward wave from (2.311) can be written in the form

$$E_y = E_0 e^{-jkx} \tag{2.312}$$

where E_0 is the magnitude of the electric field.

The corresponding plane wave is shown in Figure 2.16.

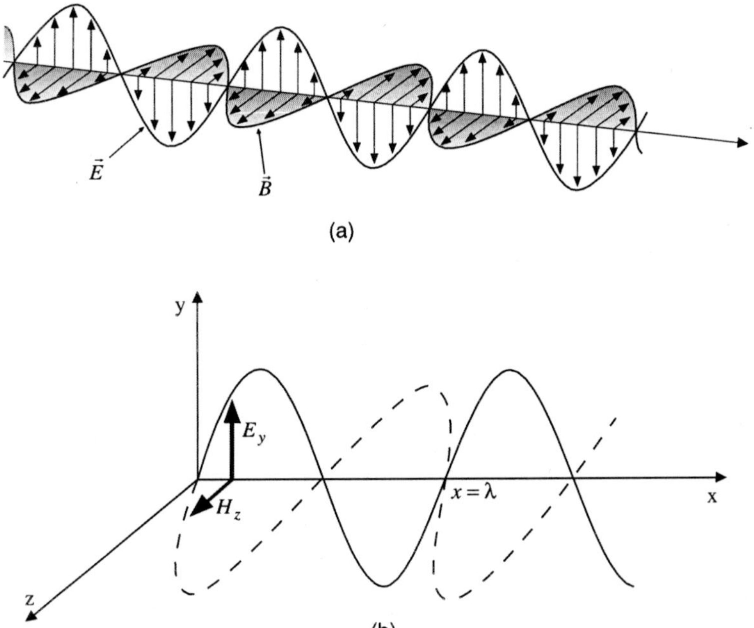

Figure 2.16 (a) The propagation of a plane wave. (b) The electric and magnetic field of an x-directed planewave.

Until now the time-harmonic variation $e^{j\omega t}$ has been assumed but omitted. The complete solution for the forward wave is thus given by

$$E_y = E_0 e^{j(\omega t - kx)} \tag{2.313}$$

and the associated magnetic field is then

$$H_z = H_0 e^{j(\omega t - kx)} = \frac{E_0}{Z_0} e^{j(\omega t - kx)} \tag{2.314}$$

where

$$Z_0 = \frac{E_0}{H_0} = \sqrt{\frac{\mu}{\varepsilon}} \tag{2.315}$$

is the wave impedance of the medium. For free space, it is approximately $Z_0 = 120\,\pi$.

2.19 RADIATION

Radiation of electromagnetic energy is an undesired leakage phenomenon or a desired process for exciting waves in space. If one deals with desired radiation, the goal is to excite electromagnetic waves from the given source in the required direction as efficiently as possible. The matching unit between the source and waves in space is known as the *radiator, antenna*, or *areal*. It is worth emphasizing that the results developed for radiating or transmitting antennas can be applied to the same antenna when used for receiving applications. The relation of the radiating case to the receiving situation can be made rigorously by means of the so-called principle of reciprocity [5].

The simplest radiating system is that of an ideal short linear element with current considered uniform over its entire length. More complex antenna structures can be considered to be made up of a large number of such elementary antennas with the corresponding magnitudes and phases of their currents.

2.19.1 Radiation Mechanism

The radiation is accomplished by a time-varying current or by an acceleration or deceleration of the charge. In other words, if a charge is oscillating in a time-motion, it creates radiation.

When the wire is initially excited, the charges are set by the source-created electric field. The acceleration of the charges is accomplished by the external source in which forces set the charges in motion and produce the associated field. The deceleration of the charges at the end of the wire is accomplished by the self (internal) forces associated with the induced field due to the buildup of charge concentration at the wire ends. These internal forces receive energy from the charge buildup as its velocity is reduced to zero at the ends of the wire.

The existence of the electromagnetic waves on the antenna is associated with the presence of charges inside the conductors. However, when the waves are radiated they form the closed loops and there are no charges to sustain their existence.

Consequently, the electric charges are required to create the fields but they are not necessary to sustain them and can exist in their absence.

2.19.2 Hertzian Dipole

The current element, usually called Hertzian dipole, is located in z-axis direction, with its location at the origin of a set of spherical coordinates in Figure 2.17.

The length h of this electrically short antenna is very small compared to wavelength. Since the wire radius a is small compared to the wavelength, the particular integral for retarded potential can be written in the form

$$\vec{A}(r) = \int_{s'} \frac{\mu \vec{J}(\vec{r}')\vec{S}e^{-jkR}}{4\pi R} d\vec{s}' \tag{2.316}$$

which results in

$$\vec{A}(z) = \int_{L} \frac{\mu I(z')e^{-jkR}}{4\pi R} \vec{e} dz' \tag{2.317}$$

and finally, if the current along the short antenna is assumed to be constant and expressed in the following phasor form:

$$I(z) = I_0 e^{j\omega t} \tag{2.318}$$

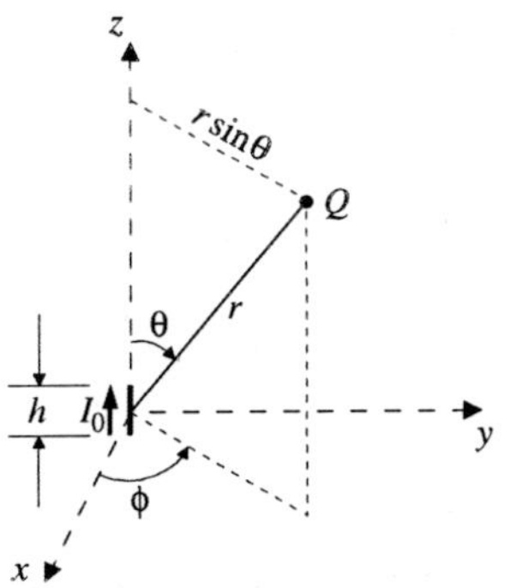

Figure 2.17 Hertzian dipole.

one obtains the final expression for the related retarded potential

$$A_z(z) = \int_L \frac{\mu I(z')e^{-jkR}}{4\pi R} dz' = \frac{\mu I_0 e^{-jkr}}{4\pi r} \int_{-h}^{h} dz' = \frac{\mu h I_0}{4\pi r} e^{-jkr} \tag{2.319}$$

If the system of spherical coordinates is considered, it follows

$$A_r = A_z \cos\theta = \mu \frac{hI_0}{4\pi r} e^{-jkr} \cos\theta \tag{2.320}$$

$$A_\theta = -A_z \sin\theta = -\mu \frac{hI_0}{4\pi r} e^{-jkr} \sin\theta \tag{2.321}$$

The electric and magnetic field components can be found directly from the magnetic vector potential components by using the time harmonic version of relations (2.121), (2.125), and (2.151).

The complex phasor form of eqns (2.125) and (2.151) is given as follows:

$$\vec{E} = -j\omega\vec{A} - \nabla\varphi \tag{2.322}$$

$$\nabla\vec{A} = -j\omega\mu\varepsilon\varphi \tag{2.323}$$

and the electric field is then given by

$$\vec{E} = -j\omega\vec{A} + \frac{1}{j\omega\mu\varepsilon}\nabla(\nabla\vec{A}) \tag{2.324}$$

Thus, the corresponding field components are

$$H_\phi = \frac{hI_0}{4\pi} e^{-jkr} \left(\frac{jk}{r} + \frac{1}{r^2}\right) \sin\theta \tag{2.325}$$

$$E_r = \frac{hI_0}{4\pi} e^{-jkr} \left(\frac{2Z_0}{r^2} + \frac{2}{j\omega\varepsilon r^3}\right) \cos\theta \tag{2.326}$$

$$E_\theta = \frac{hI_0}{4\pi} e^{-jkr} \left(\frac{j\omega\mu}{r} + \frac{1}{j\omega\varepsilon r^3} + \frac{Z_0}{r^2}\right) \sin\theta \tag{2.327}$$

At very great distances from the source ($r \gg \lambda$), the only significant terms for E and H are those varying as $1/r$. This is the region where the far field is significant and it is called the Fraunhofer region, Figure 2.18.

The components of the far field are

$$H_\phi = \frac{jkhI_0}{4\pi r} e^{-jkr} \sin\theta \tag{2.328}$$

$$E_\theta = \frac{j\omega\mu h I_0}{4\pi r} e^{-jkr} \sin\theta = Z_0 H_\phi \tag{2.329}$$

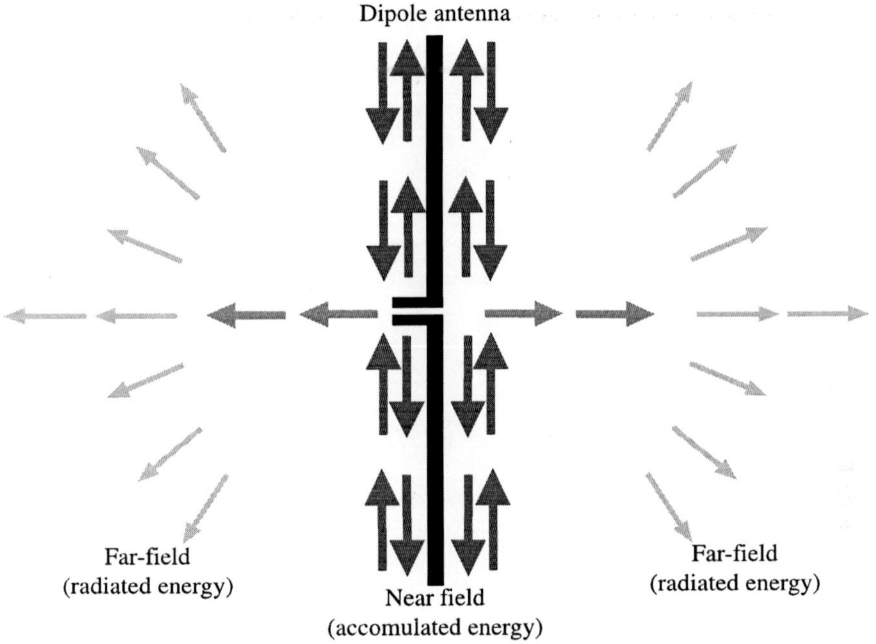

Figure 2.18 Radiation of an electrically short antenna.

The total power collected in the far field can be obtained by using expression (2.246). Namely, the total power is the integral of the time average Poyinting vector over any surrounding surface. For simplicity, this surface is a sphere of radius r,

$$P = \oint_s \frac{1}{2}\mathrm{Re}(\vec{E}x\vec{H}^*)\mathrm{d}\vec{S} = \int_0^{2\pi}\int_0^{\pi} \frac{1}{2}\mathrm{Re}(E_\theta H_\phi^*)r^2 \sin\theta \mathrm{d}\theta \mathrm{d}\phi$$
$$= \frac{Z_0 k^2 I_0^2 h^2}{16\pi}\int_0^{\pi} \sin^3\theta \mathrm{d}\theta = \frac{Z_0 \pi I_0^2}{3}\left(\frac{h}{\lambda}\right)^2 \tag{2.330}$$

It is obvious that the power radiated by the electrically short antenna is proportional to the squared value of the ratio h/λ. Consequently, it is also clear that the Hertzian dipole with the entire length h that is very small compared to wavelength λ is a radiator with a very small efficiency.

2.19.3 Fundamental Antenna Parameters

To describe antenna properties, it is necessary to define various parameters. Some of the parameter definitions specified in Ref. [6] are given in this chapter.

2.19.3.1 Radiation Power Density The power density of an antenna is equal to the real part of the time average Poyinting vector, and for the case of a free space it follows

$$\bar{P}_d = \frac{1}{2}\mathrm{Re}(E_\theta H_\phi) = \frac{E_\theta^2}{2Z_0} \tag{2.331}$$

where E_θ and H_Φ are the maximal (peak) values of the electric and magnetic fields, respectively. If the rms values are considered then eqn (2.331) becomes

$$\bar{P}_d = \frac{E_\theta^2}{Z_0} \tag{2.332}$$

In the far field region, E_θ and H_Φ vary as $1/r$ and thus P_d varies as $1/r^2$.

2.19.3.2 Radiation Intensity Radiation intensity, or the antenna power pattern, in a given direction is defined as the power radiated from an antenna per unit solid angle. The radiation intensity is a far field parameter that can be obtained by simply multiplying the radiation power density by the square distance, that is,

$$U = r^2\frac{1}{2}\mathrm{Re}(E_\theta H_\phi) = r^2\frac{1}{2}\frac{E_\theta}{Z_0} \tag{2.333}$$

The relationship between the total power and radiation intensity is given by

$$P_{rad} = \oint_\Omega U d\Omega = \int_0^{2\pi}\int_0^{\pi} U \sin\theta \, d\theta \, d\phi \tag{2.334}$$

where $d\Omega = \sin\theta \, d\theta \, d\Phi$.

In particular, for an isotropic source, eqn (2.334) becomes

$$P_{rad} = \oint_\Omega U_0 d\Omega = 4\pi U_0 \tag{2.335}$$

or the radiation intensity of an isotropic source is given by

$$U_0 = \frac{P_{rad}}{4\pi} \tag{2.336}$$

As it is obvious, the radiation intensity U is independent of the angles Φ and θ.

2.19.9.3 Directivity Directivity of an antenna is defined as the ratio of the radiation intensity in a given direction from the antenna to the radiation intensity averaged over all directions. The average radiation intensity is equal to the total power radiated by the antenna divided by 4π. If the direction is not strictly specified, the direction of maximum radiation intensity is implied.

Namely, the directivity D of a nonisotropic source is equal to the ratio of its radiation intensity in a given direction over that of an isotropic source,

$$D = \frac{U}{U_0} = \frac{4\pi U}{P_{rad}} \tag{2.337}$$

If the direction is not specified, it implies the direction of maximum radiation intensity (maximum directivity) determined as

$$D_{\max} = \frac{U_{\max}}{U_0} = \frac{4\pi U_{\max}}{P_{rad}} \tag{2.338}$$

where $D_{\max}$ is the maximum directivity, U_0 is the radiation intensity of isotropic source, and $U_{\max}$ is the maximum radiation intensity. To summarize: directivity is a measure that describes only the directional properties of the antenna and, therefore, controlled only by the pattern.

2.19.3.4 Input Impedance, Radiation, and Loss Resistance Input impedance is defined as the ratio of the voltage to current at the pair of input antenna terminals:

$$Z_a = R_a + jX_a \tag{2.339}$$

where R_a is the resistance at antenna terminals and X_s is the reactance at antenna terminals.

In addition, the resistive part of eqn (2.339) consists of two components,

$$R_a = R_r + R_L \tag{2.340}$$

where R_r is the radiation resistance and R_L is the loss resistance of the antenna.

The radiation resistance R_r is defined as the ratio of the total power radiated by the antenna P_0 to the square of the rms antenna input current I,

$$R_r = \frac{P_0}{I^2} \tag{2.341}$$

If the peak value of the current is considered, then it follows

$$R_r = \frac{P_0}{2I_m^2} \tag{2.342}$$

This is an equivalent resistance that would dissipate a power equal to the total radiated power when the current through it is equal to the antenna input current.

2.19.3.5 Gain and Radiation Efficiency Antenna gain is closely related to the directivity, but it is also a measure that takes into account the efficiency of the antenna. Gain of an antenna in a given direction is defined as the ratio of the intensity, in a given direction, to the radiation intensity that would be obtained if the power accepted by the antenna were radiated isotropically. This is called *absolute gain*. The radiation intensity corresponding to the isotropically radiated power is equal to the power input by the antenna divided by 4π. Thus, it follows

$$G(\phi, \theta) = 4\pi \frac{U(phi, \theta)}{P_{in}} \tag{2.343}$$

where P_{in} is the total input power.

However, in most of the cases, one deals with *relative gain* which is often defined as the ratio of the power gain in a given direction to the power gain of a reference in its referenced direction. The power input has to be the same for both antennas. The reference antenna is usually a dipole, horn, or any other antenna whose gain can be determined or known.

In most of the cases, however, the reference antenna is a *lossless isotropic source*.

When the direction is not stated, the power gain is usually taken in the direction of maximum radiation.

In addition, the total radiated power P_{rad} is related to the total input power P_{in}, hence, it follows

$$P_{rad} = eP_{in} \tag{2.344}$$

where e is the antenna radiation efficiency defined as the ratio of the power delivered to the radiation resistance R_r and radiation losses R_L,

$$e = \frac{R_r}{R_r + R_L} \tag{2.345}$$

Radiation efficiency can be considered as a measure of the dissipative or heat losses in an antenna. It is worth noting that impedance mismatch losses at the input port of the antenna are not included in the radiation efficiency.

Taking into account eqn (2.344), (2.343) becomes

$$G(\phi, \theta) = e\left[4\pi \frac{U(\phi, \theta)}{P_{rad}}\right] \tag{2.346}$$

which is related to the directivity by expression

$$G(\phi, \theta) = eD(\phi, \theta) \tag{2.347}$$

In many applications, partial gains G_Φ and G_θ are used. These partial gains are defined as

$$G_\phi = 4\pi \frac{U_\phi}{P_{in}} \tag{2.348}$$

$$G_\theta = 4\pi \frac{U_\theta}{P_{in}} \tag{2.349}$$

where U_Φ is the radiation intensity in a given direction contained in Φ-field component, U_θ is the radiation intensity in a given direction contained in θ-field component, and P_{in} is the total input power.

Usually, the gain is given in terms of decibels instead of the dimensionless quantity.

2.19.4 Linear Antennas

In most of the applications, practical antennas, either as a devices or EMC models (when the frequency of the fields increases to the point where the wavelength becomes comparable with the dimension of the source and victim equipment, EMC models based on the quasi-state assumptions for the fields become inaccurate and antenna models must be used), cannot be represented by the Hertz dipole approximation. However, finite length wire antennas (linear antennas) can be modeled as a superposition of infinitesimal radiation sources.

Antenna problem can be considered in two different modes

- radiation (transmitting) mode;
- scattering (receiving) mode.

Radiation as a phenomenon can be desired, if the antenna is designed as a device and used either in transmitting or scattering mode, or undesired, if electromagnetic interferences are considered via antenna model.

In either the transmission or the reception case, the key to understanding the behavior of radiated or scattered fields is the knowledge of the current distribution induced along the antenna. For the case of electrically short wires where the length is small compared to the wavelength, the current can be approximated as constant (as in the case of Hertz dipole) or linear function. On the contrary, if the wire length or frequency increases, current distribution changes significantly. There are several approaches to determine the current distribution of a linear antenna. The simplest way is to assume the current waveform. Nevertheless, such an approach could provide satisfactory results in many applications if thin wires are considered; there are situations where an accurate current distribution calculation is necessary. A more rigorous approach to this problem is to obtain the current distribution by solving the corresponding integral equation that is discussed in details in Chapters 8 and 9. This section deals with the approximate current distributions.

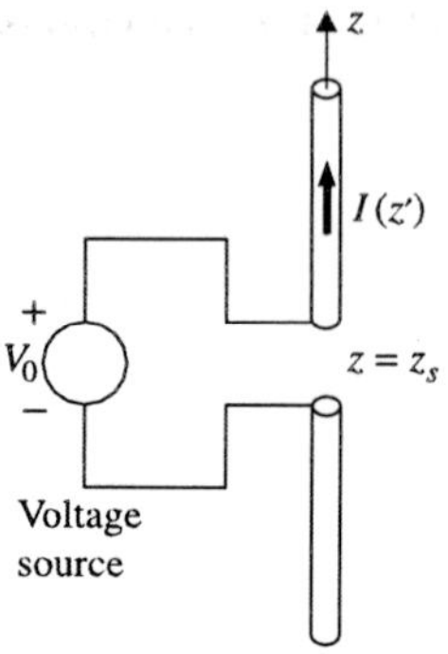

Figure 2.19 Center-fed dipole antenna.

2.19.4.1 ***Center-Fed Straight Wire Antenna*** One of the simplest antennas most commonly used in practice and also as an EMC model is the center-fed dipole antenna. This antenna as a device is shown in Figure 2.19, whereas the antenna as an undesired source of radiation in EMC is shown in Figure 2.20. The entire length of the wire is L.

At high frequencies the reasonable and traditionally widely used approximation for the antenna current is the sinusoidal distribution defined as

$$I(z) = I_0 \sin\left[k\left(\frac{L}{2} - |z|\right)\right] \tag{2.350}$$

At sufficiently low frequencies where $\lambda > \frac{L}{2}$, that is, $k\left(\frac{L}{2} - |z|\right) \ll 1$, eqn (2.350) becomes

$$I(z) = I_0\left(1 - \frac{2|z|}{L}\right) \tag{2.351}$$

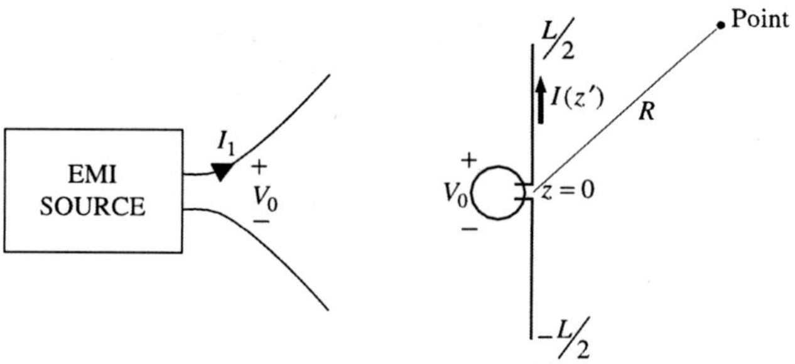

Figure 2.20 Center-fed wire antenna representation of an EMI source.

The magnetic vector potential (2.319) for the case of sinusoidal current distribution becomes

$$A_z(z) = \int_L \frac{\mu I(z')e^{-jkR}}{4\pi R} dz' = \frac{\mu I_0}{4\pi r} e^{-jkr} \int_{-L/2}^{L/2} \sin\left[k\left(\frac{L}{2} - |z'|\right)\right] e^{-jkz'\cos\theta} dz'$$

$$= \frac{\mu I_0}{2\pi r} e^{-jkr} \frac{\cos\left(k\frac{L}{2}\cos\theta\right) - \cos\left(k\frac{L}{2}\right)}{\sin\theta} \quad (2.352)$$

Note that the distance from the source to the observation point is approximated as

$$R = r - z'\cos\theta \quad (2.353)$$

For the special case of half-wave dipole, eqn (2.352) is simplified into

$$A_z(z) = \frac{\mu I_0}{2\pi r} e^{-jkr} \frac{\cos\left(\frac{\pi}{2}\cos\theta\right)}{\sin\theta} \quad (2.354)$$

The radiated far field E_θ, where the following condition is satisfied,

$$kr = 2\pi\frac{r}{\lambda} \gg 1 \quad (2.355)$$

can be determined from the magnetic vector potential featuring eqn (2.324), and it is given by

$$E_\theta = -j\omega A_\theta = j\omega A_z \sin\theta \quad (2.356)$$

Combining eqns (2.352) and (2.355) yield

$$E_\theta = jk\frac{Z_0}{4\pi} \int_{-L/2}^{+L/2} I(z') \sin\theta \frac{e^{-jkR}}{R} dz' \quad (2.357)$$

Taking into account the approximation (2.353), eqn (2.357) becomes

$$E_\theta = jk\frac{Z_0}{4\pi r} e^{-jkr} \sin\theta \int_{-L/2}^{+L/2} I(z')e^{-jkz'\cos\theta} dz' \quad (2.358)$$

Furthermore, for the case of sinusoidal distribution, one obtains

$$E_\theta = jk\frac{Z_0}{4\pi r}e^{-jkr}\int_{-L/2}^{L/2} \sin\left[k\left(\frac{L}{2}-|z'|\right)\right]\sin\theta dz'$$
$$= jZ_0\frac{\mu I_0}{2\pi r}e^{-jkr}\frac{\cos\left(k\frac{L}{2}\cos\theta\right)-\cos\left(k\frac{L}{2}\right)}{\sin\theta} \tag{2.359}$$

In the far-field region, the magnetic and electric fields are related as follows:

$$H_\phi = \frac{E}{Z_0} = j\frac{I_0}{2\pi r}e^{-jkr}\frac{\cos\left(k\frac{L}{2}\cos\theta\right)-\cos\left(k\frac{L}{2}\right)}{\sin\theta} \tag{2.360}$$

For the case of half-wave dipole, it follows

$$E_\theta = jZ_0\frac{I_0}{2\pi r}e^{-jkr}\frac{\cos\left(\frac{\pi}{2}\cos\theta\right)}{\sin\theta} \tag{2.361}$$

$$H_\phi = j\frac{I_0}{2\pi r}e^{-jkr}\frac{\cos\left(\frac{\pi}{2}\cos\theta\right)}{\sin\theta} \tag{2.362}$$

The total power radiated by the dipole antenna that can be obtained by the dipole antenna that can be obtained from integral (2.330), that is,

$$P_{rad} = \int_0^{2\pi}\int_0^{\pi}\frac{|E_\theta|^2}{2Z_0}r^2\sin\theta\, d\theta\, d\phi = Z_0\frac{I_0^2}{4\pi}\int_0^{\pi}\frac{\left[\cos\left(k\frac{L}{2}\cos\theta\right)-\cos\left(k\frac{L}{2}\right)\right]}{\sin\theta}d\theta \tag{2.363}$$

The radiation resistance is defined with relation (2.342). For the case of half-length dipole, the total power is

$$P_{rad} = Z_0\frac{I_0^2}{4\pi}\int_0^{\pi}\frac{\cos^2\left(k\frac{L}{2}\cos\theta\right)}{\sin\theta}d\theta \tag{2.364}$$

The radiation resistance of half-length dipole is approximately equal to 73Ω.

2.20 REFERENCES

[1] P. P. Silvester, R. L. Ferrari: Finite Elements for Electrical Engineers, 3rd Edition, *Cambridge University Press, Cambridge*, 2001.

[2] M. N. O. Sadiku: Numerical Techniques in Electromagnetics, 2nd Edition, *CRC Press, Boca Raton*, 2001.

[3] W. Band: Introduction to Mathematical Physics, *D. Van Nostrand Company, New York*, 1959.

[4] F. Tesche, M. Ianoz, T. Karlsson: EMC Analysis Methods and Computational Models, *John Wiley & Sons, New York*, 1997.

[5] S. Ramo, J. R. Whinnery, T. Van Duzer: Fields and Waves in Communication Electronics, 3rd Edition, *John Wiley & Sons, Inc., New York*, 1994.

[6] C. A. Balanis: Antenna Theory, 2nd Edition, *John Wiley & Sons, Inc., New York*, 1997.

3

INTRODUCTION TO NUMERICAL METHODS IN ELECTROMAGNETICS

Various approaches are used by scientists and engineers for solving electromagnetic field problems. Generally, these approaches can be classified as experimental or theoretical (computational). Nevertheless professionals and laymen strongly believe in measurements; they are always expensive, time consuming, usually do not provide parameter variation to a great extent, and sometimes are not reliable, or even hazardous. These difficulties are usually avoided using the theoretical approaches, that is, mathematical modeling.

Almost all problems arising in science and engineering can be formulated in terms of differential, integral, and variational equations. Generally, there are two basic approaches to solve problems in electromagnetics: the differential, or the field approach, and integral, or the source approach.

- The field approach deals with a solution of a corresponding differential equation with associated boundary conditions, specified at a boundary of a computational domain. Historically, this approach has been derived by Gilbert, Faraday, and Maxwell. The field approach is very useful for handling the interior (inner, closed, or bounded) field problems (Figure 3.1).
- The electromagnetic field sources' concept, or the integral approach, is based on the solution of a corresponding integral equation. Historically, this approach was promoted by Franklin, Cavendish, and Ampere, among others. It is worth noting that the source approach is convenient for the treatment of the exterior (open, outer, or unbounded) field problems (Fig. 3.2).

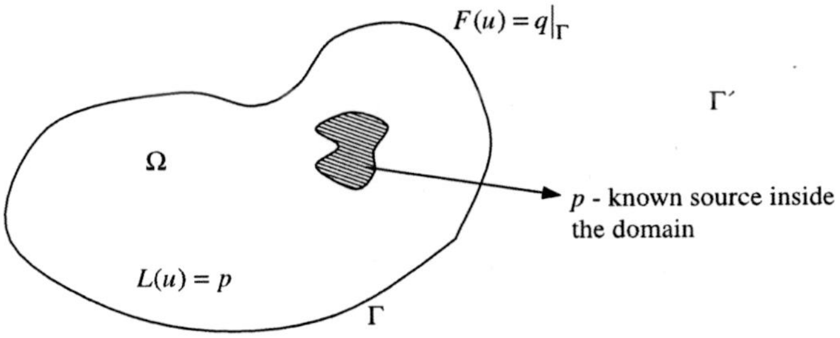

Figure 3.1 Differential approach.

Generally, the methods for solving partial differential equations or integral equations can be classified as analytical or numerical ones.

The principal drawback of the analytical methods is inability to handle problems involving complex geometries and inhomogeneous domains. This difficulty can be overcome by applying numerical methods.

The main problem arising in the application of numerical methods is related to the spurious solutions, convergence rate, and accuracy.

A classical boundary-value problem can be formulated in terms of the operator equation:

$$L(u) = p \tag{3.1}$$

on the domain Ω with conditions $F(u) = q|_\Gamma$ prescribed on the boundary Γ (Fig. 3.1), where L is linear differential operator, u is solution of the problem, and p is the excitation function representing the known sources inside the domain. Methods for the solution of the interior field problem are generally referred to as differential methods or field methods.

On the contrary, if L represents an integral operator, unknowns are related to field sources, that is, charge densities or current densities distributed along the boundary Γ' (Fig. 3.2). Once the sources are determined, field at an arbitrary point, inside or outside the domain, can be obtained by integrating the sources.

Methods of solutions for the exterior field problems are referred to as the integral methods, or method of sources. The boundary conditions used in electromagnetic field problems are usually of the Dirichlet (forced) type the, Neumann (natural) type, or their combination (mixed boundary conditions) [1,2]. These boundary conditions can be either homogeneous or inhomogeneous.

Figure 3.2 Integral approach.

3.1 ANALYTICAL VERSUS NUMERICAL METHODS

A general trade-off between analytical and numerical methods can be done, as follows:

- Analytical solution methods yield exact solutions but are limited to a narrow range of applications, mostly related to canonical problems. There are not many practical engineering problems that can be worked out using these techniques.
- Numerical techniques are applicable to almost all scientific engineering problems, but the main drawbacks are related to the approximation limit in model itself, space and time discretization. Moreover, the criteria for accuracy, stability, and convergence are not always straightforward and clear to the researcher.

The most commonly used analytical methods in electromagnetics are as follows [2]:

- Separation of variables
- Series expansion
- Conformal mapping
- Integral transforms

While the numerical methods used in electromagnetics, among others, are as follows:

- Finite difference method (FDM)
- Finite element method (FEM)
- Boundary element method (BEM)
- Method of moments (MoM)

Wide applications of numerical methods not only in electromagnetics, but also in many other continuum problems such as: hydrodynamics, thermodynamics, or acoustics are based on the advantage to model a particular problem without a requirement of high level mathematics and physics knowledge, respectively.

3.1.1 Frequency and Time Domain Modeling

The problems being analyzed can be regarded as steady state or transient, and the solution methods are usually classified as frequency or time domain. The frequency and time domain techniques for solving transient electromagnetic phenomena have been fully documented in Ref. [3].

A frequency domain solution is commonly applied for many sources but as one single frequency whereas with the time domain, it is for a single source but many frequencies [4]. While this is true for the continuous integral Fourier transform, it is often not possible in practice to obtain the equivalent solution from another domain for highly resonant structures when the discrete Fourier transform is used. Equivalence of the results in the two respective domains is true only under a precise set of conditions which are hard to meet in practice.

One important difference between the two approaches is that the time domain solution obtained is specific to the temporal variation of the excitation source. The transient response of a structure when subjected to different excitations, for example, a step-function or a Gaussian voltage source, will require the computation process to be repeated for the respective solutions, whereas in the frequency domain approach, the solution from one set of computations can be applied to obtain transient results from different sources if the geometry of the structure is unchanged. This difference is a significant factor when considering the relative merits of the two approaches.

Comparison based on computational efficiency between the two approaches [4] yields the general conclusion that the computer time required for a solution using both the integral equation frequency domain and the *implicit* time domain approaches is proportional to $(L/\Delta L)^{3(D-1)}$, where L is a characteristic dimension of the structure, ΔL the frequency dependent space discretization, and D the dimensionality, which for a wire is 2. For the *explicit* time domain approach, the time is proportional to $(L/\Delta L)^{2(D-1)+1+r}$ where $0 \leq r \leq 1$. Although in many situations, the time domain approach has better advantages; it can be said that the computational efficiency of each approach is dependent on the structure being analyzed and the form of the result being sought.

Other statements arising from the trade-off between the frequency and the time domain approach are as follows [5]:

- Better physical insight when using the time domain approach. However, an understanding of the resonant characteristic can only be obtained from the frequency spectrum.
- Nonlinearities are more conveniently handled in the time domain.
- Interactions (e.g., pulse reflection) may be isolated in the time domain using time-range gating.
- It is possible to obtain singularity expansion method poles more directly and efficiently in the time domain.
- Frequency domain formulation is relatively simpler and easier to use thus allowing more complex structures to be analyzed more conveniently.
- For complicated geometry, larger computing effort may be required for the more complex formulation of the time domain approach.

One factor of increasingly importance in favor of the frequency domain approach is in analyzing electromagnetic compatibility (EMC) properties. Accurate frequency domain information is required for such work as regulatory standards are specified in this form [6]. Due to its simpler formulation, versatile general purpose frequency domain codes are widely available and are used by professionals and amateurs alike. An efficient frequency domain transient analysis methodology is likely to find applications in many fields by users of such codes.

The problem of excessive computer time becomes critical in transient electromagnetics. In the frequency domain approach to a transient problem, the computation must be repeated over the bandwidth for hundreds or even thousands of times at a frequency sampling interval. For a highly resonant structure, this straightforward unabridged approach will simply be beyond the practical limit in terms of the computer time required to solve a problem.

Much effort in computational electromagnetics research has been directed toward reducing the computation operation count for solving problems. This is achieved through several ways, in the analytical formulation, using specialized Green's functions, approximations applied particularly to integration and matrix solving techniques. Very often improvements in computation efficiency are achieved at the expense of sacrificing accuracy and the utility of a formulation, that is, narrowing the range of applications for a particular formulation. Several techniques for reducing the computation operation count in frequency sampling when analyzing the transient behavior of resonant structures have been discussed in Ref. [3].

3.2 OVERVIEW OF NUMERICAL METHODS: DOMAIN, BOUNDARY, AND SOURCE SIMULATION

It is important to clarify some principles and ideas of how to describe field problems via partial differential or integral equations. Namely, there are some basic differences between domain methods (e.g., finite element method), boundary methods (e.g., boundary element method), and source simulation methods (charge or current simulation method). This chapter deals with the fundamentals of the FDM, FEM, and direct BEM. Implementation of static, dynamic, and transient problems is discussed in Part II and III of this book.

3.2.1 Modeling of Problems via the Domain Methods: FDM and FEM

FDM is generally one of the simplest numerical methods. Modeling via FDM is undertaken discretizing the entire domain and converting the differential operator into difference equations.

FEM modeling of partial differential equations is performed by discretizing the entire calculation domain Ω, and integrating any known sources Ω_s within the domain (Fig. 3.3a). Dirichlet and Neumann boundary conditions can be prescribed along the boundary Γ for both FDM and FEM. Modeling of differential equations with FEM results in an algebraic equation system which provides a sparse matrix. This matrix is usually banded and in many cases symmetric depending on the solution of the problem.

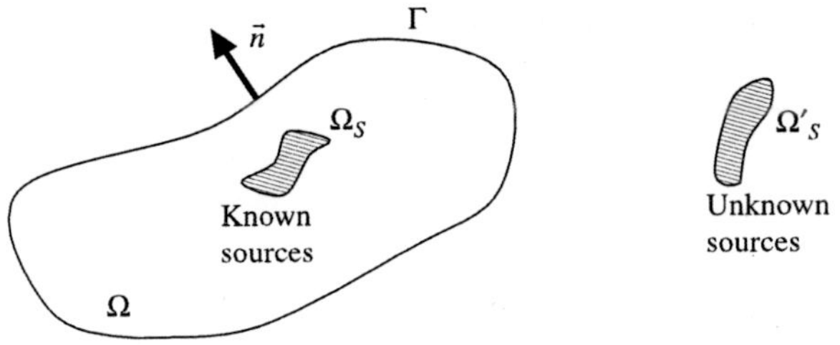

Figure 3.3 Method of fields versus method of sources.

3.2.2 Modeling of Problems via the BEM: Direct and Indirect Approach

Integral formulations of partial differential equations along the boundary are carried out using the Green integral theorem. The resulting equations are modeled discretizing only the boundary and by integrating any known sources Ω_s within a given subdomain. Modeling of partial differential equations through their boundary integral formulations results in less unknowns but dense matrices [1]. The formulation of the boundary element method using field and potential quantities rather than field sources is usually called the *direct BEM formulation*. However, there are many applications in electromagnetics of boundary integral equations for which it is more convenient to work in terms of sources. This technique is known as *direct BEM* and the modeling of the integral equations is undertaken in terms of sources distributed over the boundary. This method can be considered as a special case of the *direct BEM*. Therefore, the modeling of integral equations is related to the integration over unknown charge density or current density. Integral equations over unknown sources Ω'_s (Fig. 3.3b) can also be derived from the Green integral theorem and the solution method can be referred to as a special variant of the boundary element method—indirect boundary element method.

However, as classic BEM uses potential or field on the domain boundary and this integral equation approach deals with unknown sources, some authors use the term *finite elements for integral operators* [7]. On the contrary, in order to stress the integration over sources, the term *source element method* (SEM) or *source integration method* (SIM) is suggested [1].

3.3 THE FINITE DIFFERENCE METHOD

The finite difference method is a highly versatile method and has been used to analyze objects with an extremely wide range of size and complexity.

Historically, the finite difference method was developed in the 1920s for some applications in hydrodynamics. The finite difference method is based on the approximation of the function derivatives using the finite differences, that is, a differential equation is replaced by a finite difference equation.

3.3.1 One-Dimensional FDM

The standard definition of the first derivative of the function is given by:

$$\frac{\mathrm{d}f}{\mathrm{d}x} = \lim_{\Delta x \to 0} \frac{f(x+\Delta x) - f(x)}{\Delta x} \tag{3.2}$$

The finite difference approximation of expression (3.2) is then:

$$\frac{\mathrm{d}f}{\mathrm{d}x} \approx \frac{f(x+\Delta x) - f(x)}{\Delta x} \tag{3.3}$$

Rigorously, the approximation (3.3) arises from the Taylor series representation of the function $f(x)$:

$$f(x+\Delta x) = f(x) + \frac{\mathrm{d}f}{\mathrm{d}x}\Delta x + \frac{\mathrm{d}^2 f}{\mathrm{d}x^2}\frac{\Delta x^2}{2!} + \frac{\mathrm{d}^3 f}{\mathrm{d}x^3}\frac{\Delta x^3}{3!} \cdots \tag{3.4}$$

$$f(x-\Delta x) = f(x) - \frac{\mathrm{d}f}{\mathrm{d}x}\Delta x + \frac{\mathrm{d}^2 f}{\mathrm{d}x^2}\frac{\Delta x^2}{2!} - \frac{\mathrm{d}^3 f}{\mathrm{d}x^3}\frac{\Delta x^3}{3!} \cdots \tag{3.5}$$

Neglecting the third and higher terms of the series leads to three possible schemes for the approximation of the first derivative of function $f(x)$ via finite differences. These schemes are:

- forward scheme:

$$\frac{\mathrm{d}f}{\mathrm{d}x} \approx \frac{f(x+\Delta x) - f(x)}{\Delta x} \tag{3.6}$$

- backward scheme:

$$\frac{\mathrm{d}f}{\mathrm{d}x} \approx \frac{f(x) - f(x-\Delta x)}{\Delta x} \tag{3.7}$$

- central scheme:

$$\frac{\mathrm{d}f}{\mathrm{d}x} \approx \frac{f(x+\Delta x) - f(x-\Delta x)}{2\Delta x} \tag{3.8}$$

Figure 3.4 shows all the three schemes.

Obviously, the approximation error is proportional to the number of neglected terms in Taylor series. For expressions (3.6) and (3.7), the approximation error is of order Δx, while the expression is of order Δx^2.

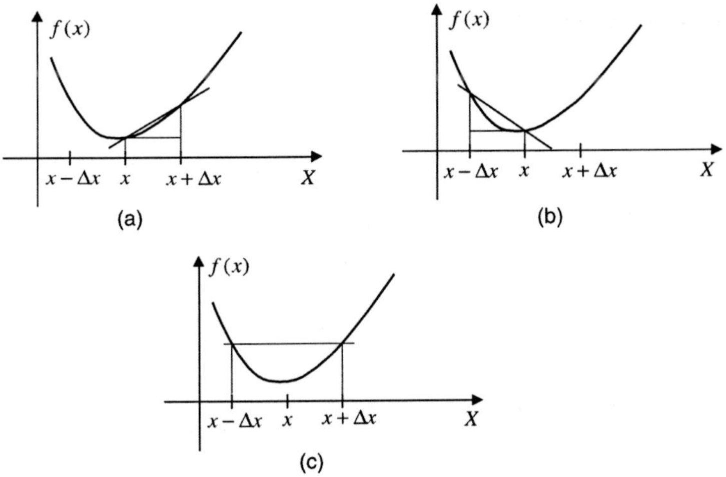

Figure 3.4 (a) "forward scheme," (b) "backward scheme," (c) "central scheme."

Second order differential operator obtained by repeating the same procedure to the first derivative of function $f(x)$ is as follows:

$$\frac{\mathrm{d}^2 f}{\mathrm{d}x^2} \approx \frac{f(x-\Delta x) - 2f(x) + f(x+\Delta x)}{\Delta x^2} \tag{3.9}$$

FDM is one of the simplest numerical methods but at the same time suffers from relatively poor convergence rate. This drawback can be overcome by increasing the number of unknowns in the corresponding matrix systems.

3.3.1.1 Computational Example Determine the approximate solution of the differential equation:

$$\frac{\mathrm{d}^2 u}{\mathrm{d}x^2} + u + x = 0; \text{ on interval } (0,1) \tag{3.10}$$

by FDM, discretizing the calculation domain to five unknowns. Prescribed boundary conditions are as follows: $u(0) = u(1) = 0$.
Discretization of the domain to five points gives: $\Delta x = 1/4$ (Fig. 3.5).

From the prescribed boundary conditions it follows:

$$u_1 = u_5 = 0 \tag{3.11}$$

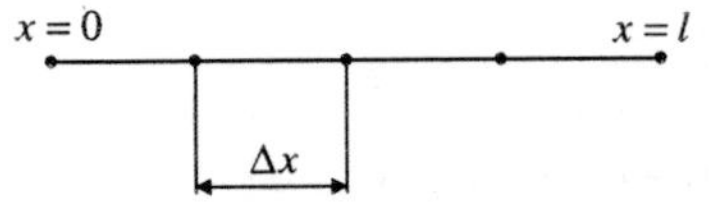

Figure 3.5 Discretization of the calculation domain.

Implementing the formula (3.9) yields:

$$\begin{aligned}\frac{u_1 - 2u_2 + u_3}{\Delta x^2} + u_2 &= -x_2\\ \frac{u_2 - 2u_3 + u_4}{\Delta x^2} + u_3 &= -x_3\\ \frac{u_3 - 2u_4 + u_5}{\Delta x^2} + u_4 &= -x_4\end{aligned} \tag{3.12}$$

Rearranging the obtained difference equations, with $u_1 = u_5 = 0$ and $\Delta x = 1/4$, leads to:

$$\begin{aligned}-31u_2 + 16u_3 &= -1/4\\ 16u_2 - 31u_3 + 16u_4 &= -1/2\\ 16u_3 - 31u_4 &= -3/4\end{aligned} \tag{3.13}$$

or in the matrix form it follows:

$$\begin{bmatrix} -31 & 16 & 0 \\ 16 & -31 & 16 \\ 0 & 16 & -31 \end{bmatrix} \begin{Bmatrix} u_2 \\ u_3 \\ u_4 \end{Bmatrix} = \begin{Bmatrix} -1/4 \\ -1/2 \\ -3/4 \end{Bmatrix} \tag{3.14}$$

Solving the matrix equations the coefficients u_2 to u_4 are obtained: $u_2 = 0.0442740; u_3 = 0.0701559; u_4 = 0.0604030$
Analytical and numerical solutions are compared in Table 3.1.

Approximate solution at arbitrary point can be obtained by means of linear interpolation.

3.3.2 Two-Dimensional FDM

Two-dimensional calculation domain in u xy-plane is considered (Figure 3.6). The domain is assumed to be source-free and if the boundary conditions in terms of potential and its normal derivative are specified, the static problem can be formulated by the Laplace equation:

$$\nabla^2 \varphi = 0 \tag{3.15}$$

TABLE 3.1 Comparison of analytical and numerical solution.

x	$\bar{u}$	u	δ (%)
0.25	0.0442740	0.0440137	0.59
0.5	0.0701559	0.069747	0.58
0.75	0.0604030	0.0600562	0.57

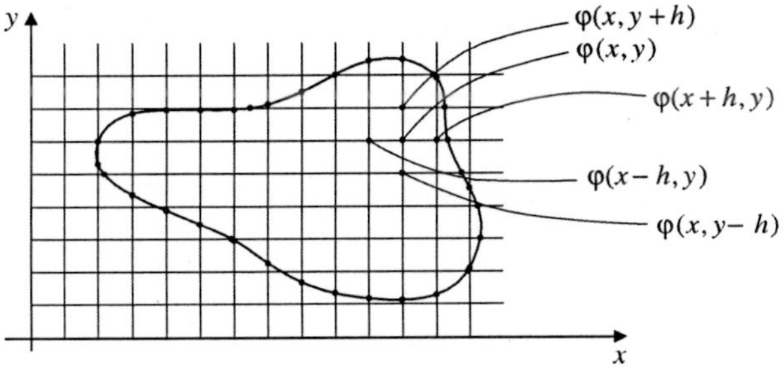

Figure 3.6 Discretization of 2D domain.

where φ stands for the unknown potential within the domain of interest.

For two-dimensional problems in rectangular coordinates, eqn (3.9) becomes:

$$\frac{\partial^2 \varphi}{\partial x^2} + \frac{\partial^2 \varphi}{\partial y^2} = 0 \tag{3.16}$$

Applying the Taylor expansions for a function of two variables in the vicinity of point (x,y) for first two terms can be written as:

$$\varphi(x+h, y) = \varphi(x, y) + h\frac{\partial \varphi(x, y)}{\partial x} \tag{3.17}$$

$$\varphi(x, y+h) = \varphi(x, y) + h\frac{\partial \varphi(x, y)}{\partial y} \tag{3.18}$$

and the potential derivatives are then:

$$\frac{\partial \varphi(x, y)}{\partial x} = \frac{\varphi(x+h, y) - \varphi(x, y)}{h} \tag{3.19}$$

$$\frac{\partial \varphi(x, y)}{\partial y} = \frac{\varphi(x, y+h) - \varphi(x, y)}{h} \tag{3.20}$$

Repeating the same procedure the second order derivatives are obtained:

$$\frac{\partial^2 \varphi}{\partial x^2} = \frac{\varphi(x+h, y) - 2\varphi(x, y) + \varphi(x-h, y)}{h^2} \tag{3.21}$$

$$\frac{\partial^2 \varphi}{\partial y^2} = \frac{\varphi(x, y+h) - 2\varphi(x, y) + \varphi(x, y-h)}{h^2} \tag{3.22}$$

Finally, substituting relations (3.21) and (3.22) in the Laplace equation (3.15) results in the following formula:

$$\varphi(x, y) = \frac{1}{4}[\varphi(x+h, y) + \varphi(x-h, y) + \varphi(x, y+h) + \varphi(x, y-h)] \tag{3.23}$$

or simply:

$$\varphi_{ij} = \frac{1}{4}[\varphi_{(i+1,j)} + \varphi_{(i-1,j)} + \varphi_{(i,j+1)} + \varphi_{(i,j-1)}] \tag{3.24}$$

where index i is related to discrete set of x variables, and j is related to discrete set of y variables.

Relation (3.24) is also known as "four-points formula" and represents the averaged value of potential at a given point.

3.3.2.1 Computational Example Consider a very long rectangular tube with dimensions $a/b = 4/3$. One plate is on the potential $\varphi = 100V$, and other plates are grounded, that is, on the potential $\varphi = 0V$, as shown in Figure 3.6. The potential inside the tube has to be determined by means of FDM.

The tube dimensions can be expressed via finite difference step h:

$$a = 4h; \quad b = 3h \tag{3.25}$$

where h is the distance between the neighboring nodes, as shown in Figure 3.7.

As the problem is two dimensional, the Laplace equation $\nabla^2\varphi = 0$ can be written as:

$$\frac{\partial^2\varphi}{\partial x^2} + \frac{\partial^2\varphi}{\partial y^2} = 0 \tag{3.26}$$

The partial differential eqn (3.26) is replaced by the following finite difference equation:

$$\varphi(x, y) = \frac{1}{4}[\varphi(x+h, y) + \varphi(x-h, y) + \varphi(x, y+h) + \varphi(x, y-h)] \tag{3.27}$$

Taking into account the symmetry of the problem, it follows:

$$\varphi_1 = \frac{1}{4}[0 + 100 + \varphi_3 + \varphi_2]$$

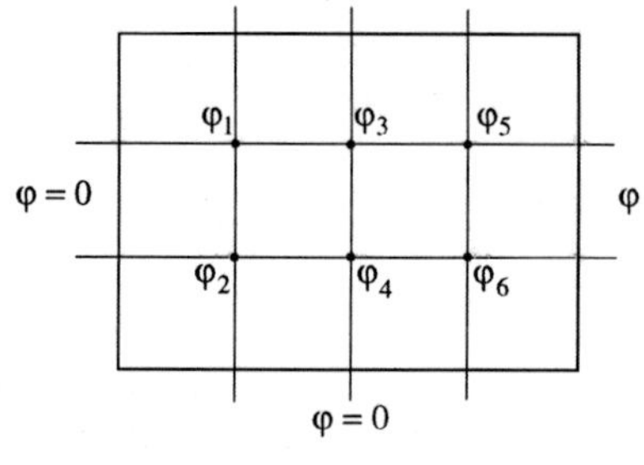

Figure 3.7 Discretization of the rectangular tube cross-section.

$$\varphi_2 = \frac{1}{4}[0 + \varphi_1 + \varphi_4 + 0]$$

$$\varphi_3 = \frac{1}{4}[2\varphi_1 + 100 + \varphi_4] \tag{3.28}$$

$$\varphi_4 = \frac{1}{4}[2\varphi_2 + \varphi_3 + 0]$$

The algebraic system of eqn (3.28) can be also written in the matrix form:

$$\begin{bmatrix} 4 & -1 & -1 & 0 \\ -1 & 4 & 0 & -1 \\ -2 & 0 & 4 & -1 \\ 0 & -2 & -1 & 4 \end{bmatrix} \begin{bmatrix} \varphi_1 \\ \varphi_2 \\ \varphi_3 \\ \varphi_4 \end{bmatrix} = \begin{bmatrix} 100 \\ 0 \\ 100 \\ 0 \end{bmatrix} \tag{3.29}$$

Solving the matrix system (3.29), the values of potential $\varphi_1, \varphi_2, \varphi_3, \varphi_4$ is obtained:

$$\begin{matrix} \varphi_1 = 41.6149 & \varphi_3 = 50.9317 \\ \varphi_2 = 15.5280 & \varphi_4 = 20.4969 \end{matrix} \tag{3.30}$$

Equation (3.26) with the prescribed boundary conditions can also be solved analytically via separation of variables method [2].

3.4 THE FINITE ELEMENT METHOD

The finite element method is one of the most commonly used numerical methods in science and engineering. The method is highly automatized and very convenient for computer implementation based on step by step algorithms. The special features of FEM are related to efficient modeling of complex shape geometries and inhomogeneous domains, and also to the relatively simple mathematical formulation providing highly banded and symmetric matrix, same as accuracy refinement by higher order approximation and automatic inclusion of natural (Neumann) boundary conditions. This method generally gives better results than the FDTD in modeling complicated boundaries and is particularly suited for problems with closed boundaries. In this method, the entire domain space to be analyzed is discretized into a grid of elements of finite size.

3.4.1 Basic Concepts of FEM

The calculation domain is discretized to sufficiently small segments—finite elements. The unknown solution over a finite element is expressed in terms of linear combination of local interpolation functions (shape functions) (Fig. 3.8).

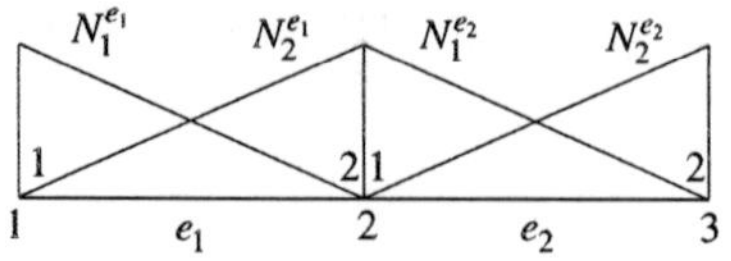

Figure 3.8 Linear shape functions.

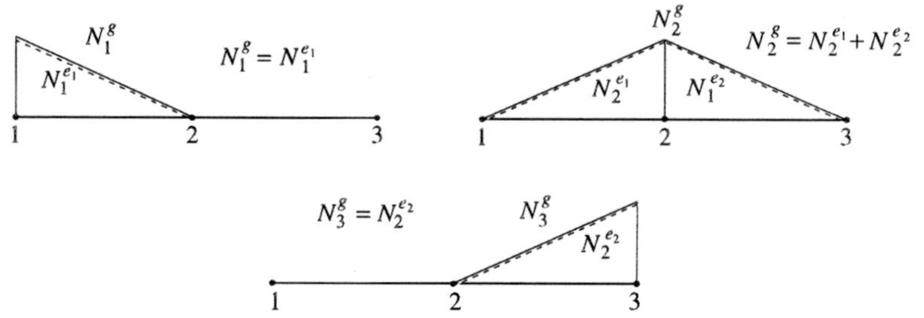

Figure 3.9 Assembling of global functions from local shape functions.

The global base functions assigned to nodes are assembled from local shape functions, assigned to elements, as it is shown in Figure 3.9 for the case of linear approximation.

The approximate solution of a problem of interest can be written as:

$$\bar{f} = \sum_{i=1}^{n} \alpha_i N_i \tag{3.31}$$

where coefficient α_i represents the solution at the global nodes, while n denotes the total number of nodes.

The approximate solution along two finite elements is shown in Figure 3.10.

The accuracy can be improved by finer discretization, or by implementation of higher order approximation.

3.4.2 One-Dimensional FEM

Many problems in science and engineering can be formulated in terms of second-order differential equations of the following form:

$$\frac{\mathrm{d}}{\mathrm{d}x}\left[\lambda \frac{\mathrm{d}u}{\mathrm{d}x}\right] + k^2 u - p = 0 \tag{3.32}$$

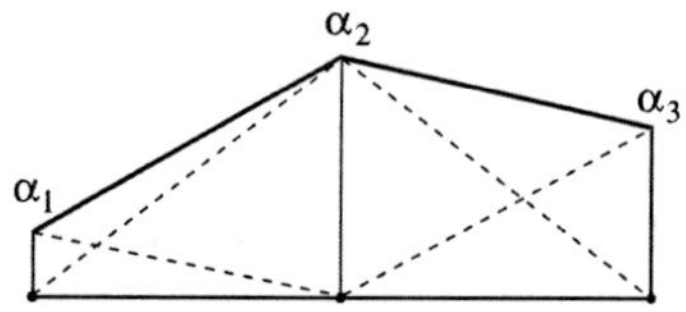

Figure 3.10 Approximate solution above two finite elements.

where λ and k are the properties of a medium and p represents the excitation function, that is, the sources within the domain of interest. Forone-dimensional case, the domain is related to interval $[a, b]$.

Substituting the approximate solution $\bar{u}$ into the differential eqn (3.32) and integrating over the calculation domain according to the weighted residual approach [7], it follows:

$$\int_a^b \left[\frac{\mathrm{d}}{\mathrm{d}x}\left(\lambda\frac{\mathrm{d}\bar{u}}{\mathrm{d}x}\right) + k^2\bar{u} - p\right] W_j dx = 0, j = 1, 2, \ldots, n \tag{3.33}$$

which can also be written as:

$$\int_a^b \frac{\mathrm{d}}{\mathrm{d}x}\left(\lambda\frac{\mathrm{d}\bar{u}}{\mathrm{d}x}\right) W_j dx + \int_a^b k^2\bar{u}W_j dx = \int_a^b pW_j dx \tag{3.34}$$

Equation (3.34) represents the *strong formulation* of the problem. Within strong formulation, base functions must be in the domain of the differential operator, and automatically satisfy the prescribed boundary conditions. The strong requirements can be avoided by moving to the *weak formulation* of the problem.

The order of differentiation can be decreased by carefully performing integration by parts. Differentiation of product of two functions can be written as:

$$\frac{\mathrm{d}}{\mathrm{d}x}\left(\lambda\frac{\mathrm{d}\bar{u}}{\mathrm{d}x}W_j\right) = \frac{\mathrm{d}}{\mathrm{d}x}\left(\lambda\frac{\mathrm{d}\bar{u}}{\mathrm{d}x}\right)W_j + \lambda\frac{\mathrm{d}\bar{u}}{\mathrm{d}x}\frac{\mathrm{d}W_j}{\mathrm{d}x} \quad ; \quad \lambda = \lambda(x) \tag{3.35}$$

Somewhat rearranging eqn (3.35) and integrating along the interval:

$$\frac{\mathrm{d}}{\mathrm{d}x}\left(\lambda\frac{\mathrm{d}\bar{u}}{\mathrm{d}x}\right)W_j = \frac{\mathrm{d}}{\mathrm{d}x}\left(\lambda\frac{\mathrm{d}\bar{u}}{\mathrm{d}x}W_j\right) - \lambda\frac{\mathrm{d}\bar{u}}{\mathrm{d}x}\frac{\mathrm{d}W_j}{\mathrm{d}x} \Big/ \int_a^b \mathrm{d}x \tag{3.36}$$

one obtains:

$$\int_a^b \frac{\mathrm{d}}{\mathrm{d}x}\left(\lambda\frac{\mathrm{d}\bar{u}}{\mathrm{d}x}\right)W_j dx = \lambda\frac{\mathrm{d}\bar{u}}{\mathrm{d}x}W_j\Big|_a^b - \int_a^b \lambda\frac{\mathrm{d}\bar{u}}{\mathrm{d}x}\frac{\mathrm{d}W_j}{\mathrm{d}x}\mathrm{d}x \tag{3.37}$$

Substituting expression (3.37) into eqn (3.34), the weak formulation is obtained:

$$\int_a^b k^2\bar{u}W_j\mathrm{d}x + \lambda\frac{\mathrm{d}\bar{u}}{\mathrm{d}x}W_j\Big|_a^b - \int_a^b \lambda\frac{\mathrm{d}\bar{u}}{\mathrm{d}x}\frac{\mathrm{d}W_j}{\mathrm{d}x}\mathrm{d}x = \int_a^b pW_j\mathrm{d}x \tag{3.38}$$

Second term on the right hand side of eqn (3.38) represents the natural boundary condition (Neumann condition) thus being directly included into the weak formulation. The physical meaning of the Neumann condition is related to the flux density at the ends of interval.

Applying the finite element algorithm, the unknown solution $\bar{u}$ is expanded into linear combination of base functions. Implementing the Galerkin Bubnov procedure (the same choice of base and test functions) yields the following matrix equation:

$$[a]\{\alpha\} = \{b\} + \{Q\} \tag{3.39}$$

where $[a]$ is the global matrix of the system, $\{\alpha\}$ is the solution vector, $\{b\}$ represents the excitation vector, and $\{Q\}$denotes the flux density.

The local approximation for the unknown function over a finite element is given by:

$$\bar{u}^e = \alpha_1^e N_1^e(x) + \alpha_2^e N_2^e(x) \tag{3.40}$$

where $N_1^e(x)$ and $N_2^e(x)$ are the linear shape functions.

Finite element matrix and vector are given by integrals:

$$a_{ji}^e = \int_{x_1^e}^{x_2^e} k^2 N_i^e N_j^e \mathrm{d}x - \int_{x_1^e}^{x_2^e} \lambda \frac{\mathrm{d}N_i^e}{\mathrm{d}x}\frac{\mathrm{d}N_j^e}{\mathrm{d}x}\mathrm{d}x \tag{3.41}$$

$$b_j^e = \int_{x_1^e}^{x_2^e} pN_j^e \mathrm{d}x \tag{3.42}$$

and the form of the local matrix equation is then:

$$\begin{bmatrix} a_{11}^e & a_{12}^e \\ a_{21}^e & a_{22}^e \end{bmatrix} \cdot \begin{bmatrix} \alpha_1^e \\ \alpha_2^e \end{bmatrix} = \begin{bmatrix} b_1^e \\ b_2^e \end{bmatrix} + \begin{bmatrix} -\lambda \left.\dfrac{du}{dx}\right|_{x=x_1^e} \\ \lambda \left.\dfrac{du}{dx}\right|_{x=x_2^e} \end{bmatrix} \tag{3.43}$$

Global matrix system is assembled from the local ones and it is given by:

$$\begin{bmatrix} a_{11}^{e_1} & a_{12}^{e_1} & 0 & \cdots & 0 \\ a_{21}^{e_1} & a_{22}^{e_1} + a_{11}^{e_2} & a_{12}^{e_2} & & 0 \\ 0 & a_{21}^{e_2} & a_{22}^{e_2} + a_{11}^{e_3} & & \vdots \\ \vdots & & & \ddots & a_{12}^{e_n} \\ 0 & 0 & \cdots & a_{21}^{e_n} & a_{22}^{e_n} \end{bmatrix} \cdot \begin{bmatrix} \alpha_1 \\ \alpha_2 \\ \alpha_3 \\ \vdots \\ \alpha_n \end{bmatrix} = \begin{bmatrix} b_1^{e_1} \\ b_2^{e_1} + b_1^{e_2} \\ b_2^{e_2} + b_1^{e_3} \\ \vdots \\ b_2^{e_n} \end{bmatrix} + \begin{bmatrix} -\lambda \left.\dfrac{du}{dx}\right|_{x=a} \\ 0 \\ 0 \\ \vdots \\ \lambda \left.\dfrac{du}{dx}\right|_{x=b} \end{bmatrix} \tag{3.44}$$

Note that the flux densities vanish at internal nodes, and only the values related to the domain boundary (in 1D case interval ends) are not equal to zero.

3.4.2.1 Incorporation of Boundary Conditions It is obvious from global matrix system (3.44) that the Neumann boundary conditions are automatically included into the weak formulation of the finite element method.

Dirichlet (forced) boundary conditions are incorporated into matrix system subsequently, that is, it follows:

$$\alpha_1 = u(a), \quad \alpha_n = u(b) \tag{3.45}$$

thus decreasing the number of unknowns, that is, the first and the last row of the matrix equations are omitted, and the global matrix equation results in matrix system of $(n-2)$ unknowns:

$$\begin{bmatrix} a_{22}^{e_1} + a_{11}^{e_2} & a_{12}^{e_2} & 0 & \cdots & 0 \\ a_{21}^{e_2} & a_{22}^{e_2} + a_{11}^{e_3} & a_{12}^{e_3} & & 0 \\ 0 & a_{21}^{e_3} & a_{22}^{e_3} + a_{11}^{e_4} & & \vdots \\ \vdots & & & \ddots & a_{12}^{e_{n-1}} \\ 0 & 0 & \cdots & a_{21}^{e_{n-1}} & a_{22}^{e_{n-1}} + a_{11}^{e_n} \end{bmatrix} \cdot \begin{bmatrix} \alpha_2 \\ \alpha_3 \\ \alpha_4 \\ \vdots \\ \alpha_{n-1} \end{bmatrix}$$
$$= \begin{bmatrix} b_2^{e_1} + b_1^{e_2} \\ b_2^{e_2} + b_1^{e_3} \\ b_2^{e_3} + b_1^{e_4} \\ \vdots \\ b_2^{e_{n-1}} + b_1^{e_n} \end{bmatrix} + \begin{bmatrix} 0 \\ 0 \\ 0 \\ \vdots \\ 0 \end{bmatrix} - \begin{bmatrix} \alpha_1 a_{21}^{e_1} \\ 0 \\ 0 \\ \vdots \\ 0 \end{bmatrix} - \begin{bmatrix} 0 \\ 0 \\ 0 \\ \vdots \\ \alpha_n a_{12}^{e_n} \end{bmatrix} \tag{3.46}$$

Once the unknown coefficients α_2 to α_{n-1}, are determined, it is possible from the first and the last row (equation) to obtain the flux densities at the ends of the interval:

$$\begin{aligned} \lambda \frac{du}{dx}\bigg|_{x=a} &= b_1^{e_1} - \alpha_1 a_{11}^{e_1} - \alpha_2 a_{12}^{e_1} \\ \lambda \frac{du}{dx}\bigg|_{x=b} &= -b_2^{e_n} + \alpha_{n-1} a_{21}^{e_n} + \alpha_n a_{22}^{e_n} \end{aligned} \tag{3.47}$$

If the Dirichlet condition is prescribed at one end of the interval and the Neumann condition at the other, then the global system consists of $n-1$ unknowns and is given by combination of system (3.44) and (3.46), respectively.

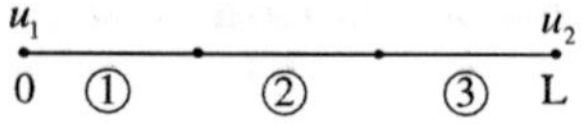

Figure 3.11 Discretization of the domain on three finite elements.

3.4.2.2 Computational Example We need to solve differential equation:

$$\frac{d^2u}{dx^2} - F = 0 \tag{3.48}$$

on the interval [0, L] with prescribed Dirichlet boundary conditions by means of FEM. The Dirichlet conditions are as follows:

$$u(0) = 0; \quad u(L) = F \tag{3.49}$$

Discretize the domain to 3 finite elements (Fig. 3.11) and use the weak formulation of the problem.

Using the weighted residual approach and carefully performing the integration by parts, the weak formulation of eqn (3.48) is obtained:

$$\frac{du}{dx}W_j\bigg|_O^L - \int_O^L \frac{du}{dx}\frac{dW_j}{dx}dx = \int_O^L FW_j dx \tag{3.50}$$

Applying the finite element algorithm, the approximate solution over a finite element is expressed as:

$$u^e = \alpha_1 N_1(x) + \alpha_2 N_2(x) \tag{3.51}$$

where N_1 and N_2 are the linear shape functions given by:

$$N_1(x) = \frac{x_2 - x}{\Delta x}; \quad \frac{\partial N_1}{\partial x} = -\frac{1}{\Delta x} \tag{3.52}$$

$$N_2(x) = \frac{x - x_1}{\Delta x}; \quad \frac{\partial N_2}{\partial x} = \frac{1}{\Delta x} \tag{3.53}$$

Furthermore, using the Galerkin–Bubnov procedure yields the finite element matrix and vector:

$$a_{ji}^e = \int_{x_1}^{x_2} \frac{dN_j^e}{dx}\frac{dN_i^e}{dx}dx \tag{3.54}$$

$$b_j^e = -\int_{x_1}^{x_2} FN_j^e dx \tag{3.55}$$

The finite element matrix and vector terms can be determined analytically as follows:

$$a_{11} = \int_{x_1}^{x_2} \frac{dN_1}{dx}\frac{dN_1}{dx}dx = \int_{x_1}^{x_2} \left(-\frac{1}{\Delta x}\right)\left(-\frac{1}{\Delta x}\right)dx = \left(\frac{1}{\Delta x}\right)^2 dx = \frac{1}{\Delta x} \tag{3.56}$$

$$a_{12} = a_{21} = \int_{x_1}^{x_2} \frac{dN_1}{dx}\frac{dN_2}{dx}dx = \int_{x_1}^{x_2} \left(-\frac{1}{\Delta x}\right)\left(\frac{1}{\Delta x}\right)dx = -\frac{1}{\Delta x} \tag{3.57}$$

$$a_{22} = \int_{x_1}^{x_2} \frac{dN_2}{dx}\frac{dN_2}{dx}dx = \int_{x_1}^{x_2} \left(\frac{1}{\Delta x}\right)\left(\frac{1}{\Delta x}\right)dx = \frac{1}{\Delta x} \tag{3.58}$$

$$b_1 = -\int_{x_1}^{x_2} F\frac{x_2 - x}{\Delta x}dx = -F\frac{\Delta x}{2} \tag{3.59}$$

$$b_2 = -\int_{x_1}^{x_2} F\frac{x - x_1}{\Delta x}dx = -F\frac{\Delta x}{2} \tag{3.60}$$

and the local matrix system (on a finite element) is then:

$$\frac{1}{\Delta x}\begin{bmatrix} 1 & -1 \\ -1 & 1 \end{bmatrix}\begin{bmatrix} \alpha_1^e \\ \alpha_2^e \end{bmatrix} = -F\frac{\Delta x}{2}\begin{Bmatrix} 1 \\ 1 \end{Bmatrix} + \begin{Bmatrix} -\left.\dfrac{du}{dx}\right|_{x=x_1} \\ \left.\dfrac{du}{dx}\right|_{x=x_2} \end{Bmatrix} \tag{3.61}$$

The global system of equations is assembled from the local ones:

$$\underbrace{\frac{1}{\Delta x}\begin{bmatrix} 1 & -1 & 0 & 0 \\ -1 & 2 & -1 & 0 \\ 0 & -1 & 2 & -1 \\ 0 & 0 & -1 & 1 \end{bmatrix}}_{\text{global-system-matrix}} \underbrace{\begin{bmatrix} \alpha_1 \\ \alpha_2 \\ \alpha_3 \\ \alpha_4 \end{bmatrix}}_{\text{solution-vector}} = \underbrace{-F\frac{\Delta x}{2}\begin{bmatrix} 1 \\ 2 \\ 2 \\ 1 \end{bmatrix}}_{\text{right-hand-side-vector}} + \underbrace{\begin{bmatrix} -\left.\dfrac{du}{dx}\right|_{x=O} \\ 0 \\ 0 \\ \left.\dfrac{du}{dx}\right|_{x=L} \end{bmatrix}}_{\text{flux-density-vector}} \tag{3.62}$$

Dirichlet conditions are inserted into the global matrix and eqn (3.62) simplifies into:

$$\frac{1}{\Delta x}\begin{bmatrix} 2 & -1 \\ -1 & 2 \end{bmatrix}\begin{bmatrix} \alpha_2 \\ \alpha_3 \end{bmatrix} = -\frac{1}{\Delta x}\begin{bmatrix} -1 & 0 \\ 0 & -1 \end{bmatrix}\begin{bmatrix} 0 \\ F \end{bmatrix} - F\frac{\Delta x}{2}\begin{bmatrix} 2 \\ 2 \end{bmatrix} + \begin{bmatrix} 0 \\ 0 \end{bmatrix} \tag{3.63}$$

Solving the linear equation system, the solution coefficients (nodes 2 and 3) are obtained:

$$\alpha_2 = -\frac{4}{3}F$$
$$\alpha_3 = -\frac{2}{3}F \tag{3.64}$$

Incorporating the calculated solution coefficients into the first and last equations of the global system, one can readily assess the flux densities at the ends of the interval (boundaries of the domain):

$$\frac{1}{\Delta x}\begin{bmatrix} -1 & 0 \\ 0 & -1 \end{bmatrix}\begin{bmatrix} \alpha_2 \\ \alpha_3 \end{bmatrix} + \frac{1}{\Delta x}\begin{bmatrix} 1 & 0 \\ 0 & 1 \end{bmatrix}\begin{bmatrix} 0 \\ F \end{bmatrix} + F\frac{\Delta x}{2}\begin{bmatrix} 1 \\ 1 \end{bmatrix} = \begin{bmatrix} -\left.\dfrac{du}{dx}\right|_{x=O} \\ \left.\dfrac{du}{dx}\right|_{x=L} \end{bmatrix} \tag{3.65}$$

that is, one obtains:

$$\left.\frac{du}{dx}\right|_{x=O} = -\frac{11}{6}F$$
$$\left.\frac{du}{dx}\right|_{x=L} = \frac{13}{6}F \tag{3.66}$$

These flux densities are consequence of the impressed Dirichlet boundary conditions.

3.4.3 Two-Dimensional FEM

The simplest discretization of a 2D domain can be performed using the so-called triangular elements (Fig. 3.12). In this case, the shape functions are given by

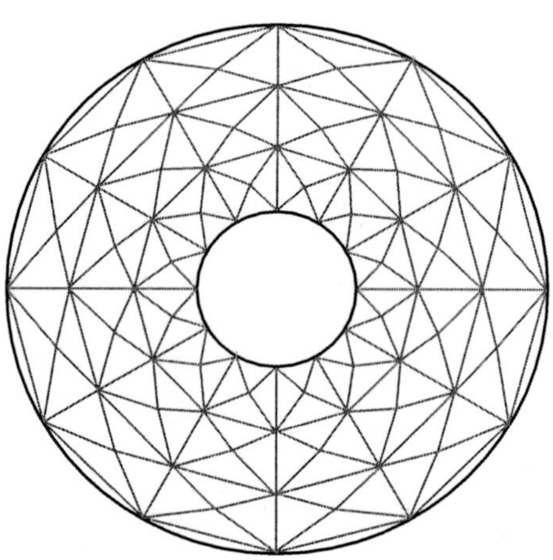

Figure 3.12 Discretization of a 2D domain via triangular elements.

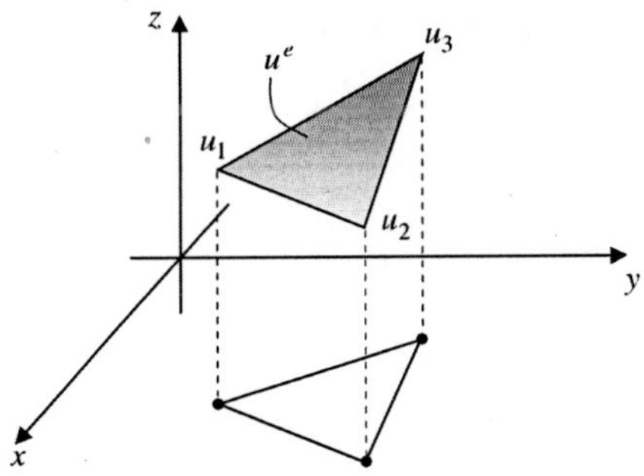

Figure 3.13 Approximate solution for 2D problem.

equations of planes in 3D space. The approximate solution is shown in Figure 3.12, while the corresponding shape functions over a triangle are shown in Figure 3.13. Figure 3.14 shows the global functions assigned to i-th node, assembled from neighboring shape functions. (Note that in 1D case, global bases always consists of two neighboring shape functions only.)

According to Figure 3.15, the solution on a triangular element is given by:

$$u^e = \alpha_1^e N_1^e(x, y) + \alpha_2^e N_2^e(x, y) + \alpha_3^e N_3^e(x, y) \tag{3.67}$$

where N_1, N_2, N_3 are 2D shape functions determined by:

$$N_1(x, y) = \frac{1}{2A}(2A_1 + b_1 x + a_1 y) \tag{3.68}$$

$$N_2(x, y) = \frac{1}{2A}(2A_2 + b_2 x + a_2 y) \tag{3.69}$$

$$N_3(x, y) = \frac{1}{2A}(2A_3 + b_3 x + a_3 y) \tag{3.70}$$

Hence, an i-th shape functions can be written as:

$$N_i(x, y) = \frac{1}{2A}(2A_i + b_i x + a_i y); \quad i = 1, 2, 3 \tag{3.71}$$

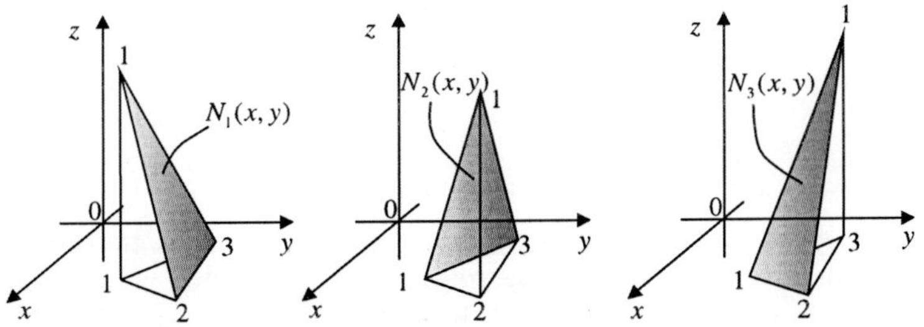

Figure 3.14 Shape functions over a triangle.

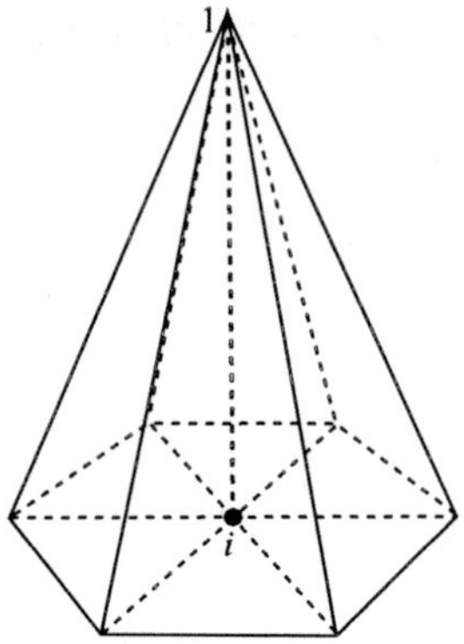

Figure 3.15 Global base function assigned to ith node.

where A denotes the area of a triangle:

$$2A = 2(A_1 + A_2 + A_3) \tag{3.72}$$

and $A_1 - A_3$, $a_1 - a_3$, and $b_1 - b_3$ are auxiliary variables:

$$2A_1 = x_2y_3 - x_3y_2 \qquad a_1 = x_3 - x_2 \qquad b_1 = y_2 - y_3 \tag{3.73}$$

$$2A_2 = x_3y_1 - x_1y_3 \qquad a_2 = x_1 - x_3 \qquad b_2 = y_3 - y_1 \tag{3.74}$$

$$2A_3 = x_1y_2 - x_2y_1 \qquad a_3 = x_2 - x_1 \qquad b_3 = y_1 - y_2 \tag{3.75}$$

or it can be simply written as follows:

$$\begin{matrix} 2A_i = x_jy_k - x_ky_j & a_i = x_k - x_j & b_i = y_j - y_k \\ i = 1, 2, 3 & j = 2, 3, 1 & k = 3, 1, 2 \end{matrix} \tag{3.76}$$

Combining relations (3.67) to (3.71), the solution on a triangle is:

$$u^e = \frac{1}{2A} \sum_{i=1}^{3} (2A_i + b_ix + a_iy)\alpha_i \tag{3.77}$$

where $n_e = 3$

3.4.3.1 ***The Weak Formulation for Generalized Helmholtz Equation*** A number of problems in science and engineering can be formulated via the generalized Helmholtz equation. As in the 1D case, FEM is implemented through the weak formulation of the problem.

The generalized inhomogeneous Helmholtz equation can be written in the following form:

$$\nabla(k\nabla u) + ru = p \tag{3.78}$$

where u is the unknown solution, k and r are the material properties, while p represents the sources inside the domain of interest.

Applying the weighted residual approach yields:

$$\int_{\Omega} [\nabla(k\nabla u) + ru - p]W_j d\Omega = 0 \tag{3.79}$$

That is, one obtains:

$$\int_{\Omega} \nabla(k\nabla u)W_j d\Omega + \int_{\Omega} ruW_j d\Omega - \int_{\Omega} pW_j d\Omega = 0 \tag{3.80}$$

Applying the simple differentiation rule:

$$\nabla(k\nabla u)W_j = \nabla\left[(k\nabla u)W_j\right] - (k\nabla u)\nabla W_j \tag{3.81}$$

and generalized Gauss integral theorem:

$$\int_{\Omega} \nabla\vec{A} d\Omega = \oint_{\Gamma} \vec{A}\vec{n} d\Gamma \tag{3.82}$$

the weak formulation of the Helmholtz equation (3.78) is obtained:

$$\int_{\Omega} \left[(k\nabla u)\nabla W_j - ruW_j\right] d\Omega = \int_{\Gamma} k\frac{\partial u}{\partial n} W_j d\Gamma - \int_{\Omega} pW_j d\Omega \tag{3.83}$$

Relation (3.83) is often called the variational equation [8].

The term on the left-hand side gives rise to the finite element matrix while the first term on the right-hand side represents the flux through the part of the domain boundary in which the Neumann boundary condition is specified. Second term on the right-hand side contains the known sources in the domain.

Applying the finite element algorithm, the unknown solution over a finite element is expressed in terms of linear combination of shape functions. In the matrix form, the approximate solution is given by:

$$u^e = \{N\}^T\{\alpha\} = [N_1 \quad N_2 \quad N_3] \begin{bmatrix} \alpha_1 \\ \alpha_2 \\ \alpha_3 \end{bmatrix} \tag{3.84}$$

where $\{\alpha\}$ denotes the unknown solution coefficients.

The gradient of scalar function u in 2D is simply determined by relation:

$$\nabla u = \frac{\partial u}{\partial x}\vec{e}_x + \frac{\partial u}{\partial y}\vec{e}_y \tag{3.85}$$

Substituting the expression (3.84) into (3.85) yields:

$$\nabla u = \begin{bmatrix} \frac{\partial u}{\partial x} \\ \frac{\partial u}{\partial y} \end{bmatrix} = \begin{bmatrix} \frac{\partial N_1}{\partial x} & \frac{\partial N_2}{\partial x} & \frac{\partial N_3}{\partial x} \\ \frac{\partial N_1}{\partial y} & \frac{\partial N_2}{\partial y} & \frac{\partial N_3}{\partial y} \end{bmatrix} \begin{bmatrix} \alpha_1 \\ \alpha_2 \\ \alpha_3 \end{bmatrix} \tag{3.86}$$

Using the Galerkin–Bubnov procedure ($W_j = N_j$) results in the following finite element matrix:

$$[a]^e = k_e \int_\Omega \begin{bmatrix} \frac{\partial N_1}{\partial x} & \frac{\partial N_1}{\partial y} \\ \frac{\partial N_2}{\partial x} & \frac{\partial N_2}{\partial y} \\ \frac{\partial N_3}{\partial x} & \frac{\partial N_3}{\partial y} \end{bmatrix} \begin{bmatrix} \frac{\partial N_1}{\partial x} & \frac{\partial N_2}{\partial x} & \frac{\partial N_3}{\partial x} \\ \frac{\partial N_1}{\partial y} & \frac{\partial N_2}{\partial y} & \frac{\partial N_3}{\partial y} \end{bmatrix} d\Omega - r_e \int_\Omega \begin{bmatrix} N_1 \\ N_2 \\ N_3 \end{bmatrix} [N_3 \quad N_2 \quad N_3] d\Omega \tag{3.87}$$

When discretizing the domain, it is necessary to choose sufficiently small elements to ensure constant properties of the medium over an element. Otherwise, parameters k and r become spatially dependent which increases the complexity of integration.

Derivatives of shape functions are simply:

$$\frac{\partial N_i}{\partial x} = \frac{b_i}{2A}; \quad \frac{\partial N_i}{\partial y} = \frac{a_i}{2A} \tag{3.88}$$

Performing certain mathematical manipulations yields:

$$[a]^e = \frac{k_e}{4A} \begin{bmatrix} a_1^2 + b_1^2 & a_1 a_2 + b_1 b_2 & a_1 a_3 + b_1 b_3 \\ a_2 a_1 + b_2 b_1 & a_2^2 + b_2^2 & a_2 a_3 + b_2 b_3 \\ a_3 a_1 + b_3 b_1 & a_3 a_2 + b_3 b_2 & a_3^2 + b_3^2 \end{bmatrix} - \frac{r_e A}{12} \begin{bmatrix} 2 & 1 & 1 \\ 1 & 2 & 1 \\ 1 & 1 & 2 \end{bmatrix} \tag{3.89}$$

The related global matrix is obtained by assembling the contributions from the local ones.

3.4.3.2 Computation of Fluxes on the Domain Boundary Once determining the complete solution for scalar potential (all coefficients αi are known), it is possible to compute Q_i which are consequence of the potentials. Total flux Q on the part of the domain boundary is determined by the integral:

$$Q = \int_{\Gamma} k\frac{\partial u}{\partial n} N_j \mathrm{d}\Gamma \tag{3.90}$$

Quantity $q = k\partial u/\partial n$ represents the flux density, that is, the Neumann (natural) boundary condition. On the finite element located on a part of the boundary, the flux can be expressed by integral:

$$Q_e = \int_{\Gamma_e} q_e N_j \mathrm{d}\Gamma \tag{3.91}$$

The flux density q is generally variable, but assumed constant on a finite element which consequently must be sufficiently small. Usually, the value of flux density on the center of the element is taken as average flux value for the whole element. Note that within the finite element algorithm, the flux Q represents the concentrated value of the flux assigned to the node.

Flux on a finite element located on a part of the boundary can be written as:

$$\{Q\}^e = \int_{\Delta\Gamma} q_e \begin{Bmatrix} N_1 \\ N_2 \end{Bmatrix} \mathrm{d}\Gamma \tag{3.92}$$

On the contrary, the concentrated flux on i-th node consists of contributions from neighboring (adjacent) elements Figure 3.16. is given by expression:

$$Q_i = \int_{\Delta\Gamma_1} q_1 N_2^{\Gamma_1} d\Gamma + \int_{\Delta\Gamma_2} q_2 N_1^{\Gamma_2} \mathrm{d}\Gamma \tag{3.93}$$

Assuming the constant densities q_1 and q_2, it follows:

$$Q_i = q_1\frac{\Delta\Gamma_1}{2} + q_2\frac{\Delta\Gamma_2}{2} \tag{3.94}$$

For the case when $\Delta\Gamma_1 = \Delta\Gamma_2 = \Delta\Gamma$ eqn (3.94) simplifies into:

$$Q_i = q_1\frac{\Delta\Gamma}{2} + q_2\frac{\Delta\Gamma}{2} = \frac{\Delta\Gamma}{2}(q_1 + q_2) \tag{3.95}$$

For example, if the boundary consists of three finite elements with constant flux density, that is, $q_1 = q_2 = q_3 = q_0$, the contributions in nodes are as follows:

$$\begin{aligned} Q_1 &= q_0\frac{\Delta\Gamma}{2}; \quad Q_2 = q_0\Delta\Gamma \\ Q_3 &= q_0\Delta\Gamma; \quad Q_4 = q_0\frac{\Delta\Gamma}{2} \end{aligned} \tag{3.96}$$

Note that only the first and last contribution represents half the value of other nodes.

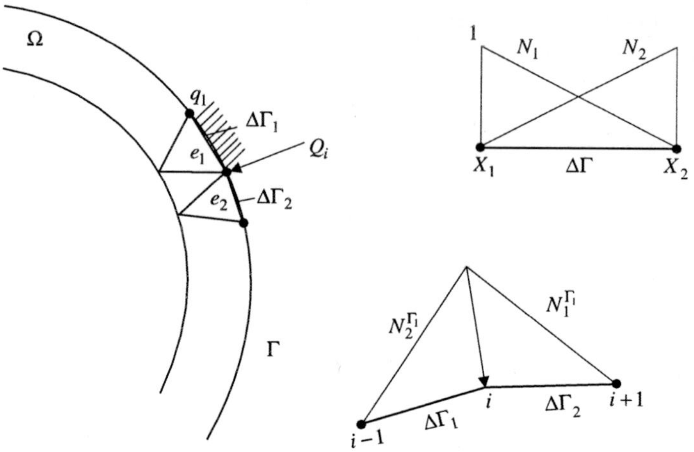

Figure 3.16 Elements on the boundary along which the flux density q is distributed, while the flux Q is concentrated in nodes.

3.4.3.3 ***Computation of Sources on a Finite Element*** Contribution of the source on a 2D finite element is related to the evaluation of the following integral:

$$\{p\}^e = \int_{\Omega e} p_e\{N\}\mathrm{d}\Omega = \int_{\Omega e} p_e \begin{Bmatrix} N_1 \\ N_2 \\ N_3 \end{Bmatrix} \mathrm{d}\Omega \tag{3.97}$$

where p denotes the source inside the domain Ω.

Ensuring the sufficiently fine domain discretization, the constant value of the source over a triangle can be assumed. According to the Galerkin procedure $W_j = N_j$, the right-hand side is of the form:

$$\{p\}^e = p_e \int_{\Omega} \{N\}\mathrm{d}\Omega \tag{3.98}$$

Integral of the shape functions over a triangular element:

$$\int_{\Omega e} N_i \mathrm{d}\Omega \quad \Rightarrow \quad \iint_{\Omega} N_i(x, y)\mathrm{d}x\mathrm{d}y \tag{3.99}$$

represents the volume of the pyramid whose height is equal to 1; the area of the base (triangle) is A (Fig. 3.17).

The right-hand side is now given by expression:

$$\{p\}^e = \frac{1}{3} p_e \begin{Bmatrix} 1 \\ 1 \\ 1 \end{Bmatrix} A \tag{3.100}$$

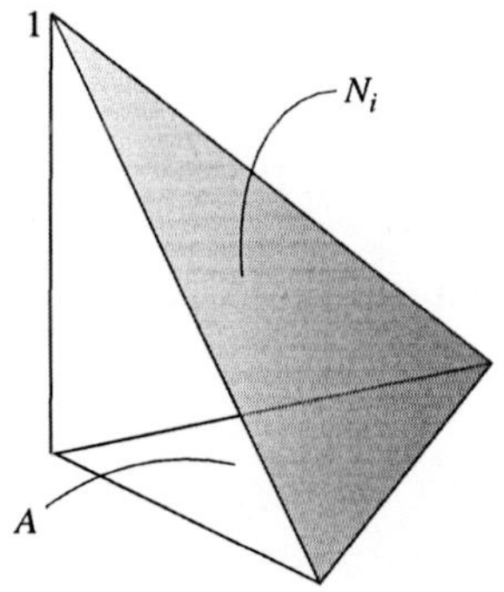

Figure 3.17 Integral of shape function over a finite element.

The assembling of the global system remains the same.

3.4.3.4. Computational Example: 2D Case Determine the distribution of the scalar potential within the domain shown in Figure 3.18 if the parameters are as follows: $\sigma_1 = 1\,\mathrm{S/m}$, $\sigma_2 = 3\,\mathrm{S/m}$, $\varphi_1 = 100\,\mathrm{V}$, $\varphi_2 = 0\,\mathrm{V}$. Discretize the domain to four finite elements.

The governing equations describing the static problem can be derived in a few steps. The electric field is determined by the gradient of scalar potential:

$$\vec{E} = -\nabla\varphi \tag{3.101}$$

Since the domain is the source-free, it follows:

$$\nabla\vec{J} = 0; \quad \vec{J} = \sigma\vec{E} \tag{3.102}$$

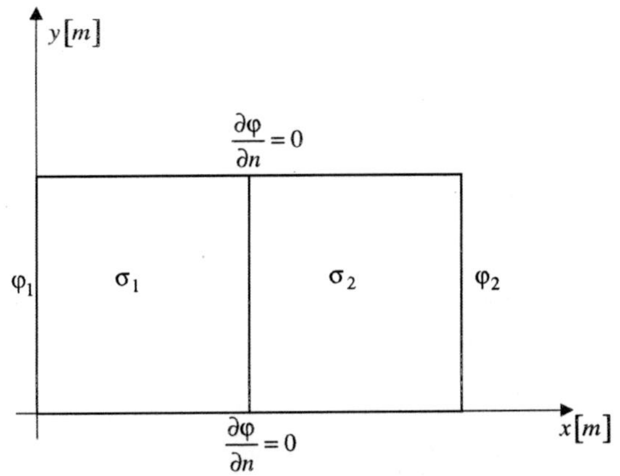

Figure 3.18 Two-dimensional calculation domain.

Combining eqns (3.101) and (3.102) results in the Laplace equation for inhomogeneous medium:

$$\nabla(\sigma\nabla\varphi) = 0 \tag{3.103}$$

Applying the weak formulation and Galerkin–Bubnov procedure leads to the following variational equation:

$$\int_{\Omega} (\sigma\nabla\varphi)\nabla N_j \mathrm{d}\Omega = \int_{\Gamma} \sigma\frac{\partial\varphi}{\partial n} N_j \mathrm{d}\Gamma \tag{3.104}$$

which can be regarded as a special case of the general variational eqn (3.83).

Taking into account the homogeneous Neumann boundary condition $\partial\varphi/\partial n = 0$ specified on a part of the boundary eqn (3.104) becomes:

$$\int_{\Omega} (\sigma\nabla\varphi)\nabla N_j \mathrm{d}\Omega = 0 \tag{3.105}$$

The approximate solution for scalar potential is:

$$\varphi = \{N\}^T\{\alpha\} \tag{3.106}$$

and the related finite element matrix, according to previous theory and expression (3.89), is given by:

$$[a]^e = \frac{\sigma_e}{4A}\begin{bmatrix} a_1^2 + b_1^2 & a_1a_2 + b_1b_2 & a_1a_3 + b_1b_3 \\ a_2a_1 + b_2b_1 & a_2^2 + b_2^2 & a_2a_3 + b_2b_3 \\ a_3a_1 + b_3b_1 & a_3a_2 + b_3b_2 & a_3^2 + b_3^2 \end{bmatrix} \tag{3.107}$$

The problem of interest is solved by discretizing the domain into four finite elements in a way shown in Figure 3.19.

The auxiliary variables are computed from the coordinates of the local nodes.

For the element e_1 (Fig. 3.20) it follows:

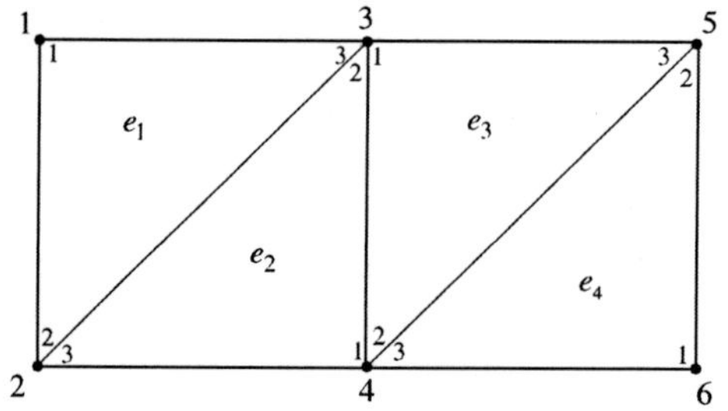

Figure 3.19 Discretization of the domain of interest.

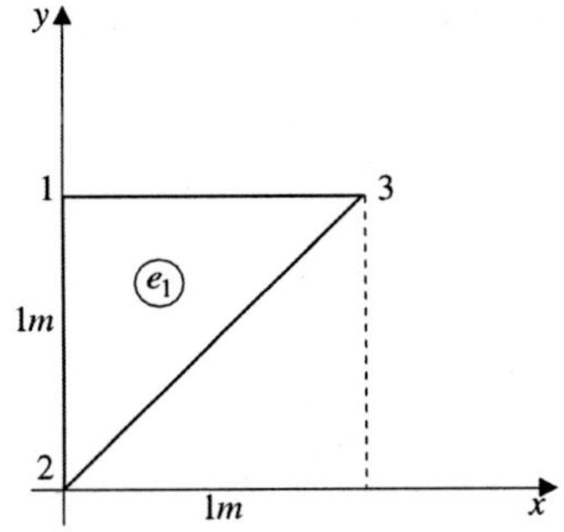

Figure 3.20 Element e_1.

TABLE 3.2 Connection of local with global nodes on e_1.

		① Loc 1	② Loc 2	③ Loc 3
①	*Loc 1*	2	−1	−1
②	*Loc 2*	−1	1	0
③	*Loc 3*	−1	0	1

Inserting the conductivity value $\sigma_1 = 1S/m$, the finite element matrix is given by:

$$[a]^{e_1} = \frac{1}{2}\begin{bmatrix} 2 & -1 & -1 \\ -1 & 1 & 0 \\ -1 & 0 & 1 \end{bmatrix} \tag{3.108}$$

The local and global nodes are related via table of connection (Table 3.2).

On element e_2, (Fig. 3.21) the value of conductivity is $\sigma_1 = 1S/m$, while auxiliary variables are given by:

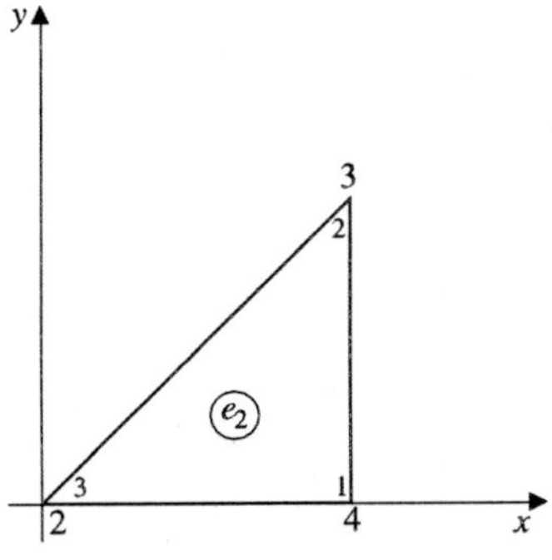

Figure 3.21 Element e_2.

TABLE 3.3 Connection of local with global nodes on e_2.

		④ *Loc 1*	③ *Loc 2*	② *Loc 3*
④	*Loc 1*	2	−1	−1
③	*Loc 2*	−1	1	0
②	*Loc 3*	−1	0	1

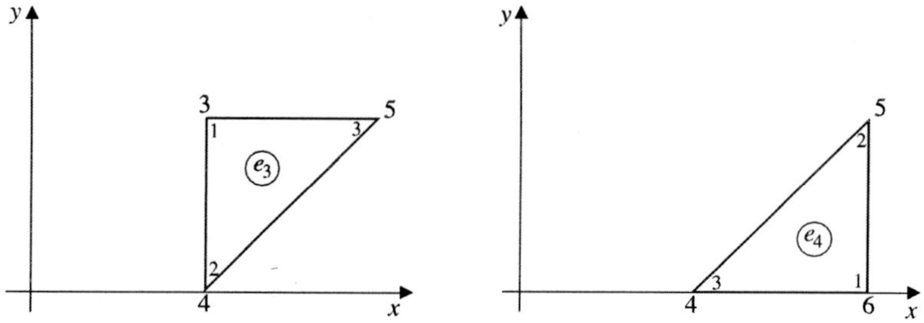

Figure 3.22 Elements e_3 and e_4.

And the finite element matrix is the same as in the previous case:

$$[a]^{e_2} = \frac{1}{2}\begin{bmatrix} 2 & -1 & -1 \\ -1 & 1 & 0 \\ -1 & 0 & 1 \end{bmatrix} \tag{3.109}$$

Connection of local with global nodes on e_2 is shown in Table 3.3:

Elements e_3 and e_4 on which the conductivity value is $\sigma_2 = 3S/m$ are shown in Figure 3.22.

The resulting finite element matrix is:

$$[a]^{e_3} = [a]^{e_4} = \frac{1}{2}\begin{bmatrix} 6 & -3 & -3 \\ -3 & 3 & 0 \\ -3 & 0 & 3 \end{bmatrix} \tag{3.110}$$

while the connection between local and global nodes is shown in Tables 3.4 and 3.5, respectively.

TABLE 3.4 Connection of local with global nodes on element e_3.

		③ *Loc 1*	④ *Loc 2*	⑤ *Loc 3*
③	*Loc 1*	6	−3	−3
④	*Loc 2*	−3	3	0
⑤	*Loc 3*	−3	0	1

TABLE 3.5 Connection of local with global nodes on element e_4.

		⑥ *Loc 1*	⑤ *Loc 2*	④ *Loc 3*
⑥	*Loc 1*	6	−3	−3
⑤	*Loc 2*	−3	3	0
④	*Loc 3*	−3	0	1

The global matrix is now assembled from the local ones and is given by:

$$\frac{1}{2}\cdot\begin{bmatrix} 2 & - & - & 0 & 0 & 0 \\ - & 2 & 0 & - & 0 & 0 \\ - & 0 & 8 & - & - & 0 \\ 0 & - & - & 8 & 0 & - \\ 0 & 0 & - & 0 & 6 & - \\ 0 & 0 & 0 & - & - & 6 \end{bmatrix}\begin{bmatrix} 100 \\ 100 \\ \alpha_3 \\ \alpha_4 \\ 0 \\ 0 \end{bmatrix} = \begin{bmatrix} 0 \\ 0 \\ 0 \\ 0 \\ 0 \\ 0 \end{bmatrix} + \begin{bmatrix} Q_1 \\ Q_2 \\ 0 \\ 0 \\ Q_5 \\ Q_6 \end{bmatrix} \tag{3.111}$$

Inserting the Dirichlet boundary conditions the coefficients α_1 and α_2 obtain value $\varphi_1 = 100\text{V}$, while coefficients α_5 and α_6 obtain value $\varphi_2 = 0\text{V}$, thus decreasing the number of unknowns from $N = 6$ to $N = 2$. Solving the global system (3.111), it follows:

$$\alpha_3 = \alpha_4 = 25V \tag{3.112}$$

Once determining the complete solution for scalar potential (all coefficients α_i are known), it is possible to compute fluxes Q_1, Q_2, Q_5, and Q_6, respectively.

3.5 THE BOUNDARY ELEMENT METHOD

3.5.1 Integral Equation Formulation

Since the middle of the eighties, the boundary element method (BEM) has been applied to a variety of problems in electromagnetics. The basic idea of BEM is to discretize the integral equation using boundary elements [1]. The well-known moment method (MoM) is equivalent to BEM when using subsectional bases and Dirac delta function as weighting functions. BEM, on the contrary, offers a more general relationship in which shape functions corresponding to subsectional bases are taking systematically as in the finite element method (FEM). Thus, BEM can be regarded as a combination of classical boundary integral equation method and the discretization concepts originated from FEM.

The first step in solving a problem via BEM is deriving the integral equation formulation of the differential equation governing the problem.

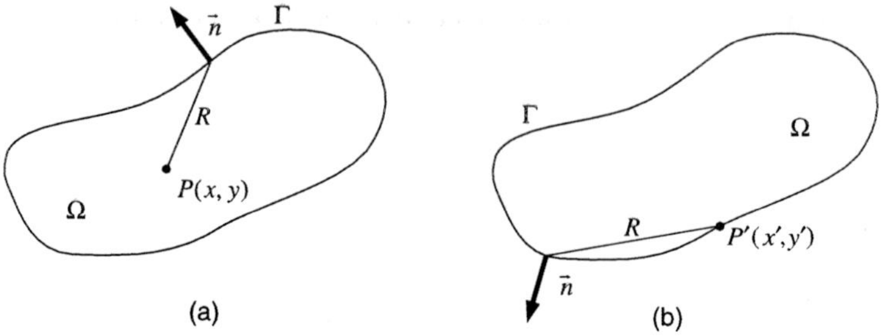

Figure 3.23 The geometry of the problem.

The governing differential equation for static field problem, either electrostatic or magnetostatic, for source-free domains is defined by the Laplace equation:

$$\nabla^2 u = 0 \tag{3.113}$$

or the Poisson equation, if sources p are present within the domain:

$$\nabla^2 u = -p \tag{3.114}$$

A typical calculation domain Ω with the related boundary Γ is shown in Figure 3.23a, where n is external normal vector to the boundary, and R denotes the distance from the source to the observation point.

It is worth mentioning that the observation point P can be also located on the boundary itself, as shown in Figure 3.23b.

The boundary conditions associated with these problems can be divided into essential condition (Dirichlet): whereas $u = u|_\Gamma$, defined on Γ_1, and natural condition (Neumann):$\partial u/\partial n|_\Gamma = q$, defined on Γ_2, as shown in Figure 3.24. The total boundary is then given by $\Gamma_1 + U\Gamma_2$.

For simplicity, the case of the Laplace equation (3.113) is considered first and afterward, the procedure can be extended to the solution of the Poisson equation (3.114).

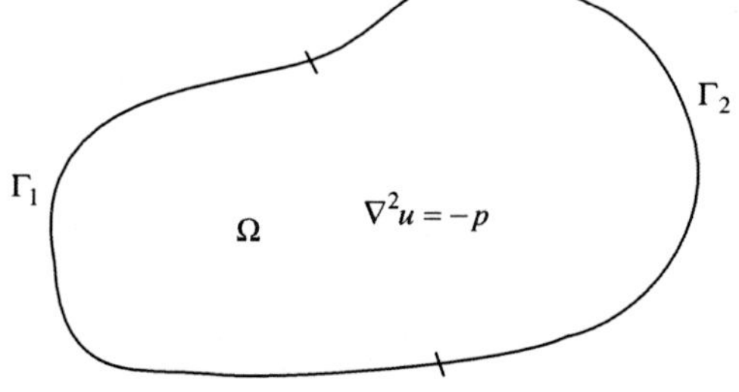

Figure 3.24 Calculation domain with boundary conditions.

Applying the weighted residual approach, eqn (3.113) can be integrated over the calculation domain Ω:

$$\int_{\Omega} \nabla^2 u \cdot W \mathrm{d}\Omega = 0 \tag{3.115}$$

where W is the weighting function.

Using the following identities:

$$\nabla(\nabla u W) = \nabla^2 u W + \nabla u \nabla W \tag{3.116}$$

$$\nabla(u \nabla W) = \nabla u \nabla W + u \nabla^2 W \tag{3.117}$$

Applying the generalized Gauss theorem, eqn (3.115) becomes:

$$\int_{\Omega} W \cdot \nabla^2 u \mathrm{d}\Omega = \int_{\Gamma} W \frac{\partial u}{\partial n} \mathrm{d}\Gamma - \int_{\Gamma} u \frac{\partial W}{\partial n} \mathrm{d}\Gamma + \int_{\Omega} u \cdot \nabla^2 W \mathrm{d}\Omega \tag{3.118}$$

Equation (3.118) can be also rewritten as:

$$\int_{\Gamma} W \frac{\partial u}{\partial n} \mathrm{d}\Gamma - \int_{\Gamma} u \frac{\partial W}{\partial n} \mathrm{d}\Gamma + \int_{\Omega} \nabla^2 W \cdot u \mathrm{d}\Omega = 0 \tag{3.119}$$

The weighting function W can be chosen to be the solution of the differential equation, that is:

$$\nabla^2 W - \delta(\vec{r} - \vec{r}') = 0 \tag{3.120}$$

where δ is the Dirac delta function, r denotes the observation points, and r' denotes the source points.

The solution of eqn (3.120) is called the fundamental solution or Green function. Thus, the domain integral in (3.119) can be written as:

$$\int_{\Omega} u \nabla^2 W \mathrm{d}\Omega = - \int_{\Omega} u \delta(\vec{r} - \vec{r}') \mathrm{d}\Omega = -u_i \tag{3.121}$$

for any point inside the domain.

Combining eqns (3.119) and (3.121), the following integral relation is obtained:

$$u_i = \int_{\Gamma} \Psi \frac{\partial u}{\partial n} \mathrm{d}\Gamma - \int_{\Gamma} u \frac{\partial \Psi}{\partial n} \mathrm{d}\Gamma \tag{3.122}$$

The integral eqn (3.122) can be regarded as the Green representation of the function u, where W can be replaced by ψ.

The function ψ is the fundamental solution of the eqn (3.120). For two-dimensional problems, it is given by the expression:

$$\Psi = -\frac{1}{2\pi} \ln R \tag{3.123}$$

where $R = |\vec{r} - \vec{r}'|$ is the distance from the source point (boundary point) to the observation point, and for three-dimensional problems by:

$$\Psi = \frac{1}{4\pi R} \tag{3.124}$$

The corresponding Green integral representation of the Poisson equation type (3.124) can be obtained in a similar way, starting from the weighted residual integral:

$$\int_{\Omega} \left[\nabla^2 u + p\right] \cdot W \mathrm{d}\Omega = 0 \tag{3.125}$$

Performing similar mathematical manipulations as it has been done before, one obtains:

$$u_i = \int_{\Gamma} \Psi \frac{\partial u}{\partial n} \mathrm{d}\Gamma - \int_{\Gamma} u \frac{\partial \Psi}{\partial n} \mathrm{d}\Gamma - \int_{\Omega} p\psi \mathrm{d}\Omega \tag{3.126}$$

Any source point inside the domain Ω is taken into account by the domain integral in eqn (3.123).

When the observation point i/P is located on the boundary Γ, as shown in Figure 3.23b, the boundary integral becomes singular as R approaches zero. In order to deal with this singularity, a small sphere for 3D or cylinder for 2D problems is considered, as shown in Figure 3.25.

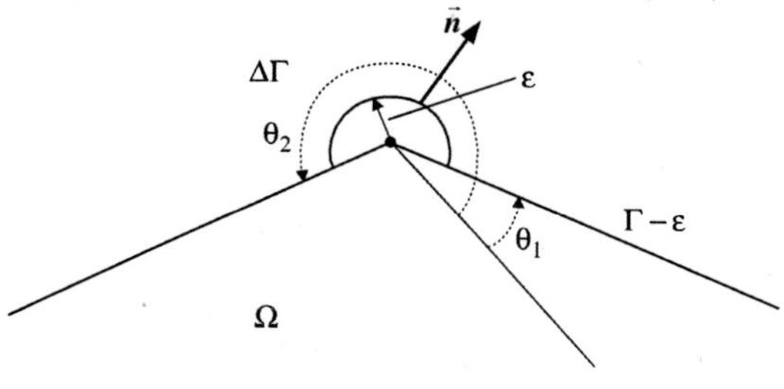

Figure 3.25 Extraction of the singularity on the contour integral.

To extract the singularity on the boundary, the first contour integral from the eqn (3.119) or (3.123) can be rewritten as follows:

$$u_i = -\frac{1}{2\pi}\int_{\Gamma-\Delta\Gamma} \frac{\partial u}{\partial n}\ln(R)\mathrm{d}\Gamma - \frac{1}{2\pi}\int_{\Delta\Gamma} \frac{\partial u}{\partial n}\ln(R)\mathrm{d}\Gamma + \frac{1}{2\pi}\int_{\Gamma-\Delta\Gamma} u\frac{\partial}{\partial n}(\ln R)\mathrm{d}\Gamma + \frac{1}{2\pi}\int_{\Delta\Gamma} u\frac{\partial}{\partial n}(\ln R)\mathrm{d}\Gamma \tag{3.127}$$

The first contour integral in eqn (3.127) then vanishes, that is:

$$\lim_{\varepsilon\to 0}\int_{\Delta\Gamma} \frac{\partial u}{\partial n}\psi \mathrm{d}\Gamma = -\lim_{\varepsilon\to 0}\int_{\theta_1}^{\theta_2} \frac{\partial u}{\partial n}\frac{1}{2\pi}\ln(\varepsilon)\varepsilon \mathrm{d}\theta = \lim_{\varepsilon\to 0}\frac{1}{2\pi}\left(\frac{\partial u}{\partial n}\right)\Big|_i .\varepsilon\ln\left(\frac{1}{\varepsilon}\right)(\theta_2-\theta_1) = 0 \tag{3.128}$$

while the second contour integral in (3.122) and (3.126) is given by:

$$\lim_{\varepsilon\to 0}\int_{\Delta\Gamma} \frac{\partial \psi}{\partial n}u \mathrm{d}\Gamma = -\lim_{\varepsilon\to 0}\frac{1}{2\pi}\int_{\theta_1}^{\theta_2} \frac{1}{\varepsilon}u\varepsilon \mathrm{d}\theta = -\frac{\theta_2-\theta_1}{2\pi}u_i \tag{3.129}$$

According to the analysis presented through the relations (3.122) to (3.129), it follows that:

$$c_i u_i = \int_{\Gamma} \Psi\frac{\partial u}{\partial n}\mathrm{d}\Gamma - \int_{\Gamma} u\frac{\partial \Psi}{\partial n}\mathrm{d}\Gamma \tag{3.130}$$

where:

$$c_i = \begin{cases} 1 & i\in\Omega \\ 1-\dfrac{\theta_2-\theta_1}{2\pi} & i\in\Gamma \\ 0 & i\notin\Omega \end{cases} \tag{3.131}$$

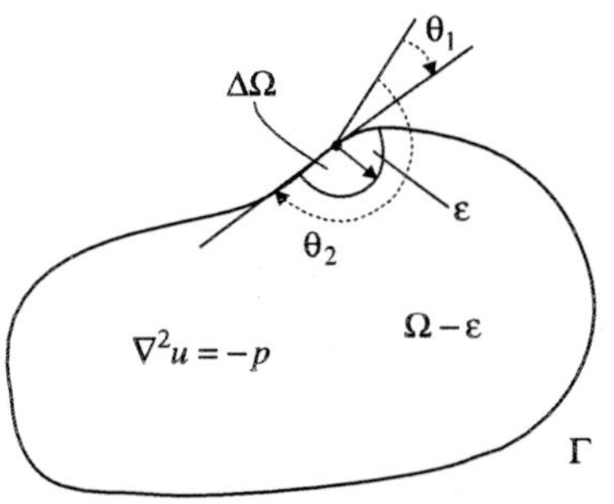

Figure 3.26 Extraction of the singularity on the domain integral.

If the integral representation (3.126) of the Poisson equation (3.124) is considered its domain term, as shown in Figure 3.26, it also vanishes as shown below:

$$\lim_{\varepsilon\to 0}\int_{\Delta\Omega}\psi p \mathrm{d}\Omega = \lim_{\varepsilon\to 0}\frac{1}{2\pi}\int_{\theta_1}^{\theta_2}\frac{1}{2\pi}\ln\left(\frac{1}{\varepsilon}\right)p_i\varepsilon \mathrm{d}\varepsilon \mathrm{d}\theta = 0 \tag{3.132}$$

Consequently, the resulting integral formulation of the Poisson equation is given by:

$$c_i u_i = \int_\Gamma \Psi\frac{\partial u}{\partial n}\mathrm{d}\Gamma - \int_\Gamma u\frac{\partial \Psi}{\partial n}\mathrm{d}\Gamma + \int_\Omega p\psi \mathrm{d}\Omega \tag{3.133}$$

Generally, for a well-posed static field problem either u or $\partial u/\partial n$ on the boundary Γ must be known, which is described by the forced (Dirichlet), natural (Neumann), or mixed (Cauchy) boundary condition.

Determining all values of potential u and its normal derivative $\partial u/\partial n$ on the boundary, the potential at an arbitrary point of the domain can be calculated.

The electrostatic potential problem ($u = \varphi$) is then defined by the following equation:

$$c_i\varphi_i = \int_\Gamma \Psi\frac{\partial \varphi}{\partial n}\mathrm{d}\Gamma - \int_\Gamma \varphi\frac{\partial \Psi}{\partial n}\mathrm{d}\Gamma + \int_\Omega \frac{\rho}{\varepsilon}\psi \mathrm{d}\Omega \tag{3.134}$$

while the magnetostatic problem for an axial component of the magnetic scalar potential $A_z(u = A_z)$ is governed by:

$$c_i A_{zi} = \int_\Gamma \Psi\frac{\partial A_z}{\partial n}\mathrm{d}\Gamma - \int_\Gamma A_z\frac{\partial \Psi}{\partial n}\mathrm{d}\Gamma + \int_\Omega \mu J_z\psi \mathrm{d}\Omega \tag{3.135}$$

In particular, the magnetostatic field of a source-free region is governed by the integral formulation of the magnetic scalar potential ($u = \varphi_\mathrm{m}$):

$$c_i\varphi_{mi} = \int_\Gamma \Psi\frac{\partial \varphi_\mathrm{m}}{\partial n}\mathrm{d}\Gamma - \int_\Gamma \varphi_\mathrm{m}\frac{\partial \Psi}{\partial n}\mathrm{d}\Gamma \tag{3.136}$$

where φ_m denotes the magnetic scalar potential.

3.5.2 Boundary Element Discretization

The basic idea of the BEM is to discretize the boundary of the domain under consideration into a set of elements. The unknown solution over each element is approximated by an interpolation function, which is associated with the values of

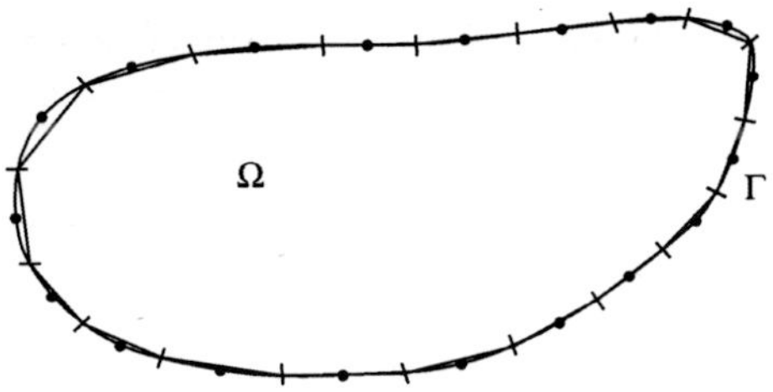

Figure 3.27 Constant element approximation.

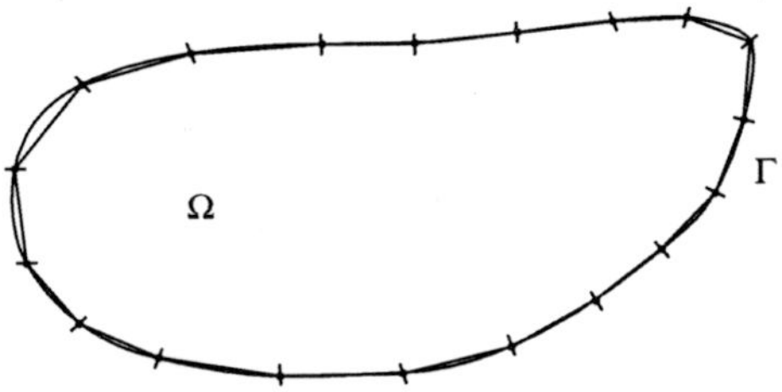

Figure 3.28 Linear element approximation.

the functions at the element nodes, so that the corresponding integral equation may be converted into a system of algebraic equations. The solution of the algebraic equation system gives the approximate solution of the original integral equation.

The boundary geometry can be discretized into a series of constant (Fig. 3.27), linear (Fig. 3.28), or quadratic elements (Fig. 3.29). The geometry of the elements is then expressed in the form of interpolation or shape functions.

3.5.2.1 Constant Boundary Elements The simplest solution can be obtained by using the constant boundary elements. The geometry of the constant boundary element for two-dimensional cases is shown in Figure 3.30.

The next step in the boundary element solution procedure is the transformation of the global coordinates of the element into the local ones (Fig. 3.31).

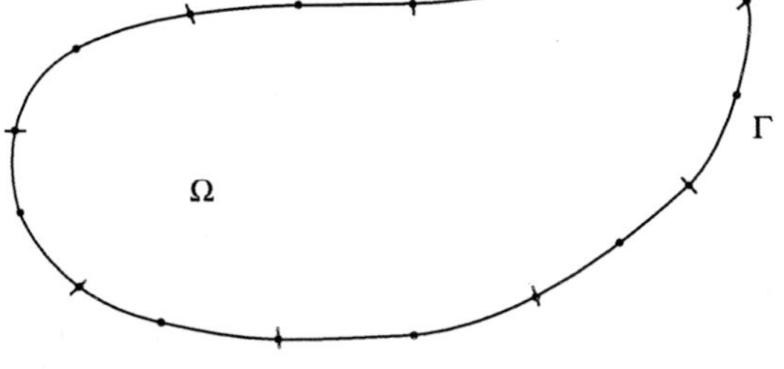

Figure 3.29 Quadratic element approximation.

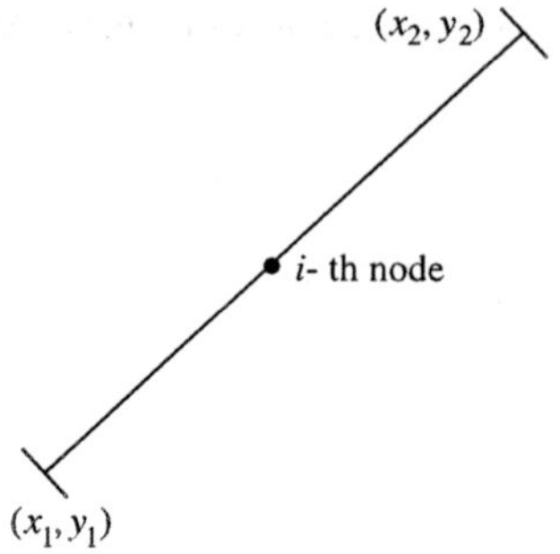

Figure 3.30 Constant boundary element approximation.

This transformation of coordinates is given by the following set of $x = x(\xi)$ and $y = y(\xi)$, that is:

$$x(\xi) = \frac{x_1 - x_2}{2}\xi + \frac{x_1 + x_2}{2} \tag{3.137}$$

$$y(\xi) = \frac{y_1 - y_2}{2}\xi + \frac{y_1 + y_2}{2} \tag{3.138}$$

where (x_1,y_1) and (x_2,y_2) are the coordinates of the element.

In addition, it follows that:

$$\mathrm{d}\Gamma = \sqrt{\mathrm{d}x^2 + \mathrm{d}y^2} = \sqrt{\left(\frac{\mathrm{d}x}{\mathrm{d}\xi}\right)^2 + \left(\frac{\mathrm{d}y}{\mathrm{d}\xi}\right)^2}\,\mathrm{d}\xi = \frac{\Delta\Gamma}{2}\mathrm{d}\xi \tag{3.139}$$

where $\Delta\Gamma$ is the segment length defined by:

$$\Delta\Gamma = \sqrt{(x_2 - x_1)^2 + (y_2 - y_1)^2} \tag{3.140}$$

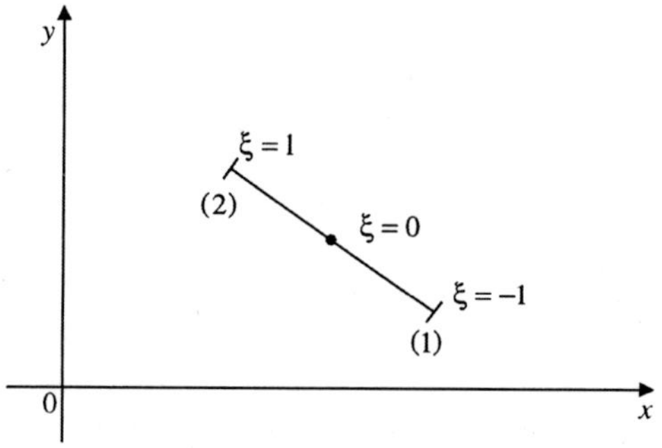

Figure 3.31 Global and local coordinates.

Using the constant boundary element approximation, the integral equation formulation of the potential problem defined by the Laplace equation becomes:

$$c_i u_i = \sum_{j=1}^{M} \left[\frac{\partial u}{\partial n}\Big|_{j\cdot} \int_{\Gamma_j} \Psi \mathrm{d}\Gamma - u_j \int_{\Gamma_j} \frac{\partial \Psi}{\partial n} \mathrm{d}\Gamma \right] \tag{3.141}$$

where i denotes the i-th boundary node and j stands for j-th boundary element.

Introducing the following notation:

$$\frac{\partial u}{\partial n}\Big|_{j\cdot} = Q_j \tag{3.142}$$

$$u = U_j \tag{3.143}$$

$$G_{ij} = \int_{\Gamma_j} \psi_{ij} \mathrm{d}\Gamma = \begin{cases} \dfrac{1}{2\pi} \displaystyle\int_{\Gamma_j} \ln\left(\frac{1}{R_{ij}}\right) \mathrm{d}\Gamma & 2D \cdots \textit{problems} \\ \displaystyle\int_{\Gamma_j} \frac{1}{4\pi R_{ij}} \mathrm{d}\Gamma & 3D \cdots \textit{problems} \end{cases} \tag{3.144}$$

$$H_{ij} = \int_{\Gamma_j} \frac{\partial \psi_{ij}}{\partial n} \mathrm{d}\Gamma = \begin{cases} \dfrac{1}{2\pi} \displaystyle\int_{\Gamma_j} \frac{\partial}{\partial n} \ln\left(\frac{1}{R_{ij}}\right) \mathrm{d}\Gamma & 2D \cdots \textit{problems} \\ \dfrac{1}{4\pi} \displaystyle\int_{\Gamma_j} \frac{\partial}{\partial n}\left(\frac{1}{R_{ij}}\right) \mathrm{d}\Gamma & 3D \cdots \textit{problems} \end{cases} \tag{3.145}$$

This results in the following system of equations for each i:

$$c_i u_i = G_{ij} Q_j - H_{ij} U_j \tag{3.146}$$

This algebraic equation system can also be written in the following matrix form:

$$[H]\{U\} = [G]\{Q\} \tag{3.147}$$

The matrix system (3.147) can be solved once the set of boundary conditions is prescribed.

If the calculation domain contains sources, then a contribution of the source term must be included into the algebraic equation system. The integral equation formulation of the potential problem defined by the Poisson equation is given by:

$$c_i u_i = \sum_{j=1}^{M} \left[Q_j \int_{\Gamma_j} \Psi \mathrm{d}\Gamma - U_j \int_{\Gamma_j} \frac{\partial \Psi}{\partial n} \mathrm{d}\Gamma \right] + \int_{\Omega_S} p \Psi \mathrm{d}\Omega \tag{3.148}$$

where p is the constant value of the source on the segment of the domain that contains sources.

The algebraic equation system is now given by:

$$P_i + \sum_{j=1}^{M} H_{ij} U_j = \sum_{j=1}^{M} Q_j G_{ij} \tag{3.149}$$

or in matrix form:

$$\{P\} + [H]\{U\} = [G]\{Q\} \tag{3.150}$$

where P_i is given by:

$$P_i = \int_{\Omega_S} p\Psi \mathrm{d}\Omega \tag{3.151}$$

If the sources inside the domain are not known, then a coupling of BEM with some domain discretization method is required.

3.5.2.2 Linear and Quadratic Elements In order to achieve a higher accuracy and faster convergence rate, the boundary of the domain can be approximated by a series of linear or quadratic elements depending on the physical nature and the geometric complexity of the problem under consideration.

In each element, the function u and the normal derivative $\partial u/\partial n$ are interpolated by means of linear or quadratic shape functions. The linear or quadratic boundary element concept implies variation for the quantities u and $\partial u/\partial n$ over the element.

As usually, the geometry of the boundary elements is also chosen to be modeled by the linear or quadratic shape functions, the elements are referred to as isoparametric elements [1].

Curvilinear elements are very well suited to modeling curved shapes so that complicated geometries can be represented with a rather small number of elements.

When using isoparametric elements, the global coordinate x is a function of the local parametric coordinate ξ on the element.

A function $x(\xi)$ for the geometry can be defined over the element in terms of the element approximating functions $f_i(\xi)$, that is:

$$x = \sum_{i=1}^{N} x_i f_i(\xi) \tag{3.152}$$

where the approximating functions $f_i(\xi)$ are usually polynomials.

Such transformation, in which the same family of approximating functions is used for the unknown quantity and to express the element shape transformation, is called isoparametric, and the related elements are referred to as isoparametric elements.

In accordance to the previous discussion, the unknowns along each element can be interpolated as follows:

$$u = \sum_{j=1}^{n_e} f_j(\xi)U_j \tag{3.153}$$

$$\frac{\partial u}{\partial n} = \sum_{j=1}^{n_e} f_j(\xi)\left.\frac{\partial u}{\partial n}\right|_{j\cdot} = \sum_{j=1}^{n_e} f_j(\xi)Q_j \tag{3.154}$$

where U_j denotes the unknown coefficients of the potential distribution, and Q_j is the value of the normal derivative at the j-th node.

3.5.2.3 ***Linear Case*** Hence, for a linear approximation, it follows:

$$u = f_1(\xi)U_1 + f_2(\xi)U_2 \tag{3.155}$$

$$\frac{\partial u}{\partial n} = f_1(\xi)Q_1 + f_2(\xi)Q_2 \tag{3.156}$$

where U_1, U_2, Q_1,Q_2 are the values of the vector potential and its normal derivative on the node $j = 1$ and $j = 2$, respectively.

The linear shape functions are given by:

$$f_1(\xi) = \frac{1}{2}(1 - \xi) \tag{3.157}$$

$$f_1(\xi) = \frac{1}{2}(1 + \xi) \tag{3.158}$$

For linear elements (Fig 3.32), the geometry is a linear function of the coordinates, that is,

$$x = f_1(\xi)x_1 + f_2(\xi)x_2 \tag{3.159}$$

$$y = f_1(\xi)y_1 + f_2(\xi)y_2 \tag{3.160}$$

3.5.2.4 ***Quadratic Case*** For a quadratic interpolation, it follows:

$$u = f_1(\xi)U_1 + f_2(\xi)U_2 + f_3(\xi)U_3 \tag{3.161}$$

$$\frac{\partial u}{\partial n} = f_1(\xi)Q_1 + f_1(\xi)Q_2 + f_3(\xi)Q_3 \tag{3.162}$$

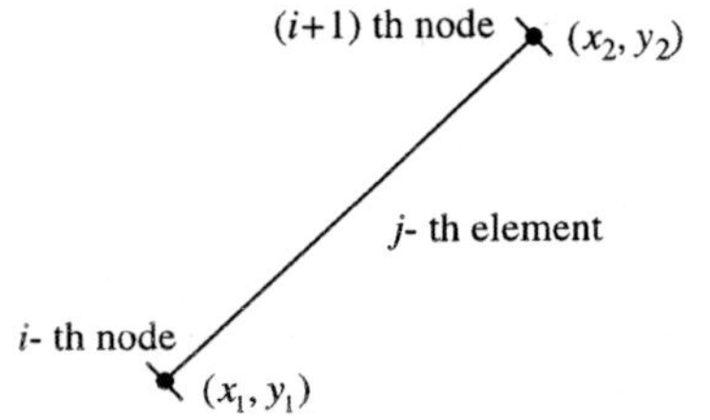

Figure 3.32 Linear boundary element approximation.

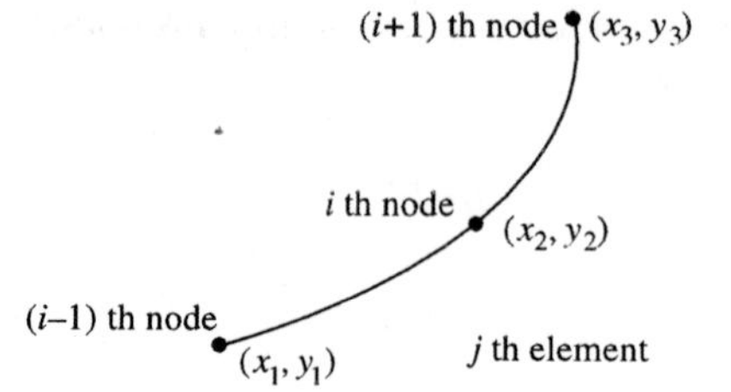

Figure 3.33 Quadratic boundary element approximation.

where the shape functions are defined as:

$$f_1(\xi) = \frac{1}{2}\xi(\xi - 1) \tag{3.163}$$

$$f_2(\xi) = \frac{1}{2}(1 + \xi)(1 - \xi) \tag{3.164}$$

$$f_3(\xi) = \frac{1}{2}\xi(\xi + 1) \tag{3.165}$$

and U_j and Q_j are the values of the vector potential and its normal derivative at the given node j, respectively.

For the case of quadratic elements (Fig. 3.33), the geometry is a function of the coordinates, that is:

$$x = f_1(\xi)x_1 + f_2(\xi)x_2 + f_3(\xi)x_3 \tag{3.166}$$

$$y = f_1(\xi)y_1 + f_2(\xi)y_2 + f_3(\xi)y_3 \tag{3.167}$$

The structure of the BEM coefficients when using linear or quadratic elements is as follows:

$$G_{ij} = \int_{\Gamma_j} f_j \psi_{ij} \mathrm{d}\Gamma = \int_{\Gamma_j} f_j \psi_{ij} \frac{\mathrm{d}\Gamma}{\mathrm{d}\xi}\mathrm{d}\xi \tag{3.168}$$

$$H_{ij} = \int_{\Gamma_j} f_j \frac{\partial \psi_{ij}}{\partial n} \mathrm{d}\Gamma = \int_{\Gamma_j} f_j \frac{\partial \psi_{ij}}{\partial n}\frac{\mathrm{d}\Gamma}{\mathrm{d}\xi}\mathrm{d}\xi \tag{3.169}$$

where $\{f\}$ denotes the corresponding linear or quadratic shape functions vector.

The algebraic equation system for the whole boundary is defined in matrix form as follows:

$$[H]\{U\} = [G]\{Q\} \tag{3.170}$$

If the source contribution needs to be taken into account, then the corresponding algebraic system is given by the following matrix equation:

$$\{P\} + [H]\{U\} = [G]\{Q\} \tag{3.171}$$

The analysis regarding the isoparametric solution presented so far is for the solution of two-dimensional potential problems.

If a three-dimensional problem is considered, then isoparametric triangular or quadrilateral surface elements have to be applied [7].

3.5.3 Computational Example for 2D Static Problem

Solving the Laplace equation $\nabla^2\varphi = 0$, it is necessary to find the distribution of the electrostatic field within the source-free quadratic domain with associated Dirichlet and Neumann boundary conditions, as shown in Figure 3.34.

The calculation domain is discretized to 12 constant elements, as shown in Figure 3.35, with $\varphi_0 = 300\text{V}$ and $d = 6$ m.

Solving the matrix system (3.170), one obtains the flux density on the boundary (nodes 4, 5, 6, 10, 11, and 12):

$$\begin{aligned} q_4 = q_5 = q_6 &= -50\frac{\text{V}}{\text{m}} \\ q_{10} = q_{11} = q_{12} &= +50\frac{\text{V}}{\text{m}} \end{aligned} \tag{3.172}$$

Knowing the values of the potential and normal derivatives in all nodes along the calculation boundary, it is possible to compute the potential inside the domain. In this particular case, the potential is calculated in the five points inside the domain (Fig. 3.35). The obtained results are:

$$\begin{aligned} \varphi_1 &= 200.28V & \varphi_4 &= 200.28V \\ \varphi_2 &= 99.74V & \varphi_5 &= 99.74V \\ \varphi_3 &= 150.01V \end{aligned}$$

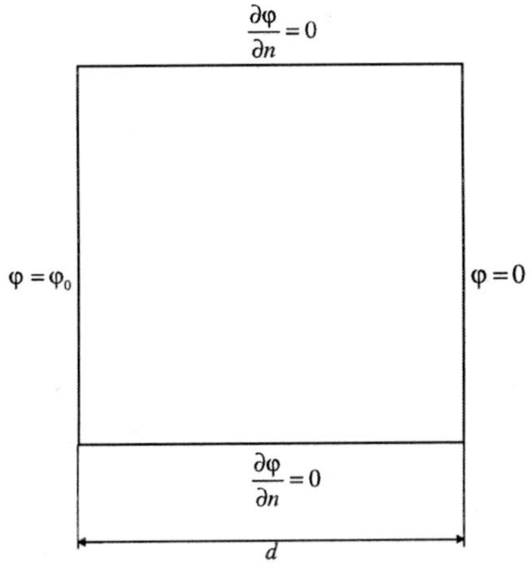

Figure 3.34 Quadratic domain with associated boundary conditions.

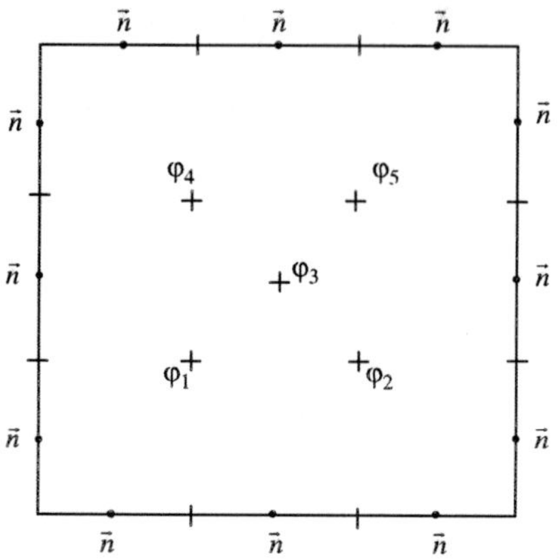

Figure 3.35 Discretization of the domain boundary.

It is worth mentioning that the obtained results are in a satisfactory agreement with the solution obtained analytically via separation of variables [8].

3.6 REFERENCES

[1] D. Poljak, C. A. Brebbia: Boundary Element Methods for Engineers, *WIT Press, Southampton-Boston*, 2005.

[2] M. N. O. Sadiku: Numerical Techniques in Electromagnetics, 2nd Edition, *CRC Press, Boca Raton*, 2001.

[3] D. Poljak, C. Y. Tham, Integral Equation Techniques in Transient Electromagnetics, *WIT Press, Southampton-Boston*, 2003.

[4] E. K. Miller: A selective survey of computational electromagnetics, *IEEE Trans. on Antennas & Propagat., Vol. 36, No. 9*, September 1988, pp 1281–1305.

[5] E K. Miller, J. A. Landt: Direct time-domain techniques for transient radiation and scattering from wires, *Proc. IEEE, Vol. 68, No. 11*, November 1980, pp 1396–1423.

[6] Computing Device, Federal Communications Commission Rules and Regulations, Vol. 2, Part 15, Subpart J, July 1981.

[7] P. P. Silvester, R. L. Ferrari: Finite Elements for Electrical Engineers, 3rd Edition, *Cambridge University Press, Cambridge* 1996.

[8] D. Poljak, N. Kovac, V. Doric: Numerical Methods in Electrical Engineering, *Lecture notes, University of Split*, 2006. (In Croatian)

4

STATIC FIELD ANALYSIS

In an electrostatic problem, it is necessary to determine the electric potential and electric field distribution. It is usually assumed that the resistive current is negligible and any magnetic effect can be neglected as well. In magnetostatics, the aim is to determine the magnetic field distribution. In this case, the frequency of the phenomenon is so low that eddy currents are negligible.

Both electrostatic and magnetostatic problems can be regarded as potential problems and analyzed using the Laplace or Poisson equation depending on the presence of a source in the region.

4.1 ELECTROSTATIC FIELDS

The electrostatic field problem is a special case of the electromagnetic field represented by Maxwell equations for which the phenomenon is not varying in time. This reduces the first Maxwell equation to:

$$\nabla x \vec{E} = 0 \tag{4.1}$$

The second and fourth Maxwell equations related to the magnetic field are not considered, while the third Maxwell equation:

$$\nabla \vec{D} = \rho \tag{4.2}$$

can be expressed:

$$\nabla (\varepsilon \vec{E}) = \rho \tag{4.3}$$

where E is the electric field intensity, ρ is the volume charge density and ε is the dielectric permittivity of the medium.

From the curl Maxwell eqn (4.1), the electric field can be expressed by a scalar potential function φ:

$$\vec{E} = -\nabla\varphi \tag{4.4}$$

Substituting eqn (4.4) into eqn (4.3) yields the Poisson equation:

$$\nabla(\varepsilon\nabla\varphi) = -\rho \tag{4.5}$$

For a linear, homogeneous and isotropic medium, eqn (4.5) can be written as:

$$\nabla^2\varphi = -\frac{\rho}{\varepsilon} \tag{4.6}$$

For the special case of a source-free region, eqn (4.5) simplifies to the Laplace equation:

$$\nabla(\varepsilon\nabla\varphi) = 0 \tag{4.7}$$

which, for a constant value of ε, gives:

$$\nabla^2\varphi = 0 \tag{4.8}$$

Many problems involving linear homogeneous and isotropic medium can be analyzed using eqn (4.8)

4.2 MAGNETOSTATIC FIELDS

The magnetostatic field is governed by the second and fourth Maxwell equations, which for static problems become:

$$\nabla x\vec{H} = \vec{J} \tag{4.9}$$

$$\nabla\vec{B} = 0 \tag{4.10}$$

For a linear medium, eqn (4.10) can be written as:

$$\nabla(\mu\vec{H}) = 0 \tag{4.11}$$

where H is the magnetic field density, J is the electric current density and μ is the permeability of the medium.

For a source-free region, the curl eqn (4.9) becomes:

$$\nabla x \vec{H} = 0 \tag{4.12}$$

and consequently, such magnetostatic field problems can be derived from a scalar potential defined via the magnetic scalar potential φ_{m}:

$$\vec{H} = -\nabla \varphi_m \tag{4.13}$$

This results in a differential equation of the Laplace type. Assuming the linear medium, eqn (4.11) can then be written as follows:

$$\nabla(\mu \nabla \varphi_m) = 0 \tag{4.14}$$

If a region contains sources in terms of current density, the magnetic vector potential is given by equation:

$$\vec{B} = \nabla x \vec{A} \tag{4.15}$$

which for linear media becomes:

$$\vec{H} = \frac{1}{\mu} \nabla x \vec{A} \tag{4.16}$$

Combining eqns (4.16) and (4.9) yields:

$$\nabla x \left(\frac{1}{\mu} \nabla x \vec{A} \right) = \vec{J} \tag{4.17}$$

For a linear, homogeneous, and isotropic media this equation can be written, as follows:

$$\nabla(\nabla \vec{A}) - \nabla^2 \vec{A} = \mu \vec{J} \tag{4.18}$$

and applying the Coulomb gauge:

$$\nabla \vec{A} = 0 \tag{4.19}$$

leads to the following Poisson equation for magnetostatic field problems:

$$\nabla^2 \vec{A} = -\mu \vec{J} \tag{4.20}$$

When $J = 0$ eqn (4.20) becomes a Laplace equation, that is:

$$\nabla^2 \vec{A} = 0 \tag{4.21}$$

It is worth mentioning that eqn (4.21) is seldom used in magnetostatics while eqn (4.8) is widely applied in electrostatic applications.

4.3 MODELING OF STATIC FIELD PROBLEMS

Static fields involving complex shape geometries and inhomogeneous domains are usually modeled via the domain methods such as Finite Difference Method (FDM) and Finite Element Method (FEM), or via boundary methods such as Boundary Element method (BEM). As the integral equation method, an alternate to the partial differential equation method, is an important approach in electrostatics and electromagnetics, this chapter deals with an analysis of static field problems via BEM. More details on the subject can be found in [1].

In accordance to the boundary element fundamentals presented in Chapter 3, the electrostatic potential problem can be defined by the following integral relationship [1]:

$$c_i\varphi_i = \int_\Gamma \Psi \frac{\partial \varphi}{\partial n} d\Gamma - \int_\Gamma \varphi \frac{\partial \Psi}{\partial n} d\Gamma + \int_\Omega \frac{\rho}{\varepsilon} \psi d\Omega \tag{4.22}$$

while the magnetostatic problem for an axial component of the magnetic scalar potential A_z is governed by [1]:

$$c_i A_{zi} = \int_\Gamma \Psi \frac{\partial A_z}{\partial n} d\Gamma - \int_\Gamma A_z \frac{\partial \Psi}{\partial n} d\Gamma + \int_\Omega \mu J_z \psi d\Omega \tag{4.23}$$

In the special case of the magnetostatic field of a source-free region, the integral formulation of the magnetic scalar potential is given by:

$$c_i {\varphi_m}_i = \int_\Gamma \Psi \frac{\partial \varphi_m}{\partial n} d\Gamma - \int_\Gamma \varphi_m \frac{\partial \Psi}{\partial n} d\Gamma \tag{4.24}$$

where φ_m denotes the magnetic scalar potential.

4.3.1 Integral Equations in Electrostatics Using Sources

If it is assumed that a volume charge density ρ within a given volume V and a surface charge density ρ_s over a given surface S are the sources of a static electric field, then the scalar potential due to such sources is given by:

$$\varphi = \int_V \frac{\rho}{\varepsilon} \psi dV + \int_{S'} \frac{\rho_s}{\varepsilon} \psi dS' \tag{4.25}$$

where ψ *is the fundamental solution, as discussed in Chapter 3.*

For two-dimensional problems, Ψ is given by the expression:

$$\Psi = -\frac{1}{2\pi}\ln R \tag{4.26}$$

where $R = |\vec{r} - \vec{r}\,'|$ is the distance from the source point (boundary point) to the observation point, and for three-dimensional problems the fundamental solution is given by:

$$\Psi = \frac{1}{4\pi R} \tag{4.27}$$

Equation (4.25) is a special case of eqn (4.22) for which the domain Ω is now replaced with V to indicate a three-dimensional volume. The boundary Γ is now called S', which represents the boundary with the unknown sources. The first term on the right hand side of eqn (4.22) is now:

$$\int_{S'} \frac{\partial \varphi}{\partial n} \psi dS' \tag{4.28}$$

where $\frac{\partial \varphi}{\partial n}$ represents the unknown charge density ρ_s, which is unknown and defined as follows:

$$\rho_s = \varepsilon \frac{\partial \varphi}{\partial n} \tag{4.29}$$

Therefore, in many electrostatic problems charge distributions are not specified. What is usually known is the potential of the conducting electrode. The surface charge on such electrodes is unknown along with the potential distribution φ in the interelectrode space. Therefore, in the case of a three-dimensional problem, given the function φ_s on S' and assuming that volume charge is absent, eqn (4.25) can be written as an integral expression in terms of the unknown surface charge density ρ_s, that is:

$$\varphi_s = \int_{S'} \frac{\rho_s}{\varepsilon} \psi \mathrm{d}S' \tag{4.30}$$

The source points and observation points are both located on the surface S'.

Function φ_s represents a known potential distribution to which a system of electrodes is charged. Potential in the space around electrodes can be calculated from eqn (4.25).

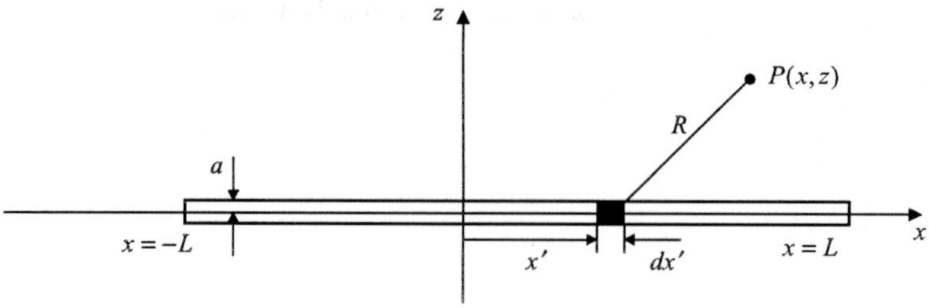

Figure 4.1 Charge distribution on a straight wire.

If a line charge density ρ_l is considered, instead of surface charge density ρ_s, then the corresponding integral equation is given by:

$$\varphi_l = \int_{S'} \frac{\rho_l}{\varepsilon} \psi \mathrm{d}l' \tag{4.31}$$

where φ_l is the potential along the line.

Determining the charge distribution along a wire is an important problem with number of applications in electrostatics.

The potential at a point $P(x,z)$ produced by the charged straight wire ρ_l, Figure 4.1, is given by the following integral expression:

$$\varphi(x,z) = \frac{1}{4\pi\varepsilon} \int_{-L}^{+L} \frac{\rho_l(x')\mathrm{d}x'}{R} \tag{4.32}$$

where $2L$ is the total length of the wire.

If the potential distribution along the wire φ_l is known, then the integral equation is given by:

$$\varphi_l = \frac{1}{4\pi\varepsilon} \int_{-L}^{+L} \frac{\rho_l(x')dx'}{R} \tag{4.33}$$

If the wire is infinitely thin, then R is:

$$R = |x - x'| \tag{4.34}$$

which becomes zero when $x = x'$, and the integral equations becomes singular.

This problem can be partially overcome by assuming any non-zero thickness of the wire:

$$R = \sqrt{(x - x')^2 + a^2} \tag{4.35}$$

where a is the wire radius.

This leads to the so-called quasisingular kernels. From a numerical point of view all integrals including $x = x'$ are almost as difficult as integrals containing true singularities; this quasisingularity problem requires special treatment through the numerical integration procedures [2]. More information on quasisingularity of integral equation kernels in antenna theory can be found in [1] and [3].

4.3.2 Computational Example: Modeling of a Lightning Rod

Electromagnetic interferences (EMI) on electrical equipment can be generated by a direct lightning strike or by an indirect discharge in the close proximity of the lightning channel [1], [3], [4]. Such lightning flash surges often cause adverse effects on electrical equipment.

The key point in the analysis of lightning protection system is the calculation of the so-called protection zone. The protection zone is a space in the vicinity of the lightning rod where the electric field intensity is much less with respect to the case where the lightning rod is absent. The geometry of such protection area is determined by the lines of the electric field in the vicinity of the rod, which is readily computed from the electric scalar potential. The calculation of the potential distribution can be carried out using the boundary element method.

A simple lightning protection system in the form of a metallic post is sited vertically on the perfectly conducting ground that is assumed to have a zero potential. A negatively charged cloud (flat model) is located at a certain distance h above ground, as shown in Figure 4.2.

The excitation is given in the form of an electrostatic field of constant value E_0 generated by the flat charged cloud. The metallic rod is assumed to be on the earth at the zero potential. The protection zone of the metallic rod is determined by the electric field distribution in its vicinity. The calculation procedure is divided in two steps. First, the calculation of the induced charge along the post is computed using the boundary element method, and then the electric field in the surrounding space is obtained using a finite difference approximation. (Finite difference approximation

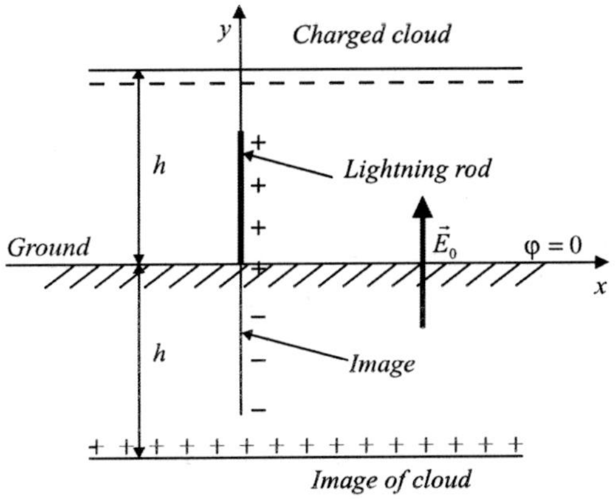

Figure 4.2 Geometry of a metallic rod.

is applied here as it is the simplest way to calculate the electric field as a gradient of the potential.)

The calculation of the induced charge along the rod is carried out using the scalar potential integral equation. The total scalar potential around the metallic rod consists of two components:

$$\varphi = \varphi_0 + \varphi_{ind} \tag{4.36}$$

where φ_0 is the potential produced by the electric field E_0, and φ_{ind} is the potential component due to the induced charge on the rod.

The potential φ_0 is given by the integral over electric field:

$$\varphi_0 = -\int_{-z/2}^{z/2} E_0 \mathrm{d}z \tag{4.37}$$

In addition, the potential component due to the induced charge is:

$$\varphi_{ind} = \frac{1}{4\pi\varepsilon}\int_{-L}^{L} \frac{\rho_l}{R}\mathrm{d}z' \tag{4.38}$$

Combining eqns (4.36) to (4.38), the total potential can be written in the form:

$$\varphi = \frac{1}{4\pi\varepsilon}\int_{-L}^{L} \frac{\rho_l}{R}\mathrm{d}z' - E_0 z \tag{4.39}$$

where ρ_l is the linear charge density, L is the entire length of the metallic post, and R is the distance from the source to the observation point, respectively.

Satisfying the boundary condition for zero potential (earth potential) at $z = 0$ results in the following integral equation:

$$\frac{1}{4\pi\varepsilon}\int_{-L}^{L} \frac{\rho_l}{R}\mathrm{d}z' = E_0 z \tag{4.40}$$

The implementation of the boundary element method to the integral eqn (4.40) can be carried out in many ways, for example by variational approach, method of average potentials, weighted residual approach, and so on.

The average value of the potential along the cylinder of length L, is given by:

$$\bar{\varphi}_{av} = \frac{1}{L}\int_{-L}^{L} \varphi(z)\mathrm{d}z = \frac{1}{4\pi\varepsilon_0 L}\int_{-L}^{L}\int_{-L}^{L} \frac{\rho_l \mathrm{d}z' \mathrm{d}z}{R} - \frac{1}{L}\int_{-L}^{L} E_0 z \mathrm{d}z \tag{4.41}$$

while applying the weighted residual approach to the integral eqn (4.40) yields:

$$\int_{-L}^{L}\left[\frac{1}{4\pi\varepsilon}\int_{-L}^{L}\frac{\rho_l}{R}\mathrm{d}z' - E_0 z\right]W_j\mathrm{d}z = 0 \tag{4.42}$$

where W_j is the set of test functions.

4.3.2.1 Boundary Element Equations The mean value of the potential along the cylinder of length L, is given by:

$$\bar{\varphi}_j = \frac{1}{\Delta l_j}\int_{-\Delta lj}^{\Delta l_{ji}}\varphi_i(z)\mathrm{d}z = \frac{1}{4\pi\varepsilon_0\Delta l_j\Delta l_i}\int_{-\Delta l_j}^{\Delta l_i}\int_{-\Delta l_i}^{\Delta l_i}\frac{\rho(z')\mathrm{d}z'\mathrm{d}z}{R}+\frac{1}{\Delta l_j}\int_{-\Delta l_j}^{\Delta l_j}E_0 z\mathrm{d}z \tag{4.43}$$

where Δl_j and Δl_i is the length of jth observation and ith source segment, respectively.

Applying the boundary element discretization, eqn (4.43) transforms into the following matrix system:

$$\{\varphi\} = [A]\{\rho_l\} - \{\varphi_0\} \tag{4.44}$$

In addition, the transformation to local coordinates results in the following BEM matrix coefficient:

$$A_{ji} = \frac{1}{4\pi\varepsilon_0\Delta l_j\Delta l_i}\int_{-\Delta l_j}^{\Delta l_i}\int_{-\Delta l_i}^{\Delta l_i}\frac{\mathrm{d}z'\mathrm{d}z}{R} = \frac{1}{16\pi\varepsilon_0}\int_{-1}^{1}\int_{-1}^{1}\frac{\mathrm{d}\xi'\mathrm{d}\xi}{R} \tag{4.45}$$

while the corresponding voltage vector is given by:

$$\varphi_{0j} = \frac{\Delta l_j}{2}\int_{-1}^{1}E_0 z(\xi)\mathrm{d}\xi \tag{4.46}$$

Taking into account the zero-potential condition on the surface of the perfectly conducting ground, the eqn (4.44) becomes:

$$[A]\{\rho_l\} = \{V\} \tag{4.47}$$

On the contrary, applying the Galerkin variant of the boundary element method to the integral eqn (4.42), the following matrix system is obtained in the form:

$$\sum_{i=1}^{M} [A]_{ji}\{q\}_i = \{V\}_j \tag{4.48}$$

where the element matrix and vector are given by:

$$[A]_{ji} = \frac{\Delta l_j \Delta l_i}{16\pi\varepsilon_0} \int_{-1}^{1} \int_{-1}^{1} \{f\}_j \{f'\}_i \frac{d\xi' d\xi}{R} \tag{4.49}$$

$$\{V\}_j = \frac{\Delta l_j}{2} \int_{-1}^{1} E_0 \{f\}_j z(\xi) d\xi \tag{4.50}$$

where $\{f\}_j$ and $\{f\}_i$ denote the set of linear shape functions.

By solving the system of eqn (4.48), the induced charges along the rod can be determined.

In chapters to follow, the indirect BEM is applied to quasistatic, scattering, and radiation problems.

4.3.2.2 Finite Difference Computation of the Electric Field The electric field at an arbitrary point in the surrounding space of the lightning protection system can be obtained as a sum of the incident field E_{inc} (due to the charged cloud) and the induced field E_{ind} (due to the charge distribution along the lightning rods):

$$E^{tot} = E^{inc} + E^{ind} \tag{4.51}$$

where the induced field is simply determined by the gradient of the scalar potential:

$$\vec{E} = -\nabla\varphi \tag{4.52}$$

The electric field from eqn (4.52) is readily computed by using the finite difference algorithm [5].

4.3.2.3 Numerical Results A single lightning rod with length $L = 10\,m$ and negligible radius is considered. An incident field due to a charged cloud is $E_0 = 1.5 \cdot 10^5 V/m$. The induced charge along the thin metallic post is shown in Figure 4.3.

It is to be observed that the positive part of the charge distribution represents the real physical solution, while the negative part is related to the post image.

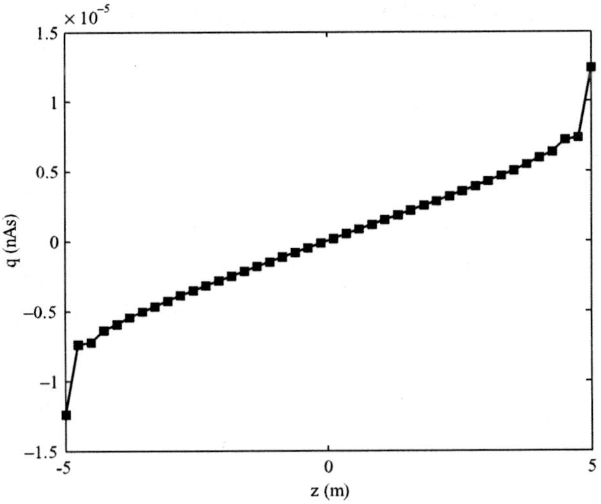

Figure 4.3 Induced charge along the post.

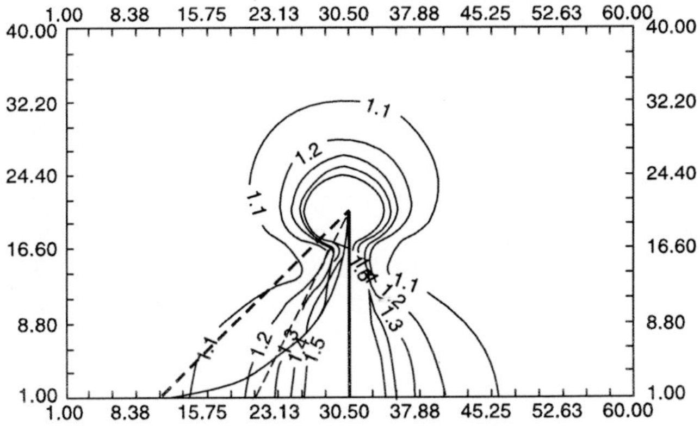

Figure 4.4 Protection zone of a single lightning rod.

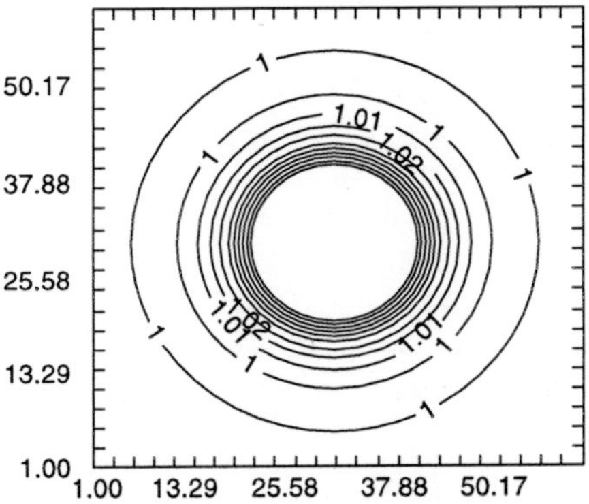

Figure 4.5 Protection zone of a single lightning rod ($y = 5$ m).

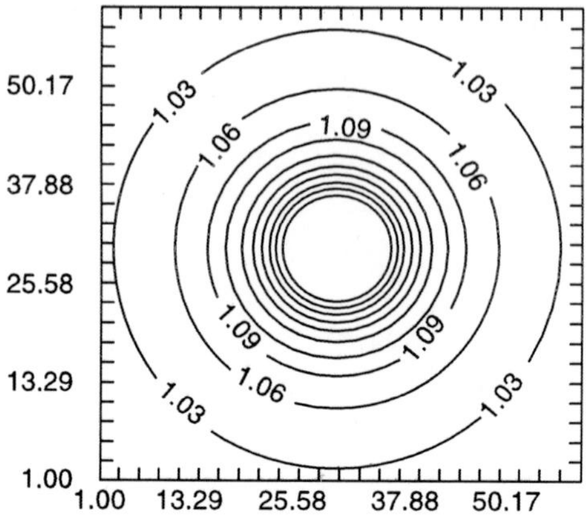

Figure 4.6 Protection zone of a single lightning rod ($y = 10\,\mathrm{m}$).

Figure 4.4 shows the protection zone of a single 20m high lightning rod in the yz plane. It can be noted that the area of very rare lightning strike is characterized by the field line with 30% greater value than the field value causing the static electricity of the cloud.

The corresponding zones in xy plane for the cases $y = 5m, y = 10m$ and $y = 20m$ are shown in Figures 4.5 to 4.7.

Numerical results presented in Figures 4.3 to 4.7 have been obtained using the piece-wise constant boundary element approximation. Computational examples related to the realistic lightning protection systems consisting of several rods have been presented in [1].

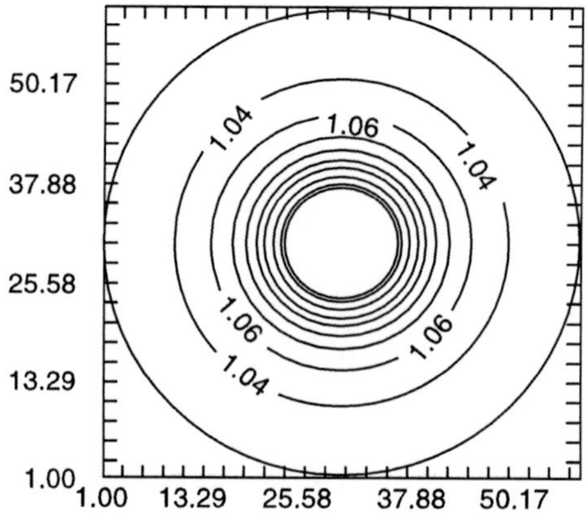

Figure 4.7 Protection zone of a single lightning rod ($y = 20\,\mathrm{m}$).

4.4 REFERENCES

[1] D. Poljak, C. A. Brebbia: The Boundary Element Method for Electrical Engineers, *WIT Press, Southampton, Boston*, 2005.

[2] P. P. Silvester, R. L. Ferrari: Finite Elements for Electrical Engineers, 3rd Edition, *Cambridge University Press, Cambridge*, 1996.

[3] D. Poljak: Electromagnetic Modelling of Wire Antenna Structures, *WIT Press, Southampton, Boston*, 2002.

[4] D. Poljak, B. Jajac: On the use of monopole antenna model in lightning protection system analysis, *International Symposium on Electromagnetic Compatibility, EMC Roma 98*, Roma, 1998.

[5] M. N. O. Sadiku: Numerical Techniques in Electromagnetics, *CRC Press, New York*, 2001.

5

QUASISTATIC FIELD ANALYSIS

5.1 INTRODUCTION

Many cases of electromagnetic fields at low frequencies including eddy currents analysis can be considered as quasistatic field problems [1–3]. For such problems diffusion processes rather than wave propagation dominate. Eddy currents are induced in any conducting medium when the magnetic field penetrating that medium changes with time, Figure 5.1. This phenomenon can be caused by a time-varying excitation function in terms of electric current or by a relative motion between the conducting medium and the exciting magnetic field. As shown in Figure 5.1, each part of such metallic structure can be referred to as a closed loop.

Although in some electromagnetic compatibility (EMC)' applications of eddy current can be useful, such as in the case of electrical machines and transformers; generally, these currents must be reduced as much as possible to avoid loss of energy and the short limit of the effective life span of electrical machinery. Thus, the determination of the eddy currents distribution in conducting media is very important in order to design more efficient and reliable electrical devices.

In general, the eddy current problems can be classified into two groups. The first one implies that the eddy current induced in a certain medium is excited by an external source. The other is referred to as a skin effect problem which is related to the calculation of the current density distribution in any arbitrary configuration of current-carrying conductors. The skin effect causes the total current density to become uneven and thus increases the losses in the system of conductors, Figure 5.2.

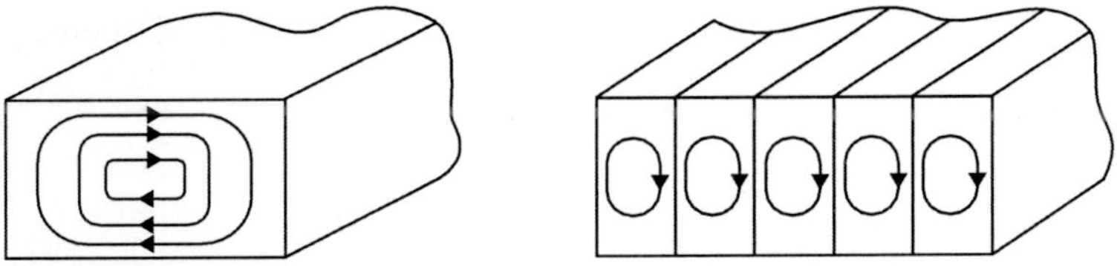

Figure 5.1 Eddy currents induced in a metallic structure exposed to time-changing magnetic field.

What is of interest in skin effect problems is to determine the total current density distribution in the arbitrary system of current-carrying conductors as the current density distribution affects the losses and the performance of conducting systems.

It is worth noting that eddy currents are induced inside the conductors inside which alternative current flows.

Hence, the skin effect phenomenon is related to the electromagnetic field in the conductors where eddy currents are to be considered.

5.2 FORMULATION OF THE QUASISTATIC PROBLEM

The eddy current problem is a low frequency phenomena for which the displacement currents are negligible with respect to the conduction currents. The problem is formulated via the diffusion equation which can be readily derived from the Maxwell curl equations.

The curl Maxwell equation for a magnetic field can be written as follows:

$$\nabla x\vec{H} = \vec{J} + \frac{\partial \vec{D}}{\partial t} \tag{5.1}$$

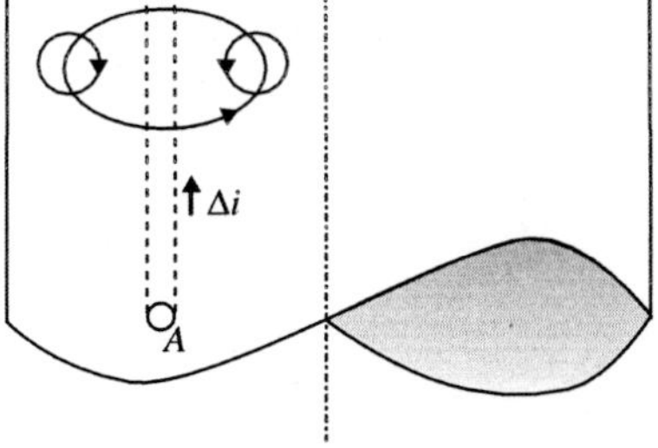

Figure 5.2 Skin effect in current carrying conductor.

where J denotes the current density, H denotes the magnetic field intensity, and D stands for the displacement current.

Taking the curl of both sides of eqn (5.1) results in

$$\nabla x \nabla x \vec{H} = \nabla x \vec{J} + \frac{\partial}{\partial t}(\nabla x \vec{D}) \tag{5.2}$$

In order to eliminate vectors J and D in favor of E, one can use the constitutive equations

$$\vec{J} = \sigma \vec{E} \tag{5.3}$$

$$\vec{D} = \varepsilon \vec{E} \tag{5.4}$$

where σ is the conductivity and ε stands for permittivity.

Assuming uniform scalar material properties, it follows that eqn (5.2) can be written as

$$\nabla x \nabla x \vec{H} = \sigma \nabla x \vec{E} + \varepsilon \frac{\partial}{\partial t}(\nabla x \vec{E}) \tag{5.5}$$

According to the curl Maxwell equation for electric field,

$$\nabla x \vec{E} = -\frac{\partial \vec{B}}{\partial t} \tag{5.6}$$

The curl of E in eqn (5.5) can be replaced by the rate of change of magnetic flux density resulting in the following relationship:

$$\nabla x \nabla x \vec{H} = -\sigma \frac{\partial \vec{B}}{\partial t} - \varepsilon \frac{\partial^2 \vec{B}}{\partial t^2} \tag{5.7}$$

In addition, if the magnetic material can be described by the following constitutive equation:

$$\vec{B} = \mu \vec{H} \tag{5.8}$$

where μ stands for medium permeability; then the wave equation for the magnetic field can be written as follows:

$$\nabla x \nabla x \vec{H} = -\mu\sigma \frac{\partial \vec{H}}{\partial t} - \mu\varepsilon \frac{\partial^2 \vec{H}}{\partial t^2} \tag{5.9}$$

Using the standard vector identity,

$$\nabla x \nabla x \vec{H} = \nabla \nabla \vec{H} - \nabla^2 \vec{H} \tag{5.10}$$

yields the general form of the magnetodynamic field equation,

$$\nabla^2 \vec{H} - \mu\sigma \frac{\partial \vec{H}}{\partial t} - \mu\varepsilon \frac{\partial^2 \vec{H}}{\partial t^2} = 0 \tag{5.11}$$

The same wave equation could be readily derived for the electric field by eliminating vector H from Maxwell equations instead of E.

The wave eqn (5.11) is valid for uniform regions and represents a significant degree of mathematical simplification as compared to the Maxwell equations. Many practical problems can be handled by solving one wave equation without reference to the other, so that only a single boundary-value problem in single vector variable remains.

For a linear, isotropic, homogeneous, and dominantly conducting medium, the displacement currents are negligible, that is, $\sigma \gg \omega\varepsilon$, and eqn (5.11) is simplified to

$$\nabla^2 \vec{H} - \mu\sigma \frac{\partial^2 \vec{H}}{\partial t^2} = 0 \tag{5.12}$$

Also, if the problem of interest is time harmonic then the complex phasor form of these electromagnetic wave equations can be used.

The complex phasor representation applied to wave eqn (5.11) results in the following equation of the Helmholtz type:

$$\nabla^2 \vec{H} - j\omega\mu\sigma \vec{H} + \omega^2 \mu\varepsilon \vec{H} = 0 \tag{5.13}$$

which is usually written in the following form:

$$\nabla^2 \vec{H} - \gamma^2 \vec{H} = 0 \tag{5.14}$$

where γ is the complex propagation constant that is simplified to

$$\gamma = \sqrt{j\omega\mu\sigma - \omega^2 \mu\varepsilon} \tag{5.15}$$

For a linear, isotropic, homogeneous, and dominantly conducting medium ($\sigma \gg \omega\varepsilon$), the propagation constant is then

$$\gamma = \sqrt{j\omega\mu\sigma} \tag{5.16}$$

and the quasistatic field problem is defined by simplified equation

$$\nabla^2 \vec{H} + j\omega\mu\sigma \vec{H} = 0 \tag{5.17}$$

Equation (5.17) is the vector Helmholtz equation (for a linear, isotropic, homogeneous, and dominantly conducting medium) and can be expressed in terms of three scalar equations, one for each component of the H vector.

5.3 INTEGRAL EQUATION REPRESENTATION OF THE HELMHOLTZ EQUATION

Applying the weighted residual approach, eqn (5.17) must be integrated over the calculation domain Ω,

$$\int_{\Omega} [\nabla^2 \vec{H} + j\omega\mu\sigma\vec{H}] W \mathrm{d}\Omega = 0 \tag{5.18}$$

where W is a weight function.

Assuming only the H_y component to be different from zero, for the sake of simplicity eqn (5.18) is simplified into

$$\int_{\Omega} [\nabla^2 H_y + j\omega\mu\sigma\vec{H}_y] W \mathrm{d}\Omega = 0 \tag{5.19}$$

Now, choosing$W = \psi$, where ψ is the fundamental solution for two-dimensional Helmholtz equation, and by using the following identities:

$$\nabla(\nabla H_y \Psi) = \nabla^2 H_y \Psi + \nabla H_y \nabla \Psi \tag{5.20}$$

$$\nabla(H_y \nabla \Psi) = \nabla H_y \nabla \Psi + H_y \nabla^2 \Psi \tag{5.21}$$

one obtains

$$\int_{\Omega} \nabla(\nabla H_y \Psi) \mathrm{d}\Omega - \int_{\Omega} \nabla H_y \nabla \Psi \mathrm{d}\Omega - \int_{\Omega} \gamma^2 H_y \Psi \mathrm{d}\Omega = 0 \tag{5.22}$$

In addition, applying the generalized Gauss theorem and identity (5.21) yields

$$\int_{\Gamma} \Psi \frac{\partial H_y}{\partial n} \mathrm{d}\Gamma - \int_{\Gamma} H_y \frac{\partial \Psi}{\partial n} \mathrm{d}\Gamma + \int_{\Omega} H_y \nabla^2 \Psi \mathrm{d}\Omega - \gamma^2 \int_{\Omega} H_y \Psi \mathrm{d}\Omega = 0 \tag{5.23}$$

which can be rewritten in the following form:

$$\int_{\Gamma} \Psi \frac{\partial H_y}{\partial n} \mathrm{d}\Gamma - \int_{\Gamma} H_y \frac{\partial \Psi}{\partial n} \mathrm{d}\Gamma + \int_{\Omega} [\nabla^2 \Psi - \gamma^2 \Psi] H_y \mathrm{d}\Omega = 0 \tag{5.24}$$

The fundamental solution of the Helmholtz partial differential equation, that is,

$$\nabla^2 \Psi - \gamma^2 \Psi + \delta(R) = 0 \tag{5.25}$$

is given by

$$\Psi = \frac{1}{2\pi} K_0(\gamma R) \tag{5.26}$$

where K_0 is the modified Bessel function of zero-order and R denotes the distance from the source to the observation point.

Taking into account the properties of the Dirac impulse, it follows that

$$\int_{\Omega} \delta(R) H_y \mathrm{d}\Omega = H_y(R) \tag{5.27}$$

and the resulting integral relation becomes

$$H_y = \int_{\Gamma} \Psi \frac{\partial H_y}{\partial n} \mathrm{d}\Gamma - \int_{\Gamma} H_y \frac{\partial \Psi}{\partial n} \mathrm{d}\Gamma \tag{5.28}$$

which is an integral (Green) representation of the function H_y.
If an interior problem is considered, $\vec{n}$ is an external normal vector to the boundary Γ, whereas H_y is related to the points inside the domain Ω as shown in Figure 5.3a.

According to the analysis already presented for potential problems in Chapter 4, the difficulties arise for the points located on the domain boundary for which $R \to 0$ (Fig. 5.3b). Thus, both integrals in eqn (5.28) must be studied carefully when the integration is performed over boundary points.

For the boundary points (x', y') it follows that

$$H_y(x', y') = \frac{1}{2\pi} \int_{\Gamma} \frac{\partial H_y}{\partial n} K_0(\gamma R) \mathrm{d}\Gamma - \frac{1}{2\pi} \int_{\Gamma} H_y \frac{\partial K_0(\gamma R)}{\partial n} \mathrm{d}\Gamma \tag{5.29}$$

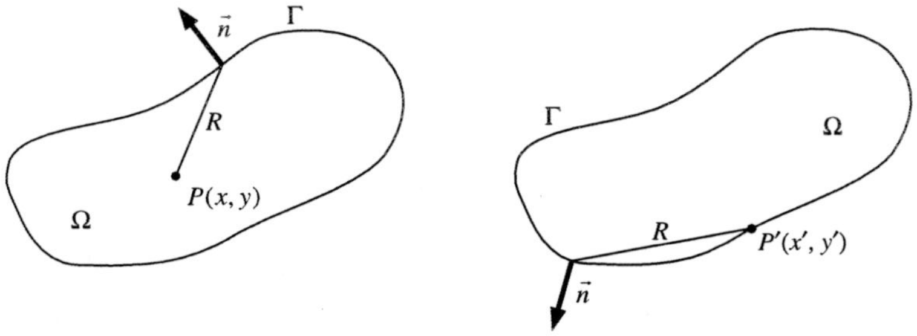

(a) Observation point inside the domain (b) Observation point at the boundary

Figure 5.3 The interior field problem.

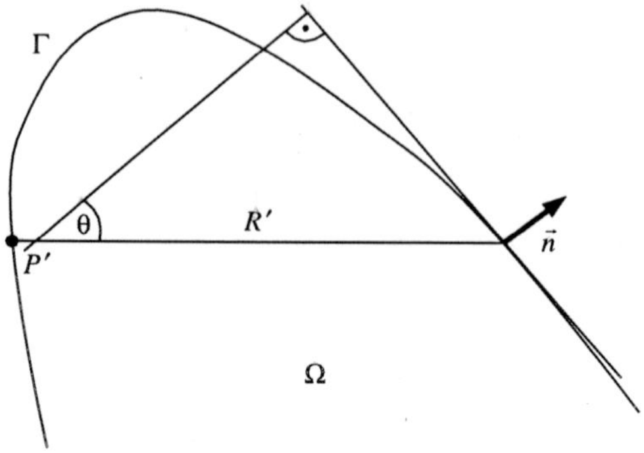

Figure 5.4 Section of a boundary.

The normal derivative of the Green function can be written in the form

$$\frac{\partial \Psi}{\partial n} = \frac{\partial \Psi}{\partial R}\frac{\partial R}{\partial n} = \frac{\partial K_0(\gamma R)}{\partial R}\frac{\partial R}{\partial n} = -\frac{\gamma}{2\pi}K_1(\gamma R)\frac{\partial R}{\partial n} \tag{5.30}$$

where the derivative of the distance R is given by

$$\frac{\partial R}{\partial n} = \cos\theta \tag{5.31}$$

θ is the angle shown in Figure 5.4.

Integral relation (5.29) can now be written as

$$H_y(x', y') = \frac{1}{2\pi}\int_\Gamma \frac{\partial H_y}{\partial n}K_0(\gamma R)\mathrm{d}\Gamma + \frac{\gamma}{2\pi}\int_\Gamma H_y K_1(\gamma R)\cos\theta\ \mathrm{d}\Gamma \tag{5.32}$$

Isolating the singularities in the two contour integrals around the semicircle $\Delta\Gamma$, as shown in Figure 5.5, one has

$$\begin{aligned} H_y(x', y') = {} & \frac{1}{2\pi}\int_{\Gamma-\Delta\Gamma} \frac{\partial H_y}{\partial n}K_0(\gamma R)\mathrm{d}\Gamma + \frac{1}{2\pi}\int_{\Delta\Gamma} \frac{\partial H_y}{\partial n}K_0(\gamma R)\mathrm{d}\Gamma \\ & -\frac{\gamma}{2\pi}\int_{\Gamma-\Delta\Gamma} H_y K_1(\gamma R)\cos\theta\ \mathrm{d}\Gamma - \frac{\gamma}{2\pi}\int_{\Delta\Gamma} H_y K_1(\gamma R)\cos\theta \mathrm{d}\Gamma \end{aligned} \tag{5.33}$$

Integrating around singularities for the case of radius $\varepsilon \to 0$ yields

$$\lim_{\varepsilon\to 0}\int_{\Delta\Gamma} \frac{\partial H_y}{\partial n}K_0(\gamma R)\mathrm{d}\Gamma = \lim_{\varepsilon\to 0}\int_0^\pi \frac{\partial H_y}{\partial n}\ln\varepsilon\cdot\varepsilon\,\mathrm{d}\Gamma = 0 \tag{5.34}$$

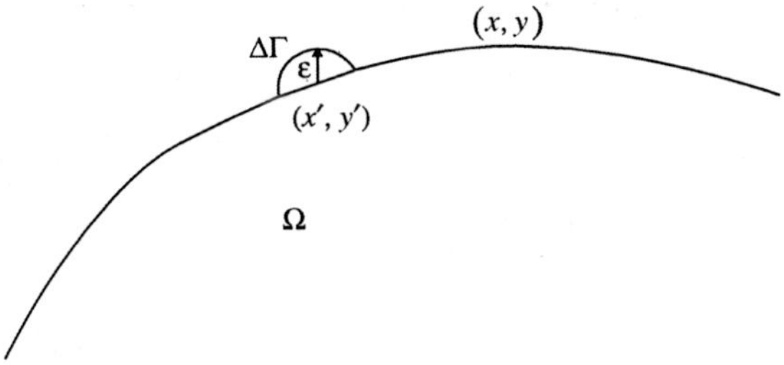

Figure 5.5 Integration around a singularity.

$$\lim_{\varepsilon\to 0}\frac{1}{2\pi}\int_{\Delta\Gamma}\frac{\partial K_0(\gamma R)}{\partial n}H_y\mathrm{d}\Gamma=\lim_{\varepsilon\to 0}\frac{1}{2\pi}\int_0^{\pi}\frac{1}{\varepsilon}H_y\varepsilon\ \mathrm{d}\Gamma=\frac{H_y}{2} \tag{5.35}$$

According to relations (5.29–5.35) where integration from 0 to π implies the smooth boundary and eqn (5.33) becomes

$$\frac{1}{2}H_y(x',y')=\frac{1}{2\pi}\int_{\Gamma}\frac{\partial H_y}{\partial n}K_0(\gamma R)\mathrm{d}\Gamma+\frac{\gamma}{2\pi}\int_{\Gamma}H_yK_1(\gamma R)\cos\theta\ \mathrm{d}\Gamma \tag{5.36}$$

Once the solution for the magnetic field and its corresponding normal derivative has been obtained, the field inside the region can be determined using formula (5.28).

The more general case of nonsmooth contour integral singularities has been discussed in Chapter 4.

5.4 COMPUTATIONAL EXAMPLE

Figure 5.6 shows the coil of a rectangular core of permeability μ located between two media with infinite permeabilities. It is assumed that $b \gg a$, the coil is of dimensions $2a \times 2h$ as shown in Figure 5.6. In this case only the y-component of the magnetic field needs to be considered.

The magnetic field to be determined on the basis of the current NI is due to the induced eddy currents. The problem can be solved analytically by using the boundary element solution and by treating the problem as two-dimensional problem.

The boundary conditions for the magnetic field arise from the current NI and can be easily posed applying the Ampere law. Because of the symmetry, it follows that

$$H_y|_{x=a}=H_y|_{x=-a} \tag{5.37}$$

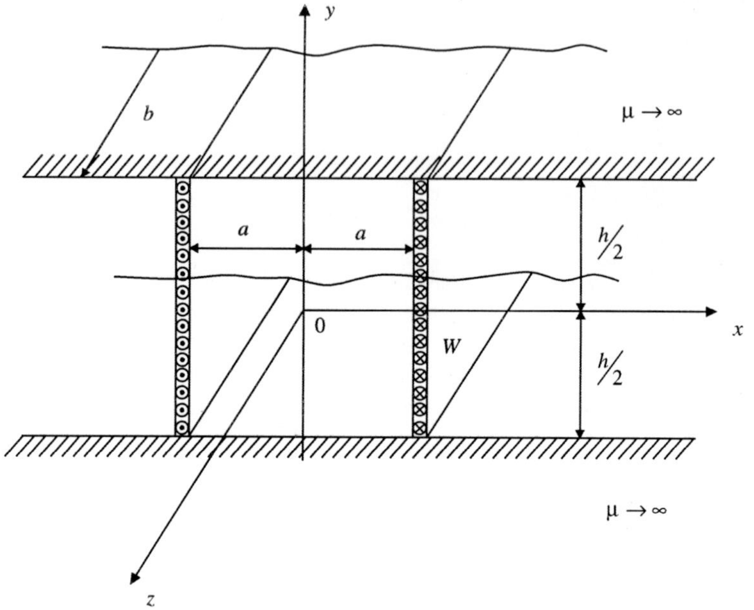

Figure 5.6 Geometry of the problem.

The Ampere law gives

$$\int_{-h/2}^{h/2} H_y \mathrm{d}y = NI \tag{5.38}$$

and

$$H_y|_{x=a} = \frac{NI}{h} \tag{5.39}$$

Furthermore, from the physical nature of the problem the magnetic field normal derivative vanishes for $y = \pm h$, that is,

$$\frac{\partial H_y}{\partial n} = 0 \tag{5.40}$$

This set of boundary conditions is represented in Figure 5.7.

The problem will be first treated analytically and then the boundary element solution will be presented.

5.4.1 Analytical Solution of the Eddy Current Problem

The analytical solution of the eddy current problem can be obtained assuming the magnetic field to vary only along the x-axis, eqn (5.17) becomes

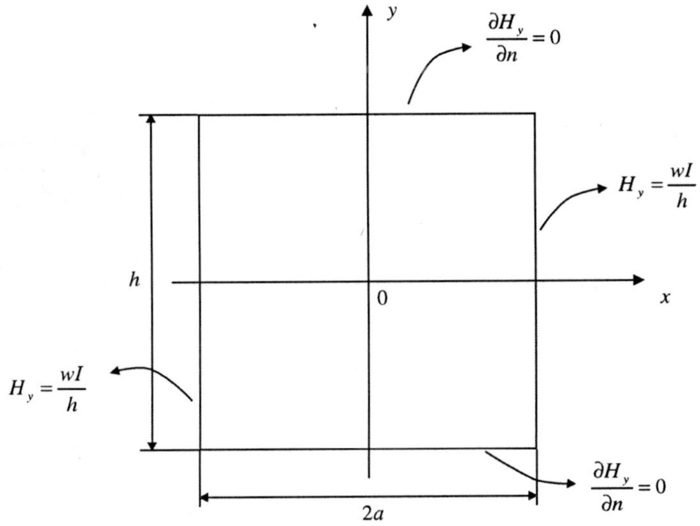

Figure 5.7 The calculation domain with the prescribed boundary conditions.

$$\frac{\partial^2 H_y}{\partial x^2} + j\omega\mu\sigma H_y = 0 \tag{5.41}$$

which can also be written as

$$\frac{\partial^2 H_y}{\partial x^2} - \gamma^2 H_y = 0 \tag{5.42}$$

The solution of the differential eqn (5.42) can be easily obtained in closed form,

$$H_y(x) = C_1 ch\gamma x + C_2 sh\gamma x \tag{5.43}$$

where C_1 and C_2 are the unknown constants to be determined from the boundary conditions.

Also, due to symmetry the solution can be written as

$$H_y(x) = 2C_1 ch\gamma x \tag{5.44}$$

Applying the prescribed boundary conditions, eqn (5.44) becomes

$$H_y(x) = \frac{NI}{h}\frac{ch\gamma x}{ch\gamma a} \tag{5.45}$$

The corresponding current density can be easily determined from the magnetic field distribution expressed by eqn (5.1) assuming the displacement currents to be negligible.

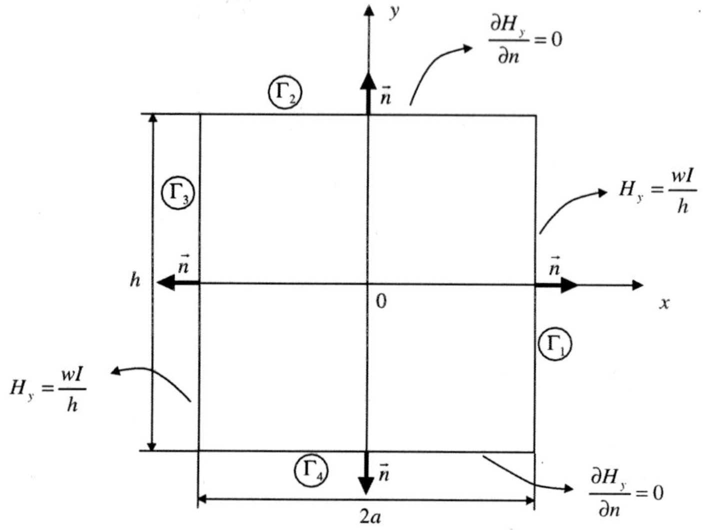

Figure 5.8 The calculation domain with prescribed boundary conditions.

5.4.2 Boundary Element Solution of the Eddy Current Problem

The set of prescribed boundary conditions with corresponding unit normal vectors is shown in Figure 5.8.

The solution of the Helmholtz equation associated with the actual eddy current problem is carried out using a constant boundary element scheme as shown in Figure 5.9.

Discretizing the boundary, the magnetic field intensity at any point in the interior of the domain (Fig. 5.10) can be expressed as

$$H_y(P) = \sum_{j=1}^{M} \left[\int_{\Gamma_j} \Psi \frac{\partial H_y}{\partial n} \mathrm{d}\Gamma - \int_{\Gamma_j} H_y \frac{\partial \Psi}{\partial n} \mathrm{d}\Gamma \right] \tag{5.46}$$

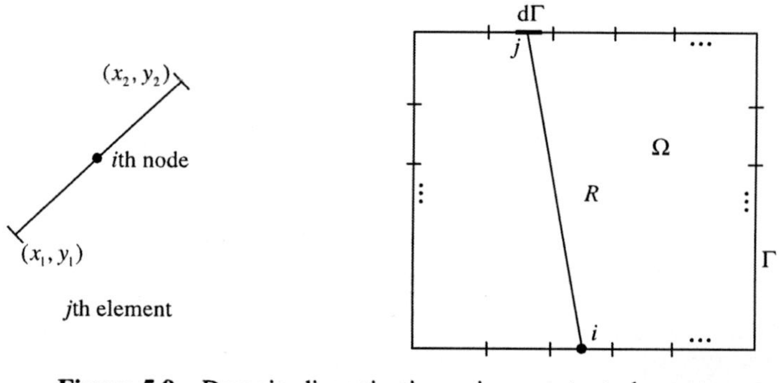

Figure 5.9 Domain discretization using constant elements.

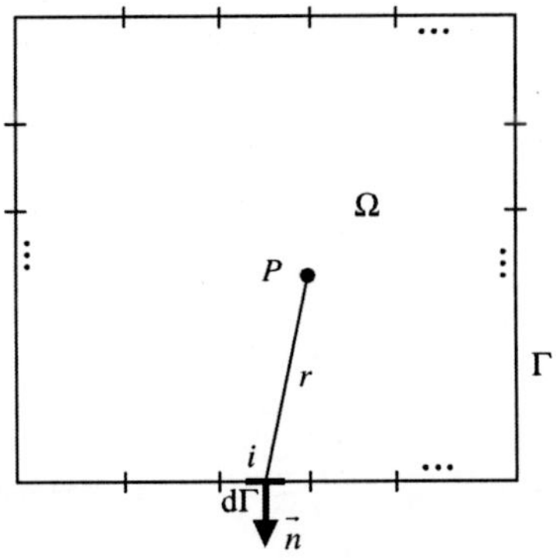

Figure 5.10 Interior of the domain.

where Ψ is the corresponding fundamental solution and M is the total number of boundary elements.

The boundary element modeling of eqn (5.32) results in the following equation for the magnetic field intensity at a point inside the domain,

$$H_y(P) = \sum_{j=1}^{M}\left[\frac{1}{2\pi}\int_{\Gamma_j}\frac{\partial H_y}{\partial n}K_0(\gamma R)\mathrm{d}\Gamma + \frac{\gamma}{2\pi}\int_{\Gamma_j}H_y K_1(\gamma R)\cos\theta\ \mathrm{d}\Gamma\right] \tag{5.47}$$

The formulation for points at the boundary points is given by the following relationship:

$$c_i H_{yi} = \sum_{j=1}^{M}\left[\int_{\Gamma_j}\Psi\frac{\partial H_y}{\partial n}\mathrm{d}\Gamma - \int_{\Gamma_j}H_y\frac{\partial \Psi}{\partial n}\mathrm{d}\Gamma\right] \tag{5.48}$$

Applying the constant boundary element approximation, eqn (5.48) can be written as

$$c_i H_{yi} = \sum_{j=1}^{M}\left(\left[\frac{\partial H_y}{\partial n}\right]_j\int_{\Gamma_j}\Psi\mathrm{d}\Gamma - H_{yj}\int_{\Gamma_j}\frac{\partial \Psi}{\partial n}\mathrm{d}\Gamma\right) \tag{5.49}$$

The resulting algebraic equation system can be expressed in matrix form as follows:

$$[H][u] = [G][q] \tag{5.50}$$

where

$$\frac{\partial H_y}{\partial n}\Big|_j = q_j \tag{5.51}$$

$$H_{y|j} = u_j \tag{5.52}$$

$$G_{ij} = \int_{\Gamma_j} \psi_{ij} = \frac{1}{2\pi}\int_{\Gamma_j} K_0(\gamma R_{ij})\mathrm{d}\Gamma \tag{5.53}$$

$$H_{ij} = \int_{\Gamma_j} \frac{\partial \psi_{ij}}{\partial n}\mathrm{d}\Gamma = -\frac{\gamma}{2\pi}\int_{\Gamma_j} K_1(\gamma R_{ij})\mathrm{d}\Gamma \tag{5.54}$$

The contribution of the element with the singularity (self-segment) of integral (5.54) vanishes, that is,

$$H_{ii} = -\frac{\gamma}{2\pi}\int_{\Gamma_j} K_1(\gamma r_{ii})\cos\theta \mathrm{d}\Gamma = 0 \tag{5.55}$$

due to the orthogonality of $\vec{r}$ and $\vec{n}$, $\cos\theta = 0$ as shown in Figure 5.11.

The boundary element matrices (5.53) and (5.54) can be computed by means of the Gauss quadrature formulas [4].

Performing the transformation of coordinates: $x = x(\xi)$, $y = y(\xi)$, as shown in Figure 5.12, the global coordinates x and y can be expressed in terms of the local coordinate ξ, that is,

$$x = \frac{x_1 - x_2}{2}\xi + \frac{x_1 + x_2}{2} \tag{5.56}$$

$$y = \frac{y_1 - y_2}{2}\xi + \frac{y_1 + y_2}{2} \tag{5.57}$$

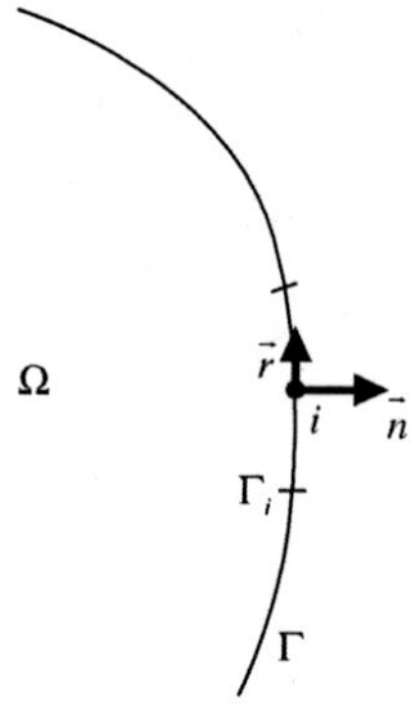

Figure 5.11 Orthogonality of vectors r and n.

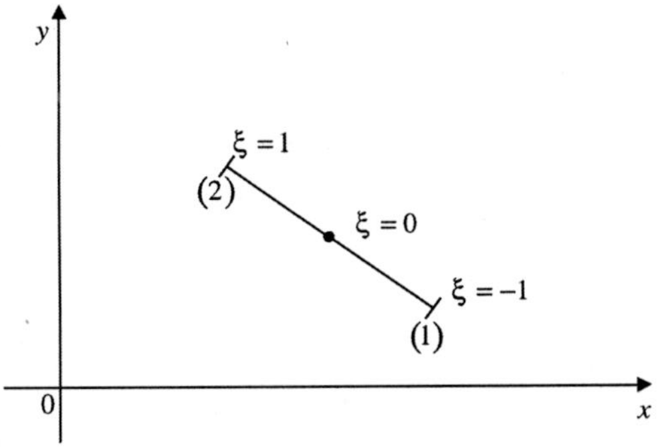

Figure 5.12 Global and local coordinates.

The differential of the line element is given by

$$\mathrm{d}\Gamma = \sqrt{\mathrm{d}x^2 + \mathrm{d}y^2} = \sqrt{\left(\frac{\mathrm{d}x}{\mathrm{d}\xi}\right)^2 + \left(\frac{\mathrm{d}y}{\mathrm{d}\xi}\right)^2}\,\mathrm{d}\xi$$
$$= \frac{\sqrt{(x_1 - x_2)^2 + (y_1 - y_2)^2}}{2}\mathrm{d}\xi = \frac{\Delta\Gamma}{2}\mathrm{d}\xi \tag{5.58}$$

and consequently the BEM matrices (5.53) and (5.54) can be written as

$$G_{ij} = \frac{1}{2\pi}\int_{\Gamma_j} K_0(\gamma R_{ij})\mathrm{d}\Gamma = \frac{\Delta\Gamma}{2}\int_{-1}^{1} K_0\left[\gamma R_{ij}(\xi)\right]\mathrm{d}\xi \tag{5.59}$$

and

$$H_{ij} = \frac{1}{2\pi}\int_{\Gamma_j} K_1(\gamma R_{ij})\cos\theta\ \mathrm{d}\Gamma = \frac{\Delta\Gamma}{2}\int_{-1}^{1} K_0\left[\gamma R_{ij}(\xi)\right]\cos\theta\,\mathrm{d}\xi \tag{5.60}$$

where

$$R_{ij}(\xi) = \sqrt{[x_i - x_j(\xi)]^2 + [y_i - y_j(\xi)]^2} \tag{5.61}$$

The example presented in Figure 5.6 is characterized by the following properties:
$\mu = 1$ Vs/Am
$\sigma = 1$ S/m
$\omega = 1$ rad/s
$NI = 1$ A

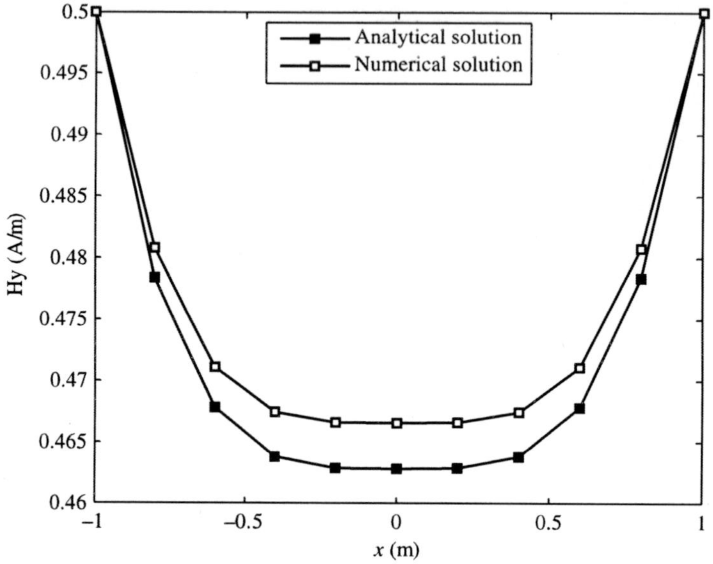

Figure 5.13 Comparison of analytical and numerical results.

The associated Dirichlet and Neumann boundary conditions are as follows:

$$H_y|_{x=a} = 0.5\ A/m;\ \frac{\partial H}{\partial n} = 0 \tag{5.62}$$

Results for the distribution of magnetic field intensity obtained here from BEM formulation using 40 constant elements are compared against the analytical solution (eqn 5.45) and presented in Figure 5.13.

Despite the use of a very crude discretization, the agreement of BEM results with the closed-form solution is satisfactory.

5.5 REFERENCES

[1] P. P. Silvester, R. L. Ferrari: Finite Elements for Electrical Engineers, 3rd Edition, *Cambridge University Press, Cambridge*, 2001.

[2] M. N. O. Sadiku: Numerical Techniques in Electromagnetics, 2nd Edition, *CRC Press, Boca Raton*, 2001.

[3] D. Poljak, C. A. Brebbia: The Boundary Element Method for Electrical Engineers, *WIT Press, Southampton, Boston*, 2005.

[4] C. A. Brebbia: The Boundary Element Method for Engineers, *Pentech Press, Plymouth*, 1978.

6

ELECTROMAGNETIC SCATTERING ANALYSIS

This chapter deals with boundary element procedures for solving electromagnetic scattering problems. The analysis considered in this chapter is limited to geometries involving very long conducting cylinders of arbitrary cross section but can be readily extended to more complex three-dimensional scattering problems.

The scattering of electromagnetic waves by systems whose dimensions are electrically small compared with a wavelength is an occurrence of great importance [1-5]. In such phenomena, the induced fields are in definite phase relationships with the incident wave and radiate electromagnetic energy in directions different from the direction of incidence. A distribution of radiated energy then generally depends on the incident wave polarization.

In contrast to electromagnetic scattering, *diffraction* traditionally deals with apertures or obstacles whose dimensions are large compared to a wavelength.

The formulation of the problem is based on the appropriate boundary integral representation of the wave equation for tangential component of the electric and magnetic field. It is worth noting that a wide range of problems can be treated using constant, linear, or quadratic elements. For the sake of simplicity, the numerical example given in this chapter is handled via constant elements.

6.1 THE ELECTROMAGNETIC WAVE EQUATIONS

Electromagnetic scattering problems can be analyzed by solving Maxwell equations. However, Maxwell equations are coupled first-order space-time partial differential equations that are very difficult to apply for the treatment of boundary-value problems. One way to overcome the difficulty is to decouple these first-order equations, thus obtaining the second-order electromagnetic wave equations.

The wave equations can be derived from the Maxwell curl equations by differentiation and substitution.

Taking curl on both sides of the Maxwell equation representing the differential form of Ampere's law, that is,

$$\nabla x\vec{H} = \vec{J} + \frac{\partial \vec{D}}{\partial t} \tag{6.1}$$

leads to the following formula:

$$\nabla x\nabla x\vec{H} = \nabla x\vec{J} + \frac{\partial}{\partial t}(\nabla x\vec{D}) \tag{6.2}$$

Vectors $\vec{J}$ and $\vec{D}$ can be expressed in terms of E using constitutive equations:

$$\vec{J} = \sigma\vec{E} \tag{6.3}$$

and

$$\vec{D} = \varepsilon\vec{E} \tag{6.4}$$

Assuming uniform scalar material properties, eqn (6.2) can be written as

$$\nabla x\nabla x\vec{H} = \sigma\nabla x\vec{E} + \varepsilon\frac{\partial}{\partial t}(\nabla x\vec{E}) \tag{6.5}$$

Furthermore, according to the curl Maxwell equation representing the differential form of the Faraday's law, that is,

$$\nabla x\vec{E} = -\frac{\partial \vec{B}}{\partial t} \tag{6.6}$$

curl of E can be replaced by the change rate of magnetic flux density leading to the following relation:

$$\nabla x\nabla x\vec{H} = -\sigma\frac{\partial \vec{B}}{\partial t} - \varepsilon\frac{\partial^2 \vec{B}}{\partial t^2} \tag{6.7}$$

If the magnetic material can be represented by the constitutive relationship, that is,

$$\vec{B} = \mu\vec{H} \tag{6.8}$$

it follows that eqn (6.7) can be written as

$$\nabla x \nabla x \vec{H} = -\mu\sigma \frac{\partial \vec{H}}{\partial t} - \mu\varepsilon \frac{\partial^2 \vec{H}}{\partial t^2} \tag{6.9}$$

Starting from the Maxwell electric curl eqn (6.6) and taking the curl on both sides and performing a similar mathematical manipulation as it has been carried out in eqns (6.1–6.9), the following equation in terms of the electric field E can be obtained:

$$\nabla x \nabla x \vec{E} = -\mu\sigma \frac{\partial \vec{E}}{\partial t} - \mu\varepsilon \frac{\partial^2 \vec{E}}{\partial t^2} \tag{6.10}$$

Using the standard vector identity, that is,

$$\nabla x \nabla x \vec{E} = \nabla \nabla \vec{E} - \nabla^2 \vec{E} \tag{6.11}$$

results in the final form of the electromagnetic wave equations for the source free region, that is,

$$\nabla^2 \vec{H} - \mu\sigma \frac{\partial \vec{H}}{\partial t} - \mu\varepsilon \frac{\partial^2 \vec{H}}{\partial t^2} = 0 \tag{6.12}$$

or

$$\nabla^2 \vec{E} - \mu\sigma \frac{\partial \vec{E}}{\partial t} - \mu\varepsilon \frac{\partial^2 \vec{E}}{\partial t^2} = 0 \tag{6.13}$$

Wave eqns (6.12) and (6.13) are valid for homogeneous regions and represent a significant degree of mathematical simplification when compared to the curl Maxwell equations. A number of practical problems can be handled by solving one wave equation without reference to the other in terms of a single boundary-value problem in a single vector variable.

If the medium of interest is linear, isotropic, homogeneous, and source free, then the set of equations (6.12) and (6.13) become:

$$\nabla^2 \vec{H} - \mu\varepsilon \frac{\partial^2 \vec{H}}{\partial t^2} = 0 \tag{6.14}$$

$$\nabla^2 \vec{E} - \mu\varepsilon \frac{\partial^2 \vec{E}}{\partial t^2} = 0 \tag{6.15}$$

Equations (6.14) and (6.15) govern the motion equations of electromagnetic waves in free space. The velocity of the electromagnetic wave in the free space is the velocity of light,

$$c = \frac{1}{\sqrt{\mu_0 \varepsilon_0}} \tag{6.16}$$

where $c = 3 \times 10^8 \mathrm{m/s}$, approximately.

6.2 COMPLEX PHASOR FORM OF THE WAVE EQUATIONS

The complex phasor representation of the electromagnetic wave equations (6.12) and (6.13) is given by the following Helmholtz type equations:

$$\nabla^2 \vec{H} - j\omega\mu\sigma\vec{H} + \omega^2\mu\varepsilon\vec{H} = 0 \tag{6.17}$$

$$\nabla^2 \vec{E} - j\omega\mu\sigma\vec{E} + \omega^2\mu\varepsilon\vec{E} = 0 \tag{6.18}$$

which is usually written in the form

$$\nabla^2 \vec{H} - \gamma^2\vec{H} = 0 \tag{6.19}$$

$$\nabla^2 \vec{E} - \gamma^2\vec{E} = 0 \tag{6.20}$$

where γ is the complex propagation constant,

$$\gamma = \sqrt{j\omega\mu\sigma - \omega^2\mu\varepsilon} \tag{6.21}$$

For a linear, isotropic, homogeneous, and source-free medium, eqns (6.17) and (6.18) become

$$\nabla^2 \vec{H} + k^2\vec{H} = 0 \tag{6.22}$$

$$\nabla^2 \vec{E} + k^2\vec{E} = 0 \tag{6.23}$$

where k is the wave number in a lossless medium given by

$$k = \omega\sqrt{\mu\varepsilon} \tag{6.24}$$

The complex form of electromagnetic potential wave equations can be derived in a similar way.

6.3 TWO-DIMENSIONAL SCATTERING FROM A PERFECTLY CONDUCTING CYLINDER OF ARBITRARY CROSS-SECTION

The electromagnetic scattering from an infinitely long, perfectly conducting (PEC) cylinder of arbitrary cross section and infinite in the z direction is considered as shown in Figure 6.1. The excitation is given in the form of a TM electromagnetic wave.

For the case of a TM wave (the case of E-field having a z-component alone), it is convenient to solve the scalar Helmholtz equation in terms of the E_z electric field component.

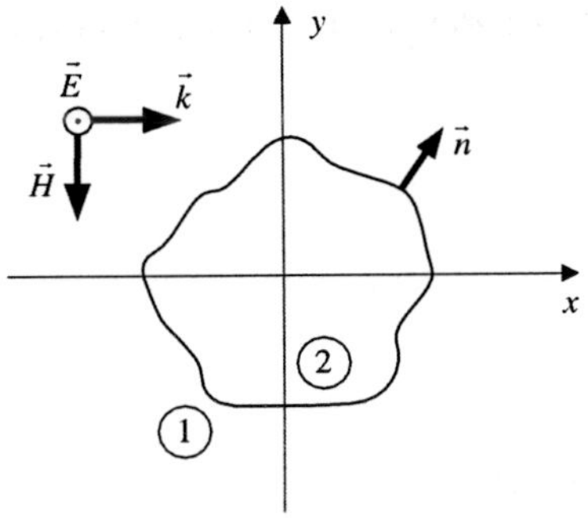

Figure 6.1 2D scattering by a cylinder of arbitrary cross section.

Thus, the problem of determining the current density induced on the surface of PEC cylinder can be formulated via the following scalar Helmholtz equation:

$$(\nabla^2 + k^2)E_z = 0 \tag{6.25}$$

Applying the Green integral theorem the following integral equation formulation is obtained:

$$E_z = \int_\Gamma \left[E_z \frac{\partial \psi}{\partial n} - \psi \frac{\partial E_z}{\partial n} \right] \mathrm{d}\Gamma \tag{6.26}$$

where ψ is the corresponding Green function that is given by

$$\psi = -\frac{j}{4} H_0^{(2)}(kR) \tag{6.27}$$

where $H_0^{(2)}(kR)$ is the zero-order Hankel function of the second kind and R is the distance from the source point to the observation point.

Furthermore, the total electric field in the vicinity of the cylinder can be expressed as a sum of the incident and scattered field components, respectively,

$$E_z^{tot} = E_z^{inc} + E_z^{sct} \tag{6.28}$$

Combining eqns (6.26), (6.27), and (6.28) yields

$$E_z = E_z^{inc} - \frac{j}{4} \int_\Gamma \left[E_z \frac{\partial H_0^{(2)}(kR)}{\partial n} - H_0^{(2)}(kR) \frac{\partial E_z}{\partial n} \right] \mathrm{d}\Gamma \tag{6.29}$$

where E_z^{inc} is the value of the incident field as a result of the known sources.

In addition, since the cylinder is perfectly conducting the total field vanishes along a metallic surface, that is,

$$E_z^{inc} + E_z^{sct} = 0 \tag{6.30}$$

Combining eqns (6.29) and (6.30) results in the following integral equation for the unknown normal derivative of the z-component electric field along the cylinder surface:

$$E_z^{inc} = -\frac{j}{4} \int_{\Gamma} H_0^{(2)}(kR) \frac{\partial E_z}{\partial n} \mathrm{d}\Gamma \tag{6.31}$$

The magnetic field on the surface of the perfectly conducting cylinder can be expressed in terms of the normal derivative of the axial electric field component,

$$\frac{\partial E_z}{\partial n} = j\omega\mu H_t \tag{6.32}$$

where H_t denotes the corresponding tangential component of the magnetic field on the domain boundary Γ.

The tangential component of the magnetic field induces the surface current density at the surface of the perfect conductor, that is,

$$J_z = H_t \tag{6.33}$$

For the actual polarization, the current density may have only axial z-component and the integral eqn (6.31) then becomes

$$E_z^{inc} = \frac{\omega \cdot \mu}{4} \int_{\Gamma} J_z(x, y) H_0^{(2)}(kR) \mathrm{d}\Gamma \tag{6.34}$$

where J_z denotes the unknown surface current density.

6.4 SOLUTION BY THE INDIRECT BOUNDARY ELEMENT METHOD

According to the boundary element algorithm, the unknown current density can be expressed as

$$J_z(x, y)_j = J_z[x(\xi), y(\xi)]_j = \sum_{m=1}^{n_e} f_{mj}(\xi) J_{mj} \tag{6.35}$$

where f_{mj} denotes the base functions, J_{mj} is the unknown coefficient of the solution, j denotes jth boundary element, and n_e is the number of local nodes per element.

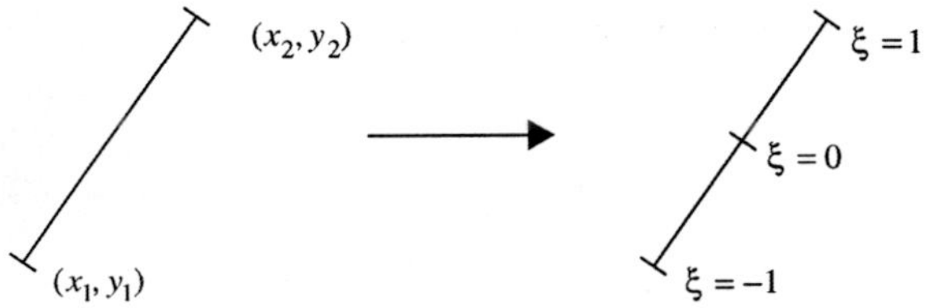

Figure 6.2 Global and local coordinates.

Global coordinates x and y become a function of dimensionless coordinate ξ using the standard transformation of coordinates,

$$x(\xi) = \frac{x_2 - x_1}{2}\xi + \frac{x_2 + x_1}{2} \tag{6.36}$$

$$y(\xi) = \frac{y_2 - y_1}{2}\xi + \frac{y_2 + y_1}{2} \tag{6.37}$$

Namely, an arbitrary boundary element defined by the global coordinates (x_1, y_1) and (x_2, y_2) is mapped into the unit element $(-1,1)$ as shown in Figure 6.2.

Furthermore in local coordinate system, the differential line element is given by

$$\mathrm{d}\Gamma = \sqrt{\left(\frac{\mathrm{d}x}{\mathrm{d}\xi}\right)^2 + \left(\frac{\mathrm{d}y}{\mathrm{d}\xi}\right)^2}\mathrm{d}\xi = \frac{1}{2}\sqrt{(x_2 - x_1)^2 + (y_2 - y_1)^2}\mathrm{d}\xi = \frac{\Delta\Gamma}{2}\mathrm{d}\xi \tag{6.38}$$

Eqn (6.35) can also be written in matrix notation,

$$J_z = \{f\}^T\{J\} \tag{6.39}$$

Discretizing the integral equation (6.34) results in the following algebraic system of equations:

$$E_{zi}^{inc} = \sum_{j=1}^{M} \int_{\Delta\Gamma_j} \{f\}_j^T \Psi_{ij}\mathrm{d}\Gamma\{J\}_j; i = 1, 2, \ldots, M \tag{6.40}$$

where $\{f\}_j$ denotes the set of shape functions and M is the total number of boundary elements.

The incident field is given in the form of a plane wave and can be expressed as

$$E_{zi}^{inc} = E_0 e^{-jkx_{mi}} \tag{6.41}$$

The fundamental solution can be written as

$$\Psi_{ij} = \frac{\omega\mu}{4}H_0^{(2)}(kR_{ij}) \tag{6.42}$$

where

$$R_{ij} = \sqrt{(x_{mi} - x_j)^2 + (y_{mi} - y_j)^2} \tag{6.43}$$

where x_{mi} and y_{mi} are the coordinates of the mth node on the jth element.

Index i denotes the observation boundary element and j is the source boundary element.

6.4.1 Constant Element Case

When using the constant boundary element approximation, the current density is considered to be constant along the particular element and hence eqn (6.40) is simplified to

$$E_{zi}^{inc} = \sum_{j=1}^{M} \int_{\Delta\Gamma_j} \Psi_{ij} d\Gamma \cdot J_i; i = 1, 2, \ldots, M \tag{6.44}$$

where E_{zi}^{inc}, Ψ_{ij}, and R_{ij} are defined by expressions (6.41), (6.42), and (6.43), respectively.

The contribution of the self-element, due to the singularity problem, is computed analytically by using the approximation of the Hankel function for small argument. Thus, for $j = i$, it follows

$$H_0^{(2)}(kR_{ii}) = j\frac{2}{\pi} \ln \frac{2}{\gamma_0(kR_{ii})} + 1 \tag{6.45}$$

and consequently, integral (6.44) can be calculated in the close form, that is,

$$\int_{\Delta\Gamma_j} \Psi_{ii} d\Gamma = \frac{\omega\mu}{4} \int_{-1}^{+1} H_0^{(2)}(kR_{ij}) d\Gamma = -\frac{j\omega\mu}{2\pi}\left[\ln\frac{\gamma_0 k}{4} + j\frac{\pi}{2} - 1\right] \tag{6.46}$$

where $\gamma_0 = 1.781072$ is the Euler number.

The discretization concept of constant element approximation is shown in Figure 6.3.

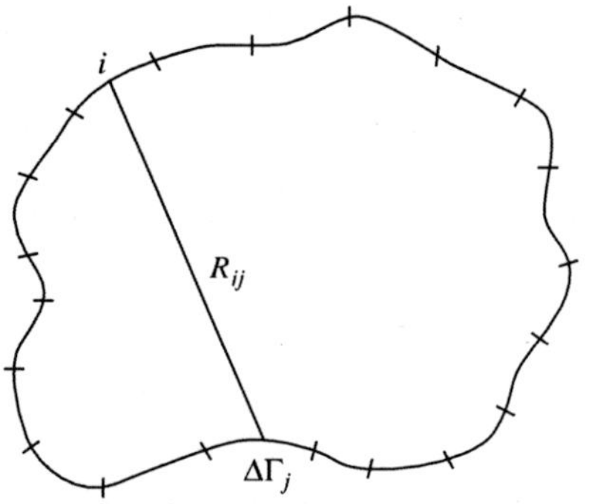

Figure 6.3 *i*th node and *j*th element — a constant element discretization.

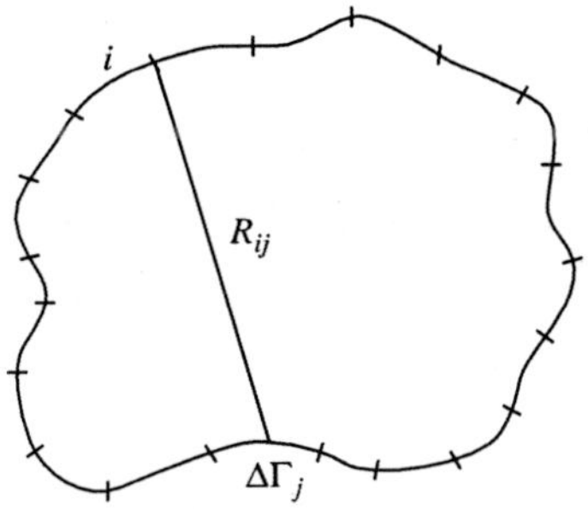

Figure 6.4 *i*th node and *j*th element — linear element discretization.

6.4.2 Linear Elements Case

When using linear element approximation, the unknown function is approximated by linear shape functions as follows:

$$J_z[x(\xi), y(\xi)]_j = \sum_{m=1}^{2} f_{mj}(\xi) J_{mj} = f_{1j}(\xi) J_{1j} + f_{2j}(\xi) J_{2j} \tag{6.47}$$

Equation (6.47) can also be rewritten in matrix form as

$$J_z[x(\xi), y(\xi)]_j = [f_{1j}(\xi)\ f_{2j}(\xi)] \begin{Bmatrix} J_{1j} \\ J_{2j} \end{Bmatrix} \tag{6.48}$$

Furthermore, the matrix eqn (6.40) in the case of linear elements becomes

$$E_{zi}^{inc} = \sum_{j=1}^{M} \int_{\Delta\Gamma_j} [f_{1j}(\xi)\ f_{2j}(\xi)] \Psi_{ij} d\Gamma \begin{Bmatrix} J_{1j} \\ J_{2j} \end{Bmatrix}; i = 1, 2, \ldots, M \tag{6.49}$$

The linear shape functions are given by

$$f_{1j}(\xi) = \frac{1}{2}(1 - \xi) \tag{6.50}$$

$$f_{2j}(\xi) = \frac{1}{2}(1 + \xi) \tag{6.51}$$

The discretization into linear elements is shown in Figure 6.4.

6.5 NUMERICAL EXAMPLE

The numerical example to follow is the calculation of the surface current density in a PEC square cylinder, Figure 6.5, for a given excitation in a form of a plane wave [6,7].

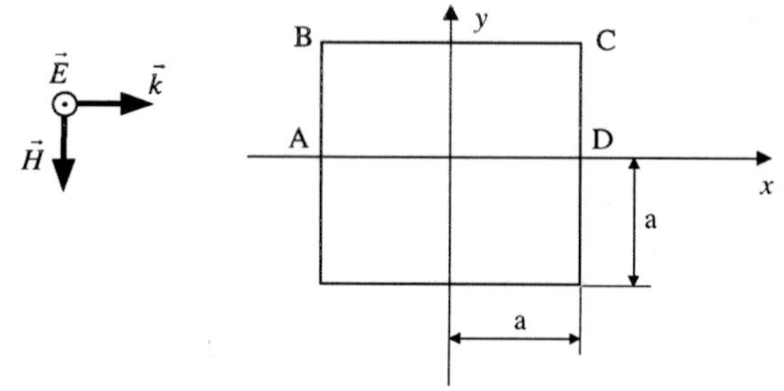

Figure 6.5 The geometry of the problem.

It is worth noting that as the current density becomes infinite at the corners, it is convenient in this case to choose constant boundary elements by which the singularity problem at these corners is avoided. The cylinder surface is discretized into 60 boundary elements.

Numerical results for induced surface current density are presented in Figures 6.6–6.8. The numerical results obtained are normalized by the value of the current density at point A as shown in Figure 6.5.

This computational example demonstrates the validity and versatility of BEM in treating electromagnetic scattering problems. In addition, considering the accuracy of the method and related computational cost, BEM seems to be the most promising technique for solving problems involving complex geometries.

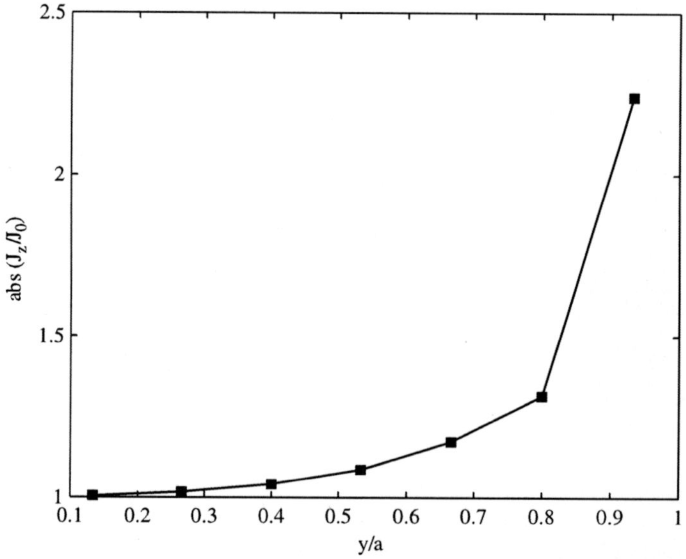

Figure 6.6 Normalized surface current density in a square cylinder between points A and B with $k_a = 1$.

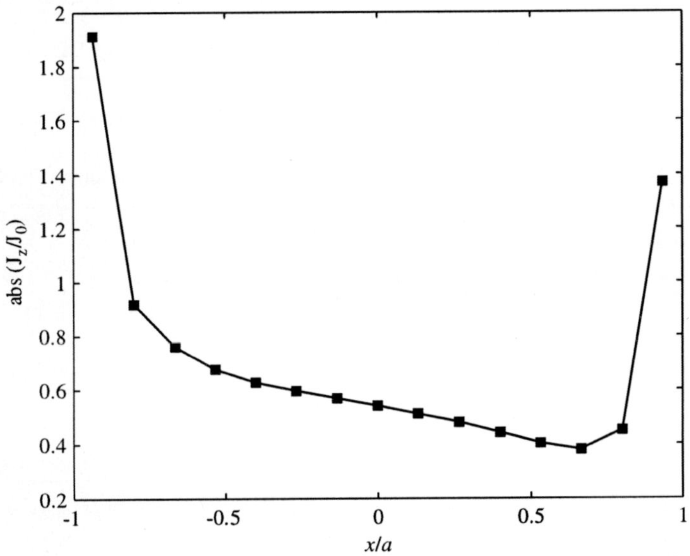

Figure 6.7 Normalized surface current density in a square cylinder between points B and C with $k_a = 1$.

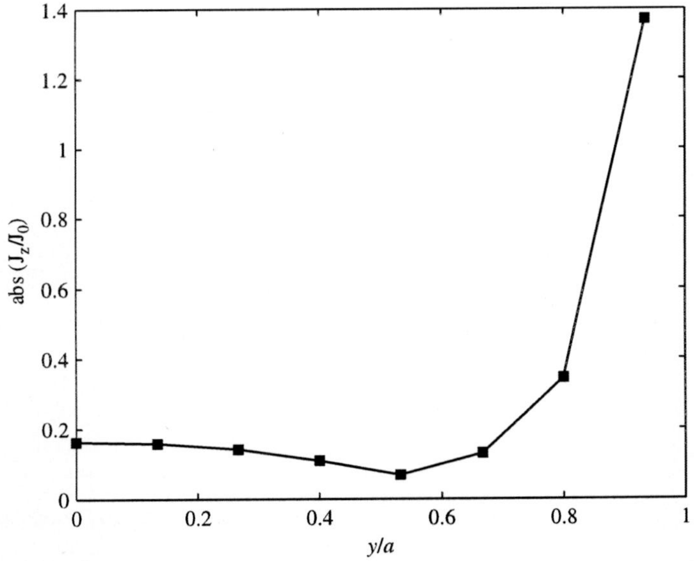

Figure 6.8 Normalized surface current density in a square cylinder between points C and D with $k_a = 1$.

6.6 REFERENCES

[1] W. Band: Introduction to Mathematical Physics, *D. Van Nostrand Company, New York*, 1959.

[2] M. N. O. Sadiku: Numerical Techniques in Electromagnetics, 2nd Edition, *CRC Press, Boca Raton*, 2001.

[3] P. P. Silvester, R. L. Ferrari: Finite Elements for Electrical Engineers, 3rd Edition, *Cambridge University Press, Cambridge*, 2001.

[4] S. Ramo, J. R. Whinnery, T. Van Duzer: Fields and Waves in Communication Electronics, 3rd Edition, *John Wiley & Sons, Inc., New York*, 1994.

[5] C. A. Balanis: Antenna Theory, 2nd Edition, *John Wiley & Sons, New York*, 1997.

[6] D. Poljak, V. Roje: Application of Boundary Element Method to Scattering from Conducting Cylinder of Arbitrary Cross-Section, *SoftCOM '95*, June 1995., pp 301–310 (in Croatian).

[7] D. Poljak, C. A. Brebbia: The Boundary Element Method for Electrical Engineers, *WIT Press, Southampton-Boston*, 2005.

PART II

ANALYSIS OF THIN WIRE ANTENNAS AND SCATTERERS

There is scarcely a subject that cannot be mathematically treated and the effect calculated beforehand, or the results determined beforehand from the available theoretical and practical data.

Nikola Tesla

7

WIRE ANTENNAS AND SCATTERERS: GENERAL CONSIDERATIONS

The analysis of radiation and scattering from the straight thin wires via integral equations is one of the oldest but still one of the most important problem in antenna theory and electromagnetic compatibility (EMC). Nevertheless, the time-harmonic or transient excitation of a straight thin wire is a standard canonical problem; this configuration is of practical interest itself, particularly, in the design of the antenna arrays and in wire-grid configurations, respectively. In a numerical sense, this relatively simple geometry is very convenient for testing newly developed numerical techniques. Since the early sixties, many numerical techniques have been tested on a thin-wire geometry. Finally, the concept of linear wire antenna has numerous applications in EMC problems. The analysis of thin wires can be carried out in the frequency and time domain. This chapter deals with general aspects of both approaches stressing out related strength and weaknesses.

7.1 FREQUENCY DOMAIN THIN WIRE INTEGRAL EQUATIONS

It was Pocklington who first formulated the frequency domain integral equation for a total current flowing along a straight thin wire antenna in 1897, having also presented a first, approximate solution of his own equation.

In the last hundred years and more, many prominent researchers have investigated both the formulation and the numerical solution of the thin wire integral equation. The most important advance in the formulation was undertaken by Hallen in the thirties. Starting from Pocklington integral equation in the frequency domain, Hallen has derived a new type of integral equation strictly related to thin wire structures. Since that time, a variety of numerical techniques for solving the integral equations of Pocklington and Hallen type in both the frequency domain (FD) and the time domain (TD) have been developed by numerous authors [1–34].

The FD numerical modeling of the wire antennas and scatterers started in the sixties with the classical paper by K.K. Mei in 1965 [1]. He has derived certain variants of Pocklington and Hallen equation as well as proposed a numerical technique for solving these equations. Today his technique would be referred to as the point-matching technique. Many important contributions to the numerical solution of Pocklington and Hallen integral equations were given in the seventies: Silvester and Chan (1972, 1973) [2,3] proposed the use of the strong finite element formulation to the Hallen and Pocklington equation, whereas Butler and Wilton (1975, 1976) [4,5] proposed several moment method techniques for solving these equations.

Besides the improvement of the numerical techniques, a significant effort has been done by many researchers to provide some advancement to the formulation itself. The most important outcome is the extension of the original Pocklington formulation to the problems including the presence of an imperfectly conducting half-space that are given by E.K. Miller et al. (1972) [6], T. Sarkar (1977) [7], and by Parhami and Mittra. Among many papers, may be the most instructive is the one from 1980 [8]. Of course, there were many other authors who participated in this field with their important contributions like the papers by Miano et al. [9,10] dealing with new numerical procedures for solving the Pocklington and Hallen integral equations.

7.2 TIME DOMAIN THIN WIRE INTEGRAL EQUATIONS

The time domain modeling of Pocklington integral equation started in the sixties with the PhD thesis of C.L. Bennett (1968) and with the paper by Sayre and Harrington (1968) [11], whereas TD Hallen integral equation was first derived by A.J. Poggio (1971) [12]. The method of solution, widely used by many researchers, is the one developed in 1973 by E.K. Miller et al. [13]. The method can be referred to as a space-time collocation technique. Even today this technique, in spite of some disadvantages concerning the stability and convergence, is one of the most commonly used. Interesting work dealing with the treatment of TD Hallen equation via method of characteristics was published by Liu and Mei (1973) [14]. Thirteen years later, S. Rao (1986) in Ref. [15] suggests the conjugate gradient method to avoid some convergence and stability problems arising from Miller technique, but due to its complexity the method has never been widely accepted. B.P. Ryne (1990) investigated the very actual problem of convergence and stability in many papers in the nineties, for example in Ref. [16], while P.J. Davies offered state-of-the-art contribution to this field (1996) [22]. F.M. Tesche (1990) [17] offers a formulation for thin wire problem with the inclusion of losses directly in the time domain.

A.R. Bretones et al. published a sequel of papers dealing with a new formulation of Pocklington equation by including the effect of the radial component of the current, for example [18]. A.G. Tijhuis et al. in numerous papers offered some significant contributions to this field, after having proposed a new formulation of the Pocklington and the Hallen equation, respectively. Moreover, they presented a variety of direct, indirect, hybrid, and iterative numerical techniques for solving these equations in both the frequency and time domains. Maybe the most instructive, among many good papers, is the one from 1992. [19]. A.R. Bretones and A.G. Tijhuis have also investigated the half-space problems directly in the time domain 1995 [20].

R. Moini et al. have developed a very useful lightning return stroke model based on the wire antenna theory. This model has been promoted in 1997 [21].

In the last decade, D. Poljak et al. have continuously worked on thin wire radiation in the presence of a half-space configuration using both the frequency and time domain approaches. The dipole antenna insulated in free space is analyzed in frequency domain by solving the FD Hallen equation via iterative and point-matching technique in Ref. [23] and also by solving the FD Pocklington equation by originally developed numerical procedures [24,25] based on Galerkin-Bubnov variant of indirect boundary element method. Time domain analysis of straight thin wire operating in the radiation mode (by solving the TD Pocklington equation via space-time finite element procedure) is given in Ref. [30] and that operating in both the radiation and scattering modes (by solving the TD Hallen equation using boundary elements/time-marching procedure) is given in Ref. [33].

Important contributions to the field of wire antennas are given in the papers dealing with the radiation and scattering of thin wire structure in the presence of a lossy half-space. This problem is modeled in the frequency domain by rigorous Sommerfeld approach. The efficient numerical procedures for evaluating the Sommerfeld integrals and solving the corresponding integral equation are developed in Refs. [26–29]. Finally the most recent contribution to this field is given in Refs. [31], [32], and [34] by proposing a new TD Hallen equation for half-space and also displaying the corresponding space-time finite element technique for solving the resulting equation. Proposed models are convenient for application to the electromagnetic compatibility (EMC) problems, particularly, those related to the lightning electromagnetics and bioelectromagnetics.

7.3 MODELING IN THE FREQUENCY AND TIME DOMAIN: COMPUTATIONAL ASPECTS

The calculation of the frequency or time domain response of thin wires has always received considerable and continuing interest in electromagnetic research and in applications.

This problem is formulated via the frequency-domain thin wire integral equation or directly in the time domain via the corresponding time domain thin wire integral equation.

It is well known that the frequency domain (FD) provides the source-independent results at a single frequency, so the set of frequency samples ought to be calculated and transformed into time domain (TD) to obtain transient response. On the contrary, the direct TD modeling provides source-dependent results over the equivalent of a band of frequencies to achieve transient result directly, but it must be repeated for every source.

Therefore, the FD or TD modeling is chosen depending on the considered problem of interest. Generally, the analysis in the time domain requires more demanding formulation of the problem and, therefore, more complex computational procedures than the frequency domain analysis does. However, the time domain modeling offers more fundamental knowledge and physical insight than the frequency domain and can also provide certain computational advantages.

The most commonly used method for solving the frequency domain thin wire integral equation (FD-TWIE) is a combination of subdomain collocation and the finite difference technique. Unfortunately, this method has been proved to suffer from a very poor convergence rate. Direct time domain techniques are more demanding than the FD techniques and unfortunately more unstable, and there are always serious problems with convergence and appearance of the spurious solutions. The known techniques for solving time domain thin wire integral equation (TD-TWIE) can be referred to as several variants of point matching techniques and marching-on-in-time, which are relatively simple, but on the contrary, suffer from a very poor convergence rate and late time rapidly growing oscillations.

Relatively recently, the thin wire integral equation (TWIE) was recently solved by the Galerkin Bubnov scheme of the indirect boundary element method (GB-IBEM). This method has shown some advantages over the point-matching techniques (avoidance of kernel quasisingularity, better convergence, easier incorporation of boundary conditions) and it is also applicable to more demanding problems including antenna arrays and to some other forms of integro-differential equation types arising in electrodynamics [Poljak-Choy 2003].

The TD-TWIE used by Poljak et al. [34] is of Hallen type since this equation does not consist of any time and space derivatives. Then the Galerkin-Bubnov finite element method is applied in space and time. This approach is shown to be more stable compared with usual marching-on-in-time schemes concerning the solution of Pocklington integral equation approach.

7.4 REFERENCES

[1] K. K. Mei: On the integral equation of thin wire antennas, *IEEE Trans. AP, 13*, 1965, pp 59–62.

[2] R. P. Silvester, K. K. Chan: Bubnov-Galerkin solutions to wire antenna problems, *Proc. IEE, 119*, 1972, pp 1095–1099.

[3] R. P. Silvester, K. K. Chan: Analysis of antenna structures assembled from arbitrarily located straight wires, *Proc. IEE, 120 (1)*, 1973.

[4] C. M. Butler, D. R. Wilton: Analysis of various numerical techniques applied to thin wire scatterers, *IEEE Trans. AP, 23*, 1975.

[5] C. M. Butler, D. R. Wilton: Efficient numerical techniques for solving Pocklington's integral equation and their relationship to other methods, *IEEE Trans. AP, 24*, 1976.

[6] E. K. Miller, A. J. Poggio, G. J. Burke, E. S. Selden: Analysis of wire antennas in the presence of a conducting half-space, Part II. The horizontal antenna in free space, *Can. J. Phys., 50*, 1972, pp 2614–2627.

[7] T. K. Sarkar: Analysis of arbitrarily oriented thin wire antennas over a plane imperfect ground, *Archiv fur elektronik und ubertragungstechnik, 31*, 1977, pp 449–457.

[8] P. Parhami, R. Mittra: Wire antennas over a lossy half-space, *IEEE Trans. AP, 28*, 1980, pp 397–403.

[9] G. Miano, L. Verolino, V. G. Vacaro: A new numerical treatment for Pocklington's integral equation, *IEEE Trans. Magnetics, 32 (3)*, 1996.

[10] G. Miano, L. Verolino, V. G. Vacaro: A hybrid procedure for solving Hallen problem, *IEEE Trans. EMC, 38 (3)*, 1996, pp 495–498.

[11] E. P. Sayre, R. F. Harrington: Transient response of a straight wire scatterers and antennas, *Proc. Intntl. Ant. Prop. Symposium*, Boston, 1968, pp 160–164.

[12] A. J. Poggio: The space-time domain magnetic vector potential integral equations, *IEEE Trans. AP, 24 (3)*, 1971, pp 702–704.

[13] E. K. Miller, A. J. Poggio, G. J. Burke: An integro-differential equation technique for the time-domain analysis of thin wire structure, Part I. The numerical method, *J.Comput. Phys., 12 (1)*, 1973, pp 24–48. Part II Numerical results, *J.Comput. Phys., 12 (2)*, 1973, pp 210–233.

[14] T. K. Liu., K. K. Mei: A time-domain integral equation solution for linear antennas and scatterers, *Radio Sci., 8 (8–9)*, 1973, pp 797–804.

[15] S. M. Rao, T. K. Sarkar, S. A. Dianat: A novel technique to the solution of transient electromagnetic scattering from thin wires, *IEEE Trans. AP, 34*, 1986, pp 630–634.

[16] B. P. Rynne, P. D. Smyth: Stability of time marching algorithms for the electric field integral equation, *J. Electr. Waves Applc., 4 (12)*, 1990, pp 1181–1205.

[17] F. M. Tesche: On the inclusion of losses in time-domain solutions of electromagnetic interaction problems, *IEEE Trans. EMC, 32 (1)*, 1990.

[18] A. R. Bretones, R. G. Martin, I. S. Garcia: Time-domain analysis of magnetic coated wire antennas and scatterers, *IEEE Trans. AP, 43 (6)*, 1995, pp 591–596.

[19] A. G. Tijhuis, Z. Q. Peng, A. R. Bretones: Transient excitation of a straight thin wire segment: A new look at an old Problem, *IEEE Trans. AP 40 (10)*, 1992, pp 1132–1146.

[20] A. R. Bretones, A. G. Tijhuis: Transient excitation of a straight thin wire segment over an interface between two dielectric half-spaces, *Radio Sci. 30 (6)*, 1995, pp 1723–1738.

[21] R. Moini, V. A. Rakov, M. A. Uman, B. A. Kordi: An antenna theory model for lightning return stroke, *Proc. EMC '97 Symposium, Zurich*, 1997, pp 149–152.

[22] P. J. Davies: On the stability of time-marching schemes for the general surface electric-field integral equation, *IEEE Trans. AP 44 (11)*, 1996, pp 1467–1473.

[23] V. Roje, D. Poljak, I. Zanchi: Current Distribution for Cylindrical Dipole and its Influence on Mutual Impedance, *Proceeding of ELMAR 34th International Symposium, Zadar*, 1992, pp. 158–160.

[24] D. Poljak, V. Roje: The weak finite element formulation for solving Pocklington's integro-differential equation, *Int. Journ. Eng. Modelling 6 (1–4)*, 1993, pp 21–25.

[25] V. Roje, D. Poljak: Various Approaches for Solving Integral Equations in Electromagnetics, *Proceedings of International Conference on Electromagnetics in Advanced Applications (ICEAA 95), Turin*, 1995, pp 329–332.

[26] D. Poljak: New numerical approach in analysis of a thin wire radiating over a lossy half-space, *Int. J. Num. Meth. Eng., 38 (22)*, 1995, pp 3803–3816.

[27] D. Poljak, V. Roje: Boundary-element approach to calculation of wire antenna parameters in the presence of dissipative half-space, *IEE Proc. Microw. Antennas Propag., 142 (6)*, 1995, pp 435–440.

[28] D. Poljak, V. Roje: Boundary Element Analysis of Resistively Loaded Wire Antenna Immersed in a Lossy Medium, *19th International Conference on Boundary Elements, BEM 19, Rome*, September 1997, pp 485–493.

[29] D. Poljak, V. Roje: The integral equation method for ground wire input impedance, in: C. Constanda, J. Saranen, S. Seikkala (Eds.): Integral Methods in Science and Engineering, Vol. I, *Longman, U.K.*, 1997, pp 139–143.

[30] D. Poljak: Finite element solution of transient radiation from thin wires, *Int. J. Eng. Modelling 8 (1–2)*, 1995, pp 31–35.

[31] D. Poljak: Transient response of wire antennas in the presence of a conducting half-space, *doctoral thesis, University of Split*, 1996.

[32] D. Poljak, V. Roje: Transient response of a thin wire in a two media configuration, *Proceedings of EMC '97 Symposium, Zurich*, 1997, pp 293–296.

[33] D. Poljak, V. Roje: Finite element technique for solving time-domain Hallen integral equation, *Proceedings of ICAP '97, Edinburgh*, 1997, pp 1.225–1.228.

[34] D. Poljak, V. Roje: Time domain calculation of the parameters of thin wire antennas and scatterers in a half-space configuration, *IEE Proc. Microw. Antennas Propag., 145 (1)*, February 1998, pp 435–440.

8

WIRE ANTENNAS AND SCATTERERS: FREQUENCY DOMAIN ANALYSIS

This chapter deals with thin wire configurations: first as individual radiating structures and second as assembled structures in wire arrays. The analysis is first undertaken by assuming the wires to be isolated in free space, while in the second part of the chapter, the influence of a dissipative half-space is taken into account using various approaches—Sommerfeld integral approach and reflection coefficient approximation. The mathematical formulations of the problems are always followed by the numerical modeling details.

8.1 THIN WIRES IN FREE SPACE

Thin wire antennas are widely used in communication systems from low to ultra high frequencies either in the form of individual elements or associated with other similar elements to form phased arrays. These antennas are most commonly used as bases for analyzing more complex configurations and as probes to explore unknown environments. In this chapter the straight wire antenna, insulated dipole antenna, and loop antenna are first studied as individual elements. Developing accurate expressions for some basic antenna types provide generalization to more complicated, composite structures. At the end of this section, an arbitrary array of straight wires is considered.

8.1.1 Single Straight Wire in Free Space

The linear dipole antenna isolated in free space is one of the simplest antenna forms. The single straight thin wire antenna having length L and of radius a insulated in free space is depicted in Figure 8.1.

The electrical properties of this antenna are determined by its axial current distribution. The current distribution along the straight thin wire is obtained as the solution of the frequency domain Pocklington integro-differential equation. This equation can be derived by expressing the time-harmonic electric field $\vec{E}$ in terms of the magnetic vector potential $\vec{A}$ and the electric scalar potential φ,

$$\vec{E} = -\nabla\varphi - j\omega\vec{A} \tag{8.1}$$

The magnetic vector and electric scalar potential for time-harmonic electric field are coupled through the well-known Lorentz Gauge equation,

$$\nabla\vec{A} = -j\,\omega\,\mu\,\varepsilon\,\varphi \tag{8.2}$$

Equation (8.2) represents the continuity equation for potential functions.

The thin wire approximation requires wire dimensions to satisfy the following conditions:

$$a \ll \lambda_0 \text{ and } a \ll L \tag{8.3}$$

where λ_0 denotes the free-space wavelength of a plane wave.

Consequently, only axial component of $\vec{A}$ exists along the cylinder. Now, combining eqns (8.1) and (8.2) yields

$$E_x = \frac{1}{j\omega\mu\varepsilon_0}\left[\frac{\partial^2 A_x}{\partial x^2} + k^2 A_x\right] \tag{8.4}$$

where ε_0 is the permittivity of the free space and ω is the applied frequency.

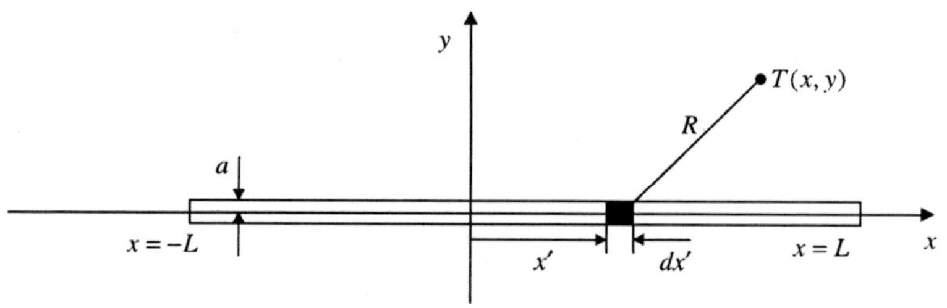

Figure 8.1 Single straight wire insulated in free space.

The magnetic vector potential A_x can be represented by an integral over the unknown axial current $I(x)$ on the cylinder:

$$A_x = -\frac{\mu}{4\pi} \int_{-L}^{L} I(x') \frac{e^{-jkR}}{R} \mathrm{d}x' \tag{8.5}$$

where R is the distance from the source point to the observation point and k is given by

$$k = \frac{2\pi}{\lambda_0} \tag{8.6}$$

and denotes the wave number of the free space.

Finally, assuming the wire to be perfectly conducting (PEC) the total tangential electric field vanishes on the metallic wire surface:

$$E_x^{inc} + E_x^{sct} = 0 \tag{8.7}$$

where E_x^{inc} is the incident field and E_x^{sct} is the scattered field due to the presence of the PEC surface.

According to eqns (8.4) and (8.5), the scattered electric field close to the antenna surface is then given by

$$E_x^{sct} = \frac{1}{j4\pi\omega\varepsilon_0} \int_{-L}^{L} \left(\frac{\partial^2}{\partial x^2} + k^2 \right) g_0(x, x') I(x') \mathrm{d}x' \tag{8.8}$$

Combining relations (8.7) and (8.8) leads to the Pocklington integro-differential equation for the unknown current distribution along the single straight wire antenna insulated in free space,

$$E_x^{inc} = -\frac{1}{j4\pi\omega\varepsilon_0} \int_{-L}^{L} \left(\frac{\partial^2}{\partial x^2} + k^2 \right) g_0(x, x') I(x') \mathrm{d}x' \tag{8.9}$$

where $g_0(x, x')$ is the free-space Green function given by

$$g_0(x, x') = \frac{e^{-jkR}}{R} \tag{8.10}$$

while the distance R from the source point to the observation point is

$$R = \sqrt{(x - x')^2 + a^2} \tag{8.11}$$

Once the axial current on the antenna is determined, other important antenna parameters, such as radiated field or input impedance, can be expressed in terms of current distribution.

The electric field tangential to the array is defined by the following equation [1]:

$$E_x(x,y) = \frac{1}{j4\pi\omega\varepsilon_0}\left[\begin{array}{l}\left(I(x')\dfrac{\partial g_0(x,x',y)}{\partial x'}\right)\Bigg|_{-L}^{L} - \displaystyle\int_{-L}^{L}\dfrac{\partial g_0(x,x',y)}{\partial x'}\dfrac{\partial I(x')}{\partial x'}\mathrm{d}x' \\ + k^2\displaystyle\int_{-L}^{L} I(x')g_0(x,x',y)\mathrm{d}x'\end{array}\right] \tag{8.12}$$

while the input impedance can be determined from the corresponding functional as presented in Ref. [2]. For rather the simple case of a single wire insulated in the free space, the expression for the input impedance is given by [3]

$$Z(I) = -\frac{1}{|I_0|^2}\int_{-L}^{L} E(x,a)I(x)^*\mathrm{d}x \tag{8.13}$$

where $E(x,a)$ is the electric field along the antenna, $I(x)^*$ is the complex conjugate of the axial current distribution, and I_0 is the antenna input current.

8.1.2 Boundary Element Solution of Thin Wire Integral Equation

In the last decade, the boundary element method (BEM) became a well established technique and is now widely used in solving electromagnetic problems [4]. BEM is a combination of classical boundary methods for solving integral equations and some numerical schemes originated from FEM. In brief, the boundary elements became a suitable technique for solving integral equations via finite elements. Till now, the free-space thin wire integral equation has been treated only by projective techniques [5, 6], usually named *strong* formulation. Within this strong formulation procedure several problems appear. Differentiation over the integral equation kernel must be performed analytically that results in quasisingularity. The method presented in this chapter is promoted in Ref. [7] and, with some modifications, also successfully used in many other papers [8–11]. The method uses some schemes from the partial differential equations modeling via weak finite element formulation and certain boundary element concepts. The established method is entitled as the Galerkin Bubnov indirect boundary element method (GB-IBEM) or the finite element integral equation method (FEIEM).

The boundary element solution of the integral eqn (8.9) has been carried out by using linear or isoparametric elements, respectively. The solution featuring linear elements can be found in Ref. [3], while the isoparametric element solution has been documented in detail in Ref. [4]. This section outlines both the methods.

8.1.2.1 Linear BEM Solution In presenting the outline of the method, it is first convenient to use an operator form of (21), which is symbolically written as

$$K(I) = E \tag{8.14}$$

where K is a linear operator, I is the unknown function to be found for a given excitation E.

The Galerkin boundary element solution starts by applying the local approximation for unknown current along the wire, that is, the unknown current $I(x)$ is then expanded into finite sum of linearly independent basis functions $\{f_i\}$ with unknown complex coefficients α_i, that is,

$$I_n(x') = \sum_{n=1}^{N_g} I_n f_n(x') \tag{8.15}$$

where $f_n(x)$ denotes the linear elements shape functions, and I_n stands for the unknown coefficients of the solution, N_g denotes the total number of basis functions.

Substituting eqn (8.15) into eqn (8.14) yields

$$K(I) = \sum_{n=1}^{N_g} I_n K(f_n) \tag{8.16}$$

The residual R can be written as follows:

$$R = \sum_{n=1}^{N_g} I_n K(f_n) - E \tag{8.17}$$

According to the definition of the scalar product of functions in Hilbert function space, the error R is weighted to zero with respect to certain weighting functions $\{W_j\}$, that is,

$$\int_\Omega R W_m{}^* d\Omega = 0; \quad m = 1, 2, \ldots, N_g \tag{8.18}$$

and Ω is the domain of interest. As the operator K is linear, performing some mathematical manipulation and by choosing $W_m = f_m$ (the Galerkin-Bubnov procedure), the system of algebraic equations is obtained;

$$\sum_{n=1}^{N_g} I_n \int_\Omega K(f_n) f_m d\Omega = \int_\Omega E f_m d\Omega; \quad m = 1, 2, \ldots, N_g \tag{8.19}$$

Featuring the weak formulation by carefully performing the integration by parts, the Galerkin formulation of the Pocklington integro-differential eqn (8.9) is given by

$$\sum_{n=1}^{N_g}\left[-\int_{-L}^{L}\int_{-L}^{L}\frac{\mathrm{d}f_m(x)}{\mathrm{d}x}\frac{\mathrm{d}f_n(x')}{\mathrm{d}x'}g_0(x,x')\mathrm{d}x'\mathrm{d}x+k^2\int_{-L}^{L}\int_{-L}^{L}f_m(x)f_{in}(x')g_0(x,x')\mathrm{d}x'\mathrm{d}x\right]I_n$$
$$=-j4\pi\omega\varepsilon\int_{-L}^{L}E_x^{inc}(x)f_m(x)\mathrm{d}x;\quad m=1,2,\ldots,N_g \tag{8.20}$$

Equation (8.20) represents the weak Galerkin-Bubnov formulation of the integral eqn (8.9). The advantage of such a formulation is obvious. The second-order differential operator is replaced by trivial differentiation over basis and test (weight) functions. The only requirement which is to be satisfied by bases and weights is that they must be chosen from the class of order-one differentiable functions.

Moreover, this formulation is convenient for implementation of BEM. Boundary conditions are subsequently incorporated into global matrix of linear equation system, which is a significant advantage over the moment methods in which basis and test functions must be chosen in a way to satisfy posed boundary conditions [8, 9].

Applying the boundary element algorithm, the domain of integration is divided into segments that are connected at nodes. Global basis functions are assigned to nodes, while shape functions are assigned to elements. A global matrix and global right-hand side vector are assembled from individual boundary element matrices and vectors, respectively, and the system of equations is given by

$$\sum_{i=1}^{M}[Z]_{ji}\{I\}_i=\{V\}_j,\quad j=1,2,\ldots,M \tag{8.21}$$

where $[Z]_{ji}$ is the mutual impedance matrix representing the interaction of the ith source to the jth observation boundary element, respectively,

$$[Z]_{ji}=-\frac{1}{j4\pi\omega\varepsilon}\int_{\Delta l_j}\int_{\Delta l_i}\{D\}_j\{D'\}_i^T g_0(x,x')\mathrm{d}x'\mathrm{d}x$$
$$+k_1^2\int_{\Delta l_j}\int_{\Delta l_i}\{f\}_j\{f'\}_i^T g_0(x,x')\mathrm{d}x'\mathrm{d}x \tag{8.22}$$

$\{V\}_j$ is the right-side voltage vector for jth observation boundary element,

$$\{V\}_j=-\int_{\Delta l_j}E_x^{inc}\{N\}_j\mathrm{d}x \tag{8.23}$$

Matrices $\{f\}$ and $\{f'\}$ contain shape functions $f_k(x)$ and $f_k(x')$, while $\{D\}$ and $\{D'\}$ contain their derivatives, where M is the total number of finite elements, n_e is the total number of local nodes per element, and Δl_i, Δl_j are the width of ith and jth finite elements, respectively.

With respect to the fact that functions $f(x)$ are required to be of class C^1 (once differentiable), a convenient choice for the shape functions over the finite elements is the family of Lagrange's polynomials given by

$$L_i(x) = \prod_{j=1}^{m} \frac{x - x_j}{x_i - x_j}, \quad j \neq i \tag{8.24}$$

For this purpose, the Pocklington's equation for free space can be treated with linear approximation and the shape functions are given as follows:

$$f_1(x) = \frac{x_2 - x}{\Delta x}, \quad f_2(x) = \frac{x - x_1}{\Delta x} \tag{8.25}$$

where x_1 and x_2 are the coordinates of the element nodes.

The evaluation of the right-hand side vector can be performed in the closed form if the delta-function voltage generator (antenna mode) or the plane wave excitation (scatterer mode) is used. In the radiation mode right-hand side vector is different from zero only in the feed-gap area. The x-component of the impressed (incident) electric field is given by

$$E_x^{inc}(x) = \frac{V_g}{\Delta l_g} \tag{8.26}$$

where V_g is the feed voltage and Δl_g is the feed-gap width.

Using the linear shape functions, it follows:

$$V_1 = \int_{-\Delta l/2}^{\Delta l/2} \frac{V_g}{\Delta l_g} \frac{x_2 - x}{\Delta l_g} \mathrm{d}x = \frac{V_g}{2} \tag{8.27}$$

$$V_2 = \int_{-\Delta l/2}^{\Delta l/2} \frac{V_g}{\Delta l_g} \frac{x - x_1}{\Delta l_g} \mathrm{d}x = \frac{V_g}{2} \tag{8.28}$$

Note that each node from the feed-gap central segment obtains half the value of the total impressed voltage.

If the scattering mode for the simple case of normal incidence is considered, the wire is illuminated by the plane wave, that is,

$$E_x^{inc}(x) = E_0 \tag{8.29}$$

The right-hand side vector differs from zero on each segment and the local voltage vector is given by following expressions:

$$V_1 = \int_{-\Delta l/2}^{\Delta l/2} E_0 \frac{x_2 - x}{\Delta l} \mathrm{d}x = E_0 \frac{\Delta l}{2} \tag{8.30}$$

$$V_2 = \int_{-\Delta l/2}^{\Delta l/2} E_0 \frac{x - x_1}{\Delta l} \mathrm{d}x = E_0 \frac{\Delta l}{2} \tag{8.31}$$

Similarly, the expressions for quadratic approximation can be obtained [7].

The full description boundary element solution procedure with linear elements can be found in Ref. [3].

8.1.2.2 Isoparametric BEM Solution Isoparametric elements use coordinate mappings derived from the element approximating functions themselves [4]. The curvilinear elements are well suited for complicated geometries which can be represented with a relatively small number of elements.

The basic idea of isoparametric transformation is to approximate the solution for the current in terms of parametric functions $f(\xi)$,

$$I_n(\xi) = \sum_{n=1} I_{ni} f_{ni}(\xi) \tag{8.32}$$

The key point of isoparametric elements lies in the fact that the global coordinate x is a function of the local parametric coordinate ξ on the element. A transformation function $x(\xi)$ defined over the element can be expressed in terms of the element approximating function, that is,

$$x = \sum_{i=1}^{N} x_i f_i(\xi) \tag{8.33}$$

where the functions $f_i(\xi)$ are polynomials.

Such transformation, in which the same family of approximation functions is used for the unknown quantity and also to express the element shape transformation, is called isoparametric and the related elements are referred to as isoparametric elements.

For the case of two-noded isoparametric linear elements used in this chapter, eqn (8.28) is simplified into

$$x = x_1 f_1(\xi) + x_2 f_2(\xi) \tag{8.34}$$

where x_1 and x_2 are the global two-noded element lengths, while the shape functions are given by

$$f_1(\xi) = \frac{1-\xi}{2}, \quad f_2(\xi) = \frac{1+\xi}{2} \tag{8.35}$$

The differentiation of eqn (8.33) yields

$$dx = \sum_{i=1}^{N} x_i \frac{f_i(\xi)}{d\xi} d\xi \tag{8.36}$$

By taking into account eqns (8.34) and (8.35), one obtains

$$\frac{dx}{d\xi} = \frac{x_2 - x_1}{2} = \frac{\Delta x}{2} \tag{8.37}$$

where Δx is the two-node element length.

Performing the boundary element discretization of the wire, the matrix equation arising from eqn (8.20) is of the form

$$\sum_{k=1}^{M} [Z]_{lk}\{I\}_k = \{V\}_l \quad l = 1, 2, \ldots, M \tag{8.38}$$

where M is the total number of elements along the wire, and $[Z]_{lk}$ is the interaction matrix representing the mutual impedance between each segment on the ith (source) wire and every segment on the jth (observation) wire,

$$\begin{aligned} [Z]^e_{lk} = &-\int_{-1}^{1}\int_{-1}^{1} \{D\}_l\{D'\}^T_k g_0(\xi,\xi')d\xi' d\xi \\ &+ k^2 \int_{-1}^{1}\int_{-1}^{1} \{f\}_l\{f'\}^T_k g_0(\xi,\xi')\left(\frac{\Delta x}{2}\right)^2 d\xi' d\xi \end{aligned} \tag{8.39}$$

Vectors $\{f\}$ and $\{f'\}$ contain shape functions $f_n(\xi)$ and $f_n(\xi')$, while $\{D\}$ and $\{D'\}$ contain their derivatives, and Δx is the segment length.

The right-hand side vector $\{V\}_l$ represents the voltage along the lth segment and it is given as follows:

$$\{V\}_l = -j4\pi\omega\varepsilon_0 \int_{-1}^{1} E_x^{inc}(\xi)\{f\}_l \frac{\Delta x}{2} d\xi \tag{8.40}$$

In the transmitting mode right-hand side vector is different from zero only in the feed-gap region and the incident field can be written as

$$E_x^{inc} = \frac{V_g}{\Delta l_g} \tag{8.41}$$

where V_g is the feed voltage and Δl_g is the feed-gap width.

8.1.3 Calculation of the Radiated Electric Field and the Input Impedance of the Wire

Once the current distribution along the wire is determined, the tangential component of the radiated electric field can be obtained using the boundary element formulation as follows:

$$E_x[x(\xi), y(\xi)] = \frac{1}{j4\pi\omega\varepsilon_0} \sum_{k=1}^{M} \left[\left(I[x'(\xi')] \frac{\partial g_0}{\partial \xi'} \right) \Bigg|_{\xi=-1}^{\xi=1} - \int_{-1}^{1} \frac{\partial g_0}{\partial \xi'} \frac{\partial I[x'(\xi')]}{\partial \xi'} \mathrm{d}\xi' \right.$$
$$\left. + k^2 \int_{-1}^{1} I[x'(\xi')] g_0[x(\xi), y(\xi), x'(\xi')] \mathrm{d}\xi' \right] \tag{8.42}$$

The current along the kth element is computed using linear interpolation,

$$I[x'(\xi')] = I_{1k} f_{1k}(\xi') + I_{2k} f_{2k}(\xi') \tag{8.43}$$

The input impedance can be readily obtained applying the boundary element algorithm to eqn (8.14) which results in the following formula:

$$Z_i = -\frac{1}{j4\pi\omega\varepsilon |I_0|^2} \left\{ I_j^* \left[\sum_{i=1}^{M} \int_{\Delta l_j} \int_{\Delta l_i} \{D\}_j \{D'\}_i^T g_0(\xi, \xi') \mathrm{d}\xi' \mathrm{d}\xi \right. \right.$$
$$\left. \left. + k^2 \int_{\Delta l_j} \int_{\Delta l_i} \{f\}_j \{f\}_i^T g_0(\xi, \xi') \left(\frac{\Delta x}{2} \right)^2 \mathrm{d}\xi' \mathrm{d}\xi \right] \alpha_i \right\}; \quad j = 1, 2, \ldots, M \tag{8.44}$$

where I_j^* denotes the complex conjugate of the current in jth node.

8.1.4 Numerical Results for Thin Wire in Free Space

Numerical results refer to unit, time-harmonic voltage excitation,

$$V = 1e^{j\omega t} \tag{8.45}$$

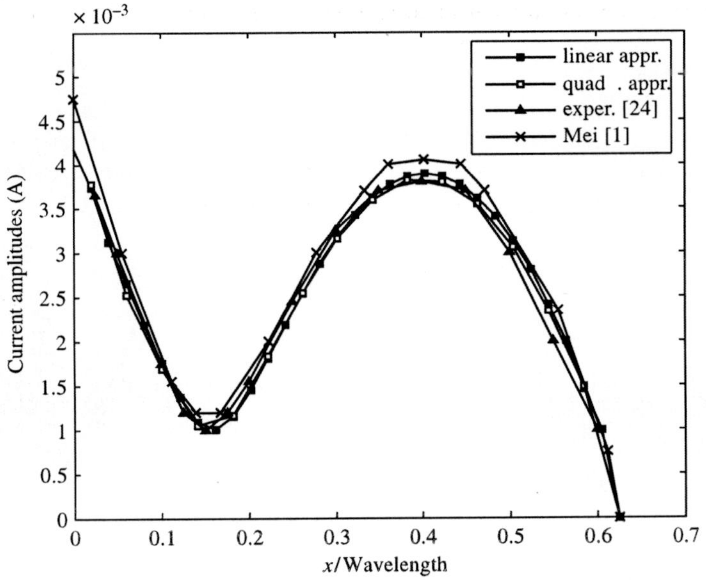

Figure 8.2 Magnitudes of current distribution on the dipole antenna insulated in free space, of half-length $L = 0.625\lambda$, radius $a = 0.007022\lambda$—comparison of various methods. (Poljak, D., New numerical approach in analysis of a thin wire radiating over a lossy half-space, *Int. J. Num. Meth. Eng.*, 38 (22), 1995, pp 3803–3816).

A single wire antenna insulated in free space ($L = 0.625\lambda$, $a = 0.00702\lambda$) is considered. The current distributions obtained by BEM (FEIEM) with linear and quadratic interpolation, respectively, are compared with Mei's theoretical results [12] (subdomain collocation), and experimental results [7] as shown in Figure 8.2.

It is obvious that BEM results with quadratic approximation are in the best agreement with experimental results, while collocation technique differs significantly from experimental results. Since there is just a slight difference between the results obtained by linear and the quadratic interpolation, linear interpolation is used in the half-space calculations because it is significantly less time-consuming. Discretization of antenna on 31 boundary elements is shown to provide convergence of the results [7].

This is presented in Figures 8.3 and 8.4 in the case of full-wave dipole antenna ($L = 0.5\lambda$, $a = 0.00702\lambda$).

8.1.5 Coated Thin Wire Antenna in Free Space

Dielectric coated wire antennas in free space are modeled via an efficient Galerkin-Bubnov boundary element scheme. The formulation is based on the Pocklington integro-differential equation for loaded wires with the dielectric coating being taken into account by means of an equivalent magnetic coating load term added to the equation.

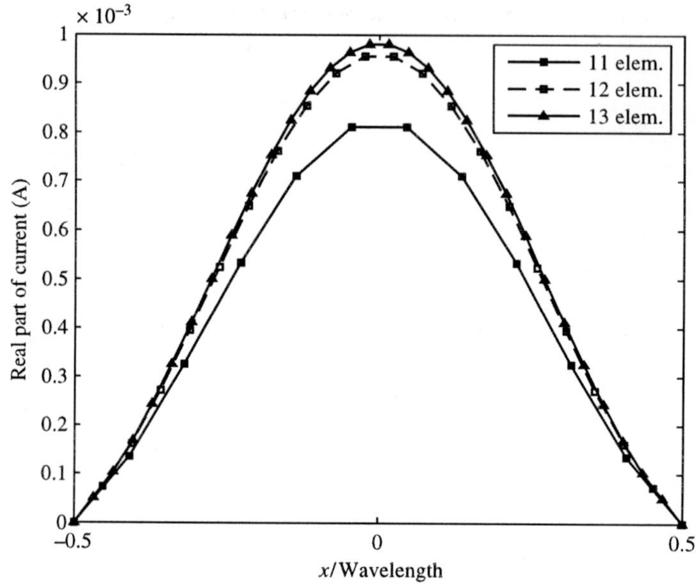

Figure 8.3 Real part of current distribution along the dipole antenna insulated in free space of half-length $L = 0.5\lambda$, radius $a = 0.007022\lambda$, for various number of finite elements (11, 21, 31)—linear interpolation. (Poljak, D., New numerical approach in analysis of a thin wire radiating over a lossy half-space, *Int. J. Num. Meth. Eng.*, 38 (22), 1995, pp 3803–3816).

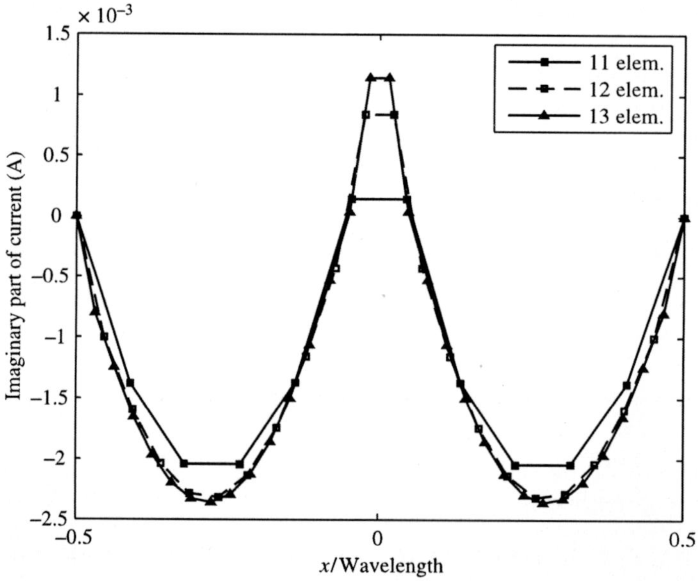

Figure 8.4 Imaginary part of current distribution along the dipole antenna insulated in free space of half-length $L = 0.5\lambda$, radius $a = 0.007022\lambda$, for various number of finite elements (11, 21, 31)—linear approximation. (Poljak, D., New numerical approach in analysis of a thin wire radiating over a lossy half-space, *Int. J. Num. Meth. Eng.*, 38 (22), 1995, pp 3803–3816).

Dielectric coated antennas are often preferable over bare wires when used in applications involving finitely conducting medium such as microwave hyperthermia, geophysical exploration, or subsurface communication. The use of these antennas avoids the often undesirable contact between them and the surrounding medium. Moreover, the radiation efficiency of the antenna can be improved when using coatings [13].

One of the principal applications of coated antennas is in microwave hyperthermia, that is, the treatment of cancer by heating the malignant tissue. The high temperature thus generated inside the tumor tissue has a cytotoxic effect and is useful in both chemotherapy and radiation therapy.

This tissue heating must be localized to maintain the temperature within the tumor tissue up to 43°C for a given time period, while the neighboring tissue temperature level is far below 43°C [13]. The insulated antennas are inserted into tumor tissue through brachytherapy catheters. Most of the heating is localized and takes place around the antenna near field.

The approximate calculation of the near field generated by an insulated dipole antenna immersed in a lossy medium has been proposed by King et al. in Ref. [14].

A more rigorous approach is applying the Moment method (MoM) approach to the near field analysis, based on the solution of the corresponding Pocklington integro-differential equation by using piecewise sinusoids was presented in Ref. [13].

The use of the bare loaded antenna approach in treating wires having dielectric, magnetic, or fixed coatings has been shown to simplify the analysis of coated antennas [3]. In Ref. [15], the Galerkin-Bubnov indirect boundary element method (GB-IBEM) [4, 16] is applied to the analysis of dielectrically coated thin wire antennas in free space.

The analysis of dielectric coated wire antennas in free space can be regarded as a starting point in more interesting case of antenna immersed in the conducting medium. The formulation of the problem is based on the corresponding Pocklington integro-differential equation. The electrically thin dielectric coating is modeled via an equivalent magnetic coating in a similar manner as presented in Ref. [17]. Using the quasistatic approximation valid for electrically thin coatings, the influence of dielectric coating can be taken into account in the Pocklington integro-differential equation via an additional impedance term.

The current distribution along the antenna is obtained by solving the Pocklington integro-differential equation by means of GB-BEM. Once the current distribution along the antenna has been obtained, the related near electromagnetic field can be computed.

A straight thin wire antenna of length $2h$ and radius a covered by a dielectric coating of outer radius b with dielectric constant ε_r in Figure 8.5 is considered.

The current along the dielectric coated antenna is governed by Pocklington type integro-differential integral equation. The integro-differential equation for the antenna current is usually derived in accordance to the well-known thin-wire approximation [3], which is valid if the antenna radius is much smaller than the wavelength of the impressed field and its length much greater then its radius.

The Pocklington integro-differential equation for loaded wire can be derived using the thin-wire approximation and enforcing the continuity condition for the

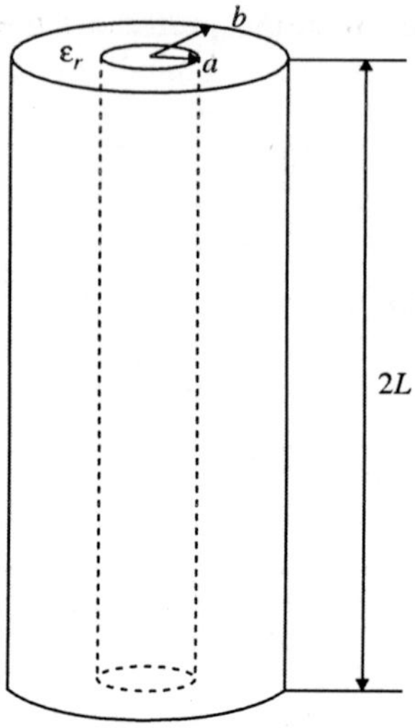

Figure 8.5 A straight thin wire with dielectric coating of thickness (b-a).

tangential components of the electric field on the thin-wire surface. The time harmonic electric field vector $\vec{E}$ can be expressed in terms of magnetic vector potential $\vec{A}$ and electric scalar potential φ as follows:

$$\vec{E} = -\nabla\varphi - j\omega\vec{A} \tag{8.46}$$

The vector and scalar potential are coupled through the commonly used Lorentz Gauge,

$$\nabla\vec{A} = -j\,\omega\,\mu\,\varepsilon\,\varphi \tag{8.47}$$

where ε and μ are the corresponding permittivity and permeability, respectively, and ω is the applied frequency.

Because of the thin wire approximation, the magnetic vector potential has only an axial component A_z.

Combining eqns (8.40) and (8.41) yields

$$E_z = \frac{1}{j\omega\mu\varepsilon}\left[\frac{\partial^2 A_z}{\partial z^2} + k^2 A_z\right] \tag{8.48}$$

where $k = \omega/c$ denotes the wave number of the free space.

The magnetic vector potential A_z can be expressed by the integral of the axial current distribution $I(z)$ flowing along the straight thin wire, that is,

$$A_z = \frac{\mu}{4\pi}\int_{-L}^{L} I(z')g(z,z')\mathrm{d}z' \tag{8.49}$$

where $g(z, z')$ is the so-called "reduced" kernel of the integral equation [3],

$$g_0(z, z') = \frac{e^{-jkR}}{R} \tag{8.50}$$

R is the distance from the source point in the antenna axis to the observation point in the surrounding medium.

The continuity conditions for the tangential electric field component at the wire antenna surface can be written as

$$E_z^{inc} + E_z^{sct} = I(z)Z_L \tag{8.51}$$

where E_z^{inc} is the incident field, E_z^{sct} is the field scattered by the coated wire, and Z_L is the corresponding impedance per unit length of the wire.

For electrically thin coatings, this effect can be taken into account through the load impedance and the dielectric coating can be expressed by means of a purely magnetic coating given by [15, 17]

$$Z_L = \frac{j\omega\mu_0}{2\pi}\frac{\varepsilon_r - 1}{\varepsilon_r}\ln\frac{b}{a} \tag{8.52}$$

The scattered field component tangential to the antenna surface can thus be expressed by the following integral:

$$E_z^{sct} = \frac{1}{j4\pi\omega\varepsilon_0}\int_{-L}^{L}\left(\frac{\partial^2}{\partial z^2} + k^2\right)g_0(z, z')I(z')dz' \tag{8.53}$$

Finally, combining relationships (8.52) and (8.53) results in the Pocklington integro-differential equation governing the current along a straight thin wire, that is,

$$E_z^{inc} = -\frac{1}{j4\pi\omega\varepsilon_0}\int_{-L}^{L}\left(\frac{\partial^2}{\partial z^2} + k^2\right)g_0(z, z')I(z')dz' + I(z)Z_L \tag{8.54}$$

where R is the distance from the source point z to the observation point z' (now both located on the thin wire antenna surface) that is given by

$$R = \sqrt{(z - z')^2 + a^2} \tag{8.55}$$

The thin wire approximation assures the zero boundary conditions for the current at the free ends of wire.

8.1.6 The Near Field of a Coated Thin Wire Antenna

The near field analysis is very important in many applications of dielectric coated antenna, such as in the case of localized heating in microwave hyperthermia. The complete electromagnetic field radiated by the coated wire can be evaluated knowing the current distribution along the antenna. However, simple analytical formulas for the near field of a coated wire are not available.

The full electric field vector can be expressed in terms of the magnetic vector potential as follows:

$$\vec{E} = \frac{1}{j\omega\mu\varepsilon_0}\nabla\left(\nabla\vec{A}\right) - j\omega\vec{A} \tag{8.56}$$

where ε_0 is the permittivity of a free space.

Because of rotational symmetry, the radiated electric field does not depend on azimuthal variable Φ and the corresponding field components are then given by

$$E_\rho = \frac{1}{j\omega\mu\varepsilon_0}\frac{\partial^2 A_z}{\partial\rho\partial z} \tag{8.57}$$

$$E_z = \frac{1}{j\omega\mu\varepsilon_0}\frac{\partial^2 A_z}{\partial z^2} - j\omega A_z \tag{8.58}$$

Inserting the expression for magnetic vector potential (8.49) into eqn (8.58), it gives

$$E_\rho = \frac{1}{j4\pi\omega\varepsilon_0}\int_{-L}^{L} I(z')\frac{\partial^2 g_0(z,z',\rho)}{\partial\rho\partial z}\mathrm{d}z' \tag{8.59}$$

and after integration by parts eqn (8.59) becomes

$$E_\rho = \frac{1}{j4\pi\omega\varepsilon_0}\int_{-L}^{L} \frac{\partial I(z')}{\partial z'}\frac{\partial g_0(z,z',\rho)}{\partial\rho}\mathrm{d}z' \tag{8.60}$$

The axial z-component of the electric field is defined by eqn (8), which after integration by parts results in the following relationship:

$$E_z = -\frac{1}{j4\pi\omega\varepsilon_0}\left[\int_{-L}^{L} \frac{\partial I(z')}{\partial z'}\frac{\partial g_0(z,z',\rho)}{\partial z}\mathrm{d}z' + k^2\int_{-L}^{L} I(z')g_0(z,z',\rho)\mathrm{d}z'\right] \tag{8.61}$$

The integrals in expressions (8.53), (8.60), and (8.61) contain quasisingular kernel due to the presence of differential operator [5]. This quasisingularity can be efficiently treated by the boundary element/finite differences approach.

8.1.7 Boundary Element Procedures for Coated Wires

The Pocklington integro-differential eqn (8.54) can be numerically solved by means of the indirect Galerkin-Bubnov boundary element method [4].

The operator form of eqn (8.54) is given by

$$K(I) = E \tag{8.62}$$

where K is a linear operator, I is the unknown current to be found for a given excitation E.

The unknown current is expressed by the sum of a finite number of linearly independent basis functions $\{f_i\}$ with unknown complex coefficients I_i, that is,

$$I(z') = \sum_{i=1}^{n} I_i f_i(z') \tag{8.63}$$

Applying the weighted residual approach and choosing the test functions to be the same as basis functions (Galerkin-Bubnov procedure), the operator eqn (8.63) transforms into a system of algebraic equation of the form:

$$\sum_{i=1}^{n} I_i \int_{-L}^{L} K(f_i) f_j \mathrm{d}z = \int_{-L}^{L} E f_j \mathrm{d}z; \quad j = 1, 2, \ldots, n \tag{8.64}$$

Performing certain mathematical manipulations the following matrix equation is obtained:

$$\sum_{i=1} [Z]_{ji} \{I\}_i = \{V\}_j \quad \text{and} \quad j = 1, 2, \ldots, M \tag{8.65}$$

where vector $\{I\}$ contains the unknown coefficients of the solution, and $\{V\}_j$ represents the local voltage vector,

$$\{V\}_j = \int_{\Delta l_j} E_z^{inc} \{f\}_j \mathrm{d}z \tag{8.66}$$

which can be evaluated in closed form if the constant and strongly localized incident electric field is assumed to be impressed along the feed-gap area,

$$E_z^{inc} = \frac{V_T}{\Delta z_g} \tag{8.67}$$

where Δz_g denotes the length of the feed-gap segment.

A linear approximation over each boundary element has been used, that is, the shape functions are given by

$$f_i(z) = \frac{z_{i+1} - z}{\Delta z} f_{i+1}(z) = \frac{z - z_i}{\Delta z} \tag{8.68}$$

and therefore it follows

$$V_{1j} = \int_{-\Delta z_g/2}^{\Delta z_g/2} \frac{V_T}{\Delta z_g} \frac{z_{j+1} - z}{\Delta z_g} \mathrm{d}z = \frac{V_T}{\Delta z_g} \frac{\Delta z_g}{2} \tag{8.69}$$

$$V_{2j} = \int_{-\Delta z_g/2}^{\Delta z_g/2} \frac{V_T}{\Delta z_g} \frac{z - z_{j+1}}{\Delta z_g} \mathrm{d}z = \frac{V_T}{\Delta z_g} \frac{\Delta z_g}{2} \tag{8.70}$$

The nonzero feed-gap element can be written as

$$\{V\}_j = \frac{V_T}{2} \begin{Bmatrix} 1 \\ 1 \end{Bmatrix} \tag{8.71}$$

where V_T is the voltage generator impressed at the feed-gap area.

Matrix $[Z]_{ji}$ represents the interaction of the ith source boundary element with the jth observation boundary element,

$$[Z]_{ji} = -\frac{1}{j4\pi\omega\varepsilon} \left(\begin{array}{c} \int_{z_j}^{z_{j+1}} \{D\}_j \int_{z_i}^{z_{i+1}} \{D\}_i^T g_0(z, z') \mathrm{d}z' \mathrm{d}z \\ + k^2 \int_{z_j}^{z_{j+1}} \{f\}_j \int_{z_i}^{z_{i+1}} \{f\}_i^T g_0(z, z') \mathrm{d}z' \mathrm{d}z \end{array} \right) \tag{8.72}$$

$$+ \int_{z_j}^{z_{j+1}} Z_L(z) \{f\}_j \{f\}_i^T \mathrm{d}z$$

Matrices $\{f\}$ and $\{f'\}$ contain the shape functions, while $\{D\}$ and $\{D'\}$ contain their derivatives, where M is the total number of segments, and z_i, z_{i+1}, z_j, and z_{j+1} are the coordinates of ith and jth wire segments, respectively.

Once the numerical results for the current distribution along the antenna have been obtained, the near electric field of the coated antenna can be determined.

The electric field components can be calculated using the boundary element/finite difference form of eqns (8.60) and (8.61) in order to avoid the problem of the quasisingularity of the Green function [4].

The radial field component defined by integral (8.60) can be computed numerically using Gaussian quadrature.

The approximation for current across the segment can be written as

$$I(z) = I_i f_i(z) + I_{i+1} f_{i+1}(z) \tag{8.73}$$

Therefore it follows

$$\frac{\partial I(z')}{\partial z}\Big|_{z=z_i} = \frac{I_{i+1} - I_i}{\Delta z} \tag{8.74}$$

and

$$E_\rho = \frac{1}{j4\pi\omega\varepsilon_0} \sum_{i=1}^{M} \frac{I_{i+1} - I_i}{\Delta z} \int_{z_i}^{z_{i+1}} \frac{\partial g_0(z, z', \rho)}{\partial \rho} dz' \tag{8.75}$$

Furthermore, the kernel can be approximated via central finite difference formula

$$\frac{\partial f(x, y)}{\partial x} = \frac{f(x + \Delta x, y) - f(x - \Delta x, y)}{2\Delta x} \tag{8.76}$$

The final formula for the radial electric field is then

$$E_\rho = \frac{1}{j4\pi\omega\varepsilon_0} \sum_{i=1}^{M} \frac{I_{i+1} - I_i}{\Delta\rho\Delta z} \int_{z_i}^{z_{i+1}} [g_0(z, z', \rho + \Delta\rho) - g_0(z, z', \rho)] dz' \tag{8.77}$$

where $\Delta\rho$ is the finite difference step.

Similarly, the axial field component is given by

$$E_z = -\frac{1}{j4\pi\omega\varepsilon_0} \left[\int_{-L}^{L} \frac{\partial I(z')}{\partial z'} \frac{\partial g_0(z, z', \rho)}{\partial z} dz' + k^2 \int_{-L}^{L} I(z') g_0(z, z', \rho) dz' \right] \tag{8.78}$$

Using linear interpolation for current over the segment it follows:

$$E_z = -\frac{1}{j4\pi\omega\varepsilon_0} \sum_{i=1}^{M} \left\{ \begin{array}{l} \dfrac{I_{i+1} - I_i}{\Delta z} \displaystyle\int_{-L}^{L} \dfrac{\partial g_0(z, z', \rho)}{\partial z} dz' \\ + k^2 \displaystyle\int_{-L}^{L} [I_i f_i(z') + I_{i+1} f_{i+1}(z')] g_0(z, z', \rho) dz' \end{array} \right\} \tag{8.79}$$

Approximating the kernel with finite differences, the final formula for the axial field component is given by

$$E_z = -\frac{1}{j4\pi\omega\varepsilon_0}\sum_{i=1}^{M}\frac{I_{i+1}-I_i}{\Delta z^2}\left\{\begin{array}{l}\int\limits_{-L}^{L}[g_0(z+\Delta z/2,z',\rho)-g_0(z-\Delta z/2,z',\rho)]\mathrm{d}z' \\ +k^2\int\limits_{-L}^{L} I(z')g_0(z,z',\rho)\mathrm{d}z'\end{array}\right\} \tag{8.80}$$

The integrals in expressions (8.79) and (8.80) can be numerically evaluated using Gaussian quadrature.

8.1.8 Numerical Results for Coated Wire

Figure 8.6 shows the real, imaginary, and absolute value of current distribution flowing along the perfectly conducting (PEC) monopole antenna of radius $a = 3.175$ mm with a dielectric coating of thickness $b - a = 3.175$ mm, and dielectric constant $\varepsilon_r = 3.2$ mounted on a PEC plane and excited at its base. Monopole length is $h = 0.125$ m and the wavelength of the excitation function ($V_T = 1$ V) is $\lambda = 0.5$ m. The related ρ-component of the electric field vs. normal distance from the antenna driving point is shown in Figure 8.7. Monopole has been modeled as a dipole of twice the length and energized at the centre by twice the voltage.

An important feature of the proposed method is its suitability for extension to more realistic cases such as a case of coated antenna radiating in the lossy medium.

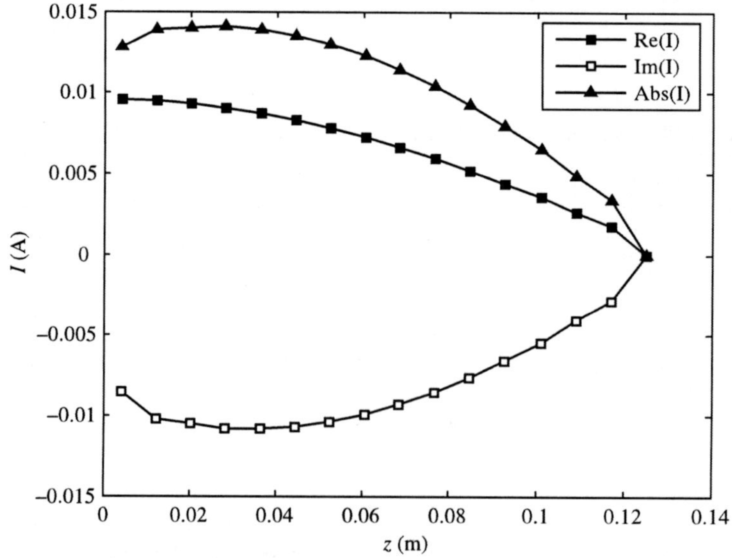

Figure 8.6 The current distribution along the monopole antenna.

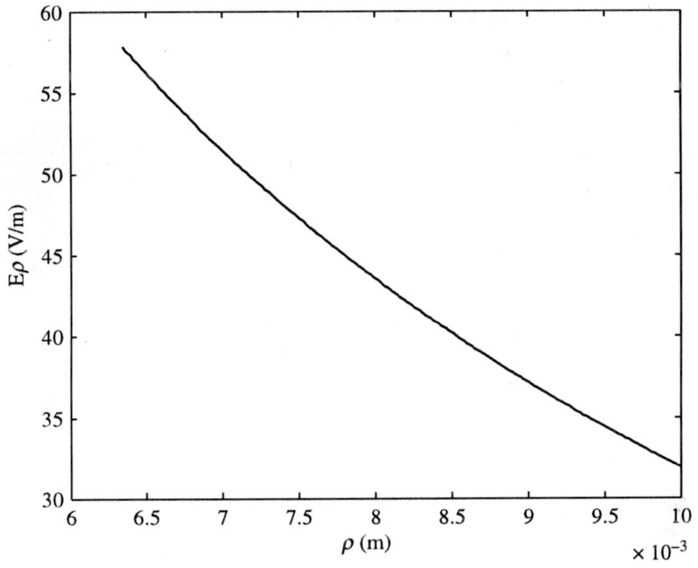

Figure 8.7 The ρ-component of the near electric field.

The coupling of the proposed thin-wire integral equation approach with the volume tensor integral equation or finite difference approach describing the surrounding space (mainly the human tissue) is also possible.

8.1.9 Thin Wire Loop Antenna

The analysis of a thin-wire loop antenna is very important in applications pertaining to direction-finding systems or UHF communications. Such antennas can also be used as probes for magnetic intensity measurements [18]. The analysis of such antenna can be simply performed by assuming constant current along the loop [19] or by solving the corresponding integral equation for the unknown current distribution along the wire [20–23]. Champagne II et al. in Ref. [20] proposed the use of curved segments within the moment method procedure applied to the set of equations for magnetic vector and electric scalar potential. Ronglin et al. [21] used parametric B-splines within the finite element technique applied to the electric field integral equation (EFIE) for curved wires [21, 22]. In contrast to these "subdomain techniques," L. Vegni et al. [23] presented a moment method analysis based on entire-domain sinusoidal basis functions applied to the integro-differential operator for thin-wire loop antenna.

A simple and highly efficient numerical procedure based on GB-IBEM is applied on the modeling of the circular loop antenna [24].

The time-harmonic behavior of a transmitting loop antenna presented in Figure 8.8 can be analyzed by solving the corresponding integro-differential equation. If the wire is assumed to be perfectly conducting, the tangential component of the total electric field on the wire vanishes.

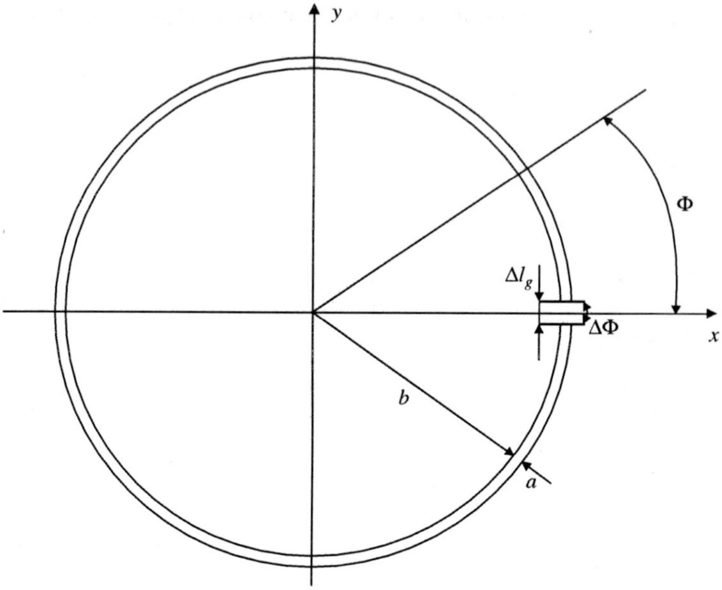

Figure 8.8 Thin wire loop antenna.

In a cylindrical coordinate system, it follows

$$E_\phi^{sct} = -E_\phi^{inc} \tag{8.81}$$

Performing some mathematical manipulations, the integro-differential equation is derived in the following form [18, 23, 24]:

$$E_\phi^{inc} = \frac{jZ_0}{4\pi}\int_{-\pi}^{\pi}\left\{kb\cos(\phi-\phi') + \frac{1}{kb}\frac{\partial^2}{\partial\phi^2}\right\}G(\phi-\phi')I(\phi')\mathrm{d}\phi' \tag{8.82}$$

where $I(\varphi)$ is the unknown current distribution, E_φ^{inc} is the tangential incident electric field, k is the free-space wave number, and $G(\varphi-\varphi')$ is the corresponding Green function given by

$$G(\phi-\phi') = \frac{e^{-jkR}}{R} \tag{8.83}$$

where the distance from the source to the observation point, respectively, is given by

$$R = b\sqrt{4\sin^2\left(\frac{\phi-\phi'}{2}\right) + \left(\frac{a}{b}\right)^2} \tag{8.84}$$

However, integro-differential eqn (8.82) is not convenient for numerical modeling due to the second-order differential operator appearing inside the integral equation kernel. This differential operator usually causes the well known quasisingularity problems [8, 9], so it is more convenient to derive an alternative version of eqn (8.82). Taking into account the symmetry property of the IE kernel,

$$\frac{\partial}{\partial \phi} G(\phi - \phi') = -\frac{\partial}{\partial \phi'} G(\phi - \phi') \tag{8.85}$$

and also performing integration by parts it follows

$$\begin{aligned} \int_{-\pi}^{\pi} I(\phi') \frac{\partial^2}{\partial \phi^2} G(\phi - \phi') \mathrm{d}\phi' &= -[I(\phi') \frac{\partial}{\partial \phi} G(\phi - \phi')]\Big|_{-\pi}^{\pi} \\ &+ \int_{-\pi}^{\pi} \frac{\partial I(\phi')}{\partial \phi'} \frac{\partial}{\partial \phi} G(\phi - \phi') \mathrm{d}\phi' \end{aligned} \tag{8.86}$$

In accordance to the circular geometry of the loop, the following condition must be imposed:

$$I(\phi = -\pi) = I(\phi = +\pi) \tag{8.87}$$

and the first term on the right-hand side of eqn (8.86) automatically disappears.

The resulting integral equation for loop antenna is then given by the expression

$$\begin{aligned} E_\phi^{inc} &= j\frac{Z_0}{4\pi} \int_{-\pi}^{\pi} kb \cos(\phi - \phi') G(\phi - \phi') I(\phi') \mathrm{d}\phi' \\ &+ j\frac{Z_0}{4\pi} \int_{-\pi}^{\pi} \frac{1}{kb} \frac{\partial I(\phi')}{\partial \phi'} \frac{\partial}{\partial \phi} G(\phi - \phi') \mathrm{d}\phi' \end{aligned} \tag{8.88}$$

It can be observed that only the first derivative of the Green function appears in the second term.

8.1.10 Boundary Element Solution of Loop Antenna Integral Equation

The boundary element approach sufficiently handles the quasisingularity problem providing fast convergence and stable numerical results. The integral eqn (8.88) can, for convenience, be rewritten in the form of a linear operator equation,

$$L(I) = E_\phi^{inc} \tag{8.89}$$

where $L(I)$ represents the left-hand side of eqn (8.88).

Applying the weighted residual approach to eqn (8.89) that gives

$$\int_{-\pi}^{\pi} [L(I) - E_{\phi}^{inc}] W_j b \mathrm{d}\phi = 0 \tag{8.90}$$

which can also be written in the form

$$j\frac{Z_0}{4\pi}\int_{-\pi}^{\pi} W_j \int_{-\pi}^{\pi} kb\cos(\phi-\phi')G(\phi-\phi')I(\phi')\mathrm{d}\phi'\mathrm{d}\phi + j\frac{Z_0}{4\pi}\int_{-\pi}^{\pi} W_j\left(\frac{\partial}{\partial\phi}\int_{-\pi}^{\pi}\frac{1}{kb}\frac{\partial I(\phi')}{\partial\phi'}G(\phi-\phi')\mathrm{d}\phi'\right)\mathrm{d}\phi = \int_{-\pi}^{\pi} E_{\phi}^{inc} W_j \mathrm{d}\phi \tag{8.91}$$

Now, in order to remove the derivative from the $G(\varphi\text{-}\varphi')$ to the test functions W_j the second term at the left-hand side should be modified.

Applying the integration by parts:

$$\int_{-\pi}^{\pi} W_j\left(\frac{\partial}{\partial\phi}\int_{-\pi}^{\pi}\frac{\partial I(\phi')}{\partial\phi'}G(\phi-\phi')\mathrm{d}\phi'\right)\mathrm{d}\phi = \left(W_j\int_{-\pi}^{\pi}\frac{\partial I(\phi')}{\partial\phi'}G(\phi-\phi')\mathrm{d}\phi'\right)\Bigg|_{-\pi}^{\pi} - \int_{-\pi}^{\pi}\frac{\partial W_j}{\partial\phi}\int_{-\pi}^{\pi}\frac{\partial I(\phi')}{\partial\phi'}G(\phi-\phi')\mathrm{d}\phi'\mathrm{d}\phi \tag{8.92}$$

the first term on the right-hand side vanishes due to the boundary condition (8.87).

Finally, the weak formulation of the integro-differential eqn (8.88) is then given in the form:

$$j\frac{Z_0}{4\pi}\int_{-\pi}^{\pi} W_j \int_{-\pi}^{\pi} kb\cos(\phi-\phi')G(\phi-\phi')I(\phi')\mathrm{d}\phi'\mathrm{d}\phi + j\frac{Z_0}{4\pi}\int_{-\pi}^{\pi}\left(\frac{\partial W_j}{\partial\phi}\int_{-\pi}^{\pi}\frac{1}{kb}\frac{\partial I(\phi')}{\partial\phi'}G(\phi-\phi')\mathrm{d}\phi'\right)\mathrm{d}\phi = \int_{-\pi}^{\pi} E_{\phi}^{inc} W_j \mathrm{d}\phi \tag{8.93}$$

Integral expression (8.93) is now convenient for the implementation of the Galerkin-Bubnov indirect boundary element method (GB-IBEM).

The unknown current over ith element can be expressed as follows:

$$I_i(\phi') = \sum_{k=1}^{n_e} f_{ik}(\phi') I_{ik} \tag{8.94}$$

where $f_{ik}(\varphi')$ denotes the kth shape function over ith finite element, while I_{ik} is the unknown value of current at the ith element.

It is suitable to rewrite the former expression into matrix notation as

$$I(\phi') = \{f\}^T\{I\} \tag{8.95}$$

where $\{f\}$ denotes the shape function vector and $\{I\}$ denotes the unknown current vector.

The test functions W_j are chosen to be the same as the shape functions,

$$W_j = f_j(\phi) \tag{8.96}$$

in accordance to the Galerkin-Bubnov procedure.

Applying the BEM discretization, integral relationship (8.93) can be written in the matrix form as

$$[Z]\{I\} = \{V\} \tag{8.97}$$

where the mutual impedance is

$$\begin{aligned}[Z]_{ji} &= j\frac{Z_0}{4\pi}\int_{\Delta\phi_j}\int_{\Delta\phi_i} kb\cos(\phi-\phi')G(\phi-\phi')\{f\}_j\{f'\}_i \mathrm{d}\phi'\mathrm{d}\phi \\ &+ j\frac{Z_0}{4\pi}\int_{\Delta\phi_j}\int_{\Delta\phi_i}\frac{1}{kb}\{D\}_j\{D'\}_i^T G(\phi-\phi')\mathrm{d}\phi')\,\mathrm{d}\phi\end{aligned} \tag{8.98}$$

where matrices $\{D\}$ and $\{D'\}$ contain the derivatives of the shape functions $f(\varphi)$ and $f(\varphi')$.

Assembling the contributions from each loop segment, the global impedance-matrix is obtained, in which nonself terms present mutual impedance of the ith source and jth observation element and self terms are related to the self-impedance contributions.

The voltage vector is given by

$$\{V\}_j = \int_{\Delta\phi_j} E_\phi^{inc}\{f\}_j \mathrm{d}\phi \tag{8.99}$$

If the receiving mode is analyzed, the incident field is given in the form of a plane wave and in the case of transmitting mode, the incident field is represented by the voltage generator in the form

$$E_\phi^{inc} = \frac{V_T}{\Delta l_g} = \frac{V_T}{b\Delta\phi_g} \tag{8.100}$$

where V_T is the feed voltage, Δl_g is the length of the feed-gap, and $\Delta\varphi_g$ denotes the angle subtended over feed-gap area.

Once the current distribution along the wire is obtained, other antenna parameters of interest (near and far field, radiation pattern, input impedance) can be calculated as indicated in Ref. [24].

8.1.11 Numerical Results for a Loop Antenna

The computational example is related to the circular loop with radius $a = 0.0027\lambda$ and loop radius $b = 0.0637\lambda$ at 3 GHz, insulated in free space. Figures 8.9 and 8.10 show the current distribution along the loop wire antenna excited by a delta-function voltage generator at $\varphi = 0°$ ($V_T = 1\,\text{V}$). All calculations are performed featuring

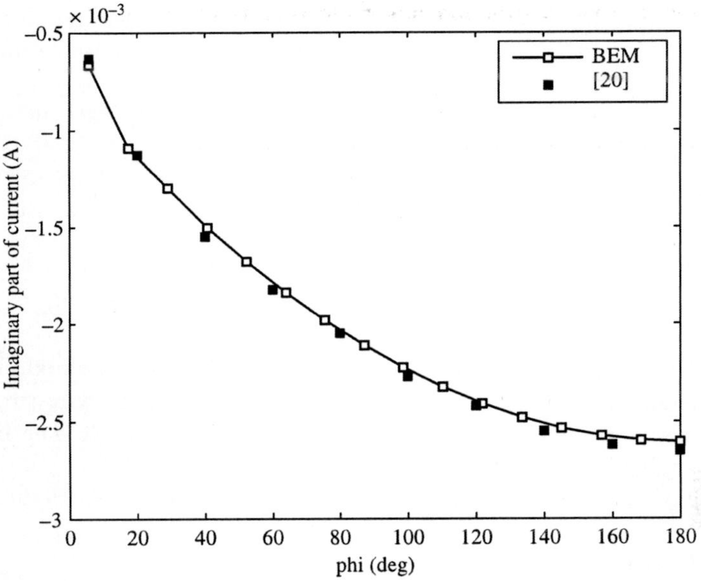

Figure 8.10 Imaginary part of current distribution along the loop with $a = 0.0027\lambda$, $b = 0.0637\lambda$, $f = 3\,\text{GHz}$. (Poljak, D., Finite Element Integral Equation Modelling of a Thin Wire Loop Antenna, *Communications in Numerical Methods in Engineering*, 14, pp 347–354, 1998).

Analysis of linear arrays by means of numerical techniques started with classical paper by Harrington [25]. It was followed by the work of many prominent researchers [5, 6, 26–32].

Straight thin-wire arrays, Figure 8.11, have been successfully treated by means of the Galerkin Bubnov scheme of the indirect boundary element method

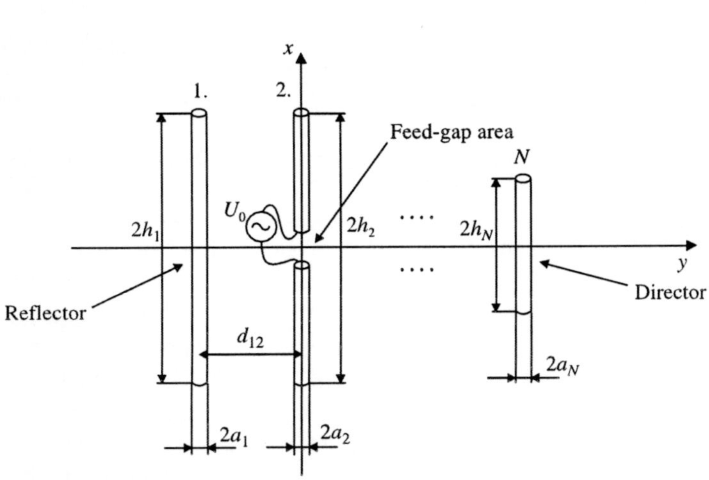

Figure 8.11 Geometry of the straight wire antenna array.

(GB-IBEM) using linear elements [2]. However, some antenna types which are very important in modern mobile communication systems, like for example helix antenna, are not convenient for numerical treatment via linear boundary elements due to curved geometries. Polynomial functions cannot approximate quantities with very rapid variations or singularities satisfactorily.

Alternative element types with more appropriate approximating functions can overcome most of the problems inherent in the simple elements. The use of curved elements can often alleviate the problems encountered in geometric modeling by shaping the elements to fit the real shapes [33].

The use of curved segments has been proposed for the MoM [20, 21] and the finite element method (FEM) [33].

The first part of this section deals with a simple and efficient Galerkin boundary element integral equation technique for the analysis of arbitrary wire structures in free space based on the isoparametric boundary element concept. The method is applied to straight wire antenna arrays.

An extension to the case of the array of straight wires, Figure 7.2, is straightforward and results in a set of coupled Pocklington integro-differential equations [2, 4]:

$$E_{xj}^{inc} = -\frac{1}{j4\pi\omega\varepsilon_0} \int_{-L}^{L} \left(\frac{\mathrm{d}^2}{\mathrm{d}x^2} + k^2\right) \sum_{j=1}^{N_w} \int_{-L_i}^{L_i} I_i(x') g_{ji}(x, x') \mathrm{d}x'; \; j = 1, 2, \ldots, N_w \quad (8.101)$$

where N_w is the total number of thin wire elements, $I_i(x')$ is the unknown current distribution along ith wire of an array to be determined, E_{xj}^{inc} is the known incident field tangential to the jth wire surface, k is the free space propagation constant, and g_{ji} is the corresponding Green function,

$$g_{ji}(x, x') = \frac{e^{-jkR_{ji}}}{R_{ji}} \quad (8.102)$$

where R_{ji} is the distance from the source point to the observation point given by

$$R_{ji} = \sqrt{(x - x')^2 + a_i^2}, j = i; \quad R_{ji} = \sqrt{(x - x')^2 + d_{ji}^2}, j \neq i \quad (8.103)$$

a_i is the radius of the ith wire and d_{ji} is the distance between ith and jth wire.

Once the currents along the wire array have been obtained, other parameters of the antenna array such as radiation pattern, input impedance, or radiation power can be obtained [2].

The near-field analysis of wire antennas is important due to numerous electromagnetic compatibility (EMC) applications require the knowledge of the field behavior in the proximity of a radiating structure.

The electric field tangential to the array is defined by the following equation [1, 2, 4]:

$$E_x(x,y) = \frac{1}{j4\pi\omega\varepsilon_0}\sum_{i=1}^{N}\left[\begin{array}{l}\left(I_i(x')\dfrac{\partial g_{ji}(x,x',y)}{\partial x'}\right)\Bigg|_{-L}^{L} - \displaystyle\int_{-L}^{L}\dfrac{\partial g_{ji}(x,x',y)}{\partial x'}\dfrac{\partial I_i(x')}{\partial x'}\mathrm{d}x' \\ +k^2\displaystyle\int_{-L}^{L} I_i(x')g_{ji}(x,x',y)\mathrm{d}x'\end{array}\right] \tag{8.104}$$

while the input impedance can be determined from the functional (8.13) as presented in Refs. [2, 4].

8.1.13 Boundary Element Analysis of Horizontal Antenna Array

The boundary element modeling of the integral equation set (8.101) by using linear elements has been documented in details in Ref. [3], while the isoparametric solution. This section outlines both the approaches.

8.1.13.1 Linear BEM Solution The Galerkin boundary element procedure starts by applying the standard representation of the unknown current along the ith wire,

$$I_n(x') = \sum_{n=1}^{N_g} I_{ni}f_{ni}(x') \tag{8.105}$$

where $f_{ni}(x)$ denotes the linear elements shape functions, I_{ni} stands for the unknown coefficients of the solution, and N_g denotes the total number of basis functions.

In addition, the Galerkin formulation of the system of integro-differential equations (8.101) for N_w wires is given by

$$\sum_{n=1}^{N_g}\sum_{i=1}^{N_w}\left[-\int_{-L_m}^{L_m}\int_{-L_n}^{L_n}\frac{\mathrm{d}f_{jm}(x)}{\mathrm{d}x}\frac{\mathrm{d}f_{in}(x')}{\mathrm{d}x'}g_{ji}(x,x')\mathrm{d}x'\mathrm{d}x + k^2\int_{-L_m}^{L_m}\int_{-L_n}^{L_n} f_{jm}(x)f_{in}(x')g_{ji}(x,x')\mathrm{d}x'\mathrm{d}x\right]I_{ni}$$
$$= -j4\pi\omega\varepsilon\int_{-L_m}^{L_m} E_{xj}^{inc}(x)f_{jm}(x)\mathrm{d}x;\quad m=1,2,\ldots,N_g;\quad j=1,2,\ldots,N_w \tag{8.106}$$

Expression (8.106) is convenient for modeling the straight wire arrays and can also be applied to wire junctions if the structure of interest is composed of straight wire elements.

The complete boundary element solution procedure with linear elements can be found in Ref. [3].

8.1.13.2 Isoparametric Boundary Element Solution The unknown current along the nth wire is expressed in terms of parametric functions $f(\xi)$,

$$I_n(\xi) = \sum_{n=1}^{M} I_{ni} f_{ni}(\xi) \tag{8.107}$$

Performing the boundary element discretization of the wire array leads to the matrix equation of the form

$$\sum_{k=1}^{M} [Z]_{lk}\{I\}_k = \{V\}_l;\; l = 1, 2, \ldots, M \tag{8.108}$$

where M is the total number of elements along the actual multiple wire configuration and $[Z]_{lk}$ is the interaction matrix representing the mutual impedance between each segment on the ith (source) wire to every segment on the jth (observation) wire,

$$\begin{aligned}[a]_{lk}^{e} = &- \int_{-1}^{1}\int_{-1}^{1} \{D\}_l\{D'\}_k^T g_{ji}(\xi, \xi')\mathrm{d}\xi'\mathrm{d}\xi \\ &+ k^2 \int_{-1}^{1}\int_{-1}^{1} \{f\}_l\{f'\}_k^T g_{ji}(\xi, \xi')\left(\frac{\Delta x}{2}\right)^2 \mathrm{d}\xi'\mathrm{d}\xi\end{aligned} \tag{8.109}$$

Vectors $\{f\}$ and $\{f'\}$ contain shape functions $f_n(\xi)$ and $f_n(\xi')$, while $\{D\}$ and $\{D'\}$ contain their derivatives, and Δx is the segment length.

The right-hand side vector $\{V\}_l$ represents the voltage along the lth segment and it is given as follows:

$$\{Z\}_l = -j4\pi\omega\varepsilon_0 \int_{-1}^{1} E_x^{inc}(\xi)\{f\}_l \frac{\Delta x}{2}\mathrm{d}\xi \tag{8.110}$$

In the transmitting mode (also called *the antenna mode*) right-hand side vector is different from zero only in the feed-gap area (Fig. 8.11) and the incident field can be written as

$$E_x^{inc} = \frac{V_g}{\Delta l_g} \tag{8.111}$$

where V_g is the feed voltage and Δl_g is the feed-gap width.

8.1.14 Radiated Electric Field of the Wire Array

Once the current distribution along the wires is determined, the tangential component of the electric field radiated by the straight wire array can be obtained using the following relation:

$$E_y[x(\xi), y(\xi)] = \frac{1}{j4\pi\omega\varepsilon_0} \sum_{i=1}^{N_w} \sum_{k=1}^{n_i} \left[\left(I[x'(\xi')] \frac{\partial G}{\partial \xi'} \right) \Bigg|_{\xi=-1}^{\xi=1} - \int_{z_{1k}}^{z_{2k}} \frac{\partial G}{\partial \xi'} \frac{\partial I[x'(\xi')]}{\partial \xi'} \mathrm{d}\xi' \right.$$
$$\left. + \mathrm{k}^2 \int_{z_{1k}}^{z_{2k}} I[x'(\xi')] G[x(\xi), y(\xi), x'(\xi')] \mathrm{d}\xi' \right] \tag{8.112}$$

where N_w is the number of wires, M is the number of elements of the ith wire representing the current along the kth element using linear interpolation

$$I[x'(\xi')] = I_{1k} f_{1k}(\xi') + I_{2k} f_{2k}(\xi') \tag{8.113}$$

8.1.15 Numerical Results for Horizontal Wire Array

A typical straight wire array in horizontal arrangement widely used in many applications is the Yagi-Uda array. Numerical results for current distribution and the electric field pattern are computed for the Yagi-Uda antenna array with radius $a = 0.0025\lambda$, reflector element length 0.479λ, fed element length 0.453λ, and director element length 0.451λ. The number of segments on all wires is 31 and generator voltage is 1 V.

Yagi-Uda arrays contain three types of wires: the fed element which contains the voltage generator, the director and the reflector elements that shape the beam in the desired direction. The distance between all wires is 0.25λ. The current distribution induced along the reflector, fed-element, and director is shown in Figures 8.12–8.14, respectively.

The BEM numerical results are in agreement with those obtained via different numerical techniques available from Refs. [1, 6].

The radiated electric field generated by this array is given in the form of the far-field pattern as shown in Figure 8.15.

The boundary element analysis presented so far can be easily extended to more complex wire configurations embedded in nonhomogeneous media.

8.1.16 Boundary Element Analysis of Vertical Antenna Array: Modeling of Radio Base Station Antennas

The radiation from GSM and UMTS base station antennas has caused rapidly growing public concern regarding potentially adverse effects on human health [34]. Consequently, the study of the intensity and form of radiated electromagnetic energy is

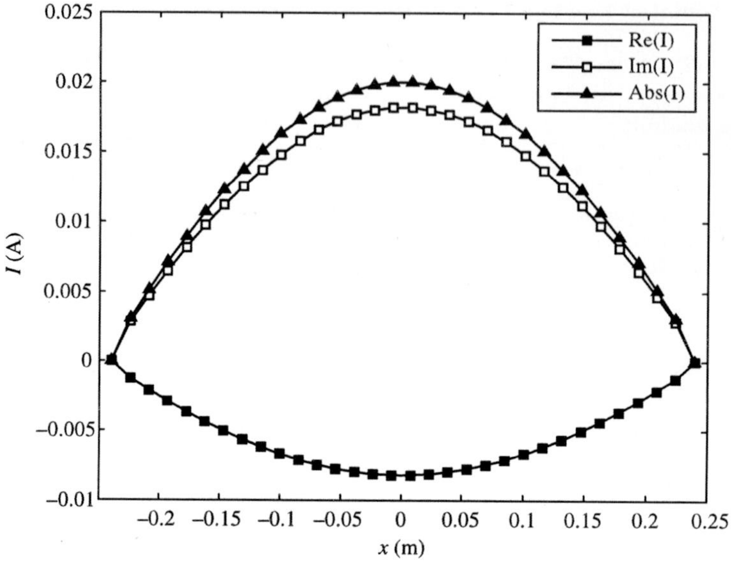

Figure 8.12 Current distribution along the reflector element.

of great interest for analyses devoted to biological effects of electromagnetic fields [35–38]. Analytical techniques for the assessment of radiation from base station antenna systems have been discussed in Refs. [35–37]. Numerical approach to this subject has been presented in Refs. [39, 40].

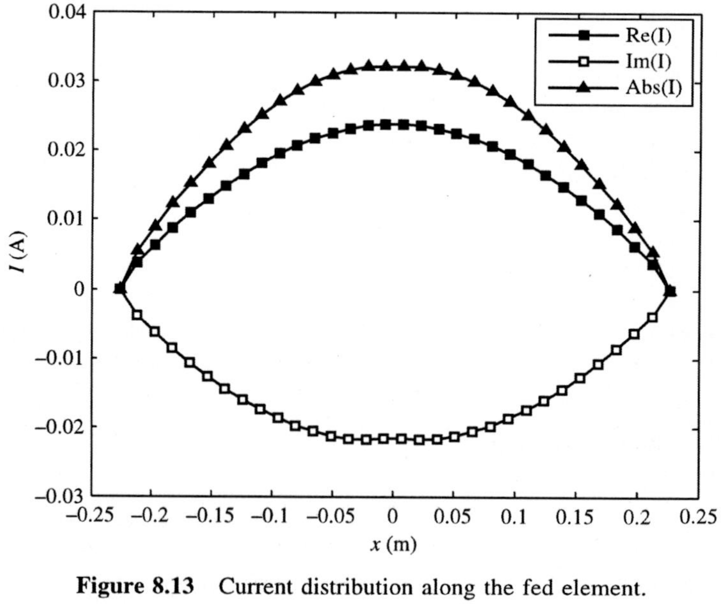

Figure 8.13 Current distribution along the fed element.

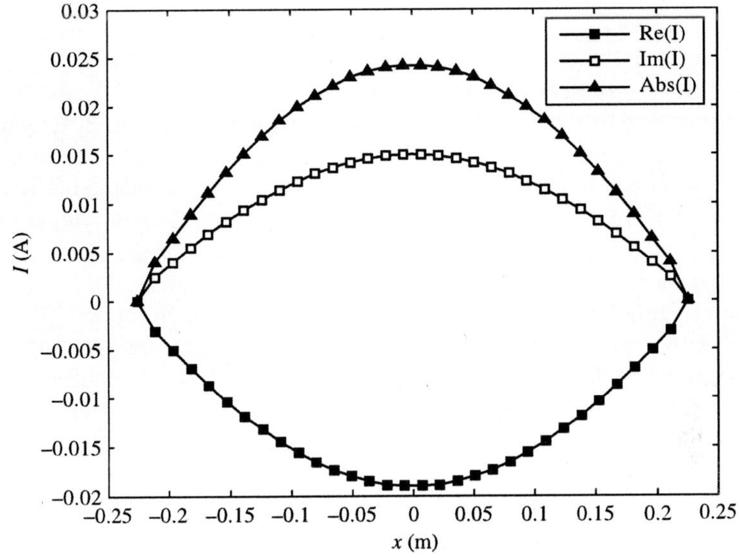

Figure 8.14 Current distribution along the director element.

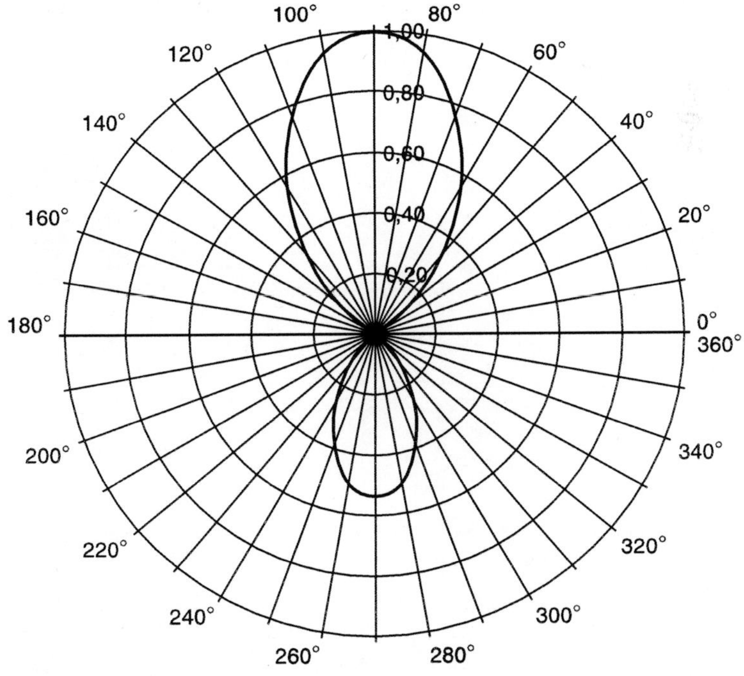

Figure 8.15 Farfield pattern.

An efficient application of boundary element modeling to analyze base station antenna systems represented by vertical array of straight wire antennas in front of a perfectly conducting (PEC) ground plane reflector has been documented in Ref. [40]. In Ref. [40], BEM results have been compared to the results obtained via numerical electromagnetic code (NEC) [41].

The formulation is based on a set of coupled Pocklington integro-differential equations for induced currents along the wires. This set of equations is numerically solved via the Galerkin-Bubnov scheme of the Indirect boundary element method (GB-IBEM).

Once the current distribution along the array has been determined, the radiated electric field can be computed. The computational results for the electric near-field and far-field have been obtained and compared with results calculated via well-known and widely adopted analytical solutions for the far field [35–37].

Figure 8.16 shows a base station antenna system represented by a vertical antenna array of M dipoles of half length L_n, distance between centers of adjacent dipoles d_c, and horizontal distance between antennas and a perfectly conducting (PEC) ground plane reflector x_{dis}.

The set of the electric field integral equations (EFIEs) for straight wire arrays can be obtained as an extension of the single wire problem. Geometry of a straight thin wire antenna of length L and radius a insulated in free space is shown in Figure 8.17.

The current distribution along the wire can be obtained as the solution of the frequency domain Pocklington integro-differential equation. This equation can be derived from Maxwell's equations by expressing the time harmonic electric field $\vec{E}$ in terms of the magnetic vector potential $\vec{A}$ and the electric scalar potential φ,

$$\vec{E} = -\nabla\varphi - j\omega\vec{A} \tag{8.114}$$

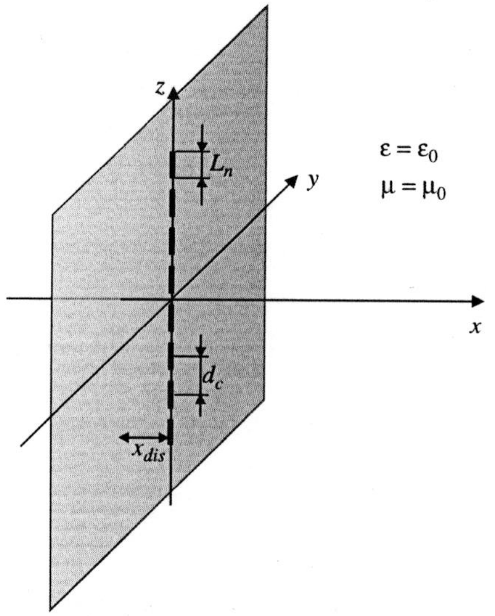

Figure 8.16 Antenna array in front of reflector.

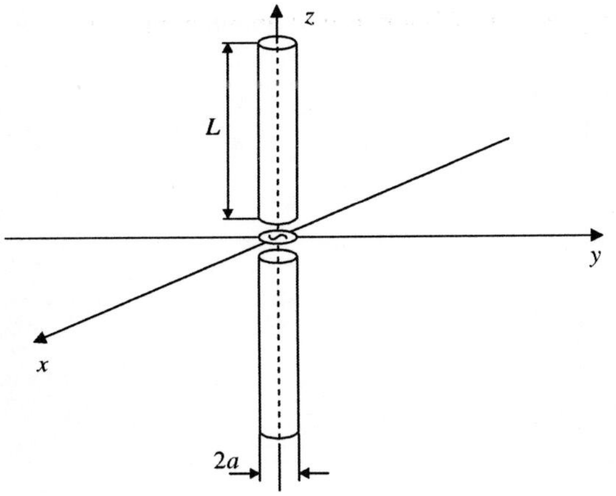

Figure 8.17 Dipole antenna insulated in free space.

The magnetic vector and electric scalar potential are coupled through the well-known Lorentz Gauge,

$$\nabla\vec{A} = -j\,\omega\,\mu\,\varepsilon\,\varphi \tag{8.115}$$

As only axial component of A along the cylinder exists in this case, combining eqns (8.114) and (8.115) yields

$$E_z = \frac{1}{j\omega\mu\varepsilon_0}\left[\frac{\partial^2 A_x}{\partial z^2} + k^2 A_z\right] \tag{8.116}$$

where ε_0 is the permittivity of the free space, ω is the applied angular frequency.

By using the electromagnetic theory, the magnetic vector potential A_x can be represented by an integral of the axial current $I(z)$ flowing along the cylinder, that is,

$$A_z = -\frac{\mu}{4\pi}\int_{-L/2}^{L/2} I(z')\frac{e^{-jkR}}{R}\mathrm{d}z' \tag{8.117}$$

where $I(z)$ is the axial current distribution, R is the distance from the source point to the observation point, and k denotes the wave number of the free space.

Finally, the condition at the air-conductor interface for the tangential electric field that vanishes on the perfectly conducting wire surface can be written as

$$E_z^{inc} + E_z^{sct} = 0 \tag{8.118}$$

where E_x^{inc} is the incident field and E_x^{sct} is the scattered field due to the presence of the perfectly conducting metallic surface.

The scattered electric field close to the antenna surface is then given by

$$E_z^{sct} = \frac{1}{j4\pi\omega\varepsilon_0} \int_{-L/2}^{L/2} \left(\frac{\partial^2}{\partial z^2} + k^2 \right) g_0(z, z') I(z') dz' \tag{8.119}$$

Combining relations (8.118) and (8.119) gives the Pocklington integro-differential equation for the current distribution along the single straight wire antenna insulated in free space,

$$E_z^{inc} = -\frac{1}{j4\pi\omega\varepsilon_0} \int_{-L/2}^{L/2} \left(\frac{\partial^2}{\partial z^2} + k^2 \right) g_0(z, z') I(z') dz' \tag{8.120}$$

where $g_0(z, z')$ is the free-space Green's function given by

$$g_0(z, z') = \frac{e^{-jkR}}{R} \tag{8.121}$$

The extension to the case of an array of straight wires as shown in Figure 8.16 is straightforward and results in a set of coupled Pocklington integro-differential equations for the currents induced along the antenna elements,

$$E_{zm}^{inc} = \frac{1}{j4\pi\omega\varepsilon_0} \sum_{n=1}^{M} \left[\frac{\partial^2}{\partial z^2} + k^2 \right] \int_{-L_n/2}^{+L_n/2} I_n(z') G_{mn}(z, z') dz'_n; \quad m = 1, 2, \ldots, M \tag{8.122}$$

where $I_n(z')$ is the current distribution along the nth wire, E_z^{inc} is the known incident field tangential to the jth wire surface, and $G_{mn}(z, z')$ is the Green's function given by

$$G_{mn}(z, z') = g_{0mn}(z, z') - g_{imn}(z, z') \tag{8.123}$$

where $g_{0mn}(z, z')$ is the free-space Green's function,

$$g_{0mn}(z, z') = \frac{e^{-jkR_{mn}}}{R_{mn}} \tag{8.124}$$

while $g_{imn}(z, z')$ is the Green's function due to the image wires,

$$g_{imn}(z, z') = \frac{e^{-jkR_{mn}^*}}{R_{mn}^*} \tag{8.125}$$

The three components of the electric field radiated by the base station antenna system of Figure 8.16 can be readily obtained by applying eqns (8.14), (8.15), and (8.17) to the geometry of vertical array. These components are given by

$$E_x = \frac{1}{j4\pi\omega\varepsilon_0}\sum_{n=1}^{M}\frac{\partial^2}{\partial x\partial z}\int_{-L_n/2}^{+L_n/2} G_{mn}(x,z')I_n(z')\mathrm{d}z'$$

$$= \frac{1}{j4\pi\omega\varepsilon_0}\sum_{n=1}^{M}\int_{-L_n/2}^{+L_n/2}\frac{\partial I_n(z')}{\partial z'}\frac{\partial G_{mn}(x,z')}{\partial x}\mathrm{d}z' \tag{8.126}$$

$$E_y = \frac{1}{j4\pi\omega\varepsilon_0}\sum_{n=1}^{M}\frac{\partial^2}{\partial x\partial z}\int_{-L_n/2}^{+L_n/2} G_{mn}(y,z')I_n(z')\mathrm{d}z'$$

$$= \frac{1}{j4\pi\omega\varepsilon_0}\sum_{n=1}^{M}\int_{-L_n/2}^{+L_n/2}\frac{\partial I_n(z')}{\partial z'}\frac{\partial G_{mn}(y,z')}{\partial y}\mathrm{d}z' \tag{8.127}$$

$$E_z = \frac{1}{j4\pi\omega\varepsilon_0}\sum_{n=1}^{M}\left[\frac{\partial^2}{\partial z^2}+k_1^2\right]\int_{-L_n/2}^{+L_n/2} G_{mn}(z,z')I_n(z')\mathrm{d}z'$$

$$= \frac{1}{j4\pi\omega\varepsilon_0}\sum_{n=1}^{M}\left[\begin{array}{l}\displaystyle\int_{-L_n/2}^{+L_n/2}\frac{\partial G_{mn}(z,z')}{\partial z}\frac{\partial I_n(z')}{\partial z'}\mathrm{d}z' \\ \displaystyle +k^2\int_{-L_n/2}^{+L_n/2} I_n(z')G_{mn}(z,z')\mathrm{d}z'\end{array}\right] \tag{8.128}$$

Solving eqn (8.122) and obtaining the equivalent current distributions along the wires, the radiated electric field can be computed.

8.1.17 Numerical Procedures for Vertical Array

Through the GB-IBEM procedure, the set of integro-differential equations (8.122) is transformed into a system of linear algebraic equations.

Applying the weighted residual approach to eqn (8.122) yields

$$\int_{-L_m/2}^{+L_m/2}\left\{\frac{1}{j4\pi\omega\varepsilon_0}\sum_{n=1}^{M}\left[\frac{\partial^2}{\partial z^2}+k^2\right]\int_{-L_n/2}^{+L_n/2} I_n(z')G_{mn}(z,z')\mathrm{d}z' - E_{zm}^{inc}\right\}W_{jm}\mathrm{d}z = 0$$

$$m = 1,2,\ldots,M; j = 1,2,\ldots,N_g \tag{8.129}$$

where N_g denotes the total number of basis functions.

Using the second-order Lagrange polynomials, that is, discretizing the array geometry into quadratic boundary elements, the current distribution along an ith wire segment (boundary element) is given by

$$
\begin{aligned}
I_{ni}(z') &= I_{1i}^{n} f_1(z') + I_{2i}^{n} f_2(z') + I_{3i}^{n} f_3(z') \\
&= I_{1i}^{n} \frac{z'-z_2}{z_1-z_2}\frac{z-z_3}{z_1-z_3} + I_{2i}^{n} \frac{z'-z_1}{z_2-z_1}\frac{z'-z_3}{z_2-z_3} + I_{3i}^{n} \frac{z'-z_1}{z_3-z_2}\frac{z'-z_2}{z_3-z_2}
\end{aligned}
\tag{8.130}
$$

while the current derivative is then

$$
\frac{\partial I_{ni}(z')}{\partial z'} = \frac{2}{\Delta l^2}\left\{ \begin{array}{l} 2(I_{1i}^{n} - I_{2i}^{n} + I_{3i}^{n})z' \\ -[(z_3+z_2)I_{1i}^{n} - (z_3+z_1)I_{2i}^{n} + (z_1+z_2)I_{3i}^{n}] \end{array} \right\}
\tag{8.131}
$$

where $\Delta l = z_3 - z_1$ denotes the segment length.

Now, deriving the weak Galerkin-Bubnov formulation of expressions (8.122) and (8.129), respectively, and taking into account expansions (8.130) and (8.131) expression (8.129) becomes

$$
\begin{aligned}
&\sum_{n=1}^{M}\sum_{i=1}^{N_n}\left[\begin{array}{l} -\int\limits_{-L_m/2}^{+L_m/2}\int\limits_{-L_n/2}^{+L_n/2} \frac{df_{jm}(z)}{dx}\frac{df_{jn}(z')}{dx'} G_{mn}(z,z')dz'dz \\ +k^2 \int\limits_{-L_m/2}^{+L_m/2}\int\limits_{-L_n/2}^{+L_n/2} f_{jm}(z) f_{in}(z') G_{mn}(z,z')dz'dz \end{array} \right] I_{ni} \\
&= -j4\pi\omega\varepsilon \int\limits_{-L_m/2}^{+L_m/2} E_{zj}^{inc}(z) f_{jm}(z)dz; \quad m = 1,2,\ldots,M; \quad j = 1,2,\ldots,N_m
\end{aligned}
\tag{8.132}
$$

where N_n is number of elements of nth antenna and N_m is number of elements of mth antenna.

Expression (8.132) can also be, for convenience, written in the matrix form

$$
\sum_{n=1}^{M}\sum_{i=1}^{N_n} [Z]_{ji}^{e}\{I_n\}_i^{e} = \{V_m\}_j^{e}, \quad m = 1,2,\ldots M; \quad j = 1,2,\ldots,N_m
\tag{8.133}
$$

$[Z]_{ji}$ is the mutual impedance matrix for jth observed boundary element on mth antenna and ith source boundary element on nth antenna,

$$
\begin{aligned}
[Z]_{ji}^{e} = &-\int\limits_{\Delta l_j}\int\limits_{\Delta l_i} \{D\}_j\{D'\}_i G_{mn}(z,z')dz'dz \\
&+k^2 \int\limits_{\Delta l_j}\int\limits_{\Delta l_i} \{f\}_j\{f'\}_i G_{mn}(z,z')dz'dz
\end{aligned}
\tag{8.134}
$$

Vectors $\{f\}$ and $\{f'\}$ contain shape functions $f_k(z)$ and $f_k(z')$, while vectors $\{D\}$ and $\{D'\}$ represent their derivatives. Vector $\{I\}_i$ contains the solution for current distribution in global nodes and $\{V_m\}_j$ is the local voltage vector given by

$$\{V_m\}_{ji}^e = -j4\pi\omega\varepsilon_0 \int_{\Delta l_j} E_{zm}^{inc}\{f\}_j \mathrm{d}z \tag{8.135}$$

where index e denotes that operations are performed over boundary element. Δl_i represents ith element length (source element) and Δl_j represents jth element length (observation element).

The voltage source expressed in the form of incident electric field is given by

$$E_{zm}^{inc} = \frac{V_g}{\Delta l_g} \tag{8.136}$$

where V_g is the impressed voltage across the feed-gap of width Δl_g.

The radiated field is also calculated using the boundary element formalism as well. The expressions for the field components are as follows:

$$E_x = \frac{1}{j4\pi\omega\varepsilon_0}\sum_{n=1}^{M}\sum_{i=1}^{N_n}\left[\begin{array}{l} \frac{4}{\Delta l_i^2}\left(I_{1i}^n - I_{2i}^n + I_{3i}^n\right)\int\limits_{-\Delta l_i/2}^{+\Delta l_i/2} z'\frac{\partial G_{mn}(x,z')}{\partial x}\mathrm{d}z' \\ -\frac{2}{\Delta l_i^2}\left[(z_3+z_2)I_{1i}^n - (z_3+z_1)I_{2i}^n + (z_1+z_2)I_{3i}^n\right]\int\limits_{-\Delta l_i/2}^{+\Delta l_i/2}\frac{\partial G_{mn}(x,z')}{\partial x}\mathrm{d}z' \end{array}\right] \tag{8.137}$$

$$E_y = \frac{1}{j4\pi\omega\varepsilon_0}\sum_{n=1}^{M}\sum_{i=1}^{N_n}\left[\begin{array}{l} \frac{4}{\Delta l_i^2}\left(I_{1i}^n - I_{2i}^n + I_{3i}^n\right)\int\limits_{-\Delta l_i/2}^{+\Delta l_i/2} z'\frac{\partial G_{mn}(x,z')}{\partial y}\mathrm{d}z' \\ -\frac{2}{\Delta l_i^2}\left[(z_3+z_2)I_{1i}^n - (z_3+z_1)I_{2i}^n + (z_1+z_2)I_{3i}^n\right]\int\limits_{-\Delta l_i/2}^{+\Delta l_i/2}\frac{\partial G_{mn}(y,z')}{\partial y}\mathrm{d}z' \end{array}\right] \tag{8.138}$$

$$E_z = \frac{1}{j4\pi\omega\varepsilon_0}\sum_{n=1}^{M}\sum_{i=1}^{N_n}\left[\begin{array}{l} \frac{4}{\Delta l_i^2}\left(I_{1i}^n - I_{2i}^n + I_{3i}^n\right)\int\limits_{-\Delta l_i/2}^{+\Delta l_i/2} z'\frac{\partial G_{mn}(z,z')}{\partial z}\mathrm{d}z' \\ +\frac{2}{\Delta l_i^2}\left[(z_3+z_2)I_{1i}^n - (z_3+z_1)I_{2i}^n + (z_1+z_2)I_{3i}^n\right]\int\limits_{-\Delta l_i/2}^{+\Delta l_i/2}\frac{\partial G_{mn}(z,z')}{\partial z}\mathrm{d}z' \\ +k^2\left(I_{1i}^n\frac{z'-z_2}{z_1-z_2}\frac{z-z_3}{z_1-z_3} + I_{2i}^n\frac{z'-z_1}{z_2-z_1}\frac{z'-z_3}{z_2-z_3} + I_{3i}^n\frac{z'-z_1}{z_3-z_2}\frac{z'-z_2}{z_3-z_2}\right)\int\limits_{-\Delta l_i/2}^{+\Delta l_i/2} G_{mn}(z,z')\mathrm{d}z' \end{array}\right] \tag{8.139}$$

where I_{1i}, I_{2i}, I_{3i} are the node currents on the ith element and nth antenna, respectively.

8.1.18 Numerical Results

The computational example presented in this section is of practical importance for the design of base station antennas. It is related to the GSM sector antenna

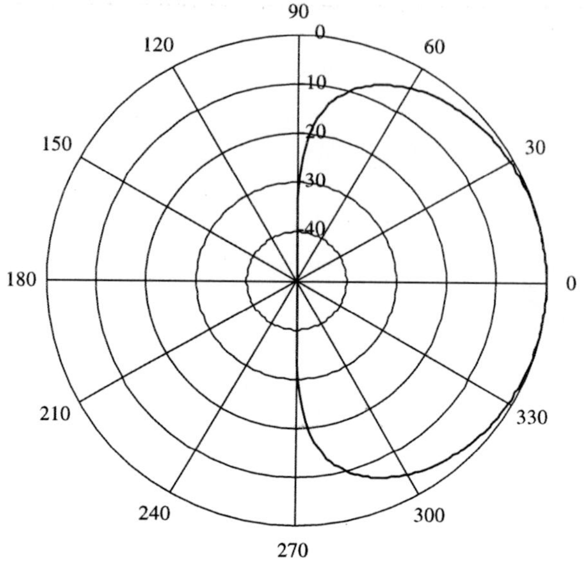

Figure 8.18 Horizontal pattern.

consisting of eight half-wave dipole antennas spaced by 0.75λ (between centers) with radius of 0.004λ. This array is distanced from the reflector at $x = -0.176\lambda$, where λ denotes the signal wavelength. The operating frequency is 900 MHz. Figures 8.18–8.22 show the results for the radiated field. For simplicity reasons, the reflector has been modeled as an infinite, perfectly conducting plane. All dipoles

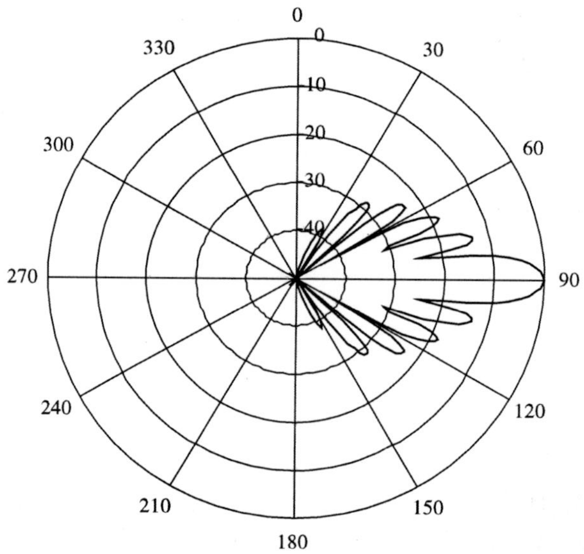

Figure 8.19 Vertical pattern.

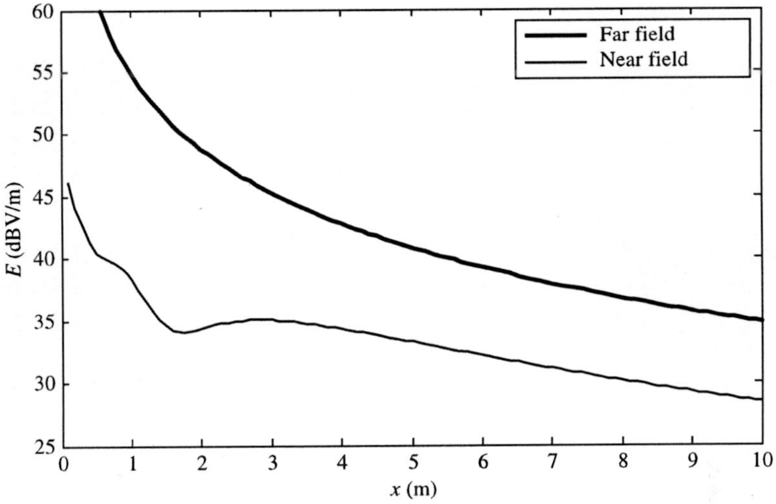

Figure 8.20 Calculated electric field $\Phi = 0°$, $z = 0\,\text{m}$.

are driven by the voltage generator placed at the center of each dipole. The total input power of each source is chosen as 30 W, which is the typical maximum power per GSM channel. The computation has been carried out by discretizing the array configuration with 88 segments.

The horizontal and vertical radiation patterns are shown in Figures 8.18 and 8.19, respectively.

Furthermore, the near-field distributions have been calculated using both exact and approximate relations.

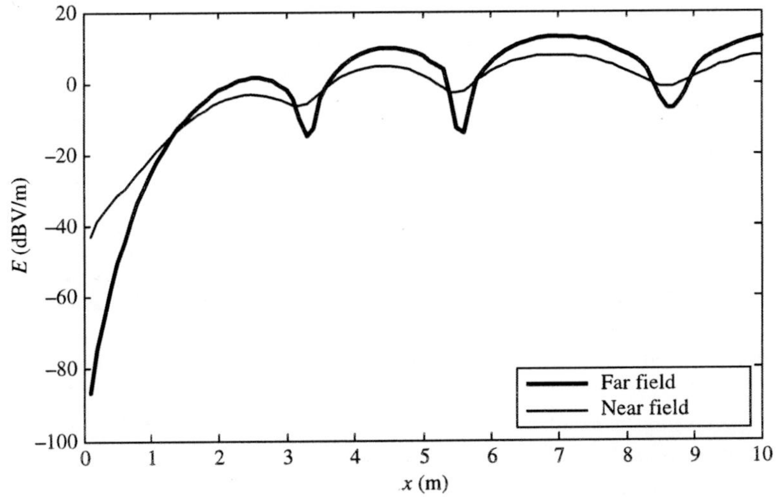

Figure 8.21 Calculated electric field $\Phi = 0°$, $z = -5\,\text{m}$.

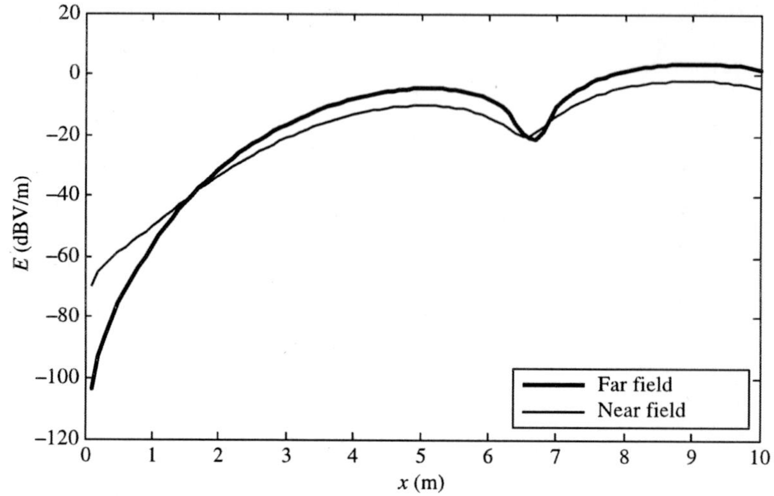

Figure 8.22 Calculated electric field $\varphi = 0°$, $z = -10\,\text{m}$.

The magnitude of the far-field radiated by the base station antenna system can be determined analytically by using the ray tracing algorithm based on the geometrical optics method. The total field is the superposition of the incident and reflected field components [36, 37],

$$E^{tot} = E^{inc} + E^{ref} \tag{8.140}$$

and their expressions are given by [3, 4]

$$E^{inc} = \frac{E_0(\phi, \vartheta)}{r} e^{-j\beta r} \tag{8.141}$$

$$E^{ref} = \Gamma_R(\phi', \vartheta') \frac{E_0(\phi', \vartheta')}{r'} e^{-j\beta r'} \tag{8.142}$$

where r and r' are the corresponding distances from the antenna system and its image, respectively, to the observation point, Γ_R is the appropriate reflection coefficient [36, 37], and E_0 is the magnitude of the incident field that is defined as

$$E_0(\phi, \vartheta) = \sqrt{30 N P_{rad} G(\phi, \vartheta)} \tag{8.143}$$

where P_{rad} is the radiated power, N is the number of carriers (antennas), and $G(\phi, \theta)$ is the radiation pattern for a particular antenna.

Comparison of the results obtained via different methods is shown in Figures 8.20–8.22.

It is obvious from Figures 8.20 to 8.22 that the use of analytical formulas for the calculation of the radiated electric field in the near zone results in significant overestimation (8–15 dB) in the main lobe, e.g., for $z = 0$ (Fig. 8.21). Beneath the main lobe, near-field values obtained by BEM could be higher than the values obtained by the far-field approximation. This is particularly visible for the nulls of the radiation pattern where the underestimation can reach the level of 12 dB (Fig. 8.21). The same can be concluded for the results presented in Figure 8.22.

Therefore, there are significant discrepancies between the results computed via different techniques. It has been shown that the analytical solutions widely used in the computation of the radiated field of wire antenna arrays can result in overestimation or underestimation of the field value computed via BEM, which is based on the rigorous integro-differential equation formulation.

Though the method presented in this section is already applicable to a wide variety of real world problems, it could be further extended using isoparametric elements to handle even more complex geometries.

8.2 THIN WIRES ABOVE A LOSSY HALF-SPACE

Analysis of radiation and scattering from horizontal wires above a conducting half-space has drawn the attention of many prominent researchers and has become of great practical importance during the last 50 years. Namely, this analysis is applicable in the field of surface wave propagation, oceanography, geophysical explorations, submarine communications, detection, and bioelectromagnetics. The model of horizontal wire antenna above dissipative half-space has numerous applications in electromagnetic compatibility (EMC).

The analysis is based on solving the corresponding integral equation for an axial current distribution along a finite length wire antenna horizontally oriented over dissipative medium [42, 43]. The effect of an imperfectly conducting half-space is taken into account via Sommerfeld integrals appearing inside the integro-differential equation kernel. It is convenient to consider two separate problems: (1) a method for solving the electric field integral equation (Pocklington's equation) and (2) calculation of Sommerfeld integrals calculation. The first problem is treated, relatively efficiently, for the case of an isolated antenna in free space, by various moment-methods variants [25–27, 44]. The most commonly used method for solving this integral equation is a combination of subdomain collocation and the finite difference technique [45, 46]. Because of its simplicity, this technique has been widely used for half-space problems, where Sommerfeld integrals cause further complexities and difficulties.

Sommerfeld integrals evaluation has drawn the attention of many prominent researchers in the last few decades and till now it has not been optimally solved due to the highly oscillating integrands that very hard to evaluate accurately. Many approximate techniques have been developed, which can be generally characterized as analytical and numerical. Good reviews are given in Refs. [42, 47]. Significant contributions are also available in Refs. [48–54].

In this, section, Pocklington's equation is solved by the Galerkin-Bubnov indirect boundary element method (GB-IBEM), while Sommerfeld integrals are treated via the exponential approximation and some analytical techniques. Solving the integral equation, the equivalent current distribution is obtained. Once the current distribution is determined, the evaluation of other parameters of interest is straightforward.

It is important to emphasize that this approach avoids the use of finite differences in the integral equation kernel, since the second-order differentiation over the IE kernel is replaced by trivial differentiation over shape and test functions outside the kernel. Single straight wire is first modeled as a separate element and then the analysis is extended to wire arrays.

8.2.1 Single Straight Wire Above a Dissipative Half-Space

The FD-TWIE for horizontal thin wire above an imperfectly conducting ground, Figure 8.23, is derived by integrating the contributions of infinitesimal dipoles over the entire length of the wire antenna [8].

The effect of a dissipative half-space can be taken into account using the rigorous Sommerfeld integral approach that requires tedious calculation of highly oscillating integrals or Fresnel reflection coefficient approximation that is computationally less demanding. This section deals with both the approaches.

8.2.1.1 Sommerfeld Integral Approach The electric field in the vicinity of a straight thin wire above real ground, Figure 8.23, can be expressed in terms of Hertz vector potential $\vec{\Pi}$,

$$\vec{E} = \nabla\left(\nabla\vec{\Pi}\right) + k^2\vec{\Pi} \tag{8.144}$$

where k is the phase constant of a free space,

$$k = \omega\sqrt{\mu\varepsilon} \tag{8.145}$$

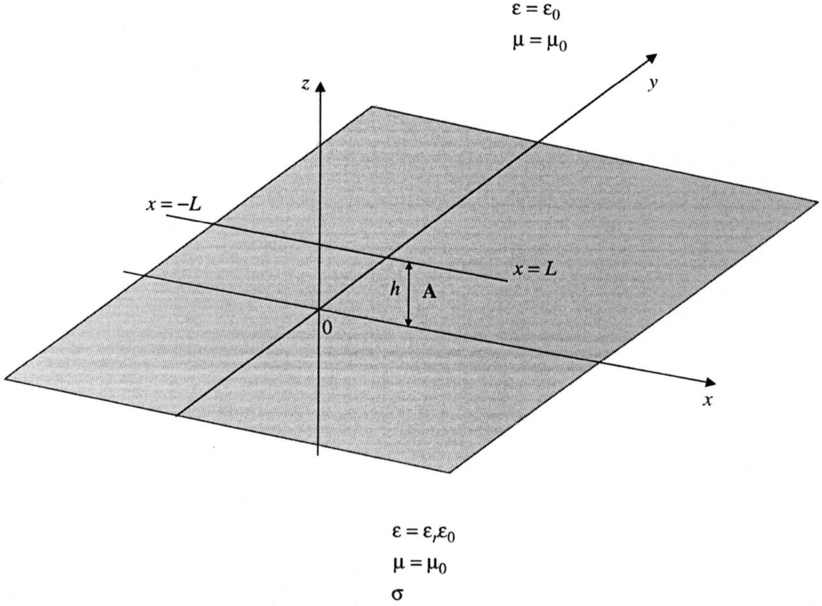

Figure 8.23 Horizontal antenna over real ground.

The scattered tangential electric field along a thin, perfectly conducting wire of length $2L$ and radius $2a$ horizontally placed above a conducting half-space is expressed in terms of Hertz vector potential components,

$$E_x^{sct}(x,z) = \left[\frac{\partial^2}{\partial x^2} + k_1{}^2\right]\Pi_x + \frac{\partial^2}{\partial x \partial z}\Pi_z \tag{8.146}$$

and Hertz potential components are given by

$$\Pi_{1x} = \frac{1}{j4\pi\omega\varepsilon}\int_{-L}^{L}\{g_0(x,x') - g_i(x,x') + U_{11}\}I(x')\mathrm{d}x' \tag{8.147}$$

$$\Pi_{1z} = \frac{1}{j4\pi\omega\varepsilon}\int_{-L}^{L}\frac{\partial W_{11}}{\partial x}I(x')\mathrm{d}x' \tag{8.148}$$

where L is half of antenna length and $I(x')$ is the unknown current to be determined. where $g_0(x,x')$ denotes the free space Green's function in the form

$$g_0(x,x') = \frac{e^{-jk_1R_1}}{R_1} \tag{8.149}$$

while $g_i(x,x')$ results from the image theory and it is given by

$$g_i(x,x') = \frac{e^{-jk_1R_2}}{R_2} \tag{8.150}$$

R_1 and R_2 are distances from the source to the observation point, respectively. If both the source and observation points are on the wire surface, then these distances are given by expressions

$$R_1 = \sqrt{(x-x')^2 + a^2}, \quad R_2 = \sqrt{(x-x')^2 + 4h^2} \tag{8.151}$$

The effect of a dissipative half-space is taken into account via *attenuation* terms in the form of Sommerfeld integrals U_{11} and W_{11}, where

$$U_{11} = 2\int_0^\infty \frac{e^{-2\mu_1(z+h)}}{\mu_1 + \mu_2}J_0(\lambda\rho)\lambda\mathrm{d}\lambda \tag{8.152}$$

$$W_{11} = 2\int_0^\infty \frac{(\mu_1 - \mu_2)e^{-2\mu_1(z+h)}}{k_2{}^2\mu_1 + k_1{}^2\mu_2}J_0(\lambda\rho)\lambda\mathrm{d}\lambda \tag{8.153}$$

where

$$\mu_1 = (\lambda^2 - k_1{}^2)^{1/2}, \quad \mu_2 = (\lambda^2 - k_2{}^2)^{1/2} \tag{8.154}$$

and λ is the integration variable, while h is the distance from the interface to the source.

Combining eqns (8.146–8.154), the scattered electric field can be expressed as follows:

$$E_x^{sct} = \frac{1}{j4\pi\omega\varepsilon} \int_{-L}^{L} \left\{ \left(\frac{\partial^2}{\partial x^2} + k_1{}^2 \right) [g_0(x,x') - g_i(x,x') + U_{11}] + \frac{\partial^2}{\partial x \partial z} [\partial W_{11}] \right\} I(x') \mathrm{d}x' \tag{8.155}$$

The excitation field component can be written as the sum of the incident field $\vec{E}^{inc}$ and field reflected from the lossy ground $\vec{E}^{ref}$ (Fig. 8.24),

$$\vec{E}^{exc} = \vec{E}^{inc} + \vec{E}^{ref} \tag{8.156}$$

The Pocklington equation for the current induced along the wire can be derived by enforcing the interface conditions for the tangential components of the electric field on the wire surface,

$$\vec{e}_x \cdot \vec{E}^{tot} = 0 \tag{8.157}$$

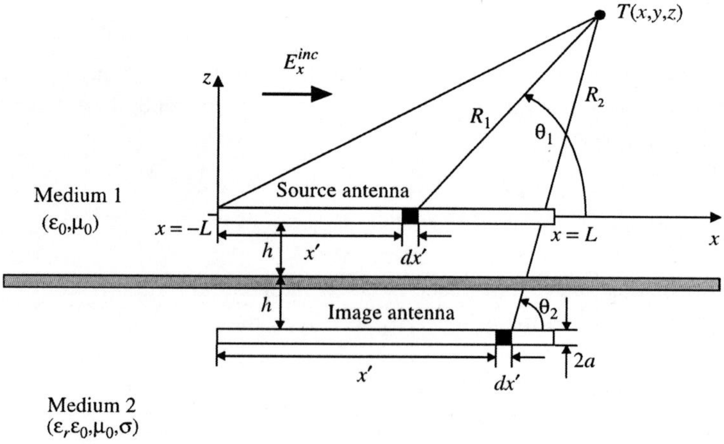

Figure 8.24 Horizontal antenna over perfectly conducting ground with source and observation point in air. (Poljak, D., Roje, V., Boundary-element approach to calculation of wire antenna parameters in the presence of dissipative half-space, *IEE Proc. Microw. Antennas Propag.*, 142 (6), 1995, pp 435–440).

where the total field $\vec{E}^{tot}$ is composed from the excitation field $\vec{E}^{exc}$ and scattered field $\vec{E}^{sct}$ components,

$$\vec{E}^{tot} = \vec{E}^{exc} + \vec{E}^{sct} \tag{8.158}$$

Assuming the conductor to be perfectly conducting, the tangential electric field vanishes on the surface of the wire, that is, it follows

$$\vec{e}_x \cdot \left(\vec{E}^{exc} + \vec{E}^{sct}\right) = 0 \tag{8.159}$$

where the excitation field component represents the sum of the incident field $\vec{E}^{inc}$ and field reflected from the lossy ground $\vec{E}^{ref}$,

$$\vec{E}^{exc} = \vec{E}^{inc} + \vec{E}^{ref} \tag{8.160}$$

$$E_x^{exc} = -\frac{1}{j4\pi\omega\varepsilon} \int_{-L}^{L} \left\{ \left(\frac{\partial^2}{\partial x^2} + k_1{}^2 \right) [g_0(x,x') - g_i(x,x') + U_{11}] + \frac{\partial^2}{\partial x \partial z} [\partial W_{11}] \right\} I(x') \mathrm{d}x' \tag{8.161}$$

If the antenna mode is considered, it follows

$$E_x^{exc} = E_x^{inc} \tag{8.162}$$

where the incident field exists only across the feed-gap area and it is given by

$$E_x^{inc} = \frac{V_g}{\Delta l_g} \tag{8.163}$$

where V_g is the equivalent voltage generator.

Furthermore, eqn (8.159) simplifies into

$$E_x^{inc} = -E_x^{sct} \tag{8.164}$$

and the corresponding Pocklington integro-differential equation is then

$$E_x^{inc} = -\frac{1}{j4\pi\omega\varepsilon} \int_{-L}^{L} \left\{ \left(\frac{\partial^2}{\partial x^2} + k_1{}^2 \right) [g_0(x,x') - g_i(x,x') + U_{11}] + \frac{\partial^2}{\partial x \partial z} [\partial W_{11}] \right\} I(x') \mathrm{d}x' \tag{8.165}$$

where E_x^{inc} denotes the incident electric field (Fig. 8.24).

If medium 2 is assumed to be perfectly conducting, terms with Sommerfeld integrals vanish and the resulting integral is of the form

$$E_x^{inc} = -\frac{1}{j4\pi\omega\varepsilon} \int_{-L}^{L} \left\{ \left(\frac{\partial^2}{\partial x^2} + k_1{}^2 \right) [g_0(x,x') - g_i(x,x')] \right\} I(x')\mathrm{d}x' \tag{8.166}$$

Finally, if an isolated cylindrical antenna in free space is considered, only the g_0 term of the total IE kernel remains, while g_i, U_{11} and W_{11} vanish, and the well-known Pocklington integro-differential equation for the free space is obtained,

$$E_x^{inc} = -\frac{1}{j4\pi\omega\varepsilon} \int_{-L}^{L} \left(\frac{\partial^2}{\partial x^2} + k_1{}^2 \right) g_0(x,x') I(x')\mathrm{d}x' \tag{8.167}$$

If the more demanding case of an arbitrarily oriented wire antenna of arbitary geometry over a lossy half-space is considered, then general curved wire electric field integral equation must be used [1].

8.2.1.2 Reflection Coefficient Approximation Besides rigorous Sommerfeld integral approach, the effects of a two-media configuration can be taken into account via the Fresnel reflection coefficient. This approach has been discussed in many papers, e.g., Refs. [43, 55, 56]. At this point, the validity of the reflection coefficient approximation (RCA) to account for the lossy interface effects should be discussed. A useful study on comparison of RCA with Sommerfeld integral approach has been carried out in Ref. [28]. The Sommerfeld integral approach has been found to be numerically stable and reliable for horizontal wire which was brought within 10^{-6} wavelengths of the interface. Furthermore, as a rough guideline, the RCA has been found to produce results generally within 10% of those obtained using rigorous Sommerfeld integral approach.

The corresponding integral equations for the current induced along the wire can be derived in a similar manner as that in the previous case, that is, by enforcing the interface conditions for the tangential components of the electric field at the wire surface,

$$\vec{e}_x \cdot \left(\vec{E}^{exc} + \vec{E}^{sct} \right) = 0 \tag{8.168}$$

where the excitation field component represents the sum of the incident field $\vec{E}^{inc}$ and the field reflected from the lossy ground $\vec{E}^{ref}$,

$$\vec{E}^{exc} = \vec{E}^{inc} + \vec{E}^{ref} \tag{8.169}$$

The scattered field component can now be written as

$$\vec{E}^{sct} = -j\omega\vec{A} - \nabla\varphi \tag{8.170}$$

where $\vec{A}$ is the magnetic vector potential and φ is the electric scalar potential.

According to the widely used thin-wire approximation [28], only the axial component of the magnetic potential differs from zero and eqn (8.170) is simplified into

$$E_x^{sct} = -j\omega A_x - \frac{\partial \varphi}{\partial x} \tag{8.171}$$

The magnetic vector potential and electric scalar potential are defined as

$$A_x = \frac{\mu}{4\pi} \int_0^L I(x')g(x,x')\mathrm{d}x' \tag{8.172}$$

$$\varphi(x) = \frac{1}{4\pi\varepsilon} \int_0^L q(x')g(x,x')\mathrm{d}x' \tag{8.173}$$

where $q(x)$ is the charge distribution along the line, $I(x')$ denotes the induced current along the line, and $g(x,x')$ stands for the Green's function that is given by

$$g(x,x') = g_0(x,x') - R_{TM}g_i(x,x') \tag{8.174}$$

where $g_0(x,x')$ is the free-space Green's function,

$$g_o(x,x') = \frac{e^{-jk_oR_o}}{R_o} \tag{8.175}$$

while $g_i(x,x')$ arises from the image theory and is given by

$$g_i(x,x') = \frac{e^{-jk_oR_i}}{R_i} \tag{8.176}$$

and R_o and R_i denote the corresponding distance from the source to the observation point, respectively.

R_{TM} is the reflection coefficient for the transverse polarization,

$$R_{TM} = \frac{n\cos\Theta - \sqrt{n - \sin^2\Theta}}{n\cos\Theta + \sqrt{n - \sin^2\Theta}} \tag{8.177}$$

which accounts for the presence of a lossy half-space.

The refraction index n is given by

$$n = \varepsilon_r - j\frac{\sigma}{\omega\varepsilon_0} \tag{8.178}$$

and argument θ is defined as

$$\Theta = arctg \frac{|x - x'|}{2h} \tag{8.179}$$

The linear charge density and the current distribution along the line are related through the equation of continuity [4]

$$q = -\frac{1}{j\omega} \frac{dI}{dx} \tag{8.180}$$

Substituting eqn (8.180) into eqn (8.173) yields

$$\varphi(x) = -\frac{1}{j4\pi\omega\varepsilon} \int_0^L \frac{\partial I(x')}{\partial x'} g(x, x') dx' \tag{8.181}$$

Combining eqns (8.171), (8.172), and (8.181) results in the following integral relationship for the scattered field:

$$E_x^{sct} = -j\omega \frac{\mu}{4\pi} \int_0^L I(x') g(x, x') dx' + \frac{1}{j4\pi\omega\varepsilon} \frac{\partial}{\partial x} \int_0^L \frac{\partial I(x')}{\partial x'} g(x, x') dx' \tag{8.182}$$

Finally, eqns (8.168) and (8.182) result in the following integral equation for the unknown current distribution induced along the wire,

$$E_x^{exc} = j\omega \frac{\mu}{4\pi} \int_0^L I(x') g(x, x') dx' - \frac{1}{j4\pi\omega\varepsilon} \frac{\partial}{\partial x} \int_0^L \frac{\partial I(x')}{\partial x'} g(x, x') dx' \tag{8.183}$$

Integral eqn (8.183) is well known in antenna theory and represents one of the most commonly used variants of the Pocklington's integro-differential equation, particularly attractive for numerical modeling, as there is no second-order differential operator under the integral sign.

8.2.2 Loaded Antenna Above a Dissipative Half-Space

Current induced on electrically short wire antennas appears usually in the form of a standing wave. The input impedance of such antennas is a function strongly dependent on frequency. On the contrary, the traveling wave antenna has input impedance which is almost frequency independent. Some decades ago Wu and King [57] proved theoretically that such a traveling wave antenna can be realized by a continuous resistive loading, if the resistance per load of the antenna varies as a function of the position along the antenna. For a certain nonuniform, resistive, frequency independent loading, the end reflections can be eliminated to a great extent, so that the equivalent traveling wave current decreases rapidly as it moves outward from the feed-gap.

These loaded antennas, due to the nonreflecting resistive loading, are convenient for broadband applications and electromagnetic pulse (EMP) simulators design. The disadvantage of resistive loadings is a reduced efficiency of the radiating antenna as radiator. This problem can be overcome by using the reactive load [58].

A comprehensive theoretical and experimental study of nonreflecting resistively loaded wire antennas in free space was given by Wu and King [57] and Shen, respectively [59]. However, in applications such as surface wave propagation, oceanography, geosounding, etc., the analysis of antenna radiating in the presence of a lossy half-space is relevant, but it is significantly less documented. One of the rare papers dealing with a loaded horizontal dipole antenna over dissipative half-space is given by Y. Rahmat-Samii et al. in Ref. [46], where the simplest variant of the moment method—pulse basis functions and point matching for solving the corresponding electric field integral equation.

The boundary element/exponential approximation approach for solving the corresponding integral equation for loaded antenna over lossy ground and input impedance assessment is presented in Ref. [60].

The dipole antenna of length $2L$ and radius a with a continuous load, horizontally located above an imperfectly conducting half-space at height h is considered. The total tangential electric field is expressed as the sum of impressed electric field generated by the source gap and the corresponding scattered field on the antenna surface,

$$E_x^{tot} = E_x^{exc} + E_x^{sct} \tag{8.184}$$

If the antenna is loaded or imperfectly conducting, the total tangential electric field can be expressed in terms of antenna current $I(x)$ and surface impedance per unit length of the antenna $Z_s(x)$,

$$E_x^{tot}(x) = Z_s(x)I(x) \tag{8.185}$$

According to the Sommerfeld integral approach to the half-space problem by combining relations (8.165), (8.183), and (8.184), the electric field integral equation for loaded horizontal wire antenna over a lossy ground becomes

$$E_x^{exc} = -\frac{1}{j4\pi\omega\varepsilon} \int_{-L}^{L} \left\{ \left(\frac{\partial^2}{\partial x^2} + k_1^2 \right) [g_0(x,x') - g_i(x,x') + U_{11}] + \frac{\partial^2}{\partial x \partial z} \left(\frac{\partial W_{11}}{\partial x} \right) \right\} I(x')\mathrm{d}x' + Z_s(x)I(x) \tag{8.186}$$

If the RCA approach is used, relations (8.183), (8.184), and (8.185) lead to the following Pocklington integro-differential variant for a loaded wire above a real ground:

$$E_x^{exc} = j\omega \frac{\mu}{4\pi} \int_0^L I(x')g(x,x')\mathrm{d}x' - \frac{1}{j4\pi\omega\varepsilon}\frac{\partial}{\partial x}\int_0^L \frac{\partial I(x')}{\partial x'} g(x,x')\mathrm{d}x' + Z_s(x)I(x) \tag{8.187}$$

Now, the character of the impedance Z_s has to be discussed.

The unloaded antenna, due to the reflections from the free ends, has a frequency sensitive input impedance. To retain the input impedance fairly constant vs. frequency, these reflections have to be reduced. This can be accomplished by using the nonreflecting resistive loading [46, 57]. Such a loading proposed in Ref. [57] is originally derived as a solution for impedance per unit length of the antenna as a function of distance from the feed point. This loading is obtained specifically when the current is represented by an outward traveling wave with no reflected wave. The expressions for such a load are then given by [46, 57]

$$Z_s(x) = \frac{Z_0}{2\pi}\frac{|\Psi|}{L - |x|} \tag{8.188}$$

where

$$Z_0 = \sqrt{\frac{\mu}{\varepsilon}}, \quad \Psi = 2\int_0^L \left(1 - \frac{z'}{L}\right) e^{-jkz'} \frac{e^{-jkr_0}}{r_0}\mathrm{d}z', \quad r_0 = \sqrt{z'^2 + a^2} \tag{8.189}$$

The main drawback of the resistive load is the reduced radiation efficiency of the antenna. This drawback can be overcome by using the reactive type of loading.

8.2.3 Electric Field and the Input Impedance of a Single Wire Above a Half-Space

The analysis of electric field scattered from single wire above a real ground has drawn the attention of many researchers, as many EMC applications require knowledge of the field behavior.

In this section, the field calculation is carried out featuring the RCA approach as the Sommerfeld integral approach is too demanding regarding analytical and numerical efforts.

The electric field components can be obtained from eqn (8.170) and they are given as follows:

$$E_x = \frac{1}{j4\pi\omega\varepsilon_0}\left[-\int_{-L}^{L}\frac{\partial I(x')}{\partial x'}\frac{\partial g(x,x')}{\partial x'}\mathrm{d}x' + k^2\int_{L}^{L} g(x,x')I(x')\mathrm{d}x'\right] \tag{8.190}$$

$$E_y = \frac{1}{j4\pi\omega\varepsilon_0}\int_{-L}^{L}\frac{\partial I(x')}{\partial x'}\frac{\partial g(x',y)}{\partial y}\mathrm{d}x' \tag{8.191}$$

$$E_z = \frac{1}{j4\pi\omega\varepsilon_0}\int_{-L}^{L}\frac{\partial I(x')}{\partial x'}\frac{\partial g(x',z)}{\partial z}\mathrm{d}x' \tag{8.192}$$

The functional for input impedance defined by relation [61]

$$Z_{in} = -\frac{1}{|I_0|^2}\int_{-L}^{L} E(x,a)I(x)^*\mathrm{d}x \tag{8.193}$$

can be determined from expression [3]

$$Z_{in}(I) = -\frac{1}{j4\pi\omega\varepsilon}\frac{1}{|I_0|^2}\int_{-L}^{L}\int_{-L}^{L}\left\{\left(\frac{\partial^2}{\partial x^2}+k_1^2\right)[g_0(x,x') - g_i(x,x') + U_{11}] + \frac{\partial^2}{\partial x\partial z}\left(\frac{\partial W_{11}}{\partial x}\right)\right\}I(x')I(x)^*\mathrm{d}x'\mathrm{d}x \tag{8.194}$$

where $E(x,a)$ is the electric field distribution of the cylindrical antenna surface, $I(x)^*$ is the complex conjugate of the axial current distribution, and I_0 is the value of the current at the antenna input.

If the loaded wire is considered, the functional for input impedance Z_{in}(I) is given in the following form:

$$Z_{in}(I) = -\frac{1}{j4\pi\omega\varepsilon}\frac{1}{|I_0|^2}\int_{-L}^{L}\int_{-L}^{L}\left\{\left(\frac{\partial^2}{\partial x^2}+k_1^2\right)[g_0(x,x') - g_i(x,x') + U_{11}] + \frac{\partial^2}{\partial x\partial z}\left(\frac{\partial W_{11}}{\partial x}\right)\right\}I(x')I(x)^*\mathrm{d}x'\mathrm{d}x + \frac{1}{j4\pi\omega\varepsilon}\frac{1}{|I_0|^2}\int_{-L}^{L} Z_s(x)I(x)I(x)^*\mathrm{d}x \tag{8.195}$$

If RCA approach is used, eqn (8.93) becomes

$$Z_{in}(I) = -\frac{1}{j4\pi\omega\varepsilon}\frac{1}{|I_0|^2}\int_{-L}^{L}\int_{-L}^{L}\left\{\left(\frac{\partial^2}{\partial x^2}+k_1^2\right)[g_0(x,x') - R_{TM}g_i(x,x')]\right\}I(x')I(x)^*\mathrm{d}x'\mathrm{d}x \tag{8.196}$$

and if the case of a loaded wire is considered, the functional for input impedance Z_{in}(I) can be written as

$$Z_{in}(I) = -\frac{1}{j4\pi\omega\varepsilon}\frac{1}{|I_0|^2}\int_{-L}^{L}\int_{-L}^{L}\left\{\left(\frac{\partial^2}{\partial x^2}+k_1^2\right)[g_0(x,x') - R_{TM}g_i(x,x')]\right\}I(x')I(x)^*\mathrm{d}x'\mathrm{d}x$$
$$+\frac{1}{j4\pi\omega\varepsilon}\frac{1}{|I_0|^2}\int_{-L}^{L} Z_s(x)I(x)I(x)^*\mathrm{d}x \tag{8.197}$$

8.2.4 Boundary Element Analysis for Single Wire Above a Real Ground

This section deals with the boundary element solution of half-space integral equations for both unloaded and loaded wire. Essentially, the numerical method is the same as in the case of a straight wire in free space. Only the differences in final expressions are presented in this section.

8.2.4.1 Boundary Element Modeling of Unloaded Wire Performing certain mathematical manipulation, the weak Galerkin-Bubnov formulation of integro-differential equation for unloaded wire yields

$$\sum_{i=1}^{n}\alpha_i\left\{-\int_{-L}^{L}\frac{\mathrm{d}f_j(x)}{\mathrm{d}x}\int_{-L}^{L}\frac{\mathrm{d}f_i(x')}{\mathrm{d}x'}g_H(x,x')\mathrm{d}x'\mathrm{d}x + {k_1}^2\int_{-L}^{L}f_j(x)\int_{-L}^{L}f_i(x')g_H(x,x')\mathrm{d}x'\mathrm{d}x\right.$$
$$\left.-\int_{-L}^{L}\frac{\mathrm{d}N_j(x)}{\mathrm{d}x}\int_{-L}^{L}f_i(x')g_V(x,x')\mathrm{d}x'\mathrm{d}x\right\} \tag{8.198}$$
$$= \frac{4\pi\omega\varepsilon}{j}\int_{-L}^{L}{E_x}^i(x)f_j(x)\mathrm{d}x,\quad j=1,2,\ldots,n$$

where

$$g_H(x,x') = g_0(x,x') - g_i(x,x') + U_{11} \tag{8.199}$$

and,

$$g_V(x, x') = \frac{\partial^2 W_{11}}{\partial x \partial z} \tag{8.200}$$

The boundary element discretization results in the matrix equation

$$\sum_{i=1}^{M} [Z]_{ji} \{I\}_i = \{V\}_j, \quad j = 1, 2, \ldots, M \tag{8.201}$$

where $[Z]_{ji}$ is the mutual impedance matrix representing the interaction of the ith source to the $j$the observation element.

$$\begin{aligned}[Z]_{ji} = -\frac{1}{j4\pi\omega\varepsilon} \Bigg(& \int_{\Delta l_j} \int_{\Delta l_i} \{D\}_j \{D'\}_i^T g_H(x, x') \mathrm{d}x' \mathrm{d}x \\ & + k_1^2 \int_{\Delta l_j} \int_{\Delta l_i} \{f\}_j \{f'\}_i^T g_H(x, x') \mathrm{d}x' \mathrm{d}x \\ & - \int_{\Delta l_j} \int_{\Delta l_i} \{D\}_j \{f'\}_i^T g_V(x, x') \mathrm{d}x' \mathrm{d}x \Bigg)\end{aligned} \tag{8.202}$$

This matrix represents, in fact, the mutual impedance of ith and jth finite element.

If the RCA is applied, the mutual impedance matrix is of the form

$$\begin{aligned}[Z]_{ji} = -\frac{1}{j4\pi\omega\varepsilon} \Bigg(& \int_{\Delta l_j} \int_{\Delta l_i} \{D\}_j \{D'\}_i^T g_H(x, x') \mathrm{d}x' \mathrm{d}x \\ & + k_1^2 \int_{\Delta l_j} \int_{\Delta l_i} \{f\}_j \{f'\}_i^T g_H(x, x') \mathrm{d}x' \mathrm{d}x \\ & - \int_{\Delta l_j} \int_{\Delta l_i} \{D\}_j \{f'\}_i^T g_V(x, x') \mathrm{d}x' \mathrm{d}x \Bigg)\end{aligned} \tag{8.203}$$

The local voltage vector $\{V\}_j$ is given by

$$\{V\}_j = -\int_{\Delta l_j} E_x^{inc} \{N\}_j \mathrm{d}x \tag{8.204}$$

Matrices $\{f\}$ and $\{f'\}$ contain shape functions $f_k(x)$ and $f_k(x')$, while $\{D\}$ and $\{D'\}$ contain their derivatives, where M is the total number of finite elements, n_e is the total number of local nodes per element, and Δl_i, Δl_j are the width of ith and jth finite elements, respectively.

In the radiation mode, right-hand side vector is different from zero only in the feed gap area. The x-component of the impressed (incident) electric field is given by

$$E_x^{inc}(x) = \frac{V_g}{\Delta l_g} \tag{8.205}$$

where V_g is the feed voltage and Δl_g is the feed-gap width.

8.2.4.2 Boundary Element Modeling of Loaded Wire The weak Galerkin-Bubnov formulation of eqn (8.187) for the loaded wire is given by

$$\sum_{i=1}^{n} \alpha_i \frac{1}{j4\pi\omega\varepsilon} \left\{ -\int_{-L}^{L} \frac{\mathrm{d}f_j(x)}{\mathrm{d}x} \int_{-L}^{L} \frac{\mathrm{d}f_i(x')}{\mathrm{d}x'} g_H(x,x')\mathrm{d}x'\mathrm{d}x + k_1^2 \int_{-L}^{L} f_j(x) \int_{-L}^{L} f_i(x') g_H(x,x')\mathrm{d}x'\mathrm{d}x \right.$$
$$\left. - \int_{-L}^{L} \frac{\mathrm{d}f_j(x)}{\mathrm{d}x} \int_{-L}^{L} f_i(x') g_V(x,x')\mathrm{d}x'\mathrm{d}x + \int_{-L}^{L} Z_s(x) f_j(x) f_i(x)\mathrm{d}x \right.$$
$$= - \int_{-L}^{L} E_x^{inc}(x) f_j(x)\mathrm{d}x; \quad j = 1,2,\ldots,n \tag{8.206}$$

The boundary element discretization results in the matrix equation

$$\sum_{i=1}^{M} [Z]_{ji} \{I\}_i = \{V\}_j; \quad j = 1,2,\ldots,M \tag{8.207}$$

For the case of loaded wire $[Z]_{ji}$ is given by

$$\begin{aligned}
[Z]_{ji} = -\frac{1}{j4\pi\omega\varepsilon} & \int_{\Delta l_j} \int_{\Delta l_i} \{D\}_j \{D'\}_i^T g_H(x,x')\mathrm{d}x'\mathrm{d}x \\
& + k_1^2 \int_{\Delta l_j} \int_{\Delta l_i} \{f\}_j \{f'\}_i^T g_H(x,x')\mathrm{d}x'\mathrm{d}x \\
& - \int_{\Delta l_j} \int_{\Delta l_i} \{D\}_j \{f'\}_i^T g_V(x,x')\mathrm{d}x'\mathrm{d}x + \int_{\Delta l_j} Z_s(x) \{f\}_j \{f\}_j^T \mathrm{d}x
\end{aligned} \tag{8.208}$$

while right-hand side vector $\{b\}_j$ remains the same as in eqn (8.204).

If the RCA is applied, the mutual impedance matrix is of the form

$$[Z]_{ji} = -\frac{1}{j4\pi\omega\varepsilon} \int\limits_{\Delta l_j} \int\limits_{\Delta l_i} \{D\}_j \{D'\}_i^T g_H(x,x') \mathrm{d}x' \mathrm{d}x + k_1^2 \int\limits_{\Delta l_j} \int\limits_{\Delta l_i} \{f\}_j \{f'\}_i^T g_H(x,x') \mathrm{d}x' \mathrm{d}x - \int\limits_{\Delta l_j} \int\limits_{\Delta l_i} \{D\}_j \{f'\}_i^T g_V(x,x') \mathrm{d}x' \mathrm{d}x + \int\limits_{\Delta l_j} Z_s(x) \{f\}_j \{f\}_j^T \mathrm{d}x \tag{8.209}$$

The computation of Sommerfeld integrals is discussed in Section 8.2.5.

8.2.5 Treatment of Sommerfeld Integrals

The Sommerfeld integrals can be solved using analytical techniques [46, 47, 52, 61, 62], asymptotic expansions [46, 63], and in recent times by various numerical techniques [48–51]. If Sommerfeld integrals are computed numerically, the results obtained are valid over a wide range of parameters, but this is a rather time-consuming process. Since these integrals are, in general, highly oscillatory and difficult to evaluate numerically, the basic intention is to avoid numerical integration.

Two different methods are outlined in this section: the exponential approximations technique and saddle-point method.

8.2.5.1 The Exponential Approximations Technique The integrands are approximated by the set of exponential functions with unknown exponents and multiplicative coefficients, so that integrals U_{11} and W_{11} can then be solved analytically.

In general, there are two integral types of interest:

$$I_1 = \int\limits_0^\infty f(\lambda) J_0(\lambda\rho) e^{-\psi(\lambda)} \mathrm{d}\lambda \tag{8.210}$$

and

$$I_2 = \frac{\partial}{\partial\rho} \int\limits_0^\infty g(\lambda) J_0(\lambda\rho) e^{-\psi(\lambda)} \mathrm{d}\lambda \tag{8.211}$$

where

$$\psi(\lambda) = 2\mu_1 h \tag{8.212}$$

$$f(\lambda) = \frac{\lambda}{\mu_1 + \mu_2} \tag{8.213}$$

$$g(\lambda) = -\frac{x - x'}{|x - x'|} \frac{4\lambda\mu_1(\mu_1 - \mu_2)}{k_1^2 \mu_2 + k_2^2 \mu_1} \tag{8.214}$$

Now, the functions $f(\lambda)$ and $g(\lambda)$ multiplied by exp(-$\psi(\lambda)$) are expressed by means of exponential sums, that is,

$$f(\lambda)e^{-\psi(\lambda)} = \sum_{k=1}^{N} a_k e^{-\lambda\xi_k} \tag{8.215}$$

$$g(\lambda)e^{-\psi(\lambda)} = \sum_{k=1}^{N} b_k e^{-\lambda\xi_k} \tag{8.216}$$

Taking into account the well-known identities [8, 9]

$$\int_0^\infty J_0(\lambda\rho)e^{-\lambda w}d\lambda = \frac{1}{(\rho^2 + w^2)^{1/2}} \tag{8.217}$$

and

$$\frac{\partial}{\partial\rho}\int_0^\infty J_0(\lambda\rho)e^{-\lambda w}d\lambda = \frac{\rho}{(\rho^2 + w^2)^{3/2}} \tag{8.218}$$

expressions (8.210) and (8.211) can obviously be presented in the form of the following sums:

$$I_1 = \sum_{k=1}^{N} a_k \frac{1}{(\rho^2 + \xi_k^2)^{1/2}} \tag{8.219}$$

$$I_2 = \sum_{k=1}^{N} b_k \frac{\rho}{(\rho^2 + \xi_k^2)^{3/2}} \tag{8.220}$$

Coefficients ξ_k must be conveniently chosen [8–11] and a_k, b_k are then calculated from the algebraic system of equation derived from eqns (8.215) and (8.216) for N values of the independent variable λ.

The choice of parameter ξ and certain values of discrete points λ is a very delicate and sensitive problem. On the basis of extensive numerical experiments and physical background of the considered half-space problem, ξ and λ are chosen to vary in the following way [8]:

$$\lambda_k = \lambda_{k-1}\frac{k_1}{\lambda_0}, \tag{8.221}$$

$$\xi_k = \xi_{k-1}\left(\frac{2h}{\lambda_0}\right)^{N-1} \tag{8.222}$$

where N is the number of members in exponential sum and λ_0 is the conveniently chosen wavelength.

8.2.5.2 Saddle-Point Method Saddle-point method is a classical procedure for analytical evaluation of the Sommerfeld integrals, but only for a distant field, ($h > \lambda/2$ and k_1R_2), and for the large ratio of the corresponding permittivities. The expressions U_{11} and g_V then become [46]

$$g_V(x,x') = 2\cos\theta\sin\theta\frac{\cos\theta - \sqrt{n-\sin^2\theta}}{n\cos\theta + \sqrt{n-\sin^2\theta}}\frac{e^{-jk_1R_2}}{R_2} \tag{8.223}$$

$$U_{11} = \frac{2\cos\theta}{\cos\theta + \sqrt{n-\sin^2\theta}}\frac{e^{-jk_1R_2}}{R_2} \tag{8.224}$$

where

$$n = \varepsilon_r - j\frac{\sigma}{\omega\varepsilon_0} \tag{8.225}$$

$$\theta = \arctan\frac{|x-x'|}{2h} \tag{8.226}$$

Using the same method, applying some further approximation, even simpler relations can be obtained which lead to the Fresnel reflection coefficient approximation [28]. The saddle-point method seems to be attractive due to the obviously simple expressions obtained for Sommerfeld integrals, but it is not accurate if the observation points are near interface. This limitation can be avoided if Sommerfeld integrals are solved directly by the convenient numerical integration technique.

8.2.6 Calculation of Electric Field and Input Impedance

The near-field analysis of wire antennas has drawn the attention of many researchers as many EMC applications require knowledge of the near-field behavior.

In order to obtain electric field value at the arbitrary point in the upper medium, the calculated current distribution $I(x')$ is substituted into the Pocklington eqn (8.30). The values of a current and its first derivative for the ith boundary element are given by

$$I(x') = I_{1i}\frac{x_{2i}-x'}{\Delta x} + I_{2i}\frac{x'-x_{1i}}{\Delta x} \tag{8.227}$$

$$\frac{\partial I(x')}{\partial x'} = \frac{I_{2i}-I_{1i}}{\Delta x} \tag{8.228}$$

where I_{1i} and I_{2i} are current values at the local nodes of the ith boundary element with coordinates x_{1i} and x_{2i}, and $\Delta x = x_{2i} - x_{1i}$ denotes dimension of the element.

Substituting current expressions (8.42) and (8.43) in eqn (8.30) results in the following equation:

$$E_x = \frac{1}{j4\pi\omega\varepsilon_0}\sum_{i=1}^{M}\left[-\frac{I_{2i}-I_{1i}}{\Delta x}\int_{x_{1i}}^{x_{2i}}\frac{\partial G(x,x')}{\partial x'}\mathrm{d}x' + k^2\int_{x_{1i}}^{x_{2i}} G(x,x')I(x')\mathrm{d}x'\right] \tag{8.229}$$

where M is the total number of wires and N_j denotes the total number of boundary elements on the jth wire and G is the Green's function given by

$$G(x,x') = g_0(x,x') - R_{TM}g_i(x,x') \tag{8.230}$$

Applying a similar procedure, it is possible to obtain the expressions for the y and z components of the electric field at an arbitrary point in the air:

$$E_y = \frac{1}{j4\pi\omega\varepsilon_0}\sum_{i=1}^{M}\frac{I_{2i}-I_{1i}}{\Delta x}\int_{x_{1i}}^{x_{2i}}\frac{\partial G(x',y)}{\partial y}\mathrm{d}x' \tag{8.231}$$

$$E_z = \frac{1}{j4\pi\omega\varepsilon_0}\sum_{i=1}^{M}\frac{I_{2i}-I_{1i}}{\Delta x}\int_{x_{1i}}^{x_{2i}}\frac{\partial G(x',z)}{\partial z}\mathrm{d}x' \tag{8.232}$$

The integrals in the above expressions are numerically evaluated using Gaussian quadrature. Because of quasisingularity of the Green's function, the first-order differential operator appearing in the integral equation kernel could cause numerical instability. In order to avoid the problem of quasisingularity of the Green's function, the first derivative of that function is approximated by means of a central finite difference formula:

$$\frac{\partial f(x,y)}{\partial x} = \frac{f(x+\Delta x,y) - f(x-\Delta x,y)}{2\Delta x} \tag{8.233}$$

where Δx stands for the size of the finite difference step.

Many procedures used for input impedance evaluation are complicated and inconvenient [64, 65]. Even for sinusoidal current approximation of an isolated antenna in free space, the expressions obtained have integral-sine and integral-cosine terms. The evaluation is, therefore, often tedious and time consuming. In the presence of inhomogeneous media, the problem becomes even more complex.

The GB-IBEM presented in Section 8.2.4.1 provides an efficient way of performing this computation. By conveniently rearranging eqn (8.194) the following expression is obtained:

$$Z_{in}(I) = -\frac{1}{j4\pi\omega\varepsilon}\frac{1}{|I_0|^2}\left\{\int_{-L}^{L}\left(\frac{\mathrm{d}}{\mathrm{d}x}+k_1^2\right)\int_{-L}^{L} g_H(x,x')I(x')I(x)^*\mathrm{d}x'\mathrm{d}x - \int_{-L}^{L}\frac{\mathrm{d}}{\mathrm{d}x}\int_{-L}^{L} g_V(x,x')I(x')I(x)^*\mathrm{d}x'\mathrm{d}x\right\} \tag{8.234}$$

If the RCA approach is used by conveniently rearranging eqn (8.196), the following expression is obtained:

$$Z_{in}(I) = -\frac{1}{j4\pi\omega\varepsilon}\frac{1}{|I_0|^2}\left\{\int_{-L}^{L}\left(\frac{\mathrm{d}}{\mathrm{d}x}+k_1^2\right)\int_{-L}^{L}[g_0(x,x') - R_{TM}g_i(x,x')]I(x')I(x)^*\mathrm{d}x'\mathrm{d}x\right\} \tag{8.235}$$

As the solution for antenna current is obtained by the GB-IBEM (FEIEM), the local expansions for current, for ith source wire segment, and jth wire segment associated with the observation point can be written in the form of linear combinations,

$$I_i(x') = \sum_{k=1}^{n_e} I_{ki}f_{ki}(x'), \quad I_j(x) = \sum_{m=1}^{n_e} I_{mj}f_{mi}(x) \tag{8.236}$$

Proceeding with the weak Galerkin-Bubnov procedure outlined in Section 8.2.4.1, eqn (8.234) becomes

$$Z_{in} = -\frac{1}{j4\pi\omega\varepsilon}\frac{1}{|I_0|^2}I_j^*\left(\sum_{i=1}^{M}\left[\int_{\Delta l_j}\int_{\Delta l_i}\{D\}_j\{D'\}_i^T g_H(x,x')\mathrm{d}x'\mathrm{d}x + k_1^2\int_{\Delta l_j}\int_{\Delta l_i}\{N\}_j\{f'\}_i^T g_H(x,x')\mathrm{d}x'\mathrm{d}x - \int_{\Delta l_j}\int_{\Delta l_i}\{D\}_j\{f'\}_i^T g_V(x,x')\right]I_i\right) \quad j = 1,2,\ldots,M \tag{8.237}$$

while eqn (8.235) becomes

$$Z_{in} = -\frac{1}{j4\pi\omega\varepsilon}\frac{1}{|I_0|^2}I_j^*\left(\sum_{i=1}^{M}\left[\int_{\Delta l_j}\int_{\Delta l_i}\{D\}_j\{D'\}_i^T[g_0(x,x') - R_{TM}g_i(x,x')]\mathrm{d}x'\mathrm{d}x\right.\right.$$
$$\left.\left. + k_1^2\int_{\Delta l_j}\int_{\Delta l_i}\{N\}_j\{f'\}_i^T[g_0(x,x') - R_{TM}g_i(x,x')]\mathrm{d}x'\mathrm{d}x \quad j = 1,2,\ldots,M\right.\right. \tag{8.238}$$

which can also be written in matrix form

$$Z_{in} = \{I\}^T[A]\{I\}^* \tag{8.239}$$

where $[A]$ is the generalized impedance matrix (global finite element matrix), which is assembled from the appropriate terms of each finite element matrices

$$[A] = [a]_{ji}; \quad i,j = 1,2,\ldots,M \tag{8.240}$$

It is obvious that relations (8.237) and (8.238), that is, the relation (8.239) provides the input impedance calculation by multiplying simply the previously formed matrices within the antenna current boundary element calculation. The algorithm presented in this way is convenient for calculating the antenna impedance, regardless whether one deals with homogeneous or inhomogeneous media.

Using relations (8.195) and (8.197), the matrix expressions for the loaded wire input impedance featuring the rigorous and RCA approach, respectively, is given by

$$\begin{aligned} Z_{in} = -\frac{1}{j4\pi\omega\varepsilon}\frac{1}{|I_0|^2}\alpha_j^*\Bigg(\sum_{i=1}^{M}\Bigg[&\int_{\Delta l_j}\int_{\Delta l_i}\{D\}_j\{D'\}_i^T g_H(x,x')\mathrm{d}x'\mathrm{d}x \\ &+ k_1^2\int_{\Delta l_j}\int_{\Delta l_i}\{f\}_j\{f'\}_i^T g_H(x,x')\mathrm{d}x'\mathrm{d}x \\ &- \int_{\Delta l_j}\int_{\Delta l_i}\{D\}_j\{f'\}_i^T g_V(x,x')\mathrm{d}x'\mathrm{d}x \\ &+ \int_{\Delta l_j} Z_s(x)\{f\}_j\{f\}_j^T\mathrm{d}x\Bigg]\alpha_i\Bigg) \quad j = 1,2,\ldots,M \end{aligned} \tag{8.241}$$

For the RCA approach, the formula for the input impedance computation is given by

$$Z_{in} = -\frac{1}{j4\pi\omega\varepsilon}\frac{1}{|I_0|^2}\alpha_j^*\left(\sum_{i=1}^{M}\left[\int_{\Delta l_j}\int_{\Delta l_i}\{D\}_j\{D'\}_i^T[g_0(x,x') - R_{TM}g_i(x,x')]\mathrm{d}x'\mathrm{d}x\right.\right.$$
$$+k_1^2\int_{\Delta l_j}\int_{\Delta l_i}\{f\}_j\{f'\}_i^T[g_0(x,x') - R_{TM}g_i(x,x')]\mathrm{d}x'\mathrm{d}x$$
$$\left.\left.+\int_{\Delta l_j}Z_s(x)\{f\}_j\{f\}_j^T\mathrm{d}x\right]\alpha_i\right) \quad j = 1,2,\ldots,M \tag{8.242}$$

This algorithm is obviously convenient for calculating the wire input impedance, regardless whether one is dealing with unloaded or loaded antenna.

8.2.7 Numerical Results for a Single Wire Above a Real Ground

All numerical results refer to unit, time-harmonic voltage excitation,

$$V = 1e^{j\omega t} \tag{8.243}$$

Figure 8.25 shows the current amplitudes along the dipole antenna radiating above lossy half-space ($2L = 10\,\mathrm{m}$, $a = 0.05\,\mathrm{m}$, $h = 5\,\mathrm{m}$, $\lambda = 5\,\mathrm{m}$, and $\varepsilon_r = 10$,

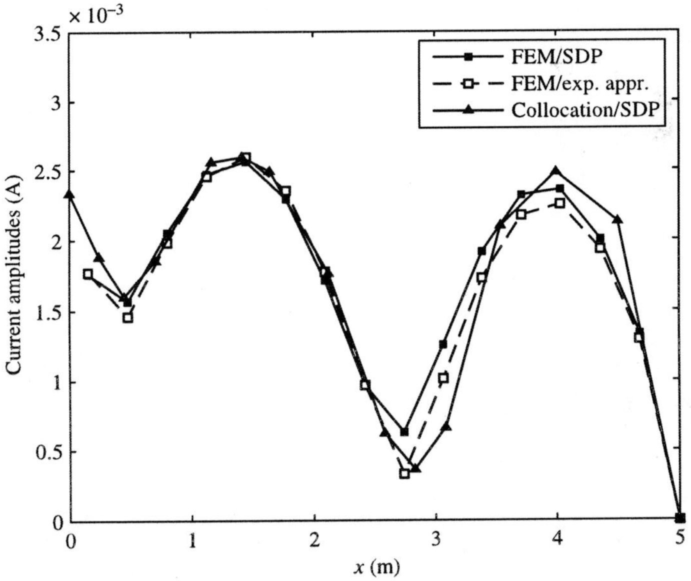

Figure 8.25 Magnitudes of current distribution along horizontal dipole antenna over lossy half-space, with $2L = 10\,\mathrm{m}$, $a = 0.05\,\mathrm{m}$, $h = 5\,\mathrm{m}$, $\lambda = 5\,\mathrm{m}$, $\varepsilon_r = 10$, $\sigma = 0.01\,\mathrm{mhos/m}$. (Poljak, D., Roje, V., Boundary-element approach to calculation of wire antenna parameters in the presence of dissipative half-space, *IEE Proc. Microw. Antennas Propag.*, 142 (6), 1995, pp 435–440).

$\sigma = 0.01$ mmho/m) obtained by different techniques; FEIEM/exponential approximations, FEIEM/steepest descent path (SDP), and collocation/SDP [3]. It also has to be pointed out that convergence of the exponential approximations for treating Sommerfeld integrals is achieved with 20 members in exponential sums (8.215) and (8.216). The convergence rate is about two times faster if one chooses parameters λ and ξ to vary with geometrical progression instead of arithmetical progression. The choice of parameter ξ and certain values of discrete points λ is a very delicate and sensitive problem. On the basis of extensive numerical experiments and physical background of the considered half-space problem, ξ and λ are defined as follows [8]:

$$\xi_k = \xi_{k-1}\left(\frac{2h}{\lambda_0}\right)^{N-1} \tag{8.244}$$

$$\lambda_k = \lambda_{k-1}\frac{k_1}{\lambda_0} \tag{8.245}$$

where N is the total number of members in exponential sum and $\lambda_0 = 1$ m is the conveniently chosen wavelength for the given single wire problem.

Figure 8.26 shows the current distribution magnitudes along the resistively loaded dipole antenna above lossy half-space ($|\psi| = 5.1083$, $a = 0.05$ m, $L = 5$ m, $h = 5$ m, $f = 60$ MHz, and $\varepsilon_r = 10$, $\sigma = 0.001$ mhos/m) obtained by FEIEM and point-matching technique [3]. It is worth mentioning that due to the resistive loading, the current distribution rapidly attenuates as it moves from the feed point.

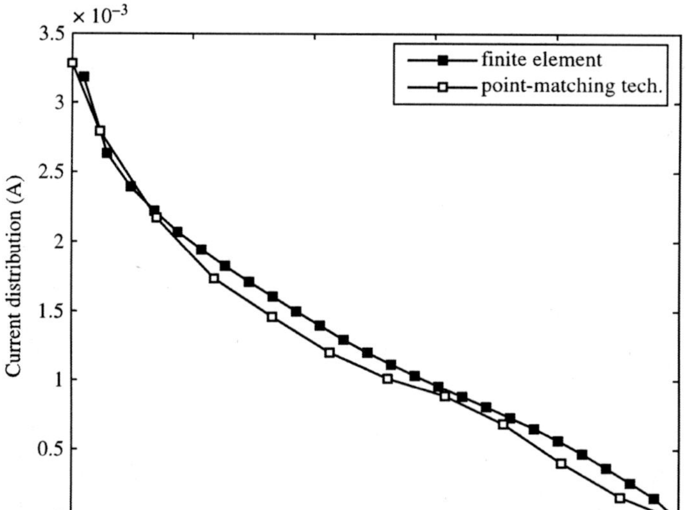

Figure 8.26 Magnitudes of current distribution along horizontal dipole antenna over lossy half-space with $|\Psi| = 5.1083$, $a = 0.05$ m, $L = 5$ m, $h = 5$ m, $f = 60$ MHz and $\varepsilon_r = 10$, $\sigma = 0.001$ mhos/m.

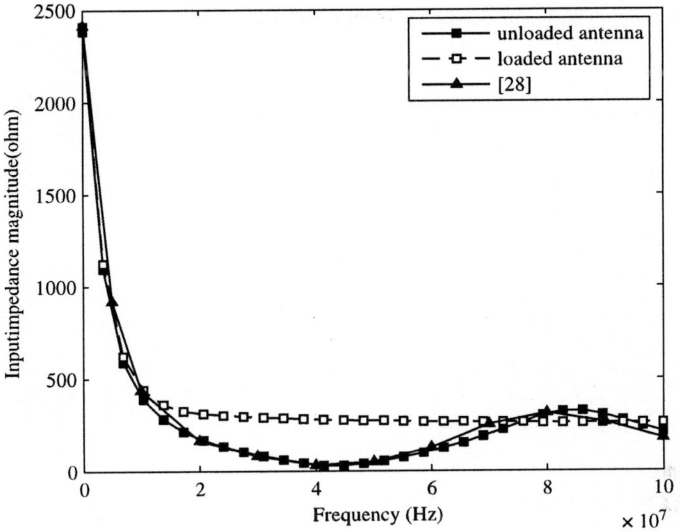

Figure 8.27 Magnitude of input impedance of loaded dipole antenna with $|\Psi| = 5.1083$, $a = 0.05\,\text{m}$, $L = 0.5\,\text{m}$ and $\varepsilon_r = 10$, $\sigma = 0.001\,\text{mhos/m}$.

Figures 8.27 and 8.28 deal with the spectrum of input impedance of unloaded and loaded dipole antenna immersed in lossy medium ($|\psi| = 5.1083$, $a = 0.05\,\text{m}$, $L = 0.5\,\text{m}$, and $\varepsilon_r = 10$, $\sigma = 0.001\,\text{mhos/m}$) and placed above lossy medium ($|\psi| = 5.1083$, $a = 0.05\,\text{m}$, $L = 5\,\text{m}$, $h = 3\,\text{m}$, and $\varepsilon_r = 40$, $\sigma = 0.001\,\text{mhos/m}$)

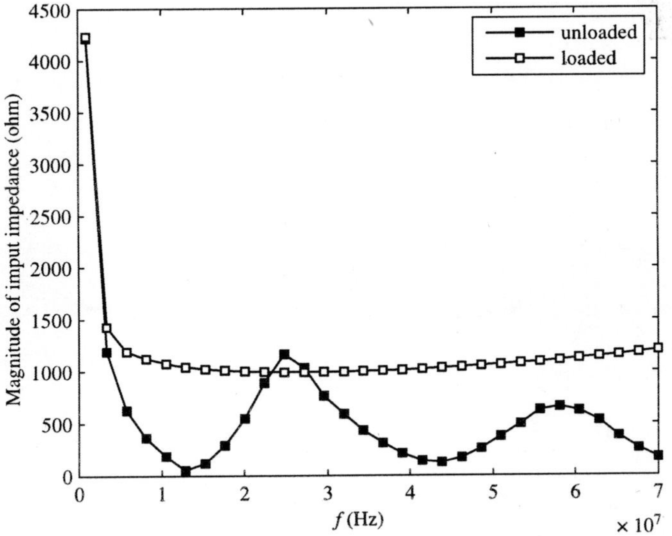

Figure 8.28 Magnitude of input impedance of horizontal loaded dipole antenna with $|\Psi| = 5.1083$, $a = 0.05\,\text{m}$, $L = 5\,\text{m}$, $h = 3\,\text{m}$, $\varepsilon_r = 40$, $\sigma = 0.001\,\text{mhos/m}$.

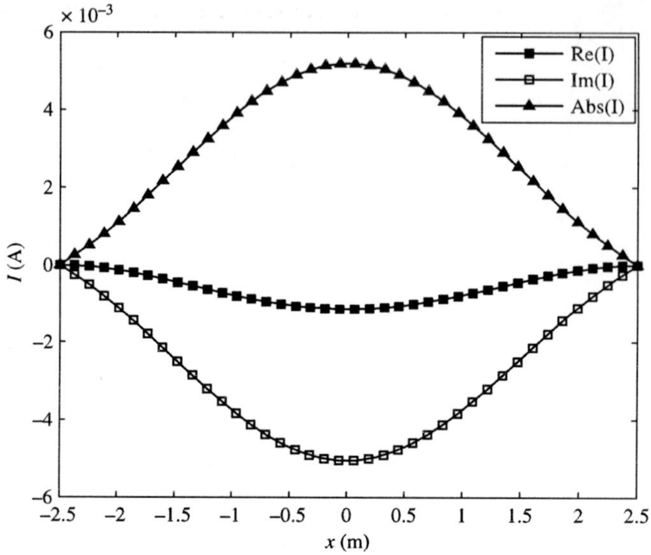

Figure 8.29 Current induced along the 5 m long scatterer illuminated by the plane wave.

respectively. The effect of resistive loading is obvious, that is, the input impedance magnitude of unloaded antenna vs frequency varies significantly, while in the case of loaded antenna it remains fairly constant in a broad frequency range.

Finally, Figures 8.29 and 8.30 are related to the RCA approximation. The current distribution induced along the 5 m and 10 m long wires illuminated by the plane wave is presented.

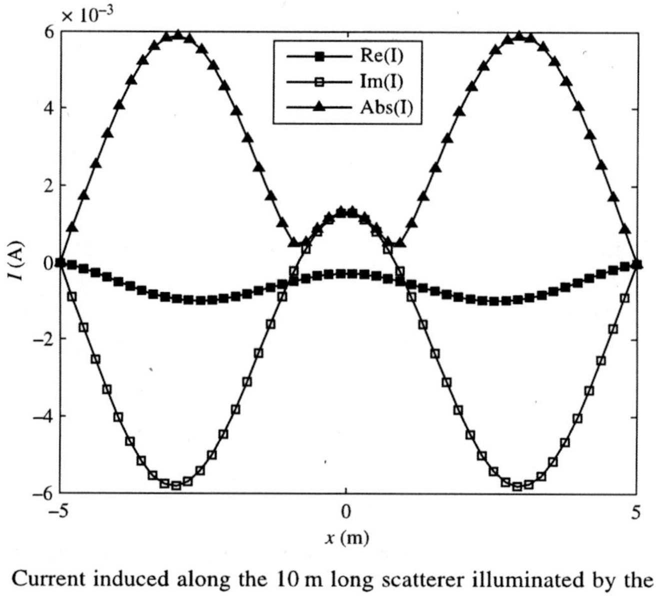

Figure 8.30 Current induced along the 10 m long scatterer illuminated by the plane wave.

8.2.8 Multiple Straight Wire Antennas Over a Lossy Half-Space

The electromagnetic modeling of multiple wire configurations in the presence of a dissipative half-space is an important part in antenna design and electromagnetic compatibility (EMC) studies. The analysis of wire antennas above a dissipative half-space is directly applicable to the area of oceanographic and geophysical researches and underwater communications. Many electromagnetic interference (EMI) sources such as underground and aboveground cables, power lines, transmission lines, mobile phones, base stations, and others can be analyzed using antenna models.

In particular, the calculation of the current induced along multiple overhead wires due to a plane wave is a key point in such analyses.

The current distribution along the multiple wire structure is governed by the set of Pocklington equation for half-space problems. The influence of lossy half-space can be taken into account via the reflection coefficient (RC) approximation [3, 4].

In this case, the geometry of interest consists of an arbitrary number of M parallel straight wires horizontally placed above an imperfect ground at height h. All wires are assumed to have the same radius a and the length of the mth wire is equal to L_m.

According to the wire antenna theory and RC approximation, the set of coupled Pocklington integral equations is given by [3, 4]

$$E_x^{exc} = -\frac{1}{j4\pi\omega\varepsilon_0}\sum_{n=1}^{M}\int_{-L_n}^{L_n}\left[\frac{\partial^2}{\partial x^2}+k_1^2\right][g_{0mn}(x,x') \\ - R'_{TM}g_{imn}(x,x')]I_n(x')\mathrm{d}x' \quad m = 1,2,\ldots,M \tag{8.246}$$

where $I_n(x')$ is the unknown current distribution induced on the nth wire axis, $g_{0mn}(x,x')$ denotes the free-space Green's function,

$$g_{0mn}(x,x') = \frac{e^{-jk_1R_{1mn}}}{R_{1mn}} \tag{8.247}$$

while in accordance to the image theory, $g_{imn}(x,x')$ is given by

$$g_{imn}(x,x') = \frac{e^{-jk_1R_{2mn}}}{R_{2mn}} \tag{8.248}$$

where k_1 is the propagation constant of free space and R_{1mn} and R_{2mn} are distances from the source point and from the corresponding image to the observation point that are defined by

$$R_{1mn} = \sqrt{(x-x')^2+a_m^2}, \\ R_{2mn} = \sqrt{(x-x')^2+4h^2}, \quad m = n \tag{8.249}$$

$$R_{1mn} = \sqrt{(x-x')^2+D_{mn}^2}, \\ R_{2mn} = \sqrt{R_{1mn}^2+4h^2}, m \neq n \tag{8.250}$$

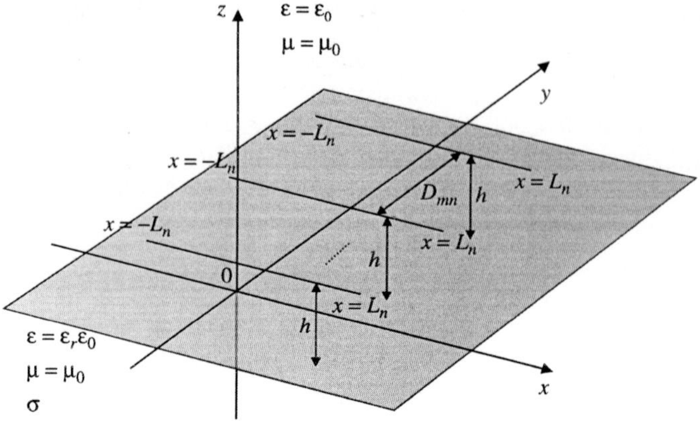

Figure 8.31 The geometry of the problem.

D_{mn} is the separation between mth and nth antenna (Fig. 8.31).

The influence of an imperfectly conducting lower medium is taken into account by means of the Fresnel plane wave reflection coefficient,

$$R'_{TM} = \frac{\underline{n}\cos\theta' - \sqrt{\underline{n} - \sin^2\theta'}}{\underline{n}\cos\theta' + \sqrt{\underline{n} - \sin^2\theta'}} \tag{8.251}$$

where θ' and $\underline{n}$ are given by

$$\theta' = arctg\frac{|x - x'|}{2h} \qquad m = n \tag{8.252}$$

$$\theta' = arctg\frac{R_{1mn}}{2h} \qquad m \neq n \tag{8.253}$$

$$n = \frac{\varepsilon_{eff}}{\varepsilon_0} \tag{8.254}$$

The wires are excited by a plane wave of arbitrary incidence (Fig. 8.32).

The tangential component of an incident plane wave can be represented in terms of its vertical E_V and horizontal E_H component as shown in Figure 8.32. Consequently, the imposed electric field at the surfaces of the conductors can be obtained as a sum of direct and reflected field components [4, 66, 67],

$$\begin{aligned} E_x^{exc} &= E_x^i + E_x^r \\ &= E_0(\sin\alpha\sin\phi - \cos\alpha\cos\theta\cos\phi)e^{-jk_1\vec{n}_i\cdot\vec{r}} \\ &\quad + E_0(R_{TE}\sin\alpha\sin\phi + R_{TM}\cos\alpha\cos\theta\cos\phi)e^{-jk_1\vec{n}_r\cdot\vec{r}} \end{aligned} \tag{8.255}$$

where α is an angle between *E-field* vector and the plane of incidence, R_{TM} and R_{TE} are the vertical and horizontal Fresnel reflection coefficients at the air-earth

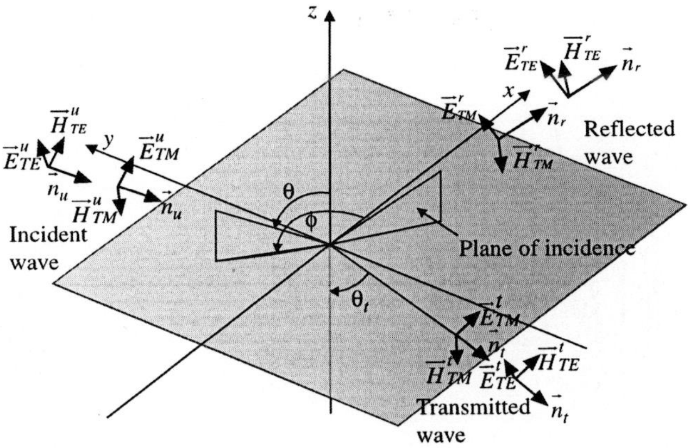

Figure 8.32 Incident, reflected and transmitted wave.

interface given by [4, 66, 67]

$$R_{TM} = \frac{\underline{n}\cos\theta - \sqrt{\underline{n} - \sin^2\theta}}{\underline{n}\cos\theta + \sqrt{\underline{n} - \sin^2\theta}} \tag{8.256}$$

$$R_{TE} = \frac{\cos\theta - \sqrt{\underline{n} - \sin^2\theta}}{\cos\theta + \sqrt{\underline{n} - \sin^2\theta}} \tag{8.257}$$

$\vec{n}_i \cdot \vec{r}$ and $\vec{n}_r \cdot \vec{r}$ are distances from the origin point to the observation point for incident and reflected waves, respectively,

$$\begin{aligned} \vec{n}_i \cdot \vec{r} &= -x\sin\theta\cos\phi - y\sin\theta\sin\phi - z\cos\theta \\ \vec{n}_r \cdot \vec{r} &= -x\sin\theta\cos\phi - y\sin\theta\sin\phi + z\cos\theta \end{aligned} \tag{8.258}$$

The set of integral equations (8.246) for straight wire array above a lossy half-space is handled by the boundary element method (BEM) featuring linear or isoparametric element approximation, respectively. The mathematical details regarding these procedures have already been presented and are outlined below for the sake of completeness.

8.2.9 Electric Field of a Wire Array Above a Lossy Half-Space

The analysis of the electric field scattered from multiple wire configuration above a real ground is carried out featuring the RCA approach, as the Sommerfeld integral approach is too demanding regarding analytical and numerical efforts.

The electric field components can be obtained from eqn (8.170) and they are given as follows:

$$E_x = \frac{1}{j4\pi\omega\varepsilon_0}\sum_{n=1}^{M}\left[-\int_{-L_n}^{+L_n}\frac{\partial I_n(x')}{\partial x'}\frac{\partial G_{nm}(x,x')}{\partial x'}\mathrm{d}x' + k^2\int_{-L_n}^{+L_n}G_{nm}(x,x')I_n(x')\mathrm{d}x'\right] \quad (8.259)$$

$$E_y = \frac{1}{j4\pi\omega\varepsilon_0}\sum_{i=1}^{M}\left[\int_{-L_2}^{+L_n}\frac{\partial I_n(x')}{\partial x'}\frac{\partial G_{nm}(y,x')}{\partial y}\mathrm{d}x'\right] \quad (8.260)$$

$$E_z = \frac{1}{j4\pi\omega\varepsilon_0}\sum_{i=1}^{M}\left[\int_{-L_n}^{+L_n}\frac{\partial I_n(x')}{\partial x'}\frac{\partial G_{nm}(z,x')}{\partial z}\mathrm{d}x'\right] \quad (8.261)$$

8.2.10 Boundary Element Analysis of Wire Array Above a Lossy Ground

The Galerkin boundary element procedure starts by applying the standard representation of the unknown current along an ith wire,

$$I_n(x') = \sum_{n=1}^{N_g} I_{ni}f_{ni}(x') \quad (8.262)$$

where $f_{ni}(x)$ denotes the linear elements shape functions, I_{ni} stands for the unknown coefficients of the solution, and N_g denotes the total number of basis functions.

The weak Galerkin-Bubnov formulation of the system of integro-differential equations (8.246) for N_w wires is given by

$$\sum_{n=1}^{N_g}\sum_{i=1}^{N_w}\left[-\int_{-L_m}^{L_m}\int_{-L_n}^{L_n}\frac{\mathrm{d}f_{jm}(x)}{\mathrm{d}x}\frac{\mathrm{d}f_{in}(x')}{\mathrm{d}x'}G_{nm}(x,x')\mathrm{d}x'\mathrm{d}x + k^2\int_{-L_m}^{L_m}\int_{-L_n}^{L_n}f_{jm}(x)f_{in}(x')G_{nm}(x,x')\mathrm{d}x'\mathrm{d}x\right]I_{ni}$$
$$-j4\pi\omega\varepsilon\int_{-L_m}^{L_m}E_{xj}^{inc}(x)f_{jm}(x)\mathrm{d}x$$
$$m = 1,2,\ldots,Ng;\quad j = 1,2,\ldots,Nw \quad (8.263)$$

Performing the boundary element discretization of the wire array leads to the matrix equation of the form:

$$\sum_{k=1}^{M}[Z]_{pk}\{I\}_k = \{V\}_p;\quad p = 1,2,\ldots,M \quad (8.264)$$

where M is the total number of elements along the actual multiple wire configuration and $[Z]_{lk}$ is the interaction matrix representing the mutual impedance between each segment on the ith (source) wire to every segment on the jth (observation) wire,

$$[Z]^e_{pk} = -\int_{-1}^{1}\int_{-1}^{1} \{D\}_l\{D'\}^T_k G_{nm}(\xi,\xi')\mathrm{d}\xi'\mathrm{d}\xi + k^2\int_{-1}^{1}\int_{-1}^{1} \{f\}_l\{f'\}^T_k G_{nm}(\xi,\xi')\left(\frac{\Delta x}{2}\right)^2 \mathrm{d}\xi'\mathrm{d}\xi \tag{8.265}$$

Vectors $\{f\}$ and $\{f'\}$ contain shape functions $f_n(\xi)$ and $f_n(\xi')$, while $\{D\}$ and $\{D'\}$ contain their derivatives, and Δx is the segment length.

The right-hand side vector $\{V\}_l$ represents the voltage along the lth segment and it is given as follows:

$$\{V\}_l = -j4\pi\omega\varepsilon_0 \int_{-1}^{1} E_x^{inc}(\xi)\{f\}_p\frac{\Delta x}{2}\mathrm{d}\xi \tag{8.266}$$

In the transmitting mode (also called *the antenna mode*) right-hand side vector is different from zero only in the feed-gap area (Fig. 8.11) and the incident field can be written as

$$E_x^{inc} = \frac{V_g}{\Delta l_g} \tag{8.267}$$

where V_g is the feed voltage and Δl_g is the feed-gap width.

8.2.11 Near-Field Calculation for Wires Above Half-Space

The near-field analysis of wire antennas has drawn the attention of many researchers as many EMC applications require knowledge of the near-field behavior.

In order to obtain electric field value at the arbitrary point in the upper medium, the calculated current distribution $I(x')$ is substituted into the Pocklington eqn (8.30). The values of a current and its first derivative for the ith boundary element are given by

$$I(x') = I_{1i}\frac{x_{2i}-x'}{\Delta x} + I_{2i}\frac{x'-x_{1i}}{\Delta x} \tag{8.268}$$

$$\frac{\partial I(x')}{\partial x'} = \frac{I_{2i}-I_{1i}}{\Delta x} \tag{8.269}$$

where I_{1i} and I_{2i} are current values at the local nodes of the ith boundary element with coordinates x_{1i} and x_{2i}, and $\Delta x = x_{2i} - x_{1i}$ denotes dimension of the element.

Substituting current expressions (8.42) and (8.43) in eqn (8.30) results in the following equation:

$$E_x = \frac{1}{j4\pi\omega\varepsilon_0}\sum_{n=1}^{M}\sum_{i=1}^{N_j}\left[-\frac{I_{i+1,n} - I_{in}}{\Delta x_j}\int_{x_{1ij}}^{x_{2ij}}\frac{\partial G_{nm}(x,x')}{\partial x'}\mathrm{d}x' + k^2\int_{x_{1ij}}^{x_{2ij}} G_{nm}(x,x')I_{in}(x')\mathrm{d}x'\right] \tag{8.270}$$

where M is the total number of wires and N_j denotes the total number of boundary elements on the jth wire, and G is the Green's function given by

$$G_{nm}(x,x') = g_{0nm}(x,x') - R_{TM}g_{inm}(x,x') \tag{8.271}$$

Applying a similar procedure, it is possible to obtain the expressions for the y and z components of the electric field at an arbitrary point in the air,

$$E_y = \frac{1}{j4\pi\omega\varepsilon_0}\sum_{n=1}^{M}\sum_{i=1}^{N_j}\frac{I_{i+1,n} - I_{in}}{\Delta x_j}\int_{x_{1ij}}^{x_{2ij}}\frac{\partial G_{nm}(x',y)}{\partial y}\mathrm{d}x' \tag{8.272}$$

$$E_z = \frac{1}{j4\pi\omega\varepsilon_0}\sum_{n=1}^{M}\sum_{i=1}^{N_j}\frac{I_{i+1,n} - I_{in}}{\Delta x_j}\int_{x_{1ij}}^{x_{2ij}}\frac{\partial G_{nm}(x',z)}{\partial z}dx' \tag{8.273}$$

The integrals in the above expressions are numerically evaluated using Gaussian quadrature. Because of quasisingularity of the Green's function, the first-order differential operator appearing in the integral equation kernel could cause numerical instability.

In order to avoid the problem of quasisingularity of the Green's function, the first derivative of that function is approximated by means of a central finite difference formula

$$\frac{\partial f(x,y)}{\partial x} = \frac{f(x+\Delta x,y) - f(x-\Delta x,y)}{2\Delta x} \tag{8.274}$$

where Δx stands for the size of the finite difference step.

8.2.12 Computational Examples for Wires Above a Lossy Half-Space

First, the behavior of two coupled dipole antennas parallel to each other are placed horizontally over tap water is analyzed. Both wires are of length $L = 0.254$ m, located at a height $h = 0.054$ m, and the distance between them is $D = 0.127$ m. The diameter of wires are $2a_1 = 1.65$ mm and $2a_2 = 6.35$ mm, respectively.

The active wire is driven by a unit time-harmonic voltage source operating at the frequency of 840 MHz.

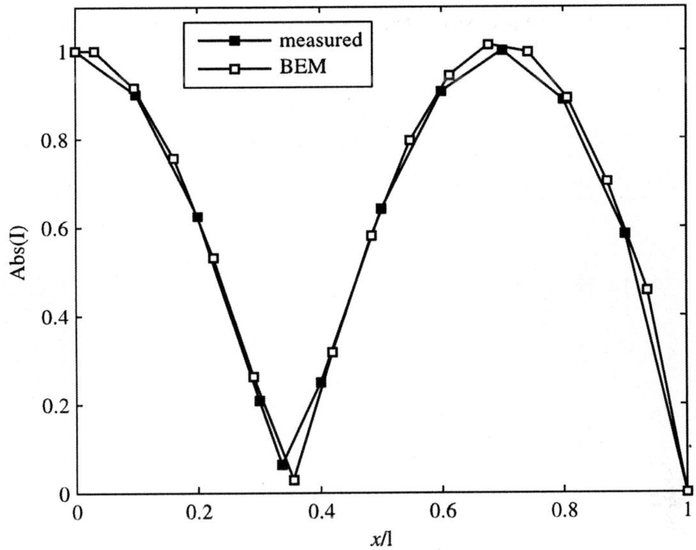

Figure 8.33 Normalized magnitudes of current distribution along the parasitic wire over tap water at f $= 840\,\mathrm{MHz}$, $\varepsilon_r = 80$, $\sigma = 0.04\,\mathrm{mho/m}$, $h = 5.4\,\mathrm{cm}$, $D = 12.7\,\mathrm{cm}$, $L = 24.5\,\mathrm{cm}$, $2a = 6.35\,\mathrm{mm}$.

Figures 8.33 and 8.34 show the normalized magnitude and phase of the current induced on the parasitic wire. The satisfactory agreement between the BEM results and the experimental results available from Ref. [68] can be noticed.

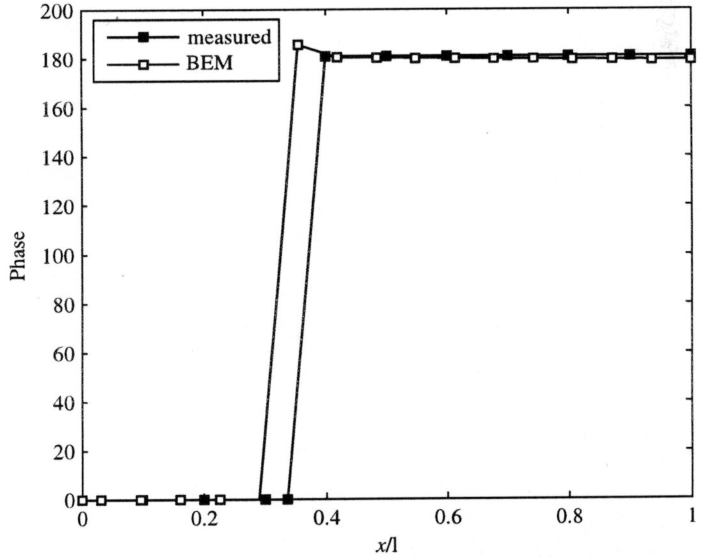

Figure 8.34 Phases of current distribution along the parasitic wire over tap water at $f = 840\,\mathrm{MHz}$, $\varepsilon_r = 80$, $\sigma = 0.04\,\mathrm{mho/m}$, $h = 5.4\,\mathrm{cm}$.

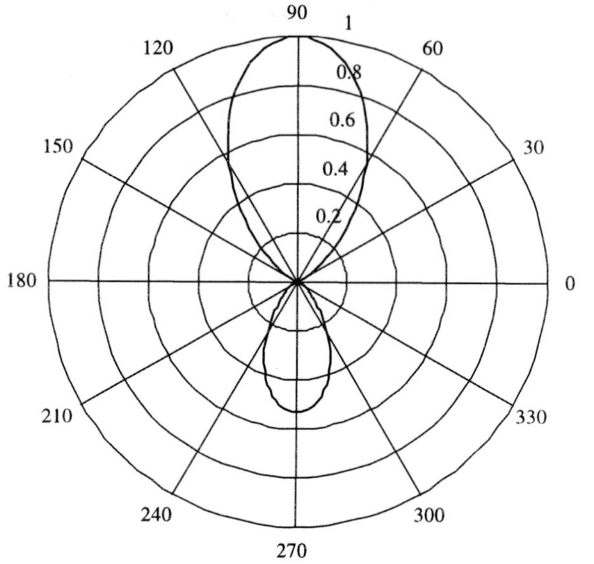

Figure 8.35 Electric field pattern for the Yagi-Uda antenna E_x component; xy plane.

The numerical results for the field pattern are related to the three element Yagi-Uda antenna. The radius of wires is $a = 0.0025\,\text{m}$ and length of wires are 0.479 m, 0.453 m, 0.451 m, respectively. The distance between the reflector and the fed element is $D_r = 0.25\,\text{m}$, whereas the distance to the director is also $D_d = 0.25\,\text{m}$. Operating frequency of the voltage generator is $f = 300\,\text{MHz}$.

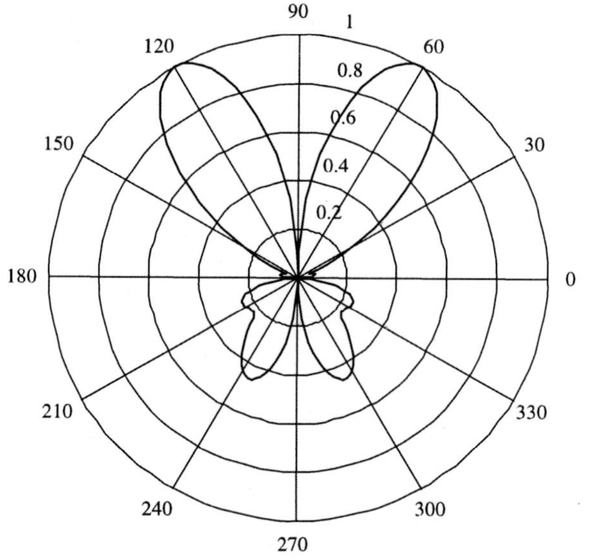

Figure 8.36 Electric field pattern for the Yagi-Uda antenna E_y component; xy plane.

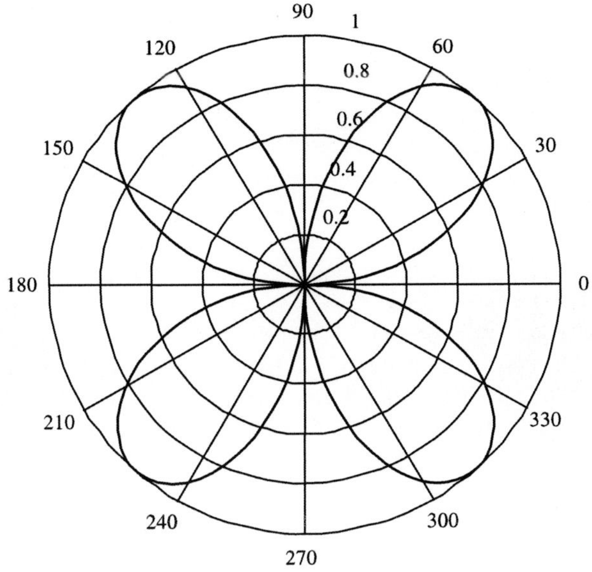

Figure 8.37 Electric field pattern for the Yagi-Uda antenna E_z component; *xz* plane.

Figures 8.35–8.38 show normalized radiation patterns for different field components. By analyzing radiation patterns for the *x* component of the electric field, a significant attenuation in the reflectors direction can be observed. This results in a higher directivity of the antenna system.

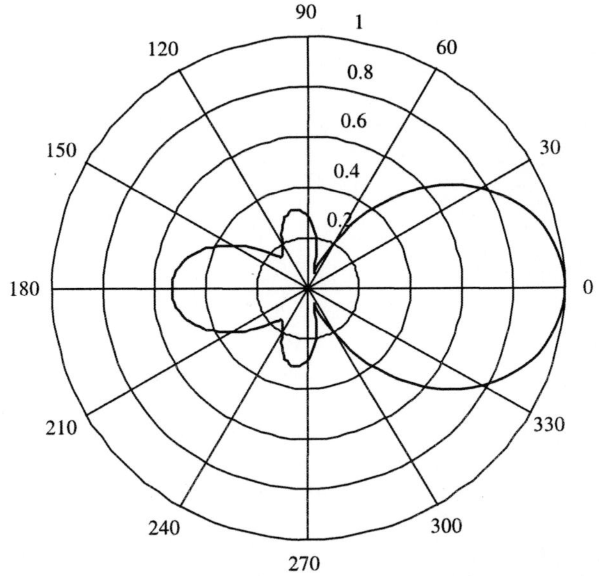

Figure 8.38 Electric field pattern for the Yagi-Uda antenna E_x component; *yz* plane.

There is no such attenuation for the case of y and z field components. However, these patterns could be neglected compared to the x component, since the amplitude of this component is lower for the order of magnitude.

8.3 REFERENCES

[1] D. Poljak, V. Doric, V. Roje: Boundary element modeling of arbitrary thin wires configuration, 22th International Conference on the Boundary Element Method, BEM XXII, *Cambridge, UK*, September 2000, pp 417–427.

[2] D. Poljak, V. Doric, V. Roje: Galerkin-Bubnov Boundary element analysis of the Yagi-Uda array, 24th International Conference on the Boundary Element Method, BEM XXIV, *Sintra, Portugal*, June 2002, pp 457–463.

[3] D. Poljak: Electromagnetic Modeling of Wire Antenna Structures, *WIT Press, Southampton, Boston*, 2002.

[4] D. Poljak, C. A. Brebbia: Boundary Element Methods for Electrical Engineers, *WIT Press, Southampton, Boston*, 2005.

[5] R. P. Silvester, K. K. Chan: Bubnov-Galerkin solutions to wire antenna problems, *Proc. IEE 119*, 1972, pp 1095–1099.

[6] R. P. Silvester, K. K. Chan: Analysis of antenna structures assembled from arbitrarily located straight wires, *Proc. IEE 120 (1)*, 1973.

[7] D. Poljak, V. Roje: The weak finite element formulation for solving Pocklington's integro-differential equation, *Int. Journ. Eng. Model. 6 (1–4)*, 1993, pp 21–25.

[8] D. Poljak: New numerical approach in analysis of a thin wire radiating over a lossy half-space, *Int. J. Num. Meth. Eng. 38 (22)*, 1995, pp 3803–3816.

[9] D. Poljak, V. Roje: Boundary-element approach to calculation of wire antenna parameters in the presence of dissipative half-space, *IEE Proc. Microw. Antennas Propag. 142 (6)*, 1995, pp. 435–440.

[10] D. Poljak, V. Roje: Boundary Element Analysis of Resistively Loaded Wire Antenna Immersed in a Lossy Medium, 19th International Conference on Boundary Elements, BEM 19, *Rome*, September 1997, pp 485–493.

[11] D. Poljak, V. Roje: The integral equation method for ground wire input impedance, in: C. Constanda, J. Saranen, S. Seikkala (Eds): Integral Methods in Science and Engineering, Vol. I, *Longman, UK*, 1997, pp 139–143.

[12] K. K. Mei: On the Integral equations of Thin Wire Antennas, *IEEE Trans. AP, 13*, 1965, pp 59–62.

[13] P. E. Atlamazoglu, N. K. Uzunoglu: A Galerkin Moment method for the Analysis of an Insulated Antenna in a Dissipative Dielectric Medium, *IEEE Trans MTT, Vol. 46, No. 7*, July 1998, pp 988–996.

[14] R. W. P. King et al.: The electromagnetic field of an insulated antenna in a conducting or dielectric medium, IEEE Trans. Microwave Theory Tech., Vol. MTT-31, July 1983, pp 574–583.

[15] D. Poljak, C. A. Brebbia: Indirect Galerkin-Bubnov boundary element analysis of coated thin wire in free space, in: Kassab, C. A. Brebbia, E. Divo, D. Poljak, (Eds), Boundary

Elements XXVII A, *Southampton, UK, WIT Press, Boston, USA, Computational Mechanics Inc.*, 2005.

[16] D. Poljak, C. A. Brebbia: Indirect Galerkin-Bubnov boundary element method for solving integral equations in electromagnetics, Engineering Analysis With Boundary Elements, EABE 28, No. 7, July 2004, pp 771–778.

[17] J. Moore, M. A. West: Simplified analysis of coated wire antennas and scatterers, *IEE Proc. Microwaves Antennas Propag., 142 (1)*, February 1995, pp 14–18.

[18] L. W. Rispin, D. C. Chang: Wire and loop antennas, in: Y. T. Lo and S. W. Lee (eds), Antenna Handbook, Volume II: Antenna Theory, Chapter 7, *Van Nostrand Reinhold, New York*, 1993.

[19] P. L. G. Overfelt: Near fields of the constant thin circular loop antenna of arbitrary radius, *IEEE Trans. AP, 44 (2)*, February 1996, pp 166–171.

[20] N. J. Champagne II, J. T. Williams, D. R. Wilton: The use of curved segments for numerically modeling thin wire antennas and scatterers, *IEEE Trans. AP, 40 (6)*, June 1992, pp 682–688.

[21] L. Ronglin, N. Guangzheng, Y. Jihui, J. Zejia: A new numerical technique for calculating current distribution on curved wire antennas – parametric B-spline finite element method, *IEEE Trans. Magnetics, 32 (3)*, May 1996, pp 906–909.

[22] L. Ronglin, N. Guangzheng: Numerical analysis of 4-arm archimedian printed spiral antenna, *IEEE Trans. Magnetics, 33* (2), March 1997, pp 1512–1515.

[23] L. Vegni, A. Toscano, D. Bonefai: Efficient moment-method analysis of a magnetic dipole, *Microwave Opt. Technol. Lett. 13 (6)*, December 1996.

[24] D. Poljak: Finite element integral equation modeling of a thin wire loop antenna, *Commun. Numerical Methods in Engineering, 14*, 1998, pp 347–354.

[25] R. F. Harrington: Matrix methods for field problems, *Proc. IEEE, 55* (2), February 1967, pp 136–149.

[26] C. M. Butler, D. R. Wilton: Analysis of various numerical techniques applied to thin wire scatterers, *IEEE Trans. AP, 23*, 1975.

[27] C. M. Butler, D. R. Wilton: Efficient numerical techniques for solving Pocklington's integral equation and their relationship to other methods, *IEEE Trans. AP, 24*, 1976.

[28] E. K. Miller, A. J. Poggio, G. J. Burke, E. S. Selden: Analysis of wire antennas in the presence of a conducting half-space, Part II. The horizontal antenna in free space, *Can. J. Phy., 50*, 1972, pp 2614–2627.

[29] T. K. Sarkar: Analysis of arbitrarily oriented thin wire antennas over a plane imperfect ground, *Archiv fur elektronik und ubertragungstechnik, 31*, 1977, pp 449–457.

[30] P. Parhami, R. Mittra: Wire antennas over a lossy half-space, *IEEE Trans. AP, 28*, 1980, pp 397–403.

[31] G. Miano, L. Verolino, V. G. Vacaro: A new numerical treatment for Pocklington's integral equation, *IEEE Trans. Magnetics, 32 (3)*, 1996.

[32] R. Y. Tay, Q. Balzano, N. Kuster: Dipole configuration with strongly improved radiation efficiency for Hand-Held transceivers, *IEEE Trans. AP, 46*, June 1998, pp 798–806.

[33] P. P. Silvester, R. L. Ferrari: Finite Elements for Electrical Engineers, 3rd edition, *Cambridge University Press, Cambridge*, 1996.

[34] E. R. Adair, R. C. Petersen: Biological effects of radio-frequency/microwave radiation, *IEEE Trans. Microwave Theory and Tech., 50 (3)*, March 2003, pp 953–962.

[35] D. Poljak: Human Exposure to Electromagnetic Fields, *WIT Press, Southampton, Boston*, 2003.

[36] P. Bernardi, M. Cavagnaro, S. Pisa, E. Piuzzi: Human exposure in the vicinity of radio base station antennas, 4th European Symposium on Electromagnetic Compatibility (EMC Europe 2000), *Brugge, Belgium*, September 11–15, 2000, pp 187–192.

[37] P. Bernardi, M. Cavagnaro, S. Pisa, E. Piuzzi: Human exposure to radio base station antennas in urban environment, *IEEE Trans. MTT, 48 (11)*, November 2000, pp 1996–2002.

[38] D. Poljak, A. Sarolic, V. Roje: Human interaction with the electromagnetic field radiated from a cellular base station antennas, Proceedings of IEEE International Symposium on EMC (EMC Europe 2002), *Sorrento, Italy*, 2002, pp 965–968.

[39] N. Herscovici, C. Christodoulou: On the design of a dual-band base station wire antenna, *IEEE Antennas and Propagation Magazine, 42 (6)*, December 2000.

[40] D. Poljak, C. A. Brebbia: Boundary element modeling of base station antennas, Engineering Analysis with Boundary Elements, EABE, 2006.

[41] G. J. Burke, A. J. Poggio: Numerical electromagnetics code (NEC)—Method of Moments, Part I, NEC Program Description—Theory, *Lawrence Livermore Laboratory, Livermore*, 1981.

[42] T. K. Sarkar: Analysis of arbitrarily oriented thin wire antennas over a plane imperfect ground, *Archiv fur elektronik und ubertragungstechnik, 31*, 1977, pp 449–457.

[43] P. Parhami, R. Mittra: Wire antennas over a lossy half-space, *IEEE Trans. AP, 28*, 1980, pp 397–403.

[44] B. H. Mc Donald, A. Wexler: Mutually constrained partial differential and integral equation field formulations, in: M. V. K. Chari and P. P. Silvester (Eds) Finite Elements in Electrical and Magnetic Field Problems, Chapter 9, *John Wiley & Sons*, 1980.

[45] L. Grcev, Z. Haznadar: A novel technique of numerical modeling of impulse current distribution in grounding systems, Proceedings of 19th International conference of lightning protection, *Graz*, 1988, paper 3.4, pp 165–169.

[46] Y. Rahmat-Samii, P. Parhami, R. Mittra: Loaded horizontal antenna over an imperfect ground, *IEEE Trans. Antennas Propagation, 26 (6)*, November 1978, pp 789–796.

[47] Y. Rahmat-Samii, P. Parhami, R. Mittra: Evaluation of Sommerfeld integrals for lossy half-space problems, *Electromagnetics 1*, January 1981, pp 1–28.

[48] W. A. Johnson, D. G. Dudley: Real axis integration of sommerfeld integrals: source and observation point in air, in: Hansen R. (Ed.), Moment Methods in Antennas and Scattering, *Artech House*, 1990, pp 313–324.

[49] B. Drachman, M. Cloud, D. P. Nyquist: Accurate evaluation of Sommerfeld integrals using the fast fourier transform, *IEEE Trans., Antennas Propagation, 37 (3)*, March 1989, pp 403–407.

[50] K. A. Michalski: On the efficient evaluation of integrals arising in the Sommerfeld half-space problem, in: Hansen R. (Ed.), Moment Methods in Antennas and Scattering, *Artech House*, 1990, pp 325–331.

[51] S. L. Dvorak: Application of the fast fourier transform to the computation of the Sommerfeld integral for a vertical electric dipole above a half-space, *IEEE Trans., Antennas Propagation, 40*, 1992, pp 798–805.

[52] R. W. P. King, S. S. Sheldon: The electromagnetic field of a vertical electric dipole over the earth or sea, *IEEE Trans. Antennas Propagation, 42 (3)*, March 1994, pp 382–389.

[53] L. Y. Liang, J. H. Jan: Error analysis for far-field approximate expression of Sommerfeld type integral, *IEEE Trans. Antennas Propagation, 42 (4)*, April 1994, pp 574.

[54] R. C. Babu: Impedance calculations for horizontal wire antennas above a lossy earth, *J. Electromagnetic Waves Appl., 11*, 1997, pp 1567–1591.

[55] V. Doric, D. Poljak, V. Roje: Electromagnetic field coupling to multiple finite length transmission lines above an imperfect ground, Proceedings of the 2003 IEEE International Symposium on Electromagnetic Compatibility (EMC), *IEEE Press*, 2003.

[56] D. Poljak, V. Roje: Induced currents and voltages along a horizontal wire above a lossy ground, 21th International Conference on Boundary Elements, BEM 21, *Oxford, England, UK*, August 25–27, 1999, pp 185–194.

[57] T. T. Wu, R. W. P. King: The cylindrical antenna with non-reflecting resistive loading, *IEEE Trans, Antennas Propagation, 13 (5)*, May 1965, pp 369–373.

[58] B. D. Popovic: Theory of cylindrical antennas with arbitrary impedance loading, *Proc. IEE, 118 (10)*, October 1971, pp 1327–1332.

[59] L. C. Shen: An experimental study of the antenna with non-reflecting resistive loading, *IEEE Trans. Antennas Propagation, 13 (5)*, 1967, pp 606–611.

[60] D. Poljak, B. Jajac, R. Simundic: Current induced along horizontal wire above an imperfectly conducting half-space, *Eng. Anal. Boundary Elements 23*, 1999, pp 835–840.

[61] M. Siegel, R. W. P. King: Radiation from linear antennas in a dissipative half-space, *IEEE Trans. Antennas Propagation, 19 (4)*, July 1971, pp 477–485.

[62] I. K. Nikoskinen, I. V. Lindell: Time-domain analysis of the Sommerfeld VMD problem based on the exact image theory, *IEEE Trans. Antennas Propagation, 38* (2), February 1990, pp 241–250.

[63] K. Sivaprasad: An asymptotic solution of dipoles in a conducting medium, *IEEE Trans. Antennas Propagation, 11*, March 1963, pp 133–142.

[64] A. W. Rudge: Input impedance of a dipole antenna above a conducting half-space, *IEEE Trans. Antennas Propagation, 20*, (*1*), January 1972, pp 86–89.

[65] R. Tiberio, G. Manara, G. Pelosi: A hybrid technique for analysing wire antennas in the presence of a plane interface, *IEEE Trans. Antennas Propagation, 33*, (*8*), August 1985.

[66] G. E. Bridges, L. Shafai: Plane wave coupling to multiple conductor transmission lines above a lossy earth, *IEEE Trans. Electromagn. Compat., EMC-31*, February 1989, pp 21–33.

[67] G. E. Bridges: Transient plane wave coupling to bare and insulated cables buried in a lossy half-space, *IEEE Trans. Electromagn. Compat., EMC-37*, February 1995, pp 62–70.

[68] K. A. Michalski, C. E. Smith, C. M. Butler: Analysis of a Horizontal Two-Element Antenna Array Above a Dielectric Half-Space, *IEE Proc. Microw. Antennas Propag., 132*, (*5*), August 1985, pp 335–338.

9

WIRE ANTENNAS AND SCATTERERS: TIME DOMAIN ANALYSIS

Time-domain (TD) modeling of transient radiation and scattering from thin wires has been found to be very efficient in revealing a number of electromagnetic phenomena that are not readily perceived in the frequency domain [1–12]. There are a number of applications of TD analysis, including short-pulse radar, lightning effects, wide-band radio communication, electromagnetic pulse, bio electromagnetics, and others.

Generally, there is no definitive advantage that could be gained by using the direct time-domain solution or using the indirect approach featuring the Inverse Fourier Transform (IFT). If the solution for a large number of incident waves arriving from various directions is of interest, then the frequency-domain approach is appropriate. Compared to the time-domain approach, frequency-domain modeling of highly resonant structures involves larger number of calculations to resolve the high frequency characteristics.

On the contrary, the time-domain approach would be a more suitable choice if the transient response is being calculated only for the early time period, as the frequency-domain approach requires computation of the frequency response up to the maximum effective frequency, and the entire range of frequency spectrum is to be transformed.

Regardless of being a rather demanding task, time-domain modeling offers a fundamental knowledge and, in general, a deeper insight into the electromagnetic phenomena [9]. Direct TD approach also provides some computational advantages [9–12].

Most of the known direct TD numerical techniques for solving various types of thin wire integral equations are often related to the wires insulated in the free space, or to the wires above a perfectly conducting ground [3,7,9]. These techniques can be referred to as several variants of point matching/marching-on-in-time procedure [9], which are relatively simple, but suffer from serious disadvantages – a very poor convergence rate and an appearance of the late time spurious oscillations [12]. The application of certain smoothing procedures, therefore, becomes necessary for obtaining any valid results. These problems can be overcome by using an efficient TD formulation for thin wires on the basis of the Hallen integral equation [5,6].

Besides providing a different perspective of electromagnetic behavior, TD modeling can be a very useful tool, and can also yield insight regarding the behavior of the energy stored in the near fields of an impulsively excited wire. In particular, the analysis of thin wire structures with a nonlinear loading is of great interest in a variety of EMC applications. Namely, the study of nonlinear phenomena is a key point in designing of protection devices since nonlinear elements attenuate the undesirable extremely strong electromagnetic fields mostly caused by a lightning or nuclear electromagnetic pulse (LEMP and NEMP). Therefore, some energy or average power measures are required for the efficiency estimation of the protection devices.

The transient response of a single thin wire with a nonlinear load, insulated in free space, has already been calculated by several authors. Liu and Tesche [13] used the Hallen integral equation approach, while Landt and Miller [14] dealt with the Pocklington integral equation approach. Some alternate treatments were also reported by Schumann [15], Liu and Tesche [16], and Landt et al. [17]. Luebbers et al. [18] have developed the finite difference time domain algorithm for treatment of nonlinear loadings, and Aygun et al. [19] have proposed a novel integral equation based technique for analyzing interaction of electromagnetic waves with nonlinear elements addressing a special concern to reducing the computational and memory requirements. Time domain variant of the Galerkin-Bubnov Indirect Boundary Element Method (GB-IBEM) for the analysis of nonlinearly loaded wire in free space and in a two media configuration has been reported in [20] and [21], respectively.

Once determining the transient response of the particular wire configuration, some other parameters of interest describing the transient behavior can be readily calculated at the considered configuration.

Therefore, time-domain energy measures for a single wire in free space have been proposed by Miller and Landt in [1], and more recently reexamined by Miller [2]. These energy measures involve integrating the squared current and charge density along a wire at each time step of a numerical solution based on a moment-method solution to the Pocklington integral equation.

As already mentioned, the Hallen integral equation has been preferred by some authors [3–6], because numerical instabilities can arise from using the Pocklington equation approach. Namely, the serious disadvantage of the Pocklington equation is the appearance of the rapidly growing nonphysical oscillations at later instants of time [12], [22]. The kernel of Pocklington equation contains the second order space-time differential operator and its numerical representation is usually not satisfactory that is found to be the main origin of the mentioned instabilities [4]. Consequently, some smoothing procedures are to be applied and the related numerical method becomes computationally less efficient [22]. The apparent advantage of the Hallen approach evidently arises from the fact that it contains neither space nor time derivatives. The TD energy measures based on Hallen integral equation solution for two-wire array and multiple wire array have been presented in [23] and [24], respectively.

This chapter deals with a direct TD modeling of a transient behavior and various thin wire configurations based on Hallen integral equation and TD scheme of the Galerkin-Bubnov Boundary Integral Equation Method (GB-BIEM). The problems analyzed include coupling between wires, coupling of external fields to wires, and effects of perfect ground plane and real ground. The time variation energy stored in the near field of wires is also analyzed.

Furthermore, the time-domain energy measures associated with the current and charge on thin wires are analyzed by spatially integrating the square of the current and charge along the wire as a function of time. Furthermore, a root-mean-square (rms) measure of the nonlinear loading effect to the transient response of the thin wire is also presented. As rms value is an energy measure of the transient current flowing along the wire, it is related to the efficiency estimation of the protection devices with nonlinear elements, thus being very important in EMC applications. A number of illustrative TD computational examples are presented at the end of each section.

9.1 THIN WIRES IN FREE SPACE

9.1.1 Single Wire in Free Space

Thin wire antenna or scatterer of length L and radius a oriented along the x-axis is considered (Fig 9.1). The wire is assumed to be perfectly conducting and excited by a time-varying voltage source, or illuminated by a plane wave.

The tangential component of the total field vanishes on the antenna surface, that is:

$$E_x^{inc} + E_x^{sct} = 0 \tag{9.1}$$

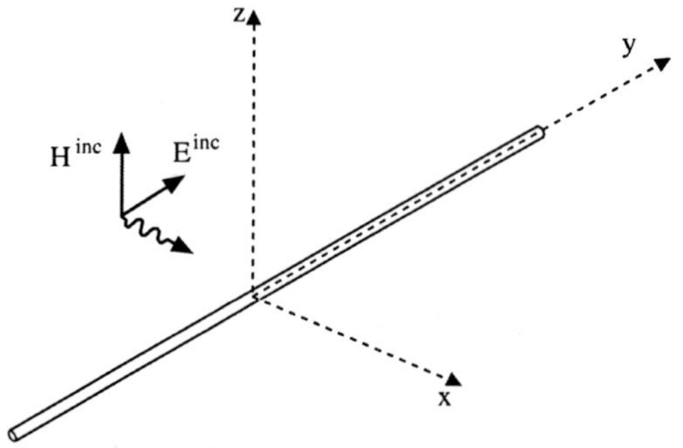

Figure 9.1 Single straight wire in free space.

where E_x^{inc} is the incident and E_x^{sct} the scattered field on the metallic wire surface. From the first Maxwell equation:

$$\nabla x\vec{E} = -\frac{\partial \vec{B}}{\partial t} \tag{9.2}$$

and using the vector magnetic potential $\vec{A}$:

$$\vec{B} = \nabla x\vec{A} \tag{9.3}$$

it follows:

$$\vec{E}^{\,inc}\Big|_{tan} = \left.\frac{\partial \vec{A}}{\partial t} + \nabla\varphi\right)\Big|_{tan} \tag{9.4}$$

where $\vec{A}$ and φ are space-time dependent magnetic vector and electric scalar potential, respectively.

These two potentials obey the Lorentz gauge:

$$\frac{\partial \varphi}{\partial t} + \frac{1}{\mu\varepsilon}\nabla\vec{A} = 0 \tag{9.5}$$

Differentiating eqn (9.4) and taking into account the Lorentz gauge (9.5) gives the wave equation for magnetic vector potential $\vec{A}$:

$$\left.\frac{\partial^2 \vec{A}}{\partial t^2} - \frac{1}{\mu\varepsilon}\nabla(\nabla\vec{A})\right)\Big|_{tan} = \frac{\partial \vec{E}^{\,inc}}{\partial t}\Big|_{tan} \tag{9.6}$$

According to the thin wire approximation, only the axial component of the vector potential exists, that is:

$$\frac{\partial^2 A_x}{\partial x^2} - \frac{1}{c^2}\frac{\partial^2 A_x}{\partial t^2} = -\frac{\partial E_x^{inc}}{\partial t} \tag{9.7}$$

where c denotes the velocity of light.

The eqn (9.7) is valid on the surface of the perfect conductor and the corresponding solution can be represented by a sum of the general solution of the homogeneous equation and a particular solution of the inhomogeneous equation:

$$A_x(x,t) = A_x^h(x,t) + A_x^p(x,t) \tag{9.8}$$

The solution of the homogeneous wave equation is given as a superposition of incident and reflected wave [4]:

$$A_x^h(x,t) = F_1(t - \frac{x}{c}) + F_2(t + \frac{x}{c}) \tag{9.9}$$

The particular solution is given by the integral [4]:

$$A_x^p(x,t) = \frac{\varepsilon c}{2} \int_{-\infty}^{x} \int_{-\infty}^{t-\frac{(x-x')}{c}} \frac{\partial E_x^{inc}(x',t')}{\partial t'} \mathrm{d}t' \mathrm{d}x' + \frac{\varepsilon c}{2} \int_{-\infty}^{x} \int_{-\infty}^{t+\frac{(x-x')}{c}} \frac{\partial E_x^{inc}(x',t')}{\partial t'} \mathrm{d}t' \mathrm{d}x' \tag{9.10}$$

Since the differential equation is related to the wire antenna surface, the eqn (9.10) simplifies into:

$$A_x^p(x,t) = \frac{1}{2Z_0} \int_0^L E_x^{inc}\left(x', t - \frac{|x-x'|}{c}\right) \mathrm{d}x' \tag{9.11}$$

where L denotes the total antenna length.

The magnetic vector potential on the perfectly conducting (PEC) wire surface may be also obtained as a solution of the following wave equation:

$$\nabla^2 \vec{A} - \mu\varepsilon \frac{\partial^2 \vec{A}}{\partial t^2} = -\mu \vec{J}(r,t) \tag{9.12}$$

where $\vec{J}$ denotes the surface current density.

The solution of differential eqn (9.12) is usually undertaken featuring the Green function theory, that is by introducing the auxiliary equation:

$$\nabla^2 g - \mu\varepsilon \frac{\partial^2 g}{\partial t^2} = -\delta(r-r', t-t') \tag{9.13}$$

The solution of eqn (9.13) is expressed in terms of the retarded Dirac impulse:

$$g(r,t,r',t') = \frac{\delta\left(t-t'-\frac{R}{c}\right)}{4\pi R} \tag{9.14}$$

where R is a distance from the source to the observation point.

The solution of eqn (9.12) may be now written in the form:

$$\vec{A}(r,t) = \frac{\mu}{4\pi} \int_{-\infty}^{t} \int_S \vec{J}_S(r',t) \frac{\delta(t-t'-R/c)}{R} \mathrm{d}S' \mathrm{d}t' \tag{9.15}$$

and performing the time domain integration, it follows:

$$\vec{A}(r,t) = \frac{\mu}{4\pi} \int_S \frac{\vec{J}_S(r', t-R/c)}{R} \mathrm{d}S' \tag{9.16}$$

According to the thin wire approximation, the equivalent current along the wire is assumed to flow in the axis, while the observation point is located on the antenna surface, that is:

$$I(z,t) = 2\pi a J_z(z,t) \tag{9.17}$$

Combining the relations (9.8)–(9.17) yields the space-time Hallen equation:

$$\int_0^L \frac{I(x',t-R/c)}{4\pi R}\,\mathrm{d}x' = F_0(t-\frac{x}{c}) + F_L(t-\frac{L-x}{c}) + \frac{1}{2Z_0}\int_0^L E_x^{inc}(x',t-\frac{|x-x'|}{c}t)\mathrm{d}x' \tag{9.18}$$

where $I(x')$ is the equivalent axial current to be determined, E_x^{inc} is the known tangential incident field, $R = [x-x')^2 + a^2]^{1/2}$ is the distance from the source point (the equivalent current in the antenna axis) to the observation point, and Z_0 is the wave impedance of a free space.

The multiple reflections of the current at the free ends of the wire are taken into account by the unknown functions $F_0(t)$ and $F_L(t)$.

In the special case of a transmitting dipole-antenna, the voltage generator differs from zero only in the feed-gap area, $x_g - \Delta l_g < x < x_g + \Delta l_g$ and can also be expressed via incident field using the Dirac function:

$$E_x^{inc}(x,t) = V_g(t)\delta(x-x_g) \tag{9.19}$$

where V_g denotes the voltage generator and $x_g = L/2$.

For sufficiently small dimensions of the gap, it follows:

$$\int_{x_g-\Delta l_g}^{x_g+\Delta l_g} E_x^{inc}\left(x',t-\frac{|x-x'|}{c}\right)\mathrm{d}x' = V_g\left(t-\frac{|x-x_g|}{c}\right) \tag{9.20}$$

and the Hallen equation for the transmitting antenna mode is given by:

$$\int_0^L \frac{I(x',t-R/c)}{4\pi R}\,\mathrm{d}x' = F_0\left(t-\frac{x}{c}\right) + F_L\left(t-\frac{L-x}{c}\right) + \frac{1}{2Z_0}V\left(t-\frac{|x-x_g|}{c}\right) \tag{9.21}$$

The unknown time functions $F_0(t)$, $F_L(t)$, $F_0(t-L/c)$, and $F_L(t-L/c)$ can be obtained by invoking the Hallen equation for $x = 0$ and $x = L$ and by applying the inverse Laplace transform. Two linear equations are obtained for the transforms $F_0(s)$ and $F_L(s)$. Solving the system of linear equations in the frequency domain and applying the inverse Laplace transform, the shifted unknown signals can be expressed in terms of the currents and excitations [3]:

$$F_0(t) = \sum_{n=0}^{\infty} K_0\left(t - \frac{2nL}{c}\right) - \sum_{n=0}^{\infty} K_L\left(t - \frac{(2n+1)L}{c}\right) \tag{9.22}$$

$$F_L(t) = \sum_{n=0}^{\infty} K_L\left(t - \frac{2nL}{c}\right) - \sum_{n=0}^{\infty} K_0\left(t - \frac{(2n+1)L}{c}\right) \tag{9.23}$$

where $K_0(t)$ and $K_L(t)$ are auxiliary functions defined by following relations:

$$K_0(t) = \int_0^L \frac{I(x', t - R_0/c)}{4\pi R_0} \mathrm{d}x' - \frac{1}{2Z_0} \int_0^L E_x^{inc}\left(x', t - \frac{x'}{c}\right) \mathrm{d}x' \tag{9.24}$$

$$K_L(t) = \int_0^L \frac{I(x', t - R_L/c)}{4\pi R_L} \mathrm{d}x' - \frac{1}{2Z_0} \int_0^\infty E_x^{inc}\left(x', t - \frac{L - x'}{c}\right) \mathrm{d}x' \tag{9.25}$$

R_0 and R_L are the distances from the wire ends to the source point.

Integral equations (9.18) and (9.21) can be solved if certain edge and initial conditions are prescribed. In accordance to the thin wire approximation, the current at the wire ends is assumed to vanish at each time, that is:

$$I(0, t) = I(L, t) = 0 \tag{9.26}$$

In addition, without any loss of generality, it can be assumed that the antenna is not excited before $t = 0$ [3], [5].

$$I(x, 0) = 0 \tag{9.27}$$

Therefore, the problem of transient radiation or scattering from straight thin wire in free space can be regarded as an initial value problem.

9.1.2 Single Wire Far Field

The formula for the far electric field of a thin wire in free space can be derived using the Maxwell equations, and it is given as follows [12]:

$$\vec{E}(r, t) = -\vec{e}_x \frac{\mu_0}{4\pi} \int_0^L \frac{\partial}{\partial t} \frac{I(x', t')}{R_1} \mathrm{d}x' + \frac{1}{4\pi\varepsilon_0} \vec{e}_{R_1} \int_0^L \frac{q(x', t')}{R_1^2} \mathrm{d}x' + \frac{1}{4\pi\varepsilon_0 c} \vec{e}_{R_1} \int_0^L \frac{\partial q(x', t')}{\partial t'} \frac{1}{R_1} \mathrm{d}x' \tag{9.28}$$

where $R_1 = [(x - x')^2 + z^2]^{1/2}$ is the distance from the wire to the observation point, the e_x and e_{R1} are the unit vectors in x and R_1 direction, respectively, and $q(x', t')$ denotes the charge distribution along the wire.

According to the well-known far-field approximation [12], the terms containing the $q(x', t')$ (near field terms) can be neglected, and eqn (9.28) becomes:

$$\vec{e}_x E_x(r,t) = -\vec{e}_x \frac{\mu_0}{4\pi} \int_0^L \frac{\partial}{\partial t} \frac{I(x', t')}{R_1} dx' \tag{9.29}$$

where the $t' = t - R_1/c$ is a retarded time.

9.1.3 Loaded Straight Thin Wire in Free Space

Antenna properties can be significantly affected by the presence of continuous or concentrated loading. This section deals with resistively and nonlinearly loaded wires, respectively.

9.1.3.1 Resistive Loading Transient response of resistively loaded thin wires has an important application in EMP simulators and lightning protection design [9,25]. The basic requirement in EMP simulators design is to generate a certain time signal having rise and fall times of order tens and hundreds of nanoseconds. On the contrary, the inconvenience in using pulse-radiating antennas of finite length is the undesirable effect caused by the reflections from the antenna open ends. These difficulties can be overcome by loading the antenna with a nonuniform resistive loading as this frequency independent loading significantly decreases the end reflections [25]. A time domain study on resistively loaded thin wire insulated in free space, based on efficient boundary element analysis, has been presented in [26]. A straight thin wire of length L and radius a, oriented along x-axis is considered. The wire is assumed to be perfectly conducting and continuously loaded with a resistive loading R_L per unit of wire length. The excitation function is a time-dependent voltage generator (antenna mode) or a plane wave (scatterer mode).

It is well known that the tangential electric field along the loaded wire is to satisfy the following condition [14]:

$$E_x^{inc} + E_x^{sct} = I(x) R_L(x) \tag{9.30}$$

where E_x^{inc} is the tangential incident field, E_x^{sct} is the scattered field, $I(x')$ is the equivalent axial current distribution along the antenna, and R_L is the resistive load per unit length of the wire.

Starting from Maxwell equations and exploiting the condition (9.28) the time-domain electric field integral equation is obtained in the form [26]:

$$-\varepsilon \frac{\partial E_x^{inc}}{\partial t} = \left[\frac{\partial^2}{\partial x^2} - \frac{1}{c} \frac{\partial^2}{\partial t^2} \right] \int_0^L \frac{I(x', t - R/c)}{4\pi R} dx' - R_L(x) \frac{\partial I(x,t)}{\partial t} \tag{9.31}$$

where c is the velocity of light, $R = [(x - x')^2 + a^2]^{1/2}$ is the distance from the source point (the equivalent current in the antenna axis) to the observation point (on the antenna surface).

Using the superposition principle [3] and integrating the eqn (9.31) the corresponding Hallen equation is obtained in the form:

$$\int_0^L \frac{I(x', t - R/c)}{4\pi R} \mathrm{d}x' = F_0(t - \frac{x}{c}) + F_L(t - \frac{L - x}{c}) + \frac{1}{2Z_0} \int_0^L E_x^{inc}(x', t - \frac{|x - x'|}{c}) \mathrm{d}x' - \frac{1}{2Z_0} \int_0^L R_L(x') I(x', t - \frac{|x - x'|}{c}) \mathrm{d}x' \tag{9.32}$$

Unknown signals F_0 and F_L are obtained using relations (9.22) and (9.23), while the auxiliary functions K_0 and K_L are defined as follows:

$$K_0(t) = \int_0^L \frac{I(x', t - R_0/c)}{4\pi R_0} \mathrm{d}x' - \frac{1}{2Z_0} \int_0^L E_x^{inc}(x', t - \frac{x'}{c}) \mathrm{d}x' + \frac{1}{2Z_0} \int_0^L R_L(x') I\left(x', t - \frac{x'}{c}\right) \mathrm{d}x' \tag{9.33}$$

$$K_L(t) = \int_0^L \frac{I(x', t - R_L/c)}{4\pi R_L} \mathrm{d}x' - \frac{1}{2Z_0} \int_0^L E_x^{inc}\left(x', t - \frac{L - x'}{c}\right) \mathrm{d}x' + \frac{1}{2Z_0} \int_0^L R_L(x') I\left(x', t - \frac{L - x'}{c}\right) \mathrm{d}x' \tag{9.34}$$

Standard nonreflecting resistive loading continuously distributed along the wire [25,26] has a following space dependence:

$$R_L(x) = \frac{R_0}{\frac{L}{2} - \left| x - \frac{L}{2} \right|} \tag{9.35}$$

The time-domain Hallen integral equation for the loaded straight thin wire (9.30) can be solved with the prescribed boundary conditions for the current at the wire ends (9.26) and the initial condition for the current along the wire at the instant $t = t_0$ (9.27).

9.1.3.2 Nonlinear Loading Analysis of nonlinear phenomena is very important for wire antennas containing semiconductor devices, integrated circuits, and voltage

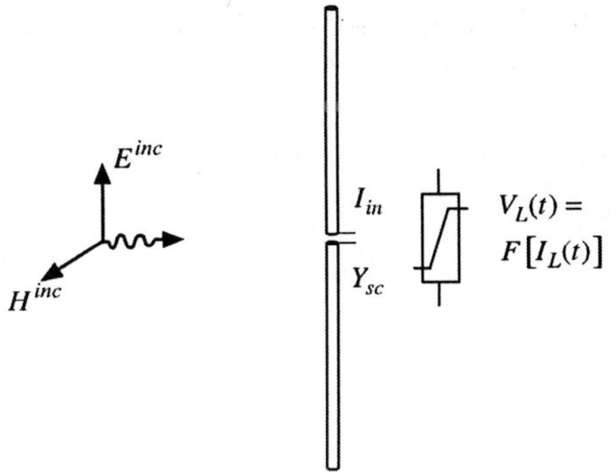

Figure 9.2 Nonlinearly loaded thin wire. (Poljak, D.; Miller, E. K.; Yoong Tham, C., Root-Mean-Square measure of Nonlinear Effects to the Transient Response of Thin Wires, *IEEE Transactions on Antennas and Propagation.* 51 (2003), 12; 3280–3283).

limiters when they are illuminated by an extremely strong signal, such as those caused by a lightning or nuclear electromagnetic pulse (LEMP and NEMP). Moreover, the analysis of nonlinear phenomena can be very useful when designing protection devices since nonlinear elements capture the undesirable and unforeseen strong electromagnetic fields.

A single thin wire of radius a and of entire length L, placed horizontally over a real ground is considered, Figure 9.2. The wire is assumed to be perfectly conducting and excited by a time-varying voltage generator or illuminated by an incident plane wave electric field. The wire also contains linear or nonlinear resistive load at its center, as shown in Figure 9.2.

If the concentrated nonlinear load is considered, then it is convenient to modify the last term on the right side of (9.321) in the following manner:

$$\frac{1}{2Z_0}\int_0^L R_L\left(x'\right)I\left(x',t-\frac{|x-x'|}{c}\right)\mathrm{d}x' = \frac{1}{2Z_0}v_L\left(t-\frac{|x-x_0|}{c}\right) \tag{9.36}$$

where v_L is the voltage on a nonlinear load placed at the wire center $x = x_0$, and a $v - I$ characteristic is determined by relation:

$$v_L(t) = F[I_L(t)] \tag{9.37}$$

and F denotes a known function and I_L is the instantaneous current flowing through the load.

9.1.4 Two Coupled Identical Wires in Free Space

Transient analysis of two identical coupled wires in free space is very important in antenna theory and EMC applications. One of the two wires which are located parallel to each other is excited by an impulsive voltage source.

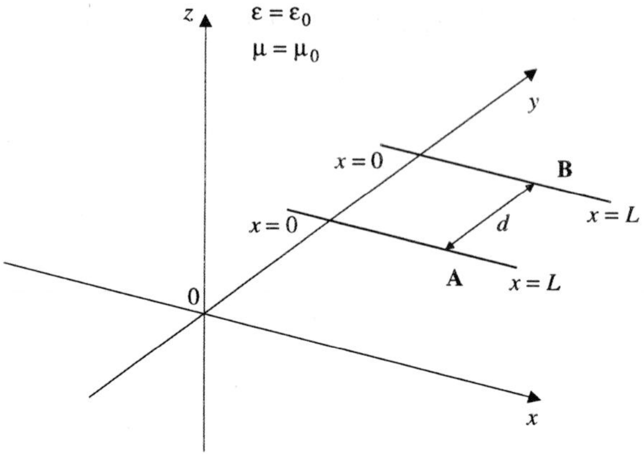

Figure 9.3 Coupled wires above real ground.

The time-domain formulation for two identical coupled wires in free space based on the set of Hallen type integral equations, has been presented in [27]. Because of the symmetry of the configuration the original geometry of two coupled wires is transformed into a two single wire modes. Each mode is represented by the corresponding Hallen integral equation.

Two identical straight and perfectly conducting thin wires A and B, Figure 9.3, of length L and radius a, parallel to each other and located along x-axis are considered. The separation between them is d. The wire A is fed by the time-dependent voltage generator $v_g(t)$.

The transient currents along the wires I_A and I_B can be calculated by solving the system of two coupled integral equations of the Pocklington or Hallen type [1]. The Hallen integral equation approach is preferred as it has proved to be more attractive from the numerical point of view [27,28].

The set of coupled Hallen equations for the thin wire structure shown in Figure 9.3, can be obtained as an extension of the Hallen equation for single wire in free space.

This extension to the problem of two coupled wires has been reported in [27] leading to the following set of coupled Hallen integral equations for unknown axial currents I_A and I_B along wires A and B, respectively:

$$\int_0^L \frac{I_A(x', t - R_a/c)}{4\pi R_a} \mathrm{d}x' + \int_0^L \frac{I_B(x', t - R_d/c)}{4\pi R_d} \mathrm{d}x'$$
$$= F_{0A}\left(t - \frac{x}{c}\right) + F_{LA}\left(t - \frac{L - x}{c}\right) + \frac{1}{2Z_0}\int_0^L E_{xA}^{inc}\left(x', t - \frac{|x - x'|}{c}\right)\mathrm{d}x' \tag{9.38}$$

$$\int_0^L \frac{I_B(x', t - R_a/c)}{4\pi R_a} dx' + \int_0^L \frac{I_A(x', t - R_d/c)}{4\pi R_d} dx'$$

$$= F_{0B}\left(t - \frac{x}{c}\right) + F_{LB}\left(t - \frac{L - x}{c}\right) + \frac{1}{2Z_0}\int_0^L E_{xB}^{inc}\left(x', t - \frac{|x - x'|}{c}\right) dx' \quad (9.39)$$

where $R_a = [(x - x')^2 + a^2]^{1/2}$, $R_d = [(x - x')^2 + d^2]^{1/2}$, E_{xA}^{inc}, E_{xB}^{inc} are the incident fields at the surface of the wire A and B, respectively, Z_0 and F_{0A}, F_{LA}, F_{0B}, F_{LB} are the unknown time signals representing the multiple reflections of the current at the ends of the wires A and B, respectively.

Invoking the integral equations for $x = 0$ and $x = L$; the set of coupled integral eqns (9.38), (9.39) are rewritten in a way that the currents I_A and I_B remain the only unknowns. Consequently, the set of resulting equations can be directly solved by the time-domain GB-IBEM (FEIEM). However, the computational cost can be significantly reduced by taking into account the symmetry of the two identical wires configuration. Using the principle of linear superposition [27], total currents I_A and I_B are simply given by:

$$I_A(x, t) = I^+(x, t) + I^-(x, t) \quad (9.40)$$

$$I_B(x, t) = I^+(x, t) - I^-(x, t) \quad (9.41)$$

where the current I^+ and I^- are the corresponding currents along wires A and B for the case of symmetric and antisymmetric excitation, respectively

Since in the original problem only wire A contains the excitation, using the superposition principle *the excitation one has* to decompose in a way so as to obtain the equal voltage sources on both the wires:

$$V^+ = \frac{V_A + V_B}{2} \quad (9.42)$$

$$V^- = \frac{V_A - V_B}{2} \quad (9.43)$$

With the $V_A = V_g$ and $V_B = 0$, it simply follows $V^+ = V^- = V_g/2$ and the original excitation can be obtained by superposition of both states:

$$V^+ + V^- = V_g \quad (9.44)$$

The incident electric field exists only in the feed-gap area Δl_g and it is simply determined by:

$$E_x^{inc} = \frac{V_g}{\Delta l_g} \quad (9.45)$$

Now, the set of decomposed Hallen integral equations can be written in the form:

$$\int_0^L \frac{I^{\pm}(x',t-R_a/c)}{4\pi R_a}\mathrm{d}x' - \int_{-\infty}^{t}\int_0^L r(\theta,\tau)\frac{I^{\pm}(x',t-R_a^*/c-\tau)}{4\pi R_a^*}\mathrm{d}x'\mathrm{d}\tau$$
$$\pm\left[\int_0^L \frac{I^{+}(x',t-R_d/c)}{4\pi R_d}\mathrm{d}x' - \int_{-\infty}^{t}\int_0^L r(\theta,\tau)\frac{I^{+}(x',t-R_d^*/c-\tau)}{4\pi R_d^*}\mathrm{d}x'\mathrm{d}\tau\right] \quad (9.46)$$
$$= F_0^{\pm}\left(t-\frac{x}{c}\right) + F_L^{\pm}\left(t-\frac{L-x}{c}\right) + \frac{1}{4Z_0}\int_0^L E_x^{inc}(x',t-\frac{|x-x'|}{c})\mathrm{d}x'$$

where

$$F_0^{\pm}(t) = \sum_{n=0}^{\infty} K_0^{\pm}\left[t-\frac{2nL}{c}\right] - \sum_{n=0}^{\infty} K_L^{\pm}\left[t-\frac{(2n+1)L}{c}\right] \quad (9.47)$$

$$F_L^{\pm}(t) = \sum_{n=0}^{\infty} K_L^{+} - \left[t-\frac{2nL}{c}\right] - \sum_{n=0}^{\infty} K_0^{+-}\left[t-\frac{(2n+1)L}{c}\right] \quad (948)$$

while the auxiliary functions $K_0^{\pm}$ and $K_L^{\pm}$ are defined by relations:

$$K_0^{\pm}(t) = \int_0^L \frac{I^{\pm}(x',t-R_{a0}/c)}{4\pi R_{a0}}\mathrm{d}x' \pm \int_0^L \frac{I^{\pm}(x',t-R_{d0}/c)}{4\pi R_{d0}}\mathrm{d}x'$$
$$-\frac{1}{4Z_0}\int_0^L E_x^{inc}\left(x',t-\frac{x'}{c}\right)\mathrm{d}x' \quad (9.49)$$

$$K_0^{\pm}(t) = \int_0^L \frac{I^{\pm}(x',t-R_{aL}/c)}{4\pi R_{aL}}\mathrm{d}x' \pm \int_0^L \frac{I^{\pm}(x',t-R_{dL}/c)}{4\pi R_{dL}}\mathrm{d}x'$$
$$-\frac{1}{4Z_0}\int_0^L E_x^{inc}\left(x',t-\frac{L-x'}{c}\right)\mathrm{d}x' \quad (9.50)$$

The set of space-time eqns (9.46) – (9.50) has to be solved twice, once for the symmetric mode (plus sign) and once for the antisymmetric mode (minus sign). The uniqueness of the solution is achieved if the zero and edge conditions at both the wires are known for $t \geq 0$, that is:

$$I^{\mp}(0,t) = I^{\pm}(L,t) = 0 \quad (9.51)$$

In addition, without any lose of generality, it can be taken that the wire is not excited before the instant $t = 0$:

$$I^{\pm}(x, 0) = 0 \tag{9.52}$$

so the set of eqns (9.46)–(9.50) can be handled as an initial value problem.

9.1.5 Measures for Postprocessing of Transient Response

Once obtaining the transient current flowing on a particular wire configuration, it is possible to determine some additional parameters providing a measure of such a transient response. Such convenient parameters presented in this section are time-domain energy measures and rms measure.

9.1.5.1 Time Domain Energy Measures for Thin Wires and Scatterers Time-domain energy measures for a single wire in free space were first presented by Miller and Landt in [1], and more recently re-examined by Miller [2]. Their energy measures involved integrating the square of the current and charge density along a wire at each time step of a numerical solution based on a moment-method solution to the thin-wire, electric-field integral equation (the Pocklington integral equation).

However, a serious disadvantage of the Pocklington equation is the appearance of the rapidly growing nonphysical oscillations at later instants of time [12,22,28]. Namely, the Pocklington equation contains the second-order space-time differential operator within its kernel and its numerical representation is not always satisfactory. This fact is claimed to be the main origin of the mentioned instabilities [12]. Consequently, relatively complex numerical procedures are to be used and the related method becomes computationally less efficient [22]. The apparent advantage of the Hallen approach evidently arises because it contains neither space nor time derivatives [3–6].

The results to follow in this book have been obtained using the Hallen integral equation (IE) for modeling various thin-wire configurations. Knowing the current distribution along the particular wire configuration, the charge is calculated from the equation of continuity:

$$\nabla \vec{J} = -\frac{\partial \rho_s}{\partial t} \tag{9.53}$$

By re-arranging the eqn (9.53) one obtains:

$$q = -\int_0^t \frac{\partial I(x', t)}{\partial z'} \mathrm{d}t \tag{9.54}$$

where q is the linear charge distribution along the wire structure.

The time-domain energy measures represented by the current and charge on an object yield insight into where and how much the object radiates as a function of time.

Having found the current and charge, measures of the H-field (kinetic) and the E-field (static) energy densities are expressed as in [1].

The H-field energy is represented by the relation [1]:

$$W_I = \frac{\mu_0}{4\pi} \int_0^L I^2(x', t) \mathrm{d}x' \tag{9.55}$$

while the E-field energy is measured by the following integral [1]:

$$W_q = \frac{1}{4\pi\varepsilon} \int_0^L q^2(x', t) \mathrm{d}x' \tag{9.56}$$

The total energy stored in the near field is proportional to the sum of W_1 and W_q [1]:

$$W_{tot} = W_I + W_q \tag{9.57}$$

9.1.5.2 ***Root-Mean-Square Measures*** It is well known from the basic theory of electric circuits that time-varying current delivers an average power to resistive load. The amount of delivered power strongly depends on a particular waveform. A measure of comparing the power delivered by different waveforms is root-mean-square (rms) or effective value of a transient current. The rms value of a time-varying current is a constant that is equal to the direct current value that would deliver the same average power to a given resistance R_L.

Instantaneous power delivered to a resistance R_L by a transient current $i(t)$ is:

$$p(t) = R_L i^2(t) \tag{9.58}$$

while the corresponding average power P_{av} is determined by the integral relation:

$$P_{av} = \frac{1}{T_0} \int_0^{T_0} p(t) \mathrm{d}t = \frac{1}{T_0} \int_0^{T_0} R_L i^2(t) \mathrm{d}t = R_L I_{\mathrm{rms}}^2 \tag{9.59}$$

from which the rms current is then:

$$I_{\mathrm{rms}} = \sqrt{\frac{1}{T_0} \int_0^{T_0} i^2(t) \mathrm{d}t} \tag{9.60}$$

Consequently, the spatial distribution of rms values of the thin wire space-time current is given by [20]:

$$I_{\text{rms}}(x) = \sqrt{\frac{1}{T_0}\int_0^{T_0} i^2(x,t)\mathrm{d}t} \tag{9.61}$$

where T_0 is the time interval of interest.

9.1.6 Computational Procedures for Thin Wires in Free Space

This chapter deals with efficient numerical procedures for handling the Hallen integral equation for unloaded and loaded single wire in free space, far field computation, and two identical coupled wires in free space. These procedures are based on time domain variant of Galerkin-Bubnov Indirect Boundary Element Method (GB-IBEM), also entitled Time Domain Finite Element Integral Equation Method (TD-FEIEM).

9.1.6.1 Numerical Solution of Time Domain Hallen Equation Most of the known direct time-domain techniques for the calculation of transient responses from thin wire structures are based on solving the time domain Pocklington integral equation, for example [7,9,12]. The most commonly used numerical technique for solving this equation is the well-known marching-on-in-time procedure combined with the method of moments (MoM) [7,9]. The main convenience of the method is that no matrix inversion is needed if the appropriate space-time discretization is performed. Yet it suffers from a serious disadvantage - the rapidly growing nonphysical oscillations appearing at later time instants. Pocklington integral equation contains the time and the space derivatives within its kernel and their numerical representations are not always satisfactory. Consequently, sophisticated and complex numerical procedures have to be used making the method computationally less efficient. On the contrary, the Hallen time-domain integral equation does not contain either space or time derivatives, which is very attractive in a numerical sense; but this equation is considered to be difficult for numerical treatment because of the presence of unknown functions which represent the solutions of the homogeneous wave equation.

An efficient procedure based on the GB-IBEM (FEIEM) for solving the single wire problem via the Hallen equation has been promoted in [29], and it is outlined in this section.

The time-domain Hallen integral equation:

$$\int_0^L \frac{I(x',t-R/c)}{4\pi R}\mathrm{d}x' = F_0\left(t-\frac{x}{c}\right) + F_L\left(t-\frac{L-x}{c}\right)$$
$$+\frac{1}{2Z_0}\int_0^L E_x^{inc}\left(x',t-\frac{|x-x'|}{c}\right)\mathrm{d}x' - \frac{1}{2Z_0}\int_0^L R_L\left(x'\right)I(x',t-\frac{|x-x'|}{c}\Big)\mathrm{d}x' \tag{9.62}$$

can be written in the operator form:

$$L(I) = E \tag{9.63}$$

where L is linear integral operator, I is the unknown function to be determined for a given excitation E.

In accordance to the weighted residual approach, eqn (9.63) is multiplied by the set of test functions and integrated over the domain of interest. Following the request for interpolation error minimization, the weighted residual integral,that is the inner product of functions $L(I) - E$ and W must vanish:

$$\int_0^L [L(I) - E] W_j \mathrm{d}x = 0; \qquad j = 1, 2, \ldots, N \tag{9.64}$$

where W_j is the set of test functions.

Applying the boundary element algorithm adjusted to space-time integral operators, the unknown solution for current is written in the form of linear combination of the base functions:

$$I(x', t') = \{f\}^T \{I\} \tag{9.65}$$

where $\{f\}$ is the vector containing the base functions and $\{I\}$ is the vector containing unknown time dependent coefficients of the solution and the local matrix system for ith source element interacting with jth observation element is obtained:

$$\int_{\Delta l_j} \int_{\Delta l_i} \{f\}_j \{f\}_i^T \frac{1}{4\pi R} \mathrm{d}x' \mathrm{d}x \{I\}_{|t-R/c} = \int_{\Delta l_j} F_0\left(t - \frac{x}{c}\right) \{f\} \mathrm{d}x$$
$$+ \int_{\Delta l_j} F_L\left(t - \frac{L-x}{c}\right) \{f\}_j \mathrm{d}x + \frac{1}{2Z_0} \int_{\Delta l_j} \int_{\Delta l_i} E_x^{inc}\left(x', t - \frac{|x - x'|}{c}\right) \mathrm{d}x' \mathrm{d}x \tag{9.66}$$

The matrix on the left-hand side of (9.66) represents the space interactions between ith source and jth observation wire elements. The first and second-term on the right-hand side contain time dependent signals F_0 and F_L expressed by the known values of current at previous instants, eqns (9.22)–(9.25). The last term on the right-hand side contains the excitation in the form of the incident field (voltage generator if the radiation mode is considered, or the plane wave if the scattering mode is considered).

If the thin wire operates in the radiation mode the excitation term can be written in the following form:

$$\frac{1}{2Z_0} \int_0^L E_x^{inc}\left(x', t - \frac{|x - x'|}{c}\right) \mathrm{d}x' = \frac{1}{2Z_0} \int_{L/2-\Delta l_g}^{L/2+\Delta l_g L} \frac{V\left(x', t - \frac{|x-x'|}{c}\right)}{\Delta l_g} \mathrm{d}x' \tag{9.67}$$

As it has been shown by numerous numerical experiments the calculation procedure becomes more accurate, stable, and efficient if the known excitation is also interpolated over wire segment [10,28]. Using the same shape functions as those used in current evaluation for the radiation mode it follows:

$$V\left(x', t - \frac{|x - x'|}{c}\right) = \{f\}^T \{V\} \tag{9.68}$$

and for the scattering mode:

$$E_x^{inc}(x', t') = \{f\}^T \{E\} \tag{9.69}$$

Both vectors {V} and {E} are time dependent vectors. In radiation problems vector {E} should be replaced with $\{V\}/l_g$ and it differs from zero only in the feed-gap area.

The local matrix form of eqn (9.66) now becomes:

$$\begin{aligned}\int_{\Delta l_j}\int_{\Delta l_i} \{f\}_j\{f\}_i^T \frac{1}{4\pi R} dx'dx\{I\}_{|t-R/c} &= \int_{\Delta l_j} F_0\left(t - \frac{x}{c}\right)\{f\}_j dx \\ &+ \int_{\Delta l_j} F_L\left(t - \frac{L-x}{c}\right)\{f\}_j dx + \frac{1}{2Z_0}\int_{\Delta l_j}\int_{\Delta l_i} \{f\}_j\{f\}_i^T dx'dx\{E\}_{|t-\frac{|x-x'|}{c}}\end{aligned} \tag{9.70}$$

The terms containing the solutions of homogeneous wave equations are of the form:

$$\begin{aligned}&\int_{\Delta l_j} F_0\left(t - \frac{x}{c}\right)\{f\}_j dx \\ &= \int_{\Delta l_j}\sum_{n=0}^{\infty} K_0\left(t - \frac{x}{c} - \frac{2nL}{c}\right)\{f\}_j dx - \int_{\Delta l_j}\sum_{n=0}^{\infty} K_L\left(t - \frac{x}{c} - \frac{2n+1}{c}L\right)\{f\}_j dx\end{aligned} \tag{9.71}$$

$$\begin{aligned}\int_{\Delta l_j} F_L\left(t - \frac{L-x}{c}\right)\{f\}_j dx &= \int_{\Delta l_j}\sum_{n=0}^{\infty} K_L\left(t - \frac{L-x}{c} - \frac{2nL}{c}\right)\{f\}_j dx \\ &- \int_{\Delta l_j}\sum_{n=0}^{\infty} K_0\left(t - \frac{L-x}{c} - \frac{2n+1}{c}L\right)\{f\}_j dx\end{aligned} \tag{9.72}$$

while the auxiliary functions $K_0(t)$ and $K_L(t)$ are determined by expressions:

$$K_0(t) = \int_{\Delta l_i} \{f\}_i^T \frac{1}{4\pi R_0} dx'\{I\}_{|t-R_0/c} = -\frac{1}{2Z_0}\int_{\Delta l_i} \{f\}_i^T dx'\{E\}_{|t-\frac{x'}{c}} \tag{9.73}$$

$$K_L(t) = \int_{\Delta l_i} \{f\}_i^T \frac{1}{4\pi R_L} dx'\{I\}_{|t-R_L/c} = -\frac{1}{2Z_0}\int_{\Delta l_i} \{f\}_i^T dx'\{E\}_{|t-\frac{L-x'}{c}} \tag{9.74}$$

The matrix form of eqn (9.70) can be written as:

$$
\begin{aligned}
[A]\{I\}\big|_{t-\frac{R}{c}} = {} & [B]\{E\}\big|_{t-\frac{|x-x'|}{c}} + [C]\left\{\sum_{n=0}^{\infty} I^n\right\}\Bigg|_{t-\frac{x}{c}\frac{2nL}{c}\frac{R_0}{c}} \\
& - [B]\left\{\sum_{n=0}^{\infty} E^n\right\}\Bigg|_{t-\frac{x}{c}\frac{2nL}{c}\frac{x'}{c}} - [D]\left\{\sum_{n=0}^{\infty} I^n\right\}\Bigg|_{t-\frac{x}{c}\frac{2n+1}{c}L\frac{R_L}{c}} \\
& - [B]\left\{\sum_{n=0}^{\infty} E^n\right\}\Bigg|_{t-\frac{x}{c}\frac{2n+1}{c}L\frac{L-x'}{c}} + [D]\left\{\sum_{n=0}^{\infty} I^n\right\}\Bigg|_{t-\frac{L-x}{c}\frac{2nL}{c}\frac{R_L}{c}} \\
& [B]\left\{\sum_{n=0}^{\infty} E^n\right\}\Bigg|_{t-\frac{L-x}{c}\frac{2nL}{c}\frac{L-x'}{c}} - [C]\left\{\sum_{n=0}^{\infty} I^n\right\}\Bigg|_{t-\frac{L-x}{c}\frac{2n+1}{c}L\frac{R_0}{c}} \\
& + [B]\left\{\sum_{n=0}^{\infty} E^n\right\}\Bigg|_{t-\frac{L-x}{c}\frac{2n+1}{c}L\frac{x'}{c}}
\end{aligned}
\tag{9.75}
$$

where the interaction matrices [A], [C], and [D] are of the form:

$$[A] = \int\limits_{\Delta l_j}\int\limits_{\Delta l_i} \frac{1}{4\pi R}\{f\}_j\{f\}_i^T \mathrm{d}x'\mathrm{d}x \tag{9.76}$$

$$[C] = \int\limits_{\Delta l_j}\int\limits_{\Delta l_i} \frac{1}{4\pi R_0}\{f\}_j\{f\}_i^T \mathrm{d}x'\mathrm{d}x \tag{9.77}$$

$$[D] = \int\limits_{\Delta l_j}\int\limits_{\Delta l_i} \frac{1}{4\pi R_L}\{f\}_j\{f\}_i^T \mathrm{d}x'\mathrm{d}x \tag{9.78}$$

where R_0 and R_L are the distances from the source elements to the wire ends.

$[B]$ matrix is given by expression:

$$[B] = \frac{1}{2Z_0}\int_{\Delta l_j}\int_{\Delta l_i} \{f\}_j\{f\}_i^T \mathrm{d}x'\mathrm{d}x \tag{9.79}$$

There are 10 time shifted terms in the global matrix system (9.75). Each of these terms contains the space dependent matrix multiplying the time dependent current vector or time dependent excitation vector. Though most of these terms contain summation from $n = 0$ to infinity, their handling is straightforward and it is related to the time interval being considered. Basically, these terms are related with the number of reflections from the ends of wire. The smaller is the time interval, the smaller is the number of terms in these sums appearing becaues of the reflections.

To obtain the time domain variation of thin wire current distribution the appropriate time discretization has to be performed. For convenience, the matrix system (9.75) is written in the form:

$$[A]\{I\}_{|_{t-R/c}} = \{g\} \tag{9.80}$$

where g is the space-time dependent vector containing the excitation and current at previous instants and represents the entire right-hand side of the eqn (9.75). The unknown time coefficients are interpolated by a linear combination of time dependent shape functions:

$$I_i(t') = \sum_{k=1}^{N_t} I_i^k T^k(t') \tag{9.81}$$

where index i denotes the ith space node and $t' = t - R/c$ is the retarded time.

Using the weighted residual approach in the time domain, for a time increment it simply follows:

$$\int_{t_k}^{t_k+\Delta t} ([A]\{I\}_{|t-R/c} - \{g\})\theta_k \mathrm{d}t = 0; \qquad k = 1, 2, \ldots, N_T \tag{9.82}$$

where N_t is the total number of time samples, and θ_k denotes the set of time dependent test functions.

Choosing the Dirac impulses as test functions from the weighted residual integral (9.82) leads to:

$$[A]\{I\}_{|_{t_k-R/c}} = \{g\}_{|_{t<t_k}} \tag{9.83}$$

If the space-time discretization is undertaken carefully in a way that within a time increment the propagation on at least one space segment occurs, that is if the well-known inequality must be satisfied:

$$\Delta t \le \frac{\Delta x}{c} \tag{9.84}$$

the following recurrence formula is obtained for the current in ith space node and at kth time instant:

$$I_{j|_{t_k}} = \frac{-\sum_{i=1}^{N_g} \overline{A_{ji}} I_{i|_{t_k-R/c}} + g_{j|_{t-t_k}}}{A_{jj}}; \qquad j = 1, 2, \ldots, N_g, k = 1, 2, \ldots, N_T \tag{9.85}$$

where N_g is the total number of global nodes, A_{ji} is the global matrix terms, g_j is the right-hand side vector containing the excitation and currents at previous instants, and the horizontal line over matrix [A] denotes the absence of diagonal terms contribution.

Therefore, the solution for transient current at present time instant for any space node is obtained as a function of currents at previous instants and not without requiring matrix inversion.

To start the stepping procedure it is necessary to prescribe the initial values of current:

$$I(x, 0) = 0 \tag{9.86}$$

and the conditions at the wire ends for all time instants:

$$I(0, t) = I(L, t) = 0 \tag{9.87}$$

It is worth noting that the presented procedure is stable for an arbitrary time interval and does not require any smoothing procedure (contrary to most of the known techniques).

9.1.6.2 Far-Field Computation The far field radiated or scattered from a thin wire insulated in free space can be obtained by integrating the space-time varying current distribution along the wire in free space from eqn (9.29). The contribution of each wire segment to the far field in the space-time observation point is given by:

$$E_x(r,t)\big|_{\Delta l_i} = -\frac{\mu_0}{4\pi}\frac{\partial}{\partial t'}\{I\}\big|_{t'} \int_{\Delta l_i} \{f\}_i^T \frac{1}{R_1}\mathrm{d}x' \tag{9.88}$$

and proceeding with the boundary element algorithm it simply follows:

$$E_x(r,t)\big|_{\Delta l_i} = -\frac{\mu_0}{4\pi}\left(\frac{I_{is}^{m+1} - I_{is}^{m}}{\Delta t}\int_{\Delta l_i} f_{is}\frac{1}{R_1}\mathrm{d}x' - \frac{\mu_0}{4\pi}\frac{I_{i,s+1}^{m+1} - I_{i,s+1}^{m}}{\Delta t}\int_{\Delta l_i} f_{i,s+1}\frac{1}{R_1}\mathrm{d}x'\right) \tag{9.89}$$

where Δt denotes the time increment, and m and m_1 are the integers of discretized times t' and t'_1, respectively.

The total tangential radiated field is then obtained by assembling all contributions along the wire:

$$E_x(r,t) = -\frac{\mu_0}{4\pi}\sum_{i=1}^{M}\left(\frac{I_{is}^{m+1} - I_{is}^{m}}{\Delta t}\int_{\Delta l_i} f_{is}\frac{1}{R_1}\mathrm{d}x' - \frac{\mu_0}{4\pi}\frac{I_{i,s+1}^{m+1} - I_{i,s+1}^{m}}{\Delta t}\int_{\Delta l_i} f_{i,s+1}\frac{1}{R_1}\mathrm{d}x'\right) \tag{9.90}$$

where M is the total number of segments.

9.1.6.3 Computational Procedures for Resistively and Nonlinearly Loaded Wire in Free Space The numerical solution procedure developed for the Hallen equation related to the unloaded antenna is also applied to the loaded antenna problem. The GB-IBEM scheme is used for space discretization, while marching-on-in-time is used for discretization in time.

According to the usual discretization procedure, the local approximation for current on a wire segment is given by relation (9.66). Using the weighted residual approach, the space boundary element discretization leads to the local equation system for ith source and jth observation wire segment:

$$
\begin{aligned}
\int_{\Delta l_j}\int_{\Delta l_i} \{f\}_j\{f\}_i^T \frac{1}{4\pi R}\mathrm{d}x'\mathrm{d}x\{I\}\Big|_{t-\frac{R}{c}} &= \int_{\Delta l_j} F_0\left(t-\frac{x}{c}\right)\{f\}_j \mathrm{d}x \\
&+ \int_{\Delta l_j} F_L\left(t-\frac{L-x}{c}\right)\{f\}_j \mathrm{d}x + \frac{1}{2Z_0}\int_{\Delta l_j}\int_{\Delta l_i} E_x^{inc}\left(x', t-\frac{|x-x'|}{c}\right)\{f\}_j \mathrm{d}x'\mathrm{d}x \\
&- \frac{1}{2Z_0}\int_{\Delta l_j}\int_{\Delta l_i} R_L(x')\{f\}_j\{f\}_i^T \mathrm{d}x'\mathrm{d}x\{I\}\Big|_{t-\frac{|x-x'|}{c}}
\end{aligned}
\tag{9.91}
$$

which can be rewritten in the matrix equation form:

$$
\begin{aligned}
[A]\{I\}\Big|_{t-\frac{R}{c}} &= [B]\{E\}\Big|_{t-\frac{|x-x'|}{c}} - [R]\{I\}\Big|_{t-\frac{|x-x'|}{c}} + [C]\left\{\sum_{n=0}^{\infty} I^n\right\}\Bigg|_{t-\frac{x}{c}-\frac{2nL}{c}-\frac{R_0}{c}} \\
&- [B]\left\{\sum_{n=0}^{\infty} E^n\right\}\Bigg|_{t-\frac{x}{c}-\frac{2nL}{c}-\frac{x'}{c}} + [R]\left\{\sum_{n=0}^{\infty} I^n\right\}\Bigg|_{t-\frac{x}{c}-\frac{2nL}{c}-\frac{x'}{c}} - [D]\left\{\sum_{n=0}^{\infty} I^n\right\}\Bigg|_{t-\frac{x}{c}-\frac{2n+1}{c}L-\frac{R_L}{c}} \\
&+ [B]\left\{\sum_{n=0}^{\infty} E^n\right\}\Bigg|_{t-\frac{x}{c}-\frac{2n+1}{c}L-\frac{L-x'}{c}} - [R]\left\{\sum_{n=0}^{\infty} I^n\right\}\Bigg|_{t-\frac{x}{c}-\frac{2n+1}{c}L-\frac{L-x'}{c}} \\
&+ [D]\left\{\sum_{n=0}^{\infty} I^n\right\}\Bigg|_{t-\frac{L-x}{c}-\frac{2nL}{c}-\frac{R_L}{c}} - [B]\left\{\sum_{n=0}^{\infty} E^n\right\}\Bigg|_{t-\frac{L-x}{c}-\frac{2nL}{c}-\frac{L-x'}{c}} \\
&+ [R]\left\{\sum_{n=0}^{\infty} I^n\right\}\Bigg|_{t-\frac{L-x}{c}-\frac{2nL}{c}-\frac{L-x'}{c}} - [C]\left\{\sum_{n=0}^{\infty} I^n\right\}\Bigg|_{t-\frac{L-x}{c}-\frac{2n+1}{c}L-\frac{R_0}{c}} \\
&+ [B]\left\{\sum_{n=0}^{\infty} E^n\right\}\Bigg|_{t-\frac{L-x}{c}-\frac{2n+1}{c}L-\frac{x'}{c}} - [R]\left\{\sum_{n=0}^{\infty} I^n\right\}\Bigg|_{t-\frac{L-x}{c}-\frac{2n+1}{c}L-\frac{x'}{c}}
\end{aligned}
\tag{9.92}
$$

where the space dependent boundary matrices are defined by relations (9.76)–(9.79) while the $[R]$ matrix is given by:

$$[R] = \frac{1}{2Z_0} \int_{\Delta l_j} \int_{\Delta l_i} R_L(x')\{f\}_j\{f\}_i^T \mathrm{d}x'\mathrm{d}x \tag{9.93}$$

When the space discretization procedure is completed, the weighted residual approach is also applied for the discretization in time. The solution in time on the ith wire segment is assumed to be in the form given by the eqn (9.81), and the time marching procedure is completely the same as the one described in section 9.1.6.1.

If the loading is nonlinear then the nonlinear $v - I$ characteristic (9.37) can be presented by an arbitrary number of piecewise linear curves. For simplicity, it is convenient to first consider a special case of two linear curves through the time scale origin, that is:

$$\begin{aligned} v_L(t) &= R_1 I_L(t); I_L(t) \geq 0 \\ v_L(t) &= R_2 I_L(t); I_L(t) < 0 \end{aligned} \tag{9.94}$$

where R_1 and R_2 are linear resistors.

This nonlinear load is included in the numerical procedure through the $[R]$ matrix, given by eqn (9.93). An extension to the case of n piecewise linear curves is straightforward.

9.1.6.4 Computational Procedures for Energy Measures Once the current is obtained using the presented boundary element procedures, the energy-measure integrals (9.55) and (9.56) can be computed.

First, the charge distribution determined by eqn (9.54) is obtained by evaluating the following integral:

$$q = -\sum_{i=1}^{M}\sum_{k=1}^{N_t} \int_{\Delta t_k} \frac{\partial}{\partial x'}\{f\}^T\{I\}\mathrm{d}t \tag{9.95}$$

for which the solution is carried out analytically and given in the form:

$$q = -\frac{1}{2c}\sum_{i=1}^{M}\sum_{m=1}^{N_t}(I_{i+1}^m + I_{i+1}^{m+1} - I_i^m - I_i^{m+1}) \tag{9.96}$$

The H-field energy is measured by evaluating the integral:

$$W_I = \frac{\mu_0}{4\pi}\sum_{i=1}^{M} \int_{\Delta l_i} (\{f\}_i^T\{I\}_i)^2 \mathrm{d}x' \tag{9.97}$$

which with closed form solution is:

$$W_I = 10^{-7}\frac{\Delta x}{3}\sum_{i=1}^{M}[(I_i^k)^2 + I_i^k I_{i+1}^k + (I_{i+1}^k)^2]; k = 1, 2, \ldots, N_t \tag{9.98}$$

The E-field energy is obtained from the integral:

$$W_q = \frac{1}{4\pi\varepsilon}\sum_{i=1}^{M}\int_{\Delta l_i}(\{f\}_i^T\{q\}_i)^2 \mathrm{d}x' \tag{9.99}$$

for which the solution is then:

$$W_q = 9x10^9\frac{\Delta x}{3}\sum_{i=1}^{M}[(q_i^k)^2 + q_i^k q_{i+1}^k + (q_{i+1}^k)^2]; k = 1, 2, \ldots, N_t \tag{9.100}$$

and the total energy measure is then given by:

$$W_{tot} = W_I + W_q \tag{9.101}$$

9.1.6.5 Computation of Spatial RMS Distribution When the current along the wire at each node and time instant is determined, the rms value of the wire current can be computed from the following relation:

$$I_{\mathrm{rms}} = \sqrt{\frac{1}{T_0}\sum_{k=1}^{N_t}\int_{t_k}^{t_{k+1}}(\{T\}^T\{I\})_i^2 \mathrm{d}t} \tag{9.102}$$

and the performing of a straight-forward integration yields:

$$I_{\mathrm{rms}} = \sqrt{\frac{\Delta t}{3T_0}\sum_{k=1}^{N_t}[(I_i^k)^2 + I_i^k I_i^{k+1} + (I_i^{k+1})^2]} \tag{9.103}$$

where $\{T\}$ is the vector containing the time domain linear shape functions.

9.1.6.6 Computational Procedures for Coupled Wires in Free Space The GB-IBEM/time marching procedure for the treatment of two identical coupled wires in free space has been presented in [27].

The local approximation for current over wire segment (for the case of symmetric and antisymmetric excitation) is given by:

$$I^{\pm}(x't') = \{f\}^T\{I\}^{\pm} \tag{9.104}$$

where $\{f\}$ denotes a vector containing space-domain shape functions, and $\{I\}^{\pm}$ is the time-dependent solution vector.

Investigating the accuracy and the convergence rate, the pulse, linear and parabolic shape functions are tested for the single wire and for the coupled wire geometry, and the linear shape functions are found to be the optimal choice regarding the accuracy and convergence vs. complexity of the algorithm and computational cost.

The spatial discretization results in the linear equation system for ith source and jth observation boundary element [27]:

$$\sum_{i=1}^{M}\left[\int_{\Delta l_j}\int_{\Delta l_i}\{f\}_j\{f\}_i^T\frac{1}{4\pi R_a}\mathrm{d}x'\mathrm{d}x\{I\}^{+-}\Big|_{t-\frac{R_a}{c}} \pm \int_{\Delta l_j}\int_{\Delta l_i}\{f\}_j\{f\}_i^T\frac{1}{4\pi R_d}\mathrm{d}x'\mathrm{d}x\{I\}^{+-}\Big|_{t-\frac{R_d}{c}}\right]$$
$$= \int_{\Delta l_j} F_0^{+-}\left(t-\frac{x}{c}\right)\{f\}_j\mathrm{d}x + \int_{\Delta l_j} F_L^{+-}\left(t-\frac{L-x}{c}\right)\{f\}_j\mathrm{d}x$$
$$+\frac{1}{4Z_0}\sum_{i=1}^{M}\int_{\Delta l_j}\int_{\Delta l_i} E_x^{inc}\left(x', t-\frac{|x-x'|}{c}\right)\{f\}_j\mathrm{d}x'\mathrm{d}x,$$
$$j = 1, 2, \ldots, N_g \tag{9.105}$$

where M is the number of space segments over the entire length of the wire. Performing further mathematical manipulation the global boundary element matrix system is obtained:

$$\sum_{i=1}^{M}[A]\{I\}\Big|_{t-\frac{R_a}{c}} = \sum_{i=1}^{M}\left([B]\{E\}\Big|_{\frac{t-|x-x'|}{c}} + [C]\left\{\sum_{n=0}^{\infty}I^n\right\}\Big|_{t-\frac{x}{c}-\frac{2nL}{c}-\frac{R_{a0}}{c}}\right.$$
$$\pm [C^d]\left\{\sum_{n=0}^{\infty}I^n\right\}\Big|_{t-\frac{x}{c}-\frac{2nL}{c}-\frac{R_{d0}}{c}} - [B]\left\{\sum_{n=0}^{\infty}E^n\right\}\Big|_{t-\frac{x}{c}-\frac{2nL}{c}-\frac{x'}{c}}$$
$$- [D]\left\{\sum_{n=0}^{\infty}I^n\right\}\Big|_{t-\frac{x}{c}-\frac{2n+1}{c}L-\frac{R_{aL}}{c}} \pm [D^d]\left\{\sum_{n=0}^{\infty}I^n\right\}\Big|_{t-\frac{x}{c}-\frac{2n+1}{c}L-\frac{R_{dL}}{c}}$$
$$+ [B]\left\{\sum_{n=0}^{\infty}E^n\right\}\Big|_{t-\frac{x}{c}-\frac{2n+1}{c}L-\frac{L-x'}{c}} + [D]\left\{\sum_{n=0}^{\infty}I^n\right\}\Big|_{t-\frac{L-x}{c}-\frac{2nL}{c}-\frac{R_{aL}}{c}}$$
$$\pm [D^d]\left\{\sum_{n=0}^{\infty}I^n\right\}\Big|_{t-\frac{L-x}{c}-\frac{2nL}{c}-\frac{R_{dL}}{c}} \pm [D]\left\{\sum_{n=0}^{\infty}I^n\right\}\Big|_{t-\frac{L-x}{c}-\frac{2nL}{c}-\frac{R^*_{dL}}{c}}$$

$$
\begin{aligned}
& -[B]\left\{\sum_{n=0}^{\infty} E^n\right\}\Bigg|_{t-\frac{L-x}{c}-\frac{2nL}{c}-\frac{L-x'}{c}} - [C]\left\{\sum_{n=0}^{\infty} I^n\right\}\Bigg|_{t-\frac{L-x}{c}-\frac{2n+1}{c}L-\frac{R_{a0}}{c}} \\
& \pm[C^d]\left\{\sum_{n=0}^{\infty} I^n\right\}\Bigg|_{t-\frac{L-x}{c}-\frac{2n+1}{c}L-\frac{R_{d0}}{c}} + [B]\left\{\sum_{n=0}^{\infty} E^n\right\}\Bigg|_{t-\frac{L-x}{c}-\frac{2n+1}{c}L-\frac{x'}{c}}\Bigg) \\
& j = 1,2,\ldots,M \qquad (9.106)
\end{aligned}
$$

where space-dependent boundary element matrices are given by relations (9.76)–(9.79), and by following relations:

$$[A^d] = \int_{\Delta l_j}\int_{\Delta l_i} \frac{1}{4\pi R_d}\{f\}_j\{f\}_i^T \mathrm{d}x'\mathrm{d}x \qquad (9.107)$$

$$[C^d] = \int_{\Delta l_j}\int_{\Delta l_i} \frac{1}{4\pi R_d 0}\{f\}_j\{f\}_i^T \mathrm{d}x'\mathrm{d}x \qquad (9.108)$$

$$[D^d] = \int_{\Delta l_j}\int_{\Delta l_i} \frac{1}{4\pi R_{dL}}\{f\}_j\{f\}_i^T \mathrm{d}x'\mathrm{d}x \qquad (9.109)$$

where $R_{a0} = [x^2 + a^2]^{1/2}$, $R_{d0} = [x^2 + d^2]^{1/2}$, $R_{aL} = [(L - x')^2 + a^2]^{1/2}$, $R_{dL} = [(L - x')^2 + d^2]^{1/2}$.

When the space discretization is performed, the weighted residual approach is used for discretization in time. The time domain solution is carried out applying the procedure described in Section 9.6.1.

9.1.7 Numerical Results for Thin Wires in Free Space

This section deals with the computational examples related to unloaded and loaded single wire and two identical coupled wires in free space.

9.1.7.1 Thin Wire Antenna and Scatterer in Free Space First, a straight single wire operating in the antenna mode (dipole antenna) is considered.

The transient current on a thin wire antenna in free space excited at its midpoint by a Gaussian voltage pulse is the most studied case in computational electromagnetics, and it is generally accepted as a canonical benchmark problem. Moreover, the Gaussian pulse is often regarded as a numerical representation of a Dirac impulse. The obtained transient response can be regarded as an impulse response of a particular linear system, and related transient response to an arbitrary incident waveform can be obtained using the convolution operator [10]. The transient current generated at the feed point of the wire because of Gaussian pulse is shown in

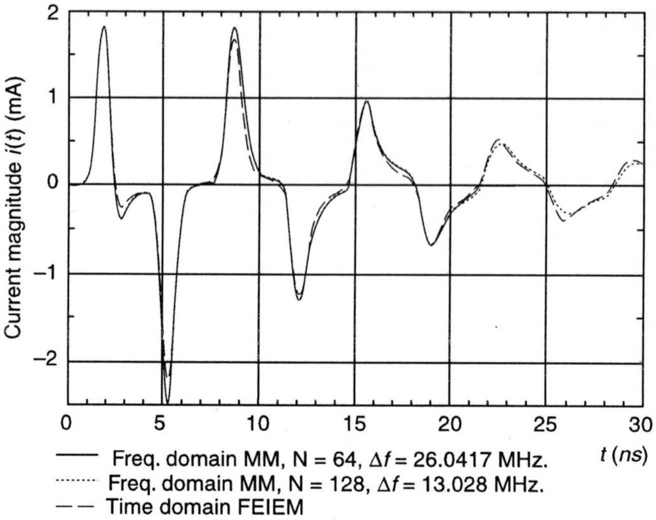

Figure 9.4 Transient current at the dipole driving point.

Figure 9.4. The length of the wire is $L = 1\,\text{m}$, and the radius is $a = 2\,\text{mm}$. The Gaussian voltage pulse is given by:

$$V(t) = V_0\, e^{-g^2(t-t_0)^2} \tag{9.110}$$

where $V_0 = 1\,\text{V}$, $g = 2^*10^9\,\text{s}^{-1}$, and $t_0 = 2\,\text{ns}$.

The transient waveform is obtained using both direct time domain approach discussed so far in this chapter and the frequency domain approach discussed in Chapter 8. Within the framework of the frequency domain approach the transient response is obtained via the Inverse Discrete Fourier Transform (IDFT). Numerical results obtained via different approaches agree satisfactorily.

$$V(t) = u(t) \tag{9.111}$$

where $u(t)$ denotes the unit step function.

Next two figures deal with a single straight wire insulated in free space operating in the scattering (receiving) mode. A thin wire scatterer of length $L = 1\,\text{m}$ and radius $a = 2\,\text{mm}$ is excited by the broadside incident Gaussian transient plane wave for the case of normal incidence:

$$E_x^{inc}(t) = E_0 e^{-g^2(t-t_0)^2} \tag{9.112}$$

where $E_0 = 1\,\text{V/m}$, $g = 2^*10^9\,\text{s}^{-1}$, and $t_0 = 2\,\text{ns}$.

The wire is illuminated along its full length simultaneously with the result that current is induced throughout the wire at the same instant. The resulting current response at the center of the wire is shown in Figure 9.7.

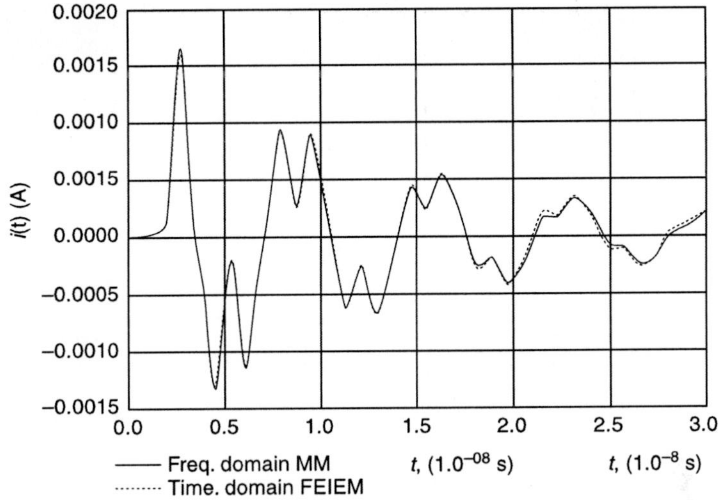

Figure 9.5 Dipole transient current at 3/4 point from one end.

Figure 9.8 shows the same wire scatterer when excited by a step function plane wave:

$$E_x^{inc}(t) = u(t) \tag{9.113}$$

It is obvious from Figures 9.4–9.8 that all numerical results obtained via direct time-domain technique are in good agreement with the results obtained via indirect

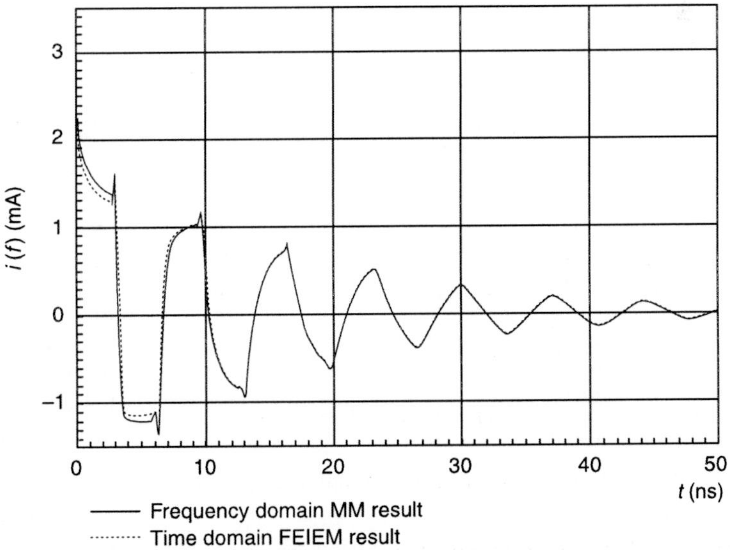

Figure 9.6 Dipole transient current for a step function input.

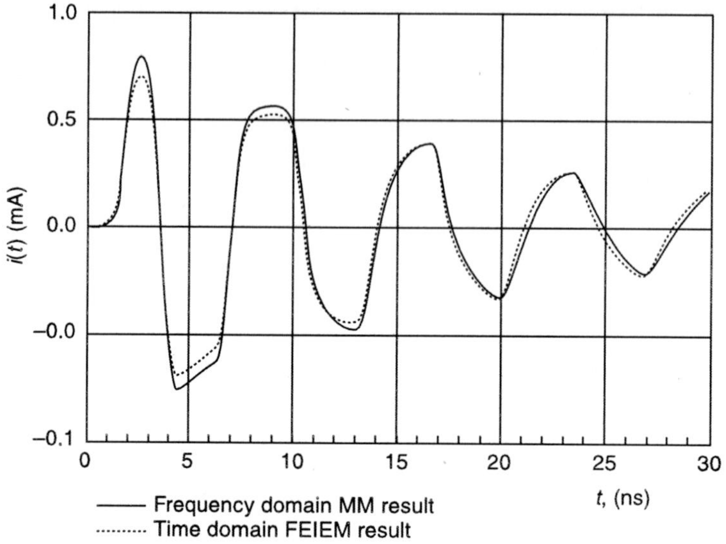

Figure 9.7 Transient current at midpoint of scatterer for Gaussian plane wave excitation.

frequency domain approach. The time domain energy variations for the single wire operating in antenna and scatterer mode are shown in Figures 9.9 and 9.10, respectively. In both the cases, the voltage pulse and the incident plane wave have the Gaussian time variation (9.110).

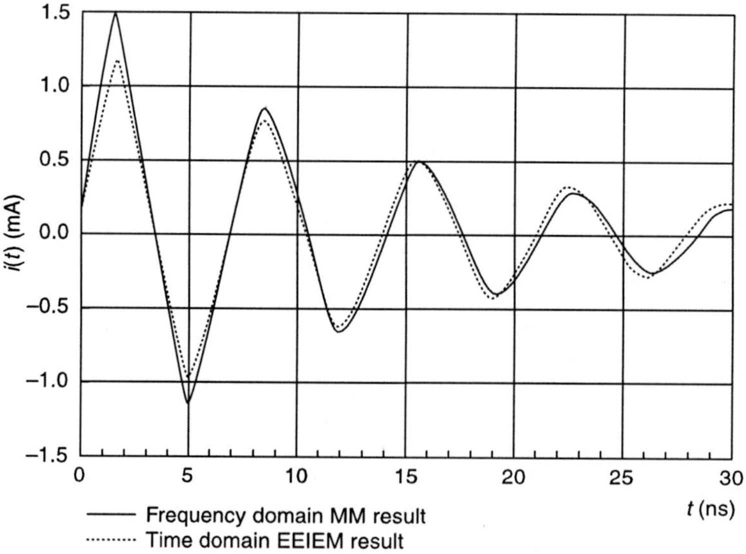

Figure 9.8 Transient current induced at the centre of the scatterer for step function plane wave excitation.

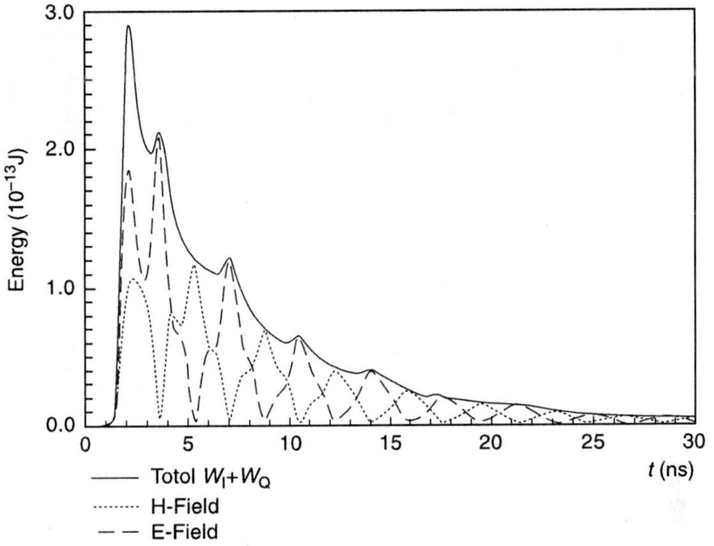

Figure 9.9 Time domain energy variation for dipole excited by a Gaussian voltage pulse.

9.1.7.2 Loaded Wires in Free Space This section deals with the influence of the resistive and nonlinear loading to the transient response of a single straight wire in free space.

Figure 9.11 shows the driving-point current against time of the single wire insulated in free space containing various resistive loadings at its center, and excited at its center by the voltage source (9.111).

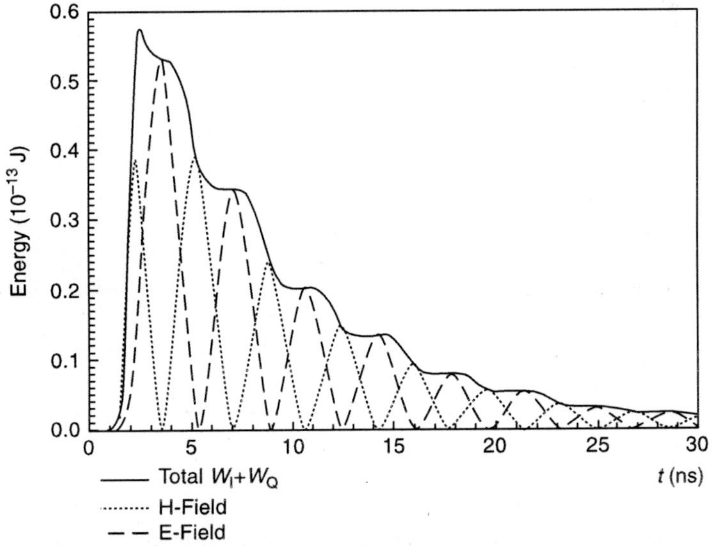

Figure 9.10 Time domain energy variation for scatterer excited by an incident plane wave.

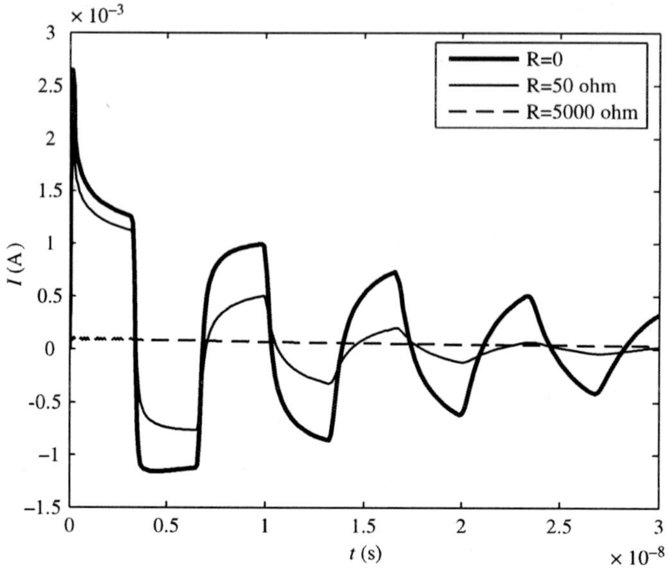

Figure 9.11 Driving-point current of the single wire in free space excited by the temporal unit step function with $L = 1$m, $a = 2$ mm. (D. Poljak, E. K. Miller, C. Y. Tham: Time Domain Energy Measures for Thin Wire Antennas and Scatterers, IEEE AP Magazine, Feb. 2002, pp 87–95).

The corresponding time-varying measure of the total energy stored in the wire near field is shown in Figure 9.12. The influence of resistive loading is obvious.

Next set of figures is related to nonlinear loading. The frequency domain method of transient analysis is based on the principle of superposition. This makes it unable to treat nonlinear load when the latter forms a part of the electromagnetic system. However, data obtained in frequency domain analysis can be transformed into the time domain and used to analyze nonlinear load. The method requires deriving a nonlinear Volterra type integral equation which is solved using input impedance and short circuit current data obtained from frequency domain analysis.

First example is related to the dipole antenna with length $L = 1$ m and radius $a = 6.74$ mm, excited at its center by Gaussian pulse voltage source (9.110) with the following parameters $V_0 = 1\ V$, $g = 3.25^{*}10^{+9}\ s^{-1}$, $t_0 = 1.389$ ns. Figure 9.13 shows the feed-point transient current of the unloaded antenna and transient load current of the antenna having a nonlinear load 0/700ohm. Obtained numerical results for the transient load current are in accordance with the results published in [14].

Figure 9.14 shows the spatial distribution of the *rms* values of the transient current along the wire clearly demonstrating the suppressing effect of the nonlinear wire loading. Figure 9.15 shows the transient current at the driving point of the thin wire containing nonlinear load at the center, and excited by the unit step voltage source (9.111).

Transient current is initially assumed to flow through 50Ω - section. Numerical results for the transient load current computed by the time domain boundary element procedure are compared with the results obtained by the frequency domain

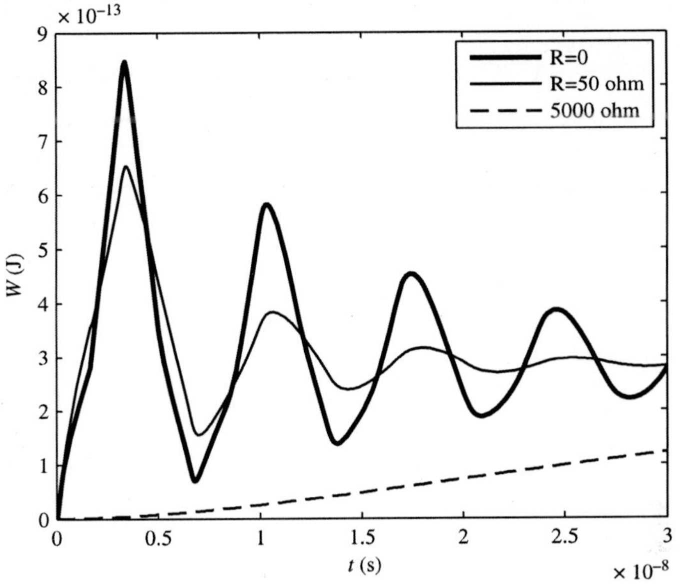

Figure 9.12 The total-energy measure as a function of time for a single wire in free space centre-excited by a unit-step voltage. (D. Poljak, E. K. Miller, C. Y. Tham: Time Domain Energy Measures for Thin Wire Antennas and Scatterers, IEEE AP Magazine, Feb. 2002, pp 87–95).

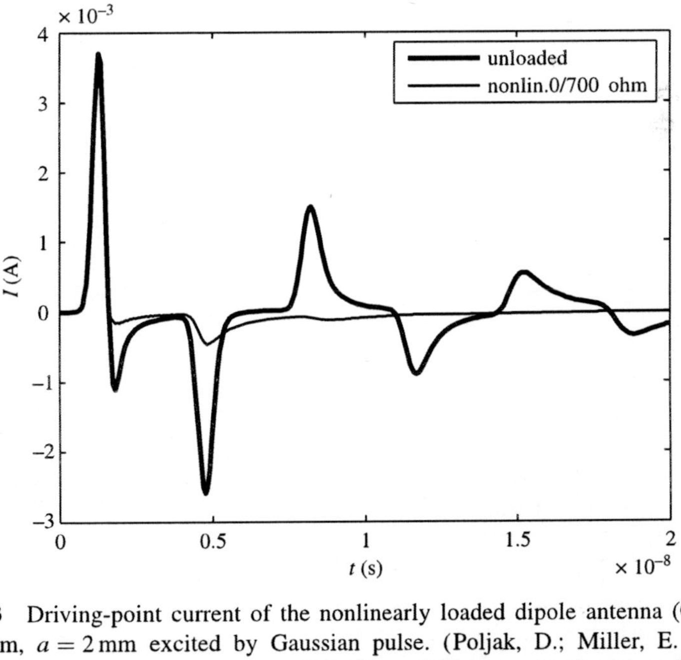

Figure 9.13 Driving-point current of the nonlinearly loaded dipole antenna (0/700 ohm) with $L = 1\,\text{m}$, $a = 2\,\text{mm}$ excited by Gaussian pulse. (Poljak, D.; Miller, E. K.; Yoong Tham, C., Root-Mean-Square measure of Nonlinear Effects to the Transient Response of Thin Wires, *IEEE Transactions on Antennas and Propagation.* 51 (2003), 12; 3280–3283).

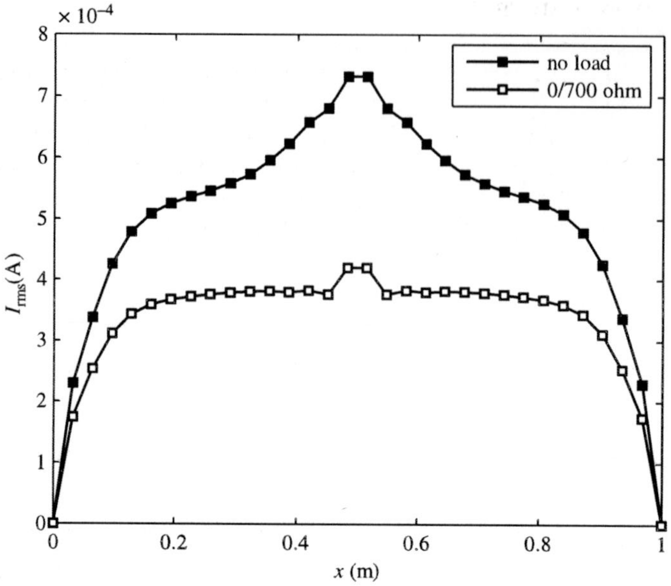

Figure 9.14 Spatial distribution of the rms value of the transient current along the dipole excited by the Gaussian pulse. (Poljak, D.; Miller, E. K.; Yoong Tham, C., Root-Mean-Square measure of Nonlinear Effects to the Transient Response of Thin Wires, *IEEE Transactions on Antennas and Propagation.* 51 (2003), 12; 3280–3283).

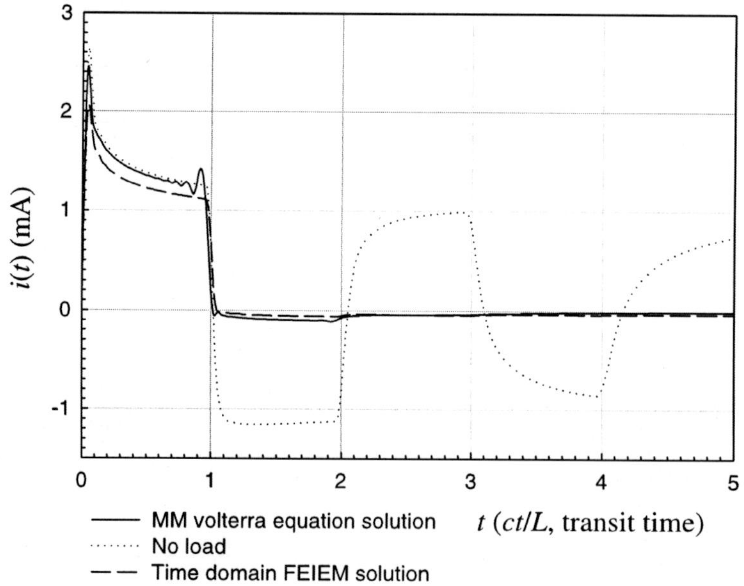

Figure 9.15 Driving-point current of the dipole antenna excited by the temporal unit step function with $L = 1\text{m}$, $a = 2\,\text{mm}$, with nonlinear load at the feed-point (50/5000 ohm). (Poljak, D.; Miller, E. K.; Yoong Tham, C., Root-Mean-Square measure of Nonlinear Effects to the Transient Response of Thin Wires, *IEEE Transactions on Antennas and Propagation.* 51 (2003), 12; 3280–3283).

solution method based on the principle of superposition. The data obtained in the frequency domain analysis can be transformed into the corresponding time domain data and used to analyze wires with nonlinear loads. The method uses a nonlinear Volterra type integral equation which is solved using input impedance and short circuit current data obtained from the frequency domain analysis and transformed into time domain using the inverse Fourier transform. Therefore, the time dependent current I_{nL} through the nonlinear load is governed by the expression [10]:

$$I_{nL} = I_{sc} - \int_0^t i\delta(t-\tau)F[i_{nL}(\tau)]d\tau \qquad (9.114)$$

where $i_{sc}(t)$ is the current flowing through the wire terminals when the nonlinear element is removed, and i_δ is the current flowing in the wire terminals because of an excitation by an impulsive voltage source.

Both of these quantities were calculated by means of the inverse Fourier transform applied to the numerical results obtained by the frequency domain method of moments (MoM) [10].

Numerical results obtained by different numerical techniques agree satisfactorily. The spatial distribution of the corresponding rms values is presented in Figure 9.16.

It can be observed that the presence of nonlinear load significantly reduces the rms value of the transient load current.

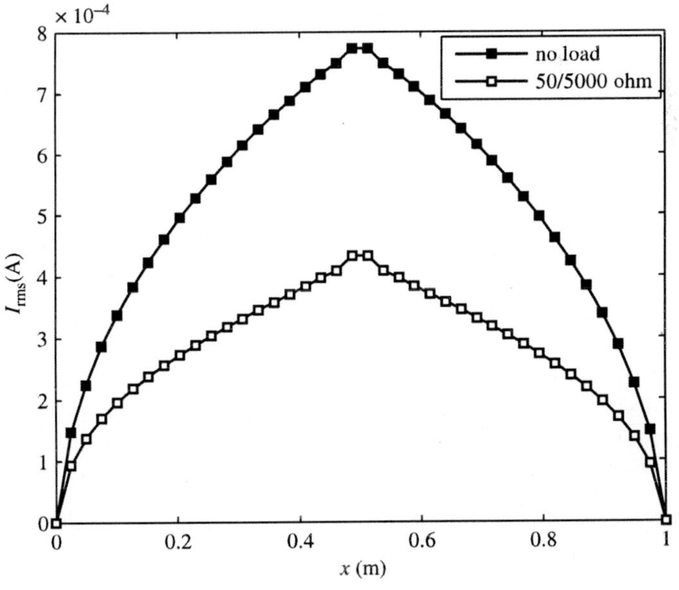

Figure 9.16 Spatial distribution of the rms value of the transient current along the dipole excited by the temporal step. (Poljak, D.; Miller, E. K.; Yoong Tham, C., Root-Mean-Square measure of Nonlinear Effects to the Transient Response of Thin Wires, *IEEE Transactions on Antennas and Propagation.* 51 (2003), 12; 3280–3283).

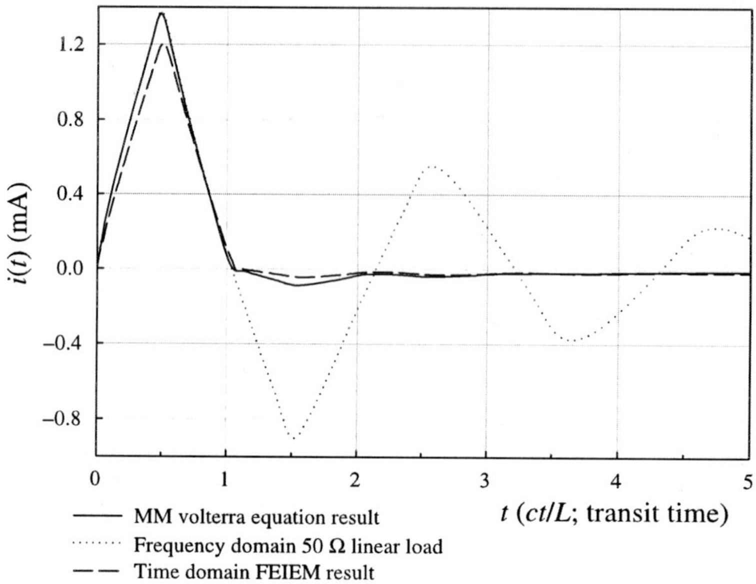

Figure 9.17 Current induced at the center of the nonlinearly loaded wire (50/5000 Ω) illuminated by the temporal unit step function with $L = 1\,\text{m}$, $a = 2\,\text{mm}$. (Poljak, D.; Miller, E. K.; Yoong Tham, C., Root-Mean-Square measure of Nonlinear Effects to the Transient Response of Thin Wires, *IEEE Transactions on Antennas and Propagation.* 51 (2003), 12; 3280–3283).

Figure 9.17 shows the transient response of the straight thin-wire scatterer with the same dimensions and with the load having a forward resistance of 50 Ω and reverse resistance of 5 kΩ. The wire is illuminated by an incident unit step function plane wave (9.113) with polarization parallel to the axial direction.

The transient current flowing through the linear 50Ω resistor is shown here in order to present the suppressing effect of the nonlinear element. Numerical results computed via different approaches agree satisfactorily yet again. The spatial distribution of the corresponding rms values is presented in Figure 9.18.

The examples also show that the oscillation at late time of both the antenna feed-point current and current induced at the wire midpoint in the scattering mode is significantly reduced with a nonlinear element. It is also seen that the magnitude of the initial current response pulse is not affected by the nonlinear nature of the load. This statement is strictly related to the given polarity of the incident field.

Observing the presented numerical results, it is obvious that the rms values of the transient current flowing through the nonlineary loaded wire is significantly attenuated with respect to the case when the nonlinear load is removed.

However, it is also obvious from the presented numerical results that the magnitude of the first pulse is not limited by the nonlinear element.

Taking into account the fact that the rms value of the time dependent current is a measure of the average power delivered to the resistive load, the rms value

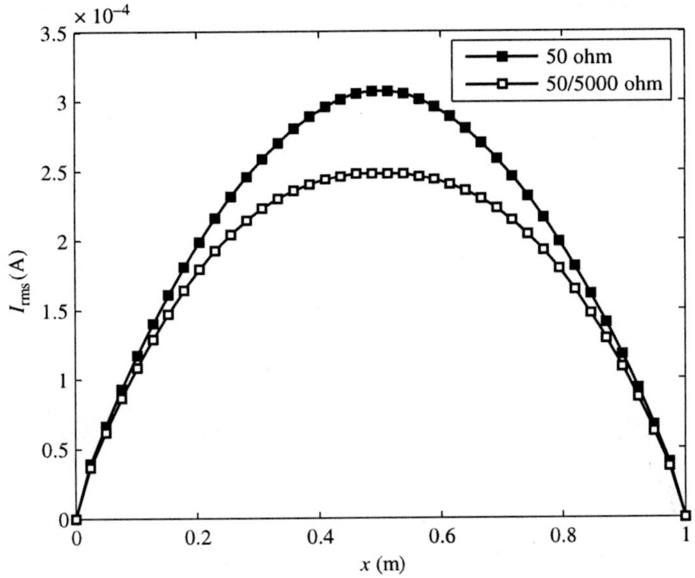

Figure 9.18 Spatial distribution of the rms value of the transient current along the straight-wire scatterer illuminated by the temporal step. (Poljak, D.; Miller, E. K.; Yoong Tham, C., Root-Mean-Square measure of Nonlinear Effects to the Transient Response of Thin Wires, *IEEE Transactions on Antennas and Propagation.* 51 (2003), 12; 3280–3283).

generally seems to be a very useful parameter in the analysis of nonlinear electromagnetic phenomena, particularly in designing the lightning protection devices.

9.1.7.3 Two Coupled Wires in Free Space This section deals with some illustrative computational examples related to coupled wires in free space. It is worth mentioning that various numerical results for two identical coupled wires in free space, on the basis of the set of Hallen equations, can be found in references [10], [23], and [27].

Two coupled identical wires in free space of length $L = 1$ m, radius $a = 2$ mm, and separated by distances $s = 1.0$ m, 0.5 m, and 0.25 m are analyzed. A Gaussian voltage pulse (9.110) is used to excite one of the wires at its midpoint. The transient current at the center of the active and passive wire, respectively, is presented in Figures 9.19 and 9.20.

The coupling effect between the wire as a result of different separation distances are obvious.

Figures 9.21 and 9.22 show the currents induced at the center of coupled wires in free space (The wire *A* is the near-side wire while the wire *B* is the far-side one) when wire *A* is exposed to the Gaussian pulse plane wave (9.112) where $V_0 = 1$ V, $g = 1.5^*10^9\, s^{-1}$ and $t_0 = 1.43\, ns$.

The length of the wires is $L = 1$ m, the radius is $a = 2$ mm, and the separation between them is $d = 0.5$ m. It is obvious from Figures 9.21 and 9.22 that the total

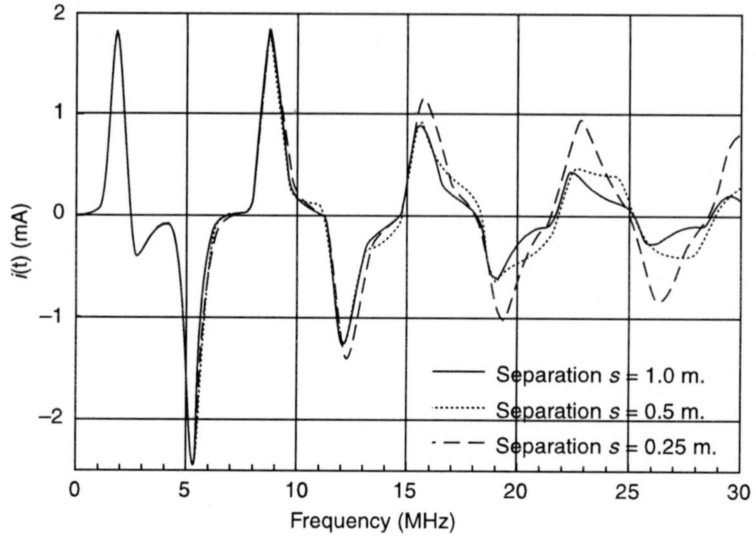

Figure 9.19 Driving point transient current at the active wire.

energy of the active wire decreases with time after the exciting pulse vanishes, while the passive wire's energy builds up.

Figure 9.23 shows the currents induced at the center of coupled wires (wire A is active wire while wire B is the passive one) when wire A is center-fed by the Gaussian pulse from Figure 4.5 with $V_0 = 1$ V, $g = 1.5 \times 10^9$ s^{-1}, and $t_0 = 1.43$ *ns*.

Figures 9.24 and 9.25 present the H-field, the E-field, and the total energy measures for the active and passive wire. It is obvious from Figures 9.24 and 9.25 that

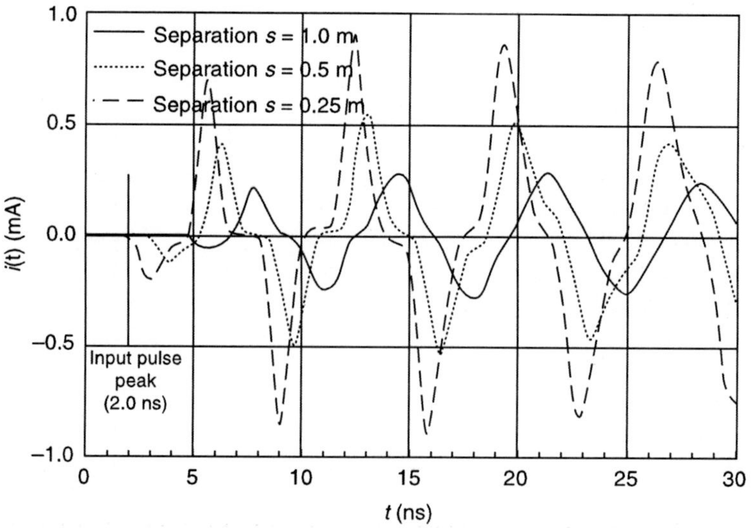

Figure 9.20 Transient current at the centre of the parasitic wire.

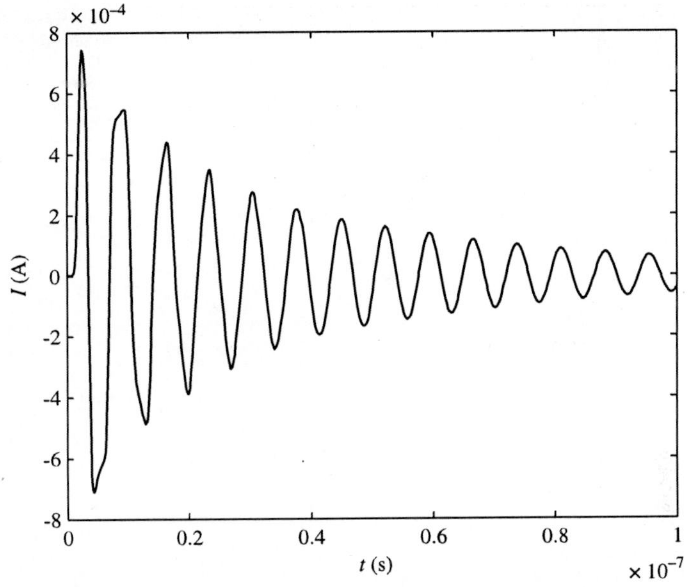

Figure 9.21 Transient response at the center of the near side wire in free space with $L = 1$ m, $a = 2$ mm, $d = 0.5$ m.

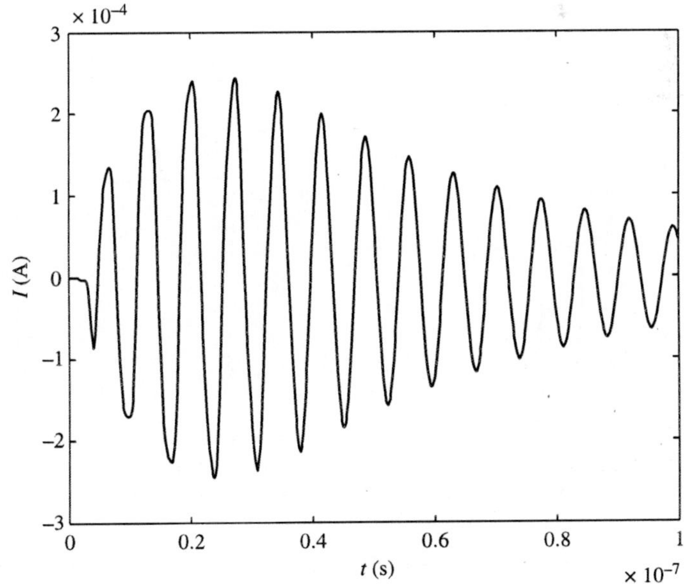

Figure 9.22 Transient response at the center of the far-side wire with $L = 1$ m, $a = 2$ mm, $d = 0.5$ m.

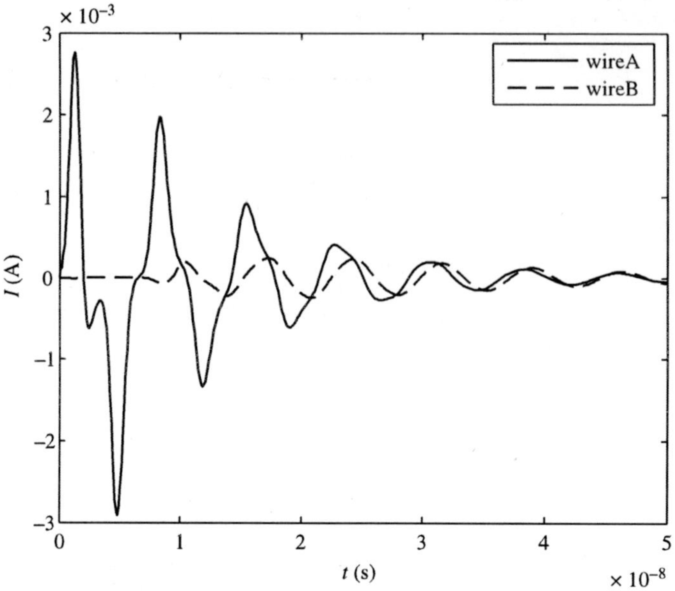

Figure 9.23 Transient current flowing on the coupled wires in free space with $L = 1$ m, $a = 2$ mm, and $d = 2$ m, when wire A is driven by a Gaussian pulse. (D. Poljak, E. K. Miller, C. Y. Tham: "Time Domain Energy Measures for Thin-Wire Antennas and Scatterers", *IEEE Antennas&Propagation Magazine*, Vol 44, No1, pp 87–95, Feb 2002).

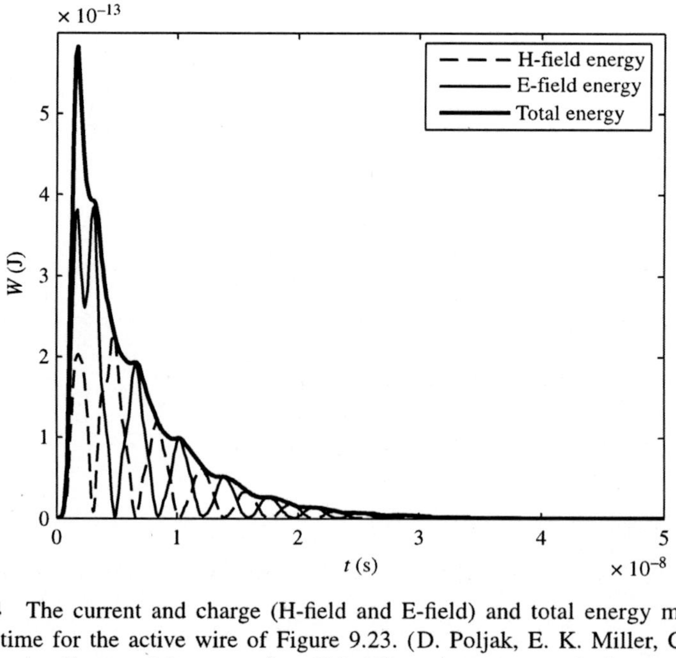

Figure 9.24 The current and charge (H-field and E-field) and total energy measure as a function of time for the active wire of Figure 9.23. (D. Poljak, E. K. Miller, C. Y. Tham: "Time Domain Energy Measures for Thin-Wire Antennas and Scatterers", *IEEE Antennas &Propagation Magazine*, Vol 44, No1, pp 87–95, Feb 2002).

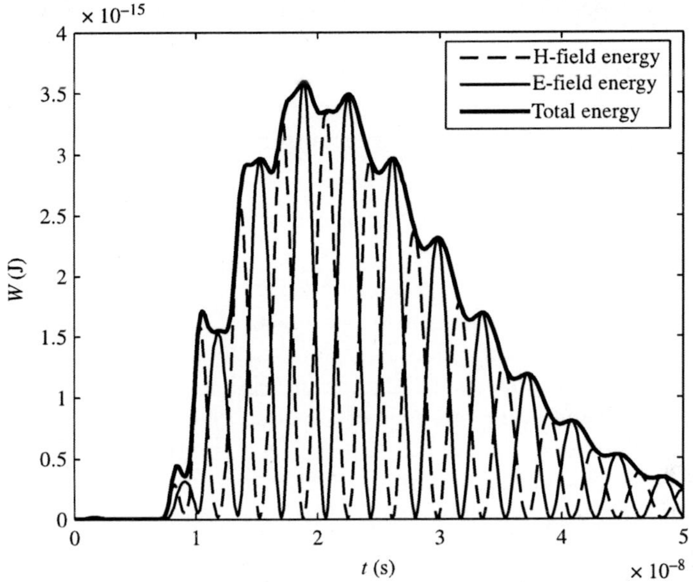

Figure 9.25 The current, charge, and total energy measures for the passive wire of Figure 9.23. (D. Poljak, E. K. Miller, C. Y. Tham: "Time Domain Energy Measures for Thin-Wire Antennas and Scatterers", *IEEE Antennas&Propagation Magazine*, Vol 44, No 1, pp 87–95, Feb 2002).

the total energy of the active wire decreases with time after the exciting pulse vanishes, while the passive wire's energy builds up.

From Figure 9.24, the charge energy can be seen to exhibit a double peak between 0 and 0.5 ns besides which it also exceeds the current energy. The first peak is normally seen for a wire excited by a pulse whose width is half or less of L/c, the transit time, but the second peak occurs because the outgoing pulses reach the wire ends before the equalization of current and charge energies, which would otherwise occur.

In addition, it is seen from Figure 9.25 that the current, charge, and total energy measures for the passive wire are all zero until the fields of the active wire have propagated across the space between them, or for about 7 ns. They increase thereafter in a nearly monotonic fashion for about 12 ns, as energy continues to be collected from the active wire, and then decline from losing energy as a result of radiation, with the maximum total energy measure of the passive wire being about 0.0067 that of the active wire.

The distinguishing feature in the coupled wire energy variation is the resonant effect. This effect causes the energy level to die off more slowly as compared with the single wire cases. The early time energy variation for the total two wire system is the same as that for the active wire alone, as the parasitic wire plays little or no part at this stage.

The energy variation in the time domain makes it possible to view the electromagnetic behavior of the two coupled wires taken together as one system. This view is not available by looking at the currents on the individual wires separately. The figures presented show that energy is transferred from the active wire to the parasitic one, some of which is then returned to the active wire later. At a certain instant, the parasitic wire becomes the more dominant of the two in terms of energy stored. The total energy dissipates very slowly when a parasitic wire is present.

9.2 THIN WIRES IN THE PRESENCE OF A TWO-MEDIA CONFIGURATION

Transient analysis of thin wires in the presence of a real ground has numerous applications in target identification, electromagnetic stimulation of biological tissue, indoor communications, and lightning protection design. However, studies of the thin wire radiating in the presence of a half-space configuration is usually performed in the frequency domain [25,30,31]. The corresponding time domain parameters can be obtained using the inverse Fourier transform. This section deals with a single wire, two coupled wires, and arbitrary wire arrays above a real ground.

9.2.1. Single Straight Wire Above a Real Ground

A direct time formulation for a straight thin wire above a dissipative half-space on the basis of the half-space Hallen equation has been proposed in [5]. This formulation is, in fact, the extension of the free-space Hallen equation [29], with the time-domain Fresnel reflection coefficient (RC) appearing in the kernel. The corresponding integral relationship equations for thin wires have been solved via efficient boundary element procedures.

A center-fed thin wire antenna or scatterer of length L and radius a, placed above a dissipative half-space along the x-axis is considered, Figure 9.26. The wire is assumed to be perfectly conducting and excited by a time-dependent voltage generator or illuminated by a plane wave. The corresponding time-domain Hallen integral equation can be derived gradually starting from the straight wire insulated in the free space.

The free-space Hallen equation is as follows:

$$\int_0^L \frac{I(x', t - R/c)}{4\pi R} dx' = F_0\left(t - \frac{x}{c}\right) + F_L\left(t - \frac{L-x}{c}\right) + \frac{1}{2Z_0} \int_0^L E_x^{inc}\left(x', t - \frac{|x-x'|}{c}\right) dx' \tag{9.115}$$

It is first transferred into the Laplace frequency domain:

$$\int_0^L \frac{I(x', s)e^{-\frac{sR}{c}}}{4\pi R} dx' = F_0(s)e^{-\frac{sx}{c}} + F_L(s)e^{-s\frac{(L-x)}{c}} + \frac{1}{2Z_0} \int_0^L E_x^{inc}(x', s)e^{-s\frac{|x-x'|}{c}} dx' \tag{9.116}$$

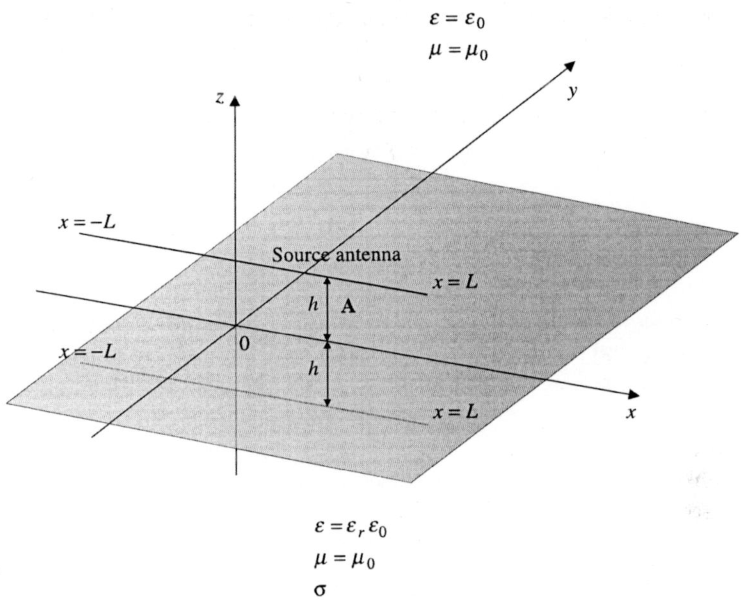

Figure 9.26 Horizontal antenna over real ground.

where $s = j\omega$ is the Laplace variable.

In accordance with the theory of images, the free space integral eqn (9.116) in the frequency domain is extended by an additional term resulting in the following integral equation:

$$\int_0^L \frac{I(x', s)e^{\frac{sR}{c}}}{4\pi R} \mathrm{d}x' - \int_0^L \frac{I(x', s)e^{\frac{sR'}{c}}}{4\pi R} \mathrm{d}x' = F_0(s)e^{-\frac{sx}{c}} + F_L(s)e^{-s\frac{(L-x)}{c}} + \frac{1}{2Z_0}\int_0^L E_x^{inc}(x', s)e^{-s\frac{|x-x'|}{c}} \mathrm{d}x' \tag{9.117}$$

where $R^* = [(x - x')^2 + 4h^2]^2$.

At the last stage, the RC approach [29,30] is applied by multiplying the Green's function of the image source by space-frequency dependent reflection coefficient $R_{TM}(\theta', s)$, defined for TM polarization. The integral equation obtained for the lossy half-space in the frequency domain is given by:

$$\int_0^L \frac{I(x, s)e^{\frac{sR}{c}}}{4\pi R} \mathrm{d}x' - \int_0^L R_{TM}(\theta, s) \frac{I(x, s)e^{\frac{sR'}{c}}}{4\pi R} \mathrm{d}x' = F_0(s)e^{-\frac{sx}{c}} + F_L(s)e^{-s\frac{L-x}{c}} + \frac{1}{2Z_0}\int_0^L E_x^{inc}(x', s)e^{-s\frac{|x-x'|}{c}} \mathrm{d}x' \tag{9.118}$$

and $R_{TM}(\theta', s)$ is determined by the expression [30,31]:

$$R_{TM}(\theta', s) = \frac{\varepsilon_r\left(1+\frac{\sigma}{\varepsilon s}\right)\cos\theta' - \sqrt{\varepsilon_r\left(1+\frac{\sigma}{\varepsilon s}\right) - \sin^2\theta'}}{\varepsilon_r\left(1+\frac{\sigma}{\varepsilon s}\right)\cos\theta' + \sqrt{\varepsilon_r\left(1+\frac{\sigma}{\varepsilon s}\right) - \sin^2\theta'}} \tag{9.119}$$

where σ and ε are the lossy medium conductivity and permittivity respectively, and $\theta' = arctg(|x - x'|/2h)$.

The reflection coefficient (RC) approach is a good approximation in half-space calculations, as long as the field is calculated far away from the source, and away from the imperfect ground to ensure $\theta' < \pi/2$. Sarkar [31] has shown that such approximation is valid as long as the dipole, which is horizontally oriented to the ground is at least $0.25\lambda/((\varepsilon_r(1+\sigma/\varepsilon s))^{1/2})$ away from interface. This condition must be satisfied for each component of the equivalent frequency spectrum. (Since the dipole antenna acts as a filter, the permissible λ is related to the corresponding $\lambda_{\max}$ of the spectrum).

Using the convolution theorem, the time-domain counterpart of eqn (9.118) is obtained in the form:

$$\int_0^L \frac{I(x', t - R/c)}{4\pi R}\,dx' - \int_{-\infty}^{t}\int_0^L r(\theta,\tau)\frac{I(x', t - R^*/c - \tau)}{4\pi R^*}\,dx'\,d\tau$$
$$= F_0\left(t - \frac{x}{c}\right) + F_L\left(t - \frac{L-x}{c}\right) + \frac{1}{2Z_0}\int_0^L E_x^{inc}\left(x', t - \frac{|x - x'|}{c}\right)dx' \tag{9.120}$$

Equation (9.120) is valid for the radiation mode in which the incident field in the excitation term from the right-hand side is expressed by means of the voltage generator (9.110). On the contrary, if the scattering mode (normal incidence, for the sake of simplicity) is considered, the excitation term is given by:

$$E_x^{exc}(t) = E_x^{inc}(t) - E_x^{ref}(t^*) \tag{9.121}$$

The transient earth-reflected field is obtained as the convolution of the incident field and the space-time reflection coefficient for the angle of incidence $\theta = 0$ (in accordance with the parallel incidence of the electric field), as is proposed in [Barnes]:

$$E_x^{ref}(t) = \int_{-\infty}^{t} E_x^{inc}(t - \tau)r(\theta = 0, \tau)d\tau \tag{9.122}$$

In this manner, the transient function $r(t)$ can be considered as the impulse response of the lower medium.

And the integral eqn (9.120) becomes:

$$\int_0^L \frac{I(x', t - R/c)}{4\pi R} \mathrm{d}x' - \int_{-\infty}^{t} \int_0^L r(\theta, \tau) \frac{I(x', t - R^*/c - \tau)}{4\pi R^*} \mathrm{d}x' \mathrm{d}\tau$$

$$= F_0\left(t - \frac{x}{c}\right) + F_L\left(t - \frac{L - x}{c}\right) + \frac{1}{2Z_0} \int_0^L E_x^{inc}\left(x', t - \frac{|x - x'|}{c}\right) \mathrm{d}x' \tag{9.123}$$

$$- \frac{1}{2Z_0} \int_{-\infty}^{t} \int_0^L E_x^{inc}\left(x', t - \frac{|x - x'|}{c} - \tau\right) r(\theta = 0, \tau) \mathrm{d}x' \mathrm{d}\tau$$

The unknown time functions $F_0(t)$, $F_L(t)$, $F_0(t - L/c)$, and $F_L(t - L/c)$ are expressed and obtained in the same manner, as in the case of free space, in terms of the auxiliary functions $K_0(t)$ and $K_L(t)$: [3,5]:

$$F_0(t) = \sum_{n=0}^{\infty} K_0\left(t - \frac{2nL}{c}\right) - \sum_{n=0}^{\infty} K_L\left(t - \frac{(2n + 1)L}{c}\right) \tag{9.124}$$

$$F_L(t) = \sum_{n=0}^{\infty} K_L\left(t - \frac{2nL}{c}\right) - \sum_{n=0}^{\infty} K_0\left(t - \frac{(2n + 1)L}{c}\right) \tag{9.125}$$

which are:

$$K_L(t) = \int_0^L \frac{I(x', t - R_0/c)}{4\pi R_0} \mathrm{d}x' - \int_{-\infty}^{t} \int_0^L r(\theta', \tau) \frac{I(x', t - R_0^*/c - \tau)}{4\pi R_0^*} \mathrm{d}x' \mathrm{d}\tau$$

$$- \frac{1}{2Z_0} \int_0^L E_x^{inc}\left(x', t - \frac{x'}{c}\right) \mathrm{d}x' \tag{9.126}$$

$$K_L(t) = \int_0^L \frac{I(x', t - R_L/c)}{4\pi R_L} \mathrm{d}x' - \int_{-\infty}^{t} \int_0^L r(\theta', \tau) \frac{I(x', t - R_L^*/c - \tau)}{4\pi R_L^*} \mathrm{d}x' \mathrm{d}\tau$$

$$- \frac{1}{2Z_0} \int_0^L E_x^{exc}\left(x', t - \frac{L - x'}{c}\right) \mathrm{d}x' \tag{9.127}$$

while R_0 and R_L are the distances from the wire ends to the source point, and R_0^*, R_L^* are the distances from the image wire ends to the image source point.

The $r(\theta, t)$ is the space-time reflection coefficient [32]:

$$r(\theta', t) = A\delta(t) + \frac{4\beta}{1 - \beta^2} \frac{e^{-at}}{t} \sum_{n=1}^{\infty} (-1)^{n+1} n A^n I_n(\alpha t) \tag{9.128}$$

where: $A = (1-\beta)/(1+\beta)$, $\beta = [\varepsilon_r - \sin^2\theta']^{1/2}/(\varepsilon_r \cos\theta')$, and $\alpha = \sigma/2\varepsilon$, $\delta(t)$ is the Dirac impulse and $I_n(t)$ is the modified Bessel function of order n.

All the approximations used in [32] for deriving the expression (9.128) are also taken into account in this work (ε_r is of order 10 or more, values for σ range from 1 to 30 mmho/m, angles of incidence ensure the condition $\theta' < \pi/2$).

If the case of a perfect (ideal) dielectric half-space is considered, the Hallen eqn (9.120) simplifies into:

$$\int_0^L \frac{I(x', t - R/c)}{4\pi R} \mathrm{d}x' - \int_0^L r(\theta) \frac{I(x', t - R^*/c)}{4\pi R^*} \mathrm{d}x'$$
$$= F_0\left(t - \frac{x}{c}\right) + F_L\left(t - \frac{L-x}{c}\right) + \frac{1}{2Z_0}\int_0^L E_x^{exc}\left(x', t - \frac{|x-x'|}{c}\right)\mathrm{d}x' \tag{9.129}$$

The unknown time functions $F_0(t)$, $F_L(t)$, $F_0(t - L/c)$, and $F_L(t - L/c)$ are expressed by relations (9.124) and (9.125), while the auxiliary functions $K_0(t)$ and $K_L(t)$ are given by:

$$\int_0^L \frac{I(x', t - R_0/c)}{4\pi R_0} \mathrm{d}x' - \int_0^L r(\theta) \frac{I(x', t - R_0^*/c)}{4\pi R_0^*} \mathrm{d}x' - \frac{1}{2Z_0}\int_0^L E_x^{exc}\left(x', t - \frac{x'}{c}\right)\mathrm{d}x' \tag{9.130}$$

$$\int_0^L \frac{I(x', t - R_L/c)}{4\pi R_L} \mathrm{d}x' - \int_0^L r(\theta) \frac{I(x', t - R_L^*/c)}{4\pi R_L^*} \mathrm{d}x' - \frac{1}{2Z_0}\int_0^L E_x^{exc}\left(x', t - \frac{L-x'}{c}\right)\mathrm{d}x' \tag{9.131}$$

Space-time integral equations (9.123) and (9.131) can be solved with the zero current at the free ends of the wire and with the initial conditions requiring the wire not to be excited before the certain instant $t = t_0$, as specified by eqns (9.86) and (9.87).

9.2.2 Far Field Equations

The formula for the far electric field of a thin horizontal wire above a lossy half-space can be derived in a similar way starting from the case of free space. The electric field of a thin wire can be expressed in the form of the TD Pocklington equation [7,12]:

$$\vec{E}(r,t) = -e_x \frac{\mu_0}{4\pi} \int_0^L \frac{\partial}{\partial t} \frac{I(x',t')}{R_1} \mathrm{d}x' + \frac{1}{4\pi\varepsilon_0} e_{R_1} \int_0^L \frac{q(x',t')}{R_1^2} \mathrm{d}x' \\ + \frac{1}{4\pi\varepsilon_0 c} e_{R_1} \int_0^L \frac{\partial q(x',t')}{\partial t'} \frac{1}{R_1} \mathrm{d}x' \tag{9.132}$$

where $R_1 = [(x-x)^2 + z^2]^{1/2}$ is the distance from the wire to the observation point, the e_x and e_{R1} are the unit vectors in x and R_1 direction, respectively and $q(x',t')$ denotes the charge distribution along the wire.

According to the well known far-field approximation [7,12] the terms containing the $q(x',t')$ (near field terms) can be neglected, and the above eqn (9.132) becomes:

$$e_x E_x(r,t) = -e_x \frac{\mu_0}{4\pi} \int_0^L \frac{\partial}{\partial t} \frac{I(x',t')}{R_1} \mathrm{d}x' \tag{9.133}$$

where $t' = t - R_1/c$ is the retarded time.

Now, using the Laplace transform it simply follows:

$$E_x(r,s) = \frac{\mu_0}{4\pi} \int_0^L s \frac{I(x',s)}{R_1} e^{-\frac{sR}{c}} \mathrm{d}x' \tag{9.134}$$

If the case of a perfectly conducting half-space is considered, additional term to take into account, that is, the image source leads to the expression:

$$E_x(r,s) = \frac{\mu_0}{4\pi} \left(\int_0^L s \frac{I(x',s)}{R_1} e^{\frac{-sR_1}{c}} \mathrm{d}x' - \int_0^L s \frac{I(x',s)}{R_1^*} e^{\frac{-sR_1^*}{c}} \mathrm{d}x' \right) \tag{9.135}$$

where $R_1^* = [(x-x')^2 + (2h+z)^2]^{1/2}$ is the distance from the image wire to the observation point.

Since the multiplication in the frequency domain implies the convolution in the time domain, the equation for the far field in the time domain is obtained by utilizing the inverse Laplace transform:

$$E_x(r,t) = -\frac{\mu_0}{4\pi} \left(\int_0^L \frac{\partial}{\partial t} \frac{I(x',t')}{R_1} \mathrm{d}x' - \int_0^L \frac{\partial}{\partial t} \frac{I(x',t_1'-\tau)}{R_1^*} \mathrm{d}x' \right) \tag{9.136}$$

where $t_1' = t - R_1^*/c$ is the retarded time.

The presence of a lossy half-space is included by extending the second term in (9.135) with the reflection coefficient. In the Laplace domain it follows:

$$E_x(r,s) = \frac{\mu_0}{4\pi}\left(\int_0^L s\frac{I(x',s)}{R_1}e^{-\frac{sR_1}{c}}\mathrm{d}x' - \int_0^L R(\theta,s')s\frac{I(x',s)}{R_1^*}e^{-\frac{sR_1^*}{c}}\mathrm{d}x'\right) \tag{9.137}$$

Using the convolution in the time domain, the time-domain far-field equation is obtained by utilizing the inverse Laplace transform:

$$E_x(r,t) = -\frac{\mu_0}{4\pi}\left(\int_0^L \frac{\partial}{\partial t}\frac{I(x',t')}{R_1}\mathrm{d}x' - \int_{-\infty}^{t} r(\theta,\tau)\int_0^L \frac{\partial}{\partial t}\frac{I(x',t_1'-\tau)}{R_1^*}\mathrm{d}x'\mathrm{d}\tau\right) \tag{9.138}$$

If the case of a dielectric half-space is considered formula (9.138) becomes:

$$E_x(r,t) = \frac{\mu_0}{4\pi}\left(\int_0^L \frac{\partial}{\partial t}\frac{I(x',t-R_1/c)}{R_1}\mathrm{d}x' - \int_0^L r(\theta)\frac{\partial}{\partial t}\frac{I(x',t-R_1^*/c)}{R_1^*}\mathrm{d}x'\right) \tag{9.139}$$

Expressions (9.136), (9.138), and (9.139) provide rather straightforward far field calculation if the transient current is already determined.

9.2.3 Loaded Straight Thin Wire Above a Lossy Half-Space

This section deals with a resistively and nonlinearly loaded wires and scatterers in the presence of a two-media configuration.

9.2.3.1 Resistively Loaded Straight Thin Wire Analysis of loaded wires in the presence of a half-space configuration is usually performed in the frequency domain [25,30,31] and the transient response is then obtained using the inverse Fourier transform.

A time-domain formulation for loaded horizontal dipole antenna radiating over a dissipative half-space, based on the extension of the half-space Hallen eqns (9.120), (9.123), or (9.129) has been reported in [6].

A straight thin wire of length L and radius a, oriented horizontally over the lossy half-space along the x-axis is considered. The wire is assumed to be perfectly conducting and continuously loaded with a resistive loading R_L per unit of wire length. The excitation function is a time-varying voltage source (antenna mode) or a plane wave (scatterer mode).

Forcing the condition for the tangential electric field along the loaded dipole antenna [6]:

$$E_x^{inc} + E_x^{sct} = IR_L \tag{9.140}$$

where E_x^{inc} is the tangential incident field, E_x^{sct} is the scattered field, $I(x')$ is the equivalent axial current distribution along the antenna, and R_L is the resistive load per unit length of the wire.

Starting from Maxwell equations and forcing the condition (9.138), the time-domain electric field integral equation is obtained in the form [23]:

$$-\varepsilon\frac{\partial E_x^{inc}}{\partial t} = \left[\frac{\partial^2}{\partial x^2} - \frac{1}{c}\frac{\partial^2}{\partial t^2}\right] - \left\{\int_0^L \frac{I(x',t-R/c)}{4\pi R}\mathrm{d}x' - \int_{-\infty}^{t} r_{TM}(\theta,t)\int_0^L \frac{I(x',t-R^*/c-\tau)}{4\pi R^*}\mathrm{d}x'\mathrm{d}\tau\right\} - R_L(x)\frac{\partial I(x,t)}{\partial t} \tag{9.141}$$

Using the superposition principle [6] and integrating eqn (9.141), the corresponding Hallen equation is obtained in the form:

$$\int_0^L \frac{I(x',t-R/c)}{4\pi R}\mathrm{d}x' - \int_{-\infty}^{t}\int_0^L r(\theta,\tau)\frac{I(x',t-R'/c-\tau)}{4\pi R'}\mathrm{d}x'\mathrm{d}\tau$$
$$= F_0\left(t-\frac{x}{c}\right) + F_L\left(t-\frac{L-x}{c}\right) + \frac{1}{2Z_0}\int_0^L E_x^{inc}\left(x',t-\frac{|x-x'|}{c}\right)\mathrm{d}x' \tag{9.142}$$
$$-\frac{1}{2Z_0}\int_0^L R_L I\left(x',t-\frac{|x-x'|}{c}\right)\mathrm{d}x'$$

$F_0(t)$ and $F_L(t)$ are the unknown time signals existing because of multiple reflections of current at wire ends given by the relations (9.124) and (9.125).

For the case of loaded wire, the auxiliary functions K_0 and K_L are defined as follows:

$$K_0(t) = \int_0^L \frac{I(x',t-R_0/c)}{4\pi R_0}\mathrm{d}x' - \int_{-\infty}^{t}\int_0^L r(\theta',\tau)\frac{I(x',t-R_0^*/c-\tau)}{4\pi R_0^*}\mathrm{d}x'\mathrm{d}\tau$$
$$-\frac{1}{2Z_0}\int_0^L E_x^{inc}\left(x',t-\frac{x'}{c}\right)\mathrm{d}x' + \frac{1}{2Z_0}\int_0^L R_L(x')I\left(x',t-\frac{x'}{c}\right)\mathrm{d}x' \tag{9.143}$$

$$K_L(t) = \int_0^L \frac{I(x',t-R_L/c)}{4\pi R_L}\mathrm{d}x' - \int_{-\infty}^{t}\int_0^L r(\theta',\tau)\frac{I(x',t-R_L^*/c-\tau)}{4\pi R_L^*}\mathrm{d}x'\mathrm{d}\tau$$
$$-\frac{1}{2Z_0}\int_0^L E_x^{inc}\left(x',t-\frac{x'}{c}\right)\mathrm{d}x' + \frac{1}{2Z_0}\int_0^L R_L(x')I\left(x',t-\frac{L-x'}{c}\right)\mathrm{d}x' \tag{9.144}$$

and the nonreflecting resistive loading is given by (9.35)

If the case of dielectric half-space is considered, eqn (9.142) simplifies into:

$$\int_0^L \frac{I(x',t-R/c)}{4\pi R}\mathrm{d}x' - \int_0^L r(\theta)\frac{I(x',t-R^*/c)}{4\pi R^*}\mathrm{d}x'$$
$$= F_0\left(t-\frac{x}{c}\right) + F_L\left(t-\frac{L-x}{c}\right) + \frac{1}{2Z_0}\int_0^L E_x^{inc}\left(x',t-\frac{|x-x'|}{c}\right)\mathrm{d}x' \tag{9.145}$$
$$-\frac{1}{2Z_0}\int_0^L R_L I\left(x',t-\frac{|x-x'|}{c}\right)\mathrm{d}x'$$

while the auxiliary functions K_0 and K_L are

$$K_0(t) = \int_0^L \frac{I(x',t-R_0/c)}{4\pi R_0}\mathrm{d}x' - \int_0^L r(\theta)\frac{I(x',t-R_0^*/c)}{4\pi R_0^*}\mathrm{d}x'$$
$$-\frac{1}{2Z_0}\int_0^L E_x^{inc}\left(x',t-\frac{x'}{c}\right)\mathrm{d}x' + \frac{1}{2Z_0}\int_0^L R_L I\left(x',t-\frac{x'}{c}\right)\mathrm{d}x' \tag{9.146}$$

$$K_L(t) = \int_0^L \frac{I(x',t-R_L/c)}{4\pi R_L}\mathrm{d}x' - \int_0^L r(\theta)\frac{I(x',t-R_L^*/c)}{4\pi R_L^*}\mathrm{d}x'$$
$$-\frac{1}{2Z_0}\int_0^L E_x^{inc}\left(x',t-\frac{L-x'}{c}\right)\mathrm{d}x' + \frac{1}{2Z_0}\int_0^L R_L I\left(x',t-\frac{L-x'}{c}\right)\mathrm{d}x' \tag{9.147}$$

Equations (9.140)–(9.147) are related to the antenna mode. If the scattering mode is of interest, then E_x^{inc} must be replaced with E_x^{exc} defined with eqn (9.121).

The time-domain Hallen integral eqns (9.142) and (9.145), can be solved with the prescribed boundary conditions for the current at the wire ends and the initial condition for the current along the wire at the instant $t = t_0$, as specified by eqns (9.86) and (9.87).

9.2.3.2 Nonlinearly Loaded Straight Thin Wire The analysis of transient behavior of a nonlinear wire antenna problem, Figure 9.27, with the presence of a half-space included, featuring the Galerkin-Bubnov Indirect Boundary Element Method (GB-IBEM) has been reported in [21].

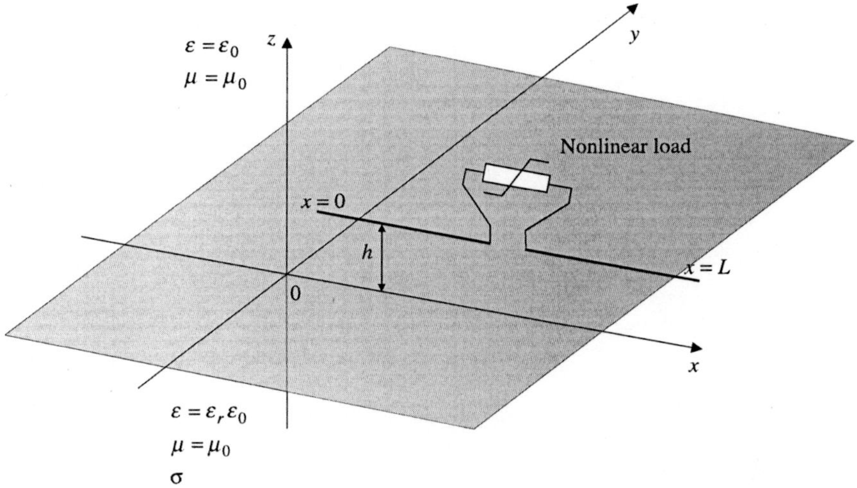

Figure 9.27 Nonlinearly loaded wire in a two-media configuration. (Poljak, D.; Yoong Tham, C.; McCowen, A. Transient Response of Nonlinearly Loaded Wires in a Two Media Configuration, *IEEE Transactions on Electromagnetic Compatibility.* 46 (2004), 1; 121–125).

The time domain Hallen equation can be obtained by modifying the loading term in Hallen eqn (9.142):

$$\int_0^L \frac{I(x',t-R/c)}{4\pi R}\mathrm{d}x' - \int_{-\infty}^{t}\int_0^L r(\theta,\tau)\frac{I(x',t-R^*/c-\tau)}{4\pi R^*}\mathrm{d}x'\mathrm{d}\tau = F_0\left(t-\frac{x}{c}\right)$$
$$+F_L\left(t-\frac{L-x}{c}\right)+\frac{1}{2Z_0}\int_0^L E_x^{exc}\left(x',t-\frac{|x-x'|}{c}\right)\mathrm{d}x' - \frac{1}{2Z_0}v_L\left(t-\frac{|x-x_0|}{c}\right) \tag{9.148}$$

F_0 and F_L expressed by the following auxiliary functions are:

$$K_0(t) = \int_0^L \frac{I(x',t-R_0/c)}{4\pi R_0}\mathrm{d}x' - \int_{-\infty}^{t}\int_0^L r(\theta,\tau)\frac{I(x',t-R_0^*/c-\tau)}{4\pi R_0^*}\mathrm{d}x'\mathrm{d}\tau$$
$$-\frac{1}{2Z_0}\int_0^L E_x^{exc}\left(x',t-\frac{x'}{c}\right)\mathrm{d}x' + \frac{1}{2Z_0}v_L\left(t-\frac{x_0}{c}\right) \tag{9.149}$$

$$K_L(t) = \int_0^L \frac{I(x',t-R_L/c)}{4\pi R_L}\mathrm{d}x' - \int_{-\infty}^{t}\int_0^L r(\theta,\tau)\frac{I(x',t-R_L^*/c-\tau)}{4\pi R_L^*}\mathrm{d}x'\mathrm{d}\tau$$
$$-\frac{1}{2Z_0}\int_0^L E_x^{exc}\left(x',t-\frac{L-x'}{c}\right)\mathrm{d}x' + \frac{1}{2Z_0}v_L\left(t-\frac{L-x_0}{c}\right) \tag{9.150}$$

If the case of a dielectric half-space is of interest, the Hallen eqn (9.148) becomes:

$$\int_0^L \frac{I(x', t - R/c)}{4\pi R} \mathrm{d}x' - \int_0^L r(\theta) \frac{I(x', t - R^*/c)}{4\pi R^*} \mathrm{d}x' \mathrm{d}\tau = F_0\left(t - \frac{x}{c}\right)$$
$$+ F_L\left(t - \frac{L - x}{c}\right) + \frac{1}{2Z_0} \int_0^L E_x^{exc}\left(x', t - \frac{|x - x'|}{c}\right) \mathrm{d}x' - \frac{1}{2Z_0} v_L\left(t - \frac{|x - x_0|}{c}\right) \tag{9.151}$$

while the auxiliary functions K_0 and K_L are:

$$K_0(t) = \int_0^L \frac{I(x', t - R_0/c)}{4\pi R_0} \mathrm{d}x' - \int_0^L r(\theta) \frac{I(x', t - R_0^*/c)}{4\pi R_0^*} \mathrm{d}x'$$
$$- \frac{1}{2Z_0} \int_0^L E_x^{exc}\left(x', t - \frac{x'}{c}\right) \mathrm{d}x' + \frac{1}{2Z_0} v_L\left(t - \frac{x_0}{c}\right) \tag{9.152}$$

$$K_L(t) = \int_0^L \frac{I(x', t - R_L/c)}{4\pi R_L} \mathrm{d}x' - \int_0^L r(\theta) \frac{I(x', t - R_L^*/c)}{4\pi R_L^*} \mathrm{d}x'$$
$$- \frac{1}{2Z_0} \int_0^L E_x^{exc}\left(x', t - \frac{L - x'}{c}\right) \mathrm{d}x' + \frac{1}{2Z_0} v_L\left(t - \frac{L - x_0}{c}\right) \tag{9.153}$$

As in all previously presented cases of TD analysis, the integral eqn (9.148) and (9.151) can be solved by prescribing certain boundary and initial conditions (9.86) and (9.87), respectively.

9.2.4 Two Coupled Horizontal Wires in a Two Media Configuration

Transient analysis of two identical coupled wires, parallel to each other and to an interface, with one of the wires excited by an impulsive voltage source is very interesting in antenna theory and EMC applications.

The knowledge of the coupling effects of such wire configuration is very useful in areas like ground-penetrating radar, mobile communication, inverse scattering, multiconductor transmission lines, printed circuit boards (PCB), or electromagnetic stimulation of biological tissue.

The problem of two coupled wires radiating in the presence of a dielectric half-space has been investigated in [33] by Bretones and Tijhuis who have established the time-domain formulation for coupling wires embedded in a two-media configuration and also have proposed the frequency domain solution method for the set of Hallen integral eqns [33,34]. However, the proposed solution method has been related to the frequency domain.

The time-domain formulation for two identical coupled wires horizontally placed over an imperfectly conducting half-space, based on the set of Hallen type integral equations, has been presented in [35]. The influence of a dissipative half-space is taken into account via the Fresnel space-time reflection coefficient appearing in the integral equation kernel. Considering the symmetry of the configuration, the original geometry of two coupled wires is transformed into two single wire modes. Each mode is represented by the corresponding Hallen integral equation.

The transient response of coupled wires is calculated using the direct time-domain code and the results are compared with those data that are computed indirectly from a frequency domain code via the Fourier transform. The Galerkin-Bubnov boundary integral equation method (GB-BIEM) combined with the marching-on-in-time algorithm is applied for solving the set of decomposed space-time Hallen integral equations, while the method of moments (MoM) is used for the solution of the frequency domain electric field integral equation (EFIE).

The transient currents along the wires I_A and I_B can be calculated by solving the system of two coupled integral equations of the Hallen type directly which is subsequently presented for the case of multiple wire configuration.

If the case of a two-wire array, the single integral eqn (9.123) is replaced by a set of two coupled integral equations for equivalent currents on both the elements.

Two identical straight, perfectly conducting loaded thin wires A and B of length L and radius a, parallel to each other and to a plane interface located along x-axis are shown in Figure 9.28. The wires are placed at height h above a real ground

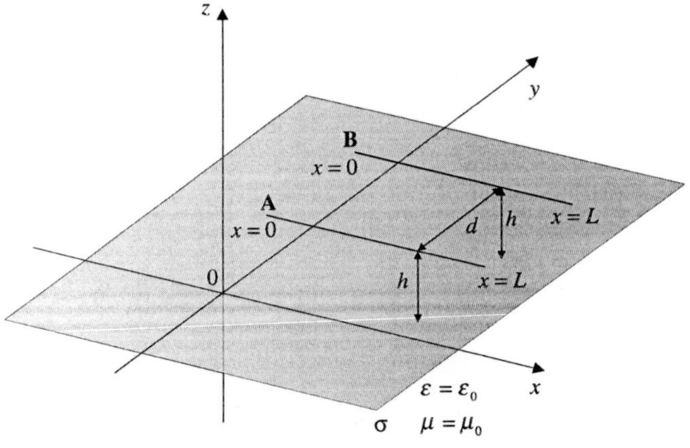

Figure 9.28 Coupled wires above real ground. (Poljak, D.; Yoong Tham, C.; Roje, V.; McCowen, Electromagnetic Pulse Excitation of Multiconductor Transmission Lines, *ICEAA'99*, pp. 789–792, Turin, Italy, Sept. 13–17, 1999).

and the separation between them is d. The wire A is assumed to be excited by the time-varying incident electric field. Within this formulation the far-side wire B is assumed to be completely shielded by the near-side wire A.

Following the similar formalism presented for single wire in [6], the set of coupled Hallen integral equations for the unknown axial currents I_A and I_B flowing along loaded wires is given by [35,36]:

$$\int_0^L \frac{I_A(x', t - R_a/c)}{4\pi R_a} \mathrm{d}x' - \int_{-\infty}^{t} \int_0^L r(\theta, \tau) \frac{I_A(x', t - R_a^*/c - \tau)}{4\pi R_a^*} \mathrm{d}x' \mathrm{d}\tau$$

$$+ \int_0^L \frac{I_B(x', t - R_d/c)}{4\pi R_d} \mathrm{d}x' - \int_{-\infty}^{t} \int_0^L r(\theta, \tau) \frac{I_B(x', t - R_d^*/c - \tau)}{4\pi R_d^*} \mathrm{d}x' \mathrm{d}\tau$$

$$= F_{0A}\left(t - \frac{x}{c}\right) + F_{LA}\left(t - \frac{L-x}{c}\right) + \frac{1}{2Z_0} \int_0^L E_{xA}^{exc}\left(x', t - \frac{|x - x'|}{c}\right) \mathrm{d}x'$$

$$- \frac{1}{2Z_0} \int_0^L Z_L(x') I_A\left(x', t - \frac{|x - x'|}{c}\right) \mathrm{d}x' \qquad (9.154)$$

$$\int_0^L \frac{I_B(x', t - R_a/c)}{4\pi R_a} \mathrm{d}x' - \int_{-\infty}^{t} \int_0^L r(\theta, \tau) \frac{I_B(x', t - R_a^*/c - \tau)}{4\pi R_a^*} \mathrm{d}x' \mathrm{d}\tau$$

$$+ \int_0^L \frac{I_A(x', t - R_d/c)}{4\pi R_d} \mathrm{d}x' - \int_{-\infty}^{t} \int_0^L r(\theta, \tau) \frac{I_A(x', t - R_d^*/c - \tau)}{4\pi R_d^*} \mathrm{d}x' \mathrm{d}\tau$$

$$= F_{0B}\left(t - \frac{x}{c}\right) + F_{LB}\left(t - \frac{L-x}{c}\right) + \frac{1}{2Z_0} \int_0^L E_{xB}^{exc}\left(x', t - \frac{|x - x'|}{c}\right) \mathrm{d}x'$$

$$- \frac{1}{2Z_0} \int_0^L Z_L(x') I_B\left(x', t - \frac{|x - x'|}{c}\right) \mathrm{d}x' \qquad (9.155)$$

where $R_d = [(x - x')^2 + d^2]^{1/2}, R_d^* = [(x - x')^2 + d^2 + 4h^2]^{1/2}, E_{xA}^{exc}, E_{xB}^{exc}$ are the total electric fields parallel to the ground plane and to the wires A and B, respectively.

If the antenna mode is considered, E_x^{exc} in eqns (9.154) and (9.155) is simply replaced with E_x^{inc}, whereas, if the scattering mode is of interest E_x^{exc} is determined with eqn (9.121).

Consequently, the set of resulting equations can be solved by the direct time-domain boundary element technique. However, the computational cost can be significantly reduced by taking into account the symmetry of the two identical wires configuration. Namely, using the principle of linear superposition [35] and [36], total currents I_A and I_B are simply given by:

$$I_A(x,t) = I^+(x,t) + I^-(x,t) \tag{9.156}$$

$$I_B(x,t) = I^+(x,t) + I^-(x,t) \tag{9.157}$$

where the currents I^+ and I^- are the corresponding currents along wires A and B for the case of symmetric and antisymmetric excitation. The treatment of the corresponding excitations is the same as already presented through relations (9.42)–(9.45).

The set of decomposed integral equations can be written in the form:

$$\begin{aligned}
&\int_0^L \frac{I^\pm(x',t-R_a/c)}{4\pi R_a}\mathrm{d}x' - \int_{-\infty}^{t}\int_0^L r(\theta,\tau)\frac{I^\pm(x',t-R_a^*/c-\tau)}{4\pi R_a^*}\mathrm{d}x'\mathrm{d}\tau \\
&\pm\left[\int_0^L \frac{I^+(x',t-R_d/c)}{4\pi R_d}\mathrm{d}x' - \int_{-\infty}^{t}\int_0^L r(\theta,\tau)\frac{I^+(x',t-R_d^*/c-\tau)}{4\pi R_d^*}\mathrm{d}x'\mathrm{d}\tau\right] \\
&= F_0^\pm\left(t-\frac{x}{c}\right) + F_L^\pm\left(t-\frac{L-x}{c}\right) + \frac{1}{4Z_0}\int_0^L E_x^{inc}\left(x',t-\frac{|x-x'|}{c}\right)\mathrm{d}x'
\end{aligned} \tag{9.158}$$

where the time signals F_0 and F_L are given by:

$$F_0^\pm(t) = \sum_{n=0}^{\infty} K_0^\pm\left(t-\frac{2nL}{c}\right) - \sum_{n=0}^{\infty} K_L^\pm\left(t-\frac{(2n+1)L}{c}\right) \tag{9.159}$$

$$F_L^\pm(t) = \sum_{n=0}^{\infty} K_L^\pm\left(t-\frac{2nL}{c}\right) - \sum_{n=0}^{\infty} K_0^\pm\left(t-\frac{(2n+1)L}{c}\right) \tag{9.160}$$

and the auxiliary functions $K_0^\pm$ and $K_L^\pm$ are:

$$\begin{aligned}
K_0^\pm(t) = &\int_0^L \frac{I^\pm(x',t-R_{a0}/c)}{4\pi R_{a0}}\mathrm{d}x' - \int_{-\infty}^{t}\int_0^L r(\theta,\tau)\frac{I^\pm(x',t-R_{a0}^*/c-\tau)}{4\pi R_{a0}^*}\mathrm{d}x'\mathrm{d}\tau \\
&\pm\left(\int_0^L \frac{I^\pm(x',t-R_{d0}/c)}{4\pi R_{d0}}\mathrm{d}x' - \int_{-\infty}^{t}\int_0^L r(\theta,\tau)\frac{I^\pm(x',t-R_{d0}^*/c-\tau)}{4\pi R_{d0}^*}\mathrm{d}x'\mathrm{d}\tau\right) \\
&-\frac{1}{4Z_0}\int_0^L E_x^{inc}\left(x',t-\frac{x'}{c}\right)\mathrm{d}x'
\end{aligned} \tag{9.161}$$

$$K_L^{\pm}(t) = \int_0^L \frac{I^{\pm}(x', t - R_{aL}/c)}{4\pi R_{aL}} \mathrm{d}x' - \int_{-\infty}^{t} \int_0^L r(\theta, \tau) \frac{I^{\pm}(x', t - R_{aL}^*/c - \tau)}{4\pi R_{aL}^*} \mathrm{d}x' \mathrm{d}\tau$$
$$\pm \left(\int_0^L \frac{I^{\pm}(x', t - R_{dL}/c)}{4\pi R_{dL}} \mathrm{d}x' - \int_{-\infty}^{t} \int_0^L r(\theta, \tau) \frac{I^{\pm}(x', t - R_{dL}^*/c - \tau)}{4\pi R_{dL}^*} \mathrm{d}x' \mathrm{d}\tau \right)$$
$$- \frac{1}{4Z_0} \int_0^L E_x^{inc}\left(x', t - \frac{L - x'}{c}\right) \mathrm{d}x' \tag{9.162}$$

The set of space-time eqns (9.158)–(9.162) for the coupled thin wires immersed in a two-media configuration must be solved twice, once for the symmetric mode (plus sign) and once for the antisymmetric mode (minus sign). The uniqueness of the solution is achieved prescribing the conditions (9.52) and (9.53).

9.2.5 Thin Wire Array Above a Real Ground

This section deals with an analysis of wire array in the presence of two-media configuration. There are numerous applications of such a study mostly related to coupling of electromagnetic pulses to antenna systems or transmission lines. Modeling of an arbitrary thin wire array is based on a set of coupled space-time integral equations of the Hallen type. The boundary element analysis of thin wire array has been carried out in [24].

All wires are located parallel to the dielectric plane at the height $z = h$, assuming dielectric plane is placed in $z = 0$, Figure 9.29. Length of the j antenna is $L_j = x_{Lj} - x_{0j}$, while distance between wires i and j is given by $d_{ij} = |y_i - y_j|$.

A set of coupled space-time integral equations for M parallel wires is given by:

$$\sum_{s=1}^{M} \int_0^{L_s} \frac{I_s(x', t - \frac{R_{vs}}{c})}{4\pi R_{vs}} \mathrm{d}x' - \sum_{s=1}^{M} \int_{-\infty}^{t} \int_0^{L_s} r_{vs}(\theta, \tau) \frac{I_s\left(x', t - \frac{R_{vs}^*}{c} - \tau\right)}{4\pi R_{vs}^*} \mathrm{d}x' \mathrm{d}\tau$$
$$= F_{0v}\left(t - \frac{x - x_{0v}}{c}\right) + F_{Lv}\left(t - \frac{x_{Lv} - x}{c}\right) + \frac{1}{2Z_0} \int_0^{L_s} E_x^{inc_v}\left(x', t - \frac{|x - x'|}{c}\right) \mathrm{d}x' \tag{9.163}$$

where v and $s = 1, 2, \ldots, M$ denote the index of the observed and source wire, respectively.

Distances between source and observation point, respectively are:

$$R_{vs}^* = \sqrt{(x - x')^2 + d_{vs}^2 + 4h^2}; \; R_{vs} = \begin{cases} \sqrt{(x - x')^2 + d_{vs}^2} : v \neq s \\ \sqrt{(x - x')^2 + a^2} : v = s \end{cases} \tag{9.164}$$

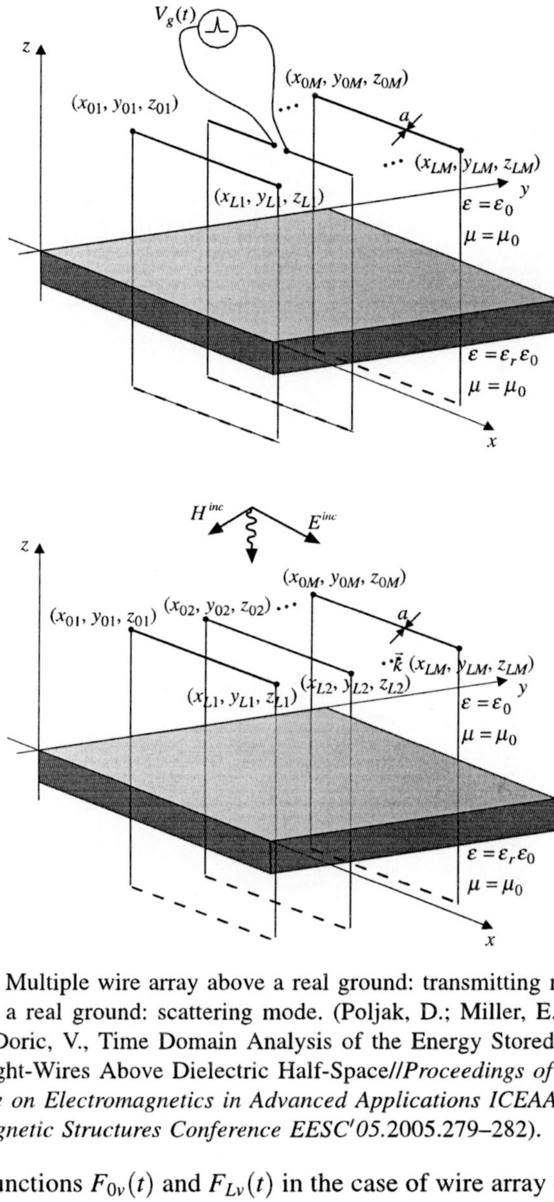

Figure 9.29 (a) Multiple wire array above a real ground: transmitting mode. (b) Multiple wire array above a real ground: scattering mode. (Poljak, D.; Miller, E. K.; Tham, C. Y.; Antonijevic, S.; Doric, V., Time Domain Analysis of the Energy Stored in the Near Field of Multiple Straight-Wires Above Dielectric Half-Space//*Proceedings of Joint 9th International Conference on Electromagnetics in Advanced Applications ICEAA'05 and 11th European Electromagnetic Structures Conference EESC'05*.2005.279–282).

The unknown functions $F_{0v}(t)$ and $F_{Lv}(t)$ in the case of wire array can be written as follows:

$$F_{0v}(t) = \sum_{n=0}^{\infty} K_{0v}(t - \frac{2nL_v}{c}) - \sum_{n=0}^{\infty} K_{Lv}(t - \frac{(2n+1)L_v}{c}) \tag{9.165}$$

$$F_{Lv}(t) = \sum_{n=0}^{\infty} K_{Lv}\left(t - \frac{2nL_v}{c}\right) - \sum_{n=0}^{\infty} K_{0v}\left(t - \frac{(2n+1)L_v}{c}\right) \tag{9.166}$$

where the auxiliary functions K_0 and K_L are:

$$K_{0v}(t) = \sum_{s=1}^{M} \int_{0}^{L_s} \frac{I_s\left(x', t - \frac{R_{vs}^{(0)}}{c}\right)}{4\pi R_{vs}^{(0)}} \mathrm{d}x' - \sum_{s=1}^{M} \int_{-\infty}^{t} \int_{0}^{L_s} r_{vs}(\theta, \tau) \frac{I_s\left(x', t - \frac{R_{vs}^{*(0)}}{c} - \tau\right)}{4\pi R_{vs}^{*(0)}} \mathrm{d}x' \mathrm{d}\tau$$
$$- \frac{1}{2Z_0} \int_{0}^{L_s} E_x^{inc_v}\left(x', t - \frac{|x - x'|}{c}\right) \mathrm{d}x' \tag{9.167}$$

$$K_{Lv}(t) = \sum_{s=1}^{M} \int_{0}^{L_s} \frac{I_s\left(x', t - \frac{R_{vs}^{(L)}}{c}\right)}{4\pi R_{vs}^{(L)}} \mathrm{d}x' - \sum_{s=1}^{M} \int_{-\infty}^{t} \int_{0}^{L_s} r_{vs}(\theta, \tau) \frac{I_s\left(x', t - \frac{R_{vs}^{*(L)}}{c} - \tau\right)}{4\pi R_{vs}^{*(L)}} \mathrm{d}x' \mathrm{d}\tau$$
$$- \frac{1}{2Z_0} \int_{0}^{L_s} E_x^{inc_v}\left(x', t - \frac{|x - x'|}{c}\right) \mathrm{d}x' \tag{9.168}$$

where $R_{vs}^{(0)}$ and $R_{vs}^{(L)}$ are distances between source point on the wire s and observation point at the ends of the wire v:

$$R_{vs}^{(0)} = R_{vs}|_{x=x_{0v}}; \quad R_{vs}^{(L)} = R_{vs}|_{x=x_{Lv}} \tag{9.169}$$

and $R_{vs}^{*(0)}$ and $R_{vs}^{*(L)}$ are the distances between source point on the image of the wire s and observation point at the ends of the wire v:

$$R_{vs}^{*(0)} = R_{vs}^{*}|_{x=x_{0v}}; \quad R_{vs}^{*(L)} = R_{vs}^{*}|_{x=x_{Lv}} \tag{9.170}$$

Space-time reflection coefficient $r_{vs}(\theta, t)$ is given by

$$r_{vs}(\theta', t) = A\delta(t) + \frac{4\beta}{1 - \beta^2} \frac{e^{-\alpha t}}{t} \sum_{n=1}^{\infty} (-1)^{n+1} n A^n I_n(\alpha t) \tag{9.171}$$

where:

$$\beta = \frac{\sqrt{\varepsilon_r - \sin^2 \theta'_{vs}}}{\varepsilon_r \cos \theta'_{vs}}; \quad \theta'_{vs} = Arctg \frac{\sqrt{(x' - x)^2 + d_{vs}^2}}{2h} \tag{9.172}$$

The set of N coupled space-time integral equations can now be solved by taking boundary conditions and initial conditions into account. The wires are not excited before the instant $t = t_0$.

Equations (9.163)–(9.172) are related to the antenna mode. If the scattering mode is of interest, then E_x^{inc} must be replaced with E_x^{exc} defined with eqn (9.121).

$$\sum_{s=1}^{M}\int_0^{L_s}\frac{I_s\left(x',t-\frac{R_{vs}}{c}\right)}{4\pi R_{vs}}\mathrm{d}x'-\sum_{s=1}^{M}\int_0^{L_s}r_{vs}(\theta)\frac{I_s\left(x',t-\frac{R_{vs}^*}{c}\right)}{4\pi R_{vs}^*}\mathrm{d}x'=F_{0v}\left(t-\frac{x-x_{0v}}{c}\right)$$
$$+F_{Lv}\left(t-\frac{x_{Lv}-x}{c}\right)+\frac{1}{2Z_0}\int_0^{L_s}E_x^{inc_v}\left(x',t-\frac{|x-x'|}{c}\right)\mathrm{d}x' \tag{9.173}$$

where the auxiliary functions K_0 and K_L are:

$$K_{0v}(t)=\sum_{s=1}^{M}\int_0^{L_s}\frac{I_s\left(x',t-\frac{R_{vs}^{(0)}}{c}\right)}{4\pi R_{vs}^{(0)}}\mathrm{d}x'-\sum_{s=1}^{M}\int_{-\infty}^{t}\int_0^{L_s}r_{vs}(\theta,\tau)\frac{I_s\left(x',t-\frac{R_{vs}^{*(0)}}{c}-\tau\right)}{4\pi R_{vs}^{*(0)}}\mathrm{d}x'\mathrm{d}\tau$$
$$-\frac{1}{2Z_0}\int_0^{L_s}E_x^{inc_v}\left(x',t-\frac{|x-x'|}{c}\right)\mathrm{d}x' \tag{9.174}$$

$$K_{Lv}(t)=\sum_{s=1}^{M}\int_0^{L_s}\frac{I_s\left(x',t-\frac{R_{vs}^{(L)}}{c}\right)}{4\pi R_{vs}^{(L)}}\mathrm{d}x'-\sum_{s=1}^{M}\int_{-\infty}^{t}\int_0^{L_s}r_{vs}(\theta,\tau)\frac{I_s\left(x',t-\frac{R_{vs}^{*(L)}}{c}-\tau\right)}{4\pi R_{vs}^{*(L)}}\mathrm{d}x'\mathrm{d}\tau$$
$$-\frac{1}{2Z_0}\int_0^{L_s}E_x^{inc_v}\left(x',t-\frac{|x-x'|}{c}\right)\mathrm{d}x' \tag{9.175}$$

where $R_{vs}^{(0)}$ and $R_{vs}^{(L)}$ are defined by relations (9.169) and (9.170).

9.2.6. Computational Procedures for Horizontal Wires Above a Dielectric Half-Space

This section deals with the numerical procedures for the solution of corresponding Hallen equations related to unloaded and loaded single wire, two coupled wires, and arbitrary wire arrays. All procedures are based on the implementation of the time domain variant of GB-IBEM, (also entitled TD FEIEM).

9.2.6.1 Numerical Solution of the Time Domain Hallen Integral Equation for Half-Space Problems The Galerkin-Bubnov Boundary Integral Equation Method (GB-BIEM) was shown to be very efficient in handling the time domain half-space Hallen eqns [10–11].

According to the usual discretization procedure, the time domain scheme of the GB-BIEM starts with the local approximation for current given in the form:

$$I(x', t') = \{f\}^T \{I\} \tag{9.176}$$

where $\{f\}$ is a vector containing shape functions, and $\{I\}$ is the time-dependent solution vector.

Utilizing the weighted residual approach, the space boundary element discretization leads to the local equation system for ith source and jth observation boundary element [10], [11]:

$$\begin{aligned}
&\int\limits_{\Delta l_j}\int\limits_{\Delta l_i} \{f\}_j \{f\}_i^T \frac{1}{4\pi R} \mathrm{d}x' \mathrm{d}x \{I\}\Big|_{t-\frac{R}{c}} - \int\limits_{\Delta l_j}\int\limits_{\Delta l_i} \{f\}_j \{f\}_i^T \frac{r(\theta')}{4\pi R} \mathrm{d}x' \mathrm{d}x \{I\}\Big|_{t-\frac{R^*}{c}} \\
&= \int\limits_{\Delta l_j} F_0\left(t - \frac{x}{c}\right)\{f\}_j \mathrm{d}x + \int\limits_{\Delta l_j} F_L\left(t - \frac{L-x}{c}\right)\{f\}_j \mathrm{d}x \\
&\quad + \frac{1}{2Z_0}\int\limits_{\Delta l_j}\int\limits_{\Delta l_i} E_x^{inc}\left(x', t - \frac{|x-x'|}{c}\right)\{f\}_j \mathrm{d}x' \mathrm{d}x
\end{aligned} \tag{9.177}$$

which can be written in the convenient matrix form:

$$\begin{aligned}
&[A]\{I\}\Big|_{t-\frac{R}{c}} - [A_1^*]\{I\}\Big|_{t-\frac{R^*}{c}} = [B]\{E\}\Big|_{t-\frac{|x-x'|}{c}} + [C]\left\{\sum_{n=0}^{\infty} I^n\right\}\Bigg|_{t-\frac{x}{c}-\frac{2nL}{c}-\frac{R_0}{c}} \\
&- [C_1^*]\left\{\sum_{n=0}^{\infty} I^n\right\}\Bigg|_{t-\frac{x}{c}-\frac{2nL}{c}-\frac{R_0^*}{c}} - -[B]\left\{\sum_{n=0}^{\infty} E^n\right\}\Bigg|_{t-\frac{x}{c}-\frac{2nL}{c}-\frac{x'}{c}} - [D]\left\{\sum_{n=0}^{\infty} I^n\right\}\Bigg|_{t-\frac{x}{c}-\frac{2n+1}{c}L-\frac{R_L}{c}} \\
&+ [D_1^*]\left\{\sum_{n=0}^{\infty} I^n\right\}\Bigg|_{t-\frac{x}{c}-\frac{2n+1}{c}L-\frac{R_L^*}{c}} + [B]\left\{\sum_{n=0}^{\infty} E^n\right\}\Bigg|_{t-\frac{x}{c}-\frac{2n+1}{c}L-\frac{L-x'}{c}} + [D]\left\{\sum_{n=0}^{\infty} I^n\right\}\Bigg|_{t-\frac{L-x}{c}-\frac{2nL}{c}-\frac{R_L}{c}} \\
&- [D_1^*]\left\{\sum_{n=0}^{\infty} I^n\right\}\Bigg|_{t-\frac{L-x}{c}-\frac{2nL}{c}-\frac{R_L^*}{c}} - [B]\left\{\sum_{n=0}^{\infty} E^n\right\}\Bigg|_{t-\frac{L-x}{c}-\frac{2nL}{c}-\frac{L-x'}{c}} - [C]\left\{\sum_{n=0}^{\infty} I^n\right\}\Bigg|_{t-\frac{L-x}{c}-\frac{2n+1}{c}L-\frac{R_0}{c}} \\
&+ [C_1^*]\left\{\sum_{n=0}^{\infty} I^n\right\}\Bigg|_{t-\frac{L-x}{c}-\frac{2n+1}{c}L-\frac{R_0^*}{c}} + [B]\left\{\sum_{n=0}^{\infty} E^n\right\}\Bigg|_{t-\frac{L-x}{c}-\frac{2n+1}{c}L-\frac{x'}{c}}
\end{aligned} \tag{9.178}$$

where the space-dependent matrices are given by:

$$[A] = \int_{\Delta l_j} \int_{\Delta l_i} \frac{1}{4\pi R} \{f\}_j \{f\}_i^T \mathrm{d}x' \mathrm{d}x \tag{9.179}$$

$$[A] = \int_{\Delta l_j} \int_{\Delta l_i} \frac{r(\theta)}{4\pi R^*} \{f\}_j \{f\}_i^T \mathrm{d}x' \mathrm{d}x \tag{9.180}$$

$$[B] = \frac{1}{2Z_0} \int_{\Delta l_j} \int_{\Delta l_i} \{f\}_j \{f\}_i^T \mathrm{d}x' \mathrm{d}x \tag{9.181}$$

$$[C] = \int_{\Delta l_j} \int_{\Delta l_i} \frac{1}{4\pi R_0} \{f\}_j \{f\}_i^T \mathrm{d}x' \mathrm{d}x \tag{9.182}$$

$$[C^*] = \int_{\Delta l_j} \int_{\Delta l_i} \frac{r(\theta)}{4\pi R_0^*} \{f\}_j \{f\}_i^T \mathrm{d}x' \mathrm{d}x \tag{9.183}$$

$$[D] = \int_{\Delta l_j} \int_{\Delta l_i} \frac{1}{4\pi R_L} \{f\}_j \{f\}_i^T \mathrm{d}x' \mathrm{d}x \tag{9.184}$$

$$[C^*] = \int_{\Delta l_j} \int_{\Delta l_i} \frac{r(\theta)}{4\pi R_L^*} \{f\}_j \{f\}_i^T \mathrm{d}x' \mathrm{d}x \tag{9.185}$$

and the $\{E\}$ vector denotes the excitation function.

It is obvious for the case of a dielectric lower medium that the reflection coefficient is only space dependent, and only the first impulse term from (9.128) exists, that is only the influence of ε_r is considered. This algorithm can generally be used in any case when the wire is far enough from the interface. (The bigger is the distance from the source antenna to the interface, the smaller is the effect of the image antenna on the total radiation in the upper medium. The exponential term decaying with the increase of time can therefore be neglected.) Such approximation significantly improves the algorithm efficiency providing a faster evaluation, and much lower memory requirements.

The solution in time on the ith wire segment can be expressed:

$$I_i(t') = \sum_{k=1}^{N_t} I_i^k T^k(t') \tag{9.186}$$

where I_i^k are the unknown coefficients and T^k are the time domain shape functions. For convenience, the matrix system (9.175) is written in the form:

$$[A]\{I\}|_{t-\frac{R}{c}} = \{g\} \tag{9.187}$$

where g is the space-time dependent vector containing the entire right-hand side of the expression (9.178).

Using the weighted residual approach in the time domain, it simply follows for a time increment:

$$\int_{t_k}^{t_k+\Delta t} [A]\{I\}_{|t-\frac{R}{c}} - \{g\}\theta_k \mathrm{d}t = 0; \quad k = 1, 2, \ldots, N_T \tag{9.188}$$

where N_t is the total number of time samples, and θ_k denotes the set of time dependent test functions.

Choosing the Dirac impulses as test functions from the weighted residual integral (9.185) it follows:

$$[A]\{I\}_{-t_k-R/c} = \{g\}_{-all\ discrete\ previous\ instants} \tag{9.189}$$

If the space-time discretization is performed carefully so that within one time increment the propagation on at least one space segment should be considered, that is if the well-known inequality is satisfied:

$$\Delta t \le \frac{\Delta x}{c} \tag{9.190}$$

the recurrence formula for the space-time dependent current can be written in the form:

$$I_j|_{t_k} = \frac{-\sum_{i=1}^{N}\left(\overline{A_{ji}I_i|_{t_k-\frac{R_{vs}}{c}}} + A_{ji}^* I_i|_{t_k-\frac{R_{vs}^*}{c}}\right) + g_j|_{all\ previous\ discrete\ instants}}{A_{jj}} \tag{9.191}$$

where A_{ji} is the global matrix term, g_{j1}^* is the whole right-side of the expression (9.178) containing the excitation and the currents at previous instants, and the overbar line denotes that the self term is omitted.

By marching-on-in-time procedure, it is possible to obtain the solution for current at present time instant for each space node as a function of currents at previous instants. To start the stepping procedure, it is necessary to prescribe the already mentioned initial conditions for the wire current.

9.2.6.2 The Far Field Computation The far field generated by the wire located over the lossy half-space is computed by using the impulse term approximation for the reflection coefficient, same as in the current calculation procedure. The boundary element contribution to the far field in the space-time observation point is given by:

$$\begin{aligned} E_x(r,t)_{|\Delta l_i} &= -\frac{\mu_0}{4\pi}\frac{\partial}{\partial t'}\{I\}_{|t'} \int_{\Delta l_i} \{f\}_i^T \frac{1}{R_1}\mathrm{d}x' \\ &+ \frac{\mu_0}{4\pi}\frac{\partial}{\partial t_1'}\{I\}_{|t_1'} \int_{\Delta l} r(\theta')\{f\}\frac{1}{R_1^*}\mathrm{d}x' \end{aligned} \tag{9.192}$$

and further applying the boundary element formalism it follows:

$$
\begin{aligned}
E_x(r,t)|_{\Delta l_i} = & -\frac{\mu_0}{4\pi}\frac{I_{i,s}^{m+1}-I_{i,s}^{m}}{\Delta t}\int\limits_{\Delta l_i} f_{i,s}\frac{1}{R_1}\mathrm{d}x' - \frac{\mu_0}{4\pi}\frac{I_{i,s+1}^{m+1}-I_{i,s+1}^{m}}{\Delta t}\int\limits_{\Delta l_i} f_{i,s+1}\frac{1}{R_1}\mathrm{d}x' \\
& + \frac{\mu_0}{4\pi}\frac{I_{i,s}^{m+1}-I_{i,s}^{m}}{\Delta t}\int\limits_{\Delta l_i} r(\theta')f_{i,s}\frac{1}{R_1^*}\mathrm{d}x' + \frac{\mu_0}{4\pi}\frac{I_{i,s+1}^{m+1}-I_{i,s+1}^{m}}{\Delta t}\int\limits_{\Delta l_i} r(\theta')f_{i,s+1}\frac{1}{R_1^*}\mathrm{d}x'
\end{aligned}
\tag{9.193}
$$

where Δt denotes the time increment and m and m_1 are the integers of discretized times t' and t_1', respectively.

The total field is then:

$$
E_x(r,t) = \sum_{i=1}^{M} E_x(r,t)_{|\Delta l_i} \tag{9.194}
$$

where M denotes the number of wire segments.

9.2.6.3 Computational Procedures for Resistively and Nonlinearly Loaded Wires Above a Dielectric Half-Space The boundary element discretization of eqn (9.151) yields to the local equation system for ith source and jth observation wire segment:

$$
\begin{aligned}
& \int\limits_{\Delta l_j}\int\limits_{\Delta l_i} \{f\}_j\{f\}_i^T\frac{1}{4\pi R}\mathrm{d}x'\mathrm{d}x\{I\}_{|t-\frac{R}{c}} - \int\limits_{\Delta l_j}\int\limits_{\Delta l_i} \{f\}_j\{f\}_i^T\frac{r(\theta)}{4\pi R^*}\mathrm{d}x'\mathrm{d}x\{I\}_{|t-\frac{R^*}{c}} \\
& = \int\limits_{\Delta l_j} F_0\left(t-\frac{x}{c}\right)\{f\}_j\mathrm{d}x + \int\limits_{\Delta l_j} F_L\left(t-\frac{L-x}{c}\right)\{f\}_j\mathrm{d}x \\
& + \frac{1}{2Z_0}\int\limits_{\Delta l_j}\int\limits_{\Delta l_i} E_x^{inc}\left(x', t-\frac{|x-x'|}{c}\right)\{f\}_j\mathrm{d}x'\mathrm{d}x \\
& \frac{1}{2Z_0}\int\limits_{\Delta l_j}\int\limits_{\Delta l_i} R_L(x')\{f\}_j\{f\}_i^T\mathrm{d}x'\mathrm{d}x\{I\}_{|t-\frac{|x-x'|}{c}}
\end{aligned}
\tag{9.195}
$$

which can be written in the matrix form as:

$$
\begin{aligned}
& [A]\{I\}\Big|_{t-\frac{R}{c}} - [A_1^*]\{I\}\Big|_{t-\frac{R^*}{c}} = [B]\{E\}\Big|_{t-\frac{|x-x'|}{c}} - [R]\{I\}\Big|_{t-\frac{|x-x'|}{c}} + [C]\left\{\sum_{n=0}^{\infty} I^n\right\}\Bigg|_{t-\frac{x}{c}-\frac{2nL}{c}-\frac{R_0}{c}} \\
& - [C_1^*]\left\{\sum_{n=0}^{\infty} I^n\right\}\Bigg|_{t-\frac{x}{c}-\frac{2nL}{c}-\frac{R_0^*}{c}} - [B]\left\{\sum_{n=0}^{\infty} E^n\right\}\Bigg|_{t-\frac{x}{c}-\frac{2nL}{c}-\frac{x'}{c}} + [R]\left\{\sum_{n=0}^{\infty} I^n\right\}\Bigg|_{t-\frac{x}{c}-\frac{2nL}{c}-\frac{x'}{c}}
\end{aligned}
$$

$$
\begin{aligned}
&- [D]\left\{\sum_{n=0}^{\infty} I^n\right\}\Bigg|_{t-\frac{x}{c}-\frac{2n+1}{c}L-\frac{R_L}{c}} + [D_1^*]\left\{\sum_{n=0}^{\infty} I^n\right\}\Bigg|_{t-\frac{x}{c}-\frac{2n+1}{c}L-\frac{R_L^*}{c}} + [B]\left\{\sum_{n=0}^{\infty} E^n\right\}\Bigg|_{t-\frac{x}{c}-\frac{2n+1}{c}L-\frac{L-x'}{c}} \\
&- [R]\left\{\sum_{n=0}^{\infty} I^n\right\}\Bigg|_{t-\frac{x}{c}-\frac{2n+1}{c}L-\frac{L-x'}{c}} + [D]\left\{\sum_{n=0}^{\infty} I^n\right\}\Bigg|_{t-\frac{L-x}{c}-\frac{2nL}{c}-\frac{R_L}{c}} - [D_1^*]\left\{\sum_{n=0}^{\infty} I^n\right\}\Bigg|_{t-\frac{L-x}{c}-\frac{2nL}{c}-\frac{R_L^*}{c}} \\
&- [B]\left\{\sum_{n=0}^{\infty} E^n\right\}\Bigg|_{t-\frac{L-x}{c}-\frac{2nL}{c}-\frac{L-x'}{c}} + [R]\left\{\sum_{n=0}^{\infty} I^n\right\}\Bigg|_{t-\frac{L-x}{c}-\frac{2nL}{c}-\frac{L-x'}{c}} - [C]\left\{\sum_{n=0}^{\infty} I^n\right\}\Bigg|_{t-\frac{L-x}{c}-\frac{2n+1}{c}L-\frac{R_0}{c}} \\
&+ [C_1^*]\left\{\sum_{n=0}^{\infty} I^n\right\}\Bigg|_{t-\frac{L-x}{c}-\frac{2n+1}{c}L-\frac{R_0^*}{c}} + [B]\left\{\sum_{n=0}^{\infty} E^n\right\}\Bigg|_{t-\frac{L-x}{c}-\frac{2n+1}{c}L-\frac{x'}{c}} - [R]\left\{\sum_{n=0}^{\infty} I^n\right\}\Bigg|_{t-\frac{L-x}{c}-\frac{2n+1}{c}L-\frac{x'}{c}}
\end{aligned}
\tag{9.196}
$$

where the space dependent boundary matrices are defined by relations (9.179)–(9.185), while the $[R]$ matrix is given by:

$$
[R] = \frac{1}{2Z_0} \int_{\Delta l_j} \int_{\Delta l_i} R_L(x') \{f\}_j \{f\}_i^T \,\mathrm{d}x'\,\mathrm{d}x \tag{9.197}
$$

and the corresponding time domain procedure is completely the same as the one described in section 9.2.6.1.

The nonlinear $v - I$ characteristic (9.37) can be presented by an arbitrary number of piecewise linear curves. For simplicity, it is convenient to first consider a special case of two linear curves through the time scale origin, that is:

$$
\begin{aligned}
v_L(t) &= R_1 I_L(t); \quad I_L(t) \geq 0 \\
v_L(t) &= R_2 I_L(t); \quad I_L(t) < 0
\end{aligned}
\tag{9.198}
$$

where R_1 and R_2 are linear resistors.

This nonlinear load is taken into account within the numerical procedure through the $[R]$ matrix. An extension to the case of n piecewise linear curves is straightforward and can be carried out easily.

9.2.6.4 Computational Procedures for Coupled Wires Above a Real Ground The local approximation for current over wire segment is given by:

$$
I^{\pm}(x', t') = \{f\}^T \{I\}^{\pm} \tag{9.199}
$$

where $\{f\}$ denotes a vector containing space-domain shape functions, and $\{I\}^{\pm}$ is the time-dependent solution vector for the symmetric and antisymmetric arrangement, respectively.

The space boundary element discretization gives the linear equation system for ith source and jth observation wire segment, respectively [35,36]:

$$\sum_{i=1}^{M}\left[\begin{array}{l}\int\limits_{\Delta l_j}\int\limits_{\Delta l_i}\{f\}_j\{f\}_i^T\frac{1}{4\pi R_a}\mathrm{d}x'\mathrm{d}x\{I\}_{\left|t-\frac{R_a}{c}\right.}^{\pm}\\ -\int\limits_{\Delta l_j}\int\limits_{\Delta l_i}r(\theta)\{f\}_j\{f\}_i^T\frac{1}{4\pi R_a^*}\mathrm{d}x'\mathrm{d}x\{I\}_{\left|t-\frac{R_a^*}{c}\right.}^{\pm}\\ \pm\left(\begin{array}{l}\int\limits_{\Delta l_j}\int\limits_{\Delta l_i}\{f\}_j\{f\}_i^T\frac{1}{4\pi R_d}\mathrm{d}x'\mathrm{d}x\{I\}_{\left|t-\frac{R_d}{c}\right.}^{\pm}\\ -\int\limits_{\Delta l_j}\int\limits_{\Delta l_i}\{f\}_j\{f\}_i^T\frac{1}{4\pi R_d^*}\mathrm{d}x'\mathrm{d}x\{I\}_{\left|t-\frac{R_d^*}{c}\right.}^{\pm}\end{array}\right)\end{array}\right] \tag{9.200}$$

$$= \int\limits_{\Delta l_j} F_0^+ - \left(t-\frac{x}{c}\right)\{f\}_j\mathrm{d}x + \int\limits_{\Delta l_j} F_L^+ - \left(t-\frac{L-x}{c}\right)\{f\}_j\mathrm{d}x;$$

$$+\frac{1}{4Z_0}\sum_{i=1}^{M}\int\limits_{\Delta l_j}\int\limits_{\Delta l_i} E_x^{inc}\left(x', t-\frac{|x-x'|}{c}\right)\{f\}_j\mathrm{d}x'\mathrm{d}x$$

$$j = 1, 2, \ldots, N_8$$

where M is the number of wire segments.

The corresponding matrix notation is given as follows:

$$\sum_{i=1}^{M}\left([A]\{I\}_{\left|t-\frac{R_a}{c}\right.} - [A^*]\{I\}_{\left|t-\frac{R_a^*}{c}\right.} \pm [A^d]\{I\}_{\left|t-\frac{R_d}{c}\right.} - [A^{d*}]\{I\}_{\left|t-\frac{R_d^*}{c}\right.}\right)$$

$$= \sum_{i=1}^{M}\left(\begin{array}{l}[B]\{E\}_{\left|t-\frac{|x-x'|}{c}\right.} + [C]\left\{\sum\limits_{n=0}^{\infty}I^n\right\}_{\left|t-\frac{x}{c}-\frac{2nL}{c}-\frac{R_{a0}}{c}\right.} - [C^*]\left\{\sum\limits_{n=0}^{\infty}I^n\right\}_{\left|t-\frac{x}{c}-\frac{2nL}{c}-\frac{R_{a0}}{c}\right.}\\ \pm[C^d]\left\{\sum\limits_{n=0}^{\infty}I^n\right\}_{\left|t-\frac{x}{c}-\frac{2nL}{c}-\frac{R_{d0}}{c}\right.} \pm[C^{d*}]\left\{\sum\limits_{n=0}^{\infty}I^n\right\}_{\left|t-\frac{x}{c}-\frac{2nL}{c}-\frac{R_{d0}^*}{c}\right.} - [B]\left\{\sum\limits_{n=0}^{\infty}E^n\right\}_{\left|t-\frac{x}{c}-\frac{2nL}{c}-\frac{x'}{c}\right.}\\ -[D]\left\{\sum\limits_{n=0}^{\infty}I^n\right\}_{\left|t-\frac{x}{c}-\frac{2n+1}{c}L-\frac{R_{aL}}{c}\right.} + [D^*]\left\{\sum\limits_{n=0}^{\infty}I^n\right\}_{\left|t-\frac{x}{c}-\frac{2n+1}{c}L-\frac{R_{aL}^*}{c}\right.} \pm [D^d]\left\{\sum\limits_{n=0}^{\infty}I^n\right\}_{\left|t-\frac{x}{c}-\frac{2n+1}{c}L-\frac{R_{dL}}{c}\right.}\\ \pm[D^{d*}]\left\{\sum\limits_{n=0}^{\infty}I^n\right\}_{\left|t-\frac{x}{c}-\frac{2n+1}{c}L-\frac{R_{dL}^*}{c}\right.} + [B]\left\{\sum\limits_{n=0}^{\infty}E^n\right\}_{\left|t-\frac{x}{c}-\frac{2n+1}{c}L-\frac{L-x'}{c}\right.} + [D]\left\{\sum\limits_{n=0}^{\infty}I^n\right\}_{\left|t-\frac{L-x}{c}-\frac{2nL}{c}-\frac{R_{aL}}{c}\right.}\\ -[D^*]\left\{\sum\limits_{n=0}^{\infty}I^n\right\}_{\left|t-\frac{L-x}{c}-\frac{2nL}{c}-\frac{R_{aL}^*}{c}\right.} \pm [D^d]\left\{\sum\limits_{n=0}^{\infty}I^n\right\}_{\left|t-\frac{L-x}{c}-\frac{2nL}{c}-\frac{R_{dL}}{c}\right.} \pm [D]\left\{\sum\limits_{n=0}^{\infty}I^n\right\}_{\left|t-\frac{L-x}{c}-\frac{2nL}{c}-\frac{R_{dL}^*}{c}\right.}\\ -[B]\left\{\sum\limits_{n=0}^{\infty}E^n\right\}_{\left|t-\frac{L-x}{c}-\frac{2nL}{c}-\frac{L-x'}{c}\right.} - [C]\left\{\sum\limits_{n=0}^{\infty}I^n\right\}_{\left|t-\frac{L-x}{c}-\frac{2n+1}{c}L-\frac{R_{a0}}{c}\right.} + [C^*]\left\{\sum\limits_{n=0}^{\infty}I^n\right\}_{\left|t-\frac{L-x}{c}-\frac{2n+1}{c}L-\frac{R_{a0}^*}{c}\right.}\\ \pm[C^d]\left\{\sum\limits_{n=0}^{\infty}I^n\right\}_{\left|t-\frac{L-x}{c}-\frac{2n+1}{c}L-\frac{R_{d0}}{c}\right.} \pm [C^{d*}]\left\{\sum\limits_{n=0}^{\infty}I^n\right\}_{\left|t-\frac{L-x}{c}-\frac{2n+1}{c}L-\frac{R_{d0}^*}{c}\right.} + [B]\left\{\sum\limits_{n=0}^{\infty}E^n\right\}_{\left|t-\frac{L-x}{c}-\frac{2n+1}{c}L-\frac{x'}{c}\right.}\end{array}\right);$$

$$j = 1, 2, \ldots, M \tag{9.201}$$

where space dependent matrices are given by relations (9.44)–(9.47) and by relations (9.199)–(9.204):

$$[A^d] = \int_{\Delta l_j} \int_{\Delta l_i} \frac{1}{4\pi R_d} \{f\}_j \{f\}_i^T \mathrm{d}x' \mathrm{d}x \tag{9.202}$$

$$[A^{d*}] = \int_{\Delta l_j} \int_{\Delta l_i} \frac{r(\theta)}{4\pi R_d^*} \{f\}_j \{f\}_i^T \mathrm{d}x' \mathrm{d}x \tag{9.203}$$

$$[C^d] = \int_{\Delta l_j} \int_{\Delta l_i} \frac{1}{4\pi R_{d0}} \{f\}_j \{f\}_i^T \mathrm{d}x' \mathrm{d}x \tag{9.204}$$

$$[C^{d*}] = \int_{\Delta l_j} \int_{\Delta l_i} \frac{r(\theta)}{4\pi R_{d0}^*} \{f\}_j \{f\}_i^T \mathrm{d}x' \mathrm{d}x \tag{9.205}$$

$$[D^d] = \int_{\Delta l_j} \int_{\Delta l_i} \frac{1}{4\pi R_{dL}} \{f\}_j \{f\}_i^T \mathrm{d}x' \mathrm{d}x \tag{9.206}$$

$$[D^{d*}] = \int_{\Delta l_j} \int_{\Delta l_i} \frac{r(\theta)}{4\pi R_{dL}^*} \{f\}_j \{f\}_i^T \mathrm{d}x' \mathrm{d}x \tag{9.207}$$

where: $R_{a0} = [x^2 + a^2]^{1/2}$, $R_{d0} = [x^2 + d^2]^{1/2}$, $R_{aL} = [(L - x')^2 + a^2]^{1/2}$, $R_{dL} = [(L - x')^2 + d^2]^{1/2}$, $R_{a0}^* = [x^2 + 4h^2]^{1/2}$, $R_{d0}^* = [x^2 + d^2 + 4h^2]^{1/2}$, $R_{aL}^* = [(L - x')^2 + 4h^2]^{1/2}$ and $R_{dL}^* = [(L - x')^2 + d^2 + 4h^2]^{1/2}$.
There are 25 time-shifted terms in the global matrix system (9.201). Each of these terms contains the space dependent matrix multiplying the time dependent current vector or time dependent excitation vector.

9.2.6.5 Space-Time Numerical Procedure for Wire Array Above a Real Ground Solutions of eqns (9.163)–(9.171) are obtained using the time domain version of GB BIEM. Local approximation for the wire segment is given by:

$$I(x', t') = \{f\}^T \{I\} \tag{9.208}$$

where $\{I\}$ denotes space-time dependant solution vector, and $\{f\}^T$ denotes vector containing a space-domain shape functions given by:

$$\begin{aligned} f_r(x') &= \frac{x_{r+1} - x'}{x_{r+1} - x_r} \\ f_{r+1}(x') &= \frac{x' - x_r}{x_{r+1} - x_r} \end{aligned} \tag{9.209}$$

where r is the global node index of the observed boundary element.

Applying boundary element discretization to eqns (9.163) and (9.171) yields the local system of linear equations for the νth wire:

$$
\begin{aligned}
&\sum_{s=1}^{M}\left[\int_{\Delta l_i}\int_{\Delta l_j}\frac{1}{4\pi R_{vs}}\{f\}_j\{f\}_i^T \mathrm{d}x'\mathrm{d}x\{I_s\}\bigg|_{t-\frac{R_{vs}}{c}}\right. \\
&\left.-\int_{\Delta l_i}\int_{\Delta l_j}\frac{r_{vs}(\theta)}{4\pi R_{vs}^*}\{f\}_j\{f\}_i^T \mathrm{d}x'\mathrm{d}x\{I_s\}\bigg|_{t-\frac{R_{vs}^*}{c}}\right] \\
&=\frac{1}{2Z_0}\int_{\Delta l_i}\int_{\Delta l_j}E_x^{inc_v}\left(x',t-\frac{|x-x'|}{c}\right)\{f\}_j \mathrm{d}x'\mathrm{d}x \\
&+\int_{\Delta l_j}F_0\left(t-\frac{x-x_{0v}}{c}\right)\{f\}_j \mathrm{d}x+\int_{\Delta l_j}F_L\left(t-\frac{x_{Lv}-x}{c}\right)\{f\}_j \mathrm{d}x
\end{aligned}
\tag{9.210}
$$

where i is the index of the source element located on the sth wire, while j denotes the index of the observed element located on the νth wire.

Performing certain mathematical manipulation yields the following matrix system:

$$
\begin{aligned}
&\sum_{s=1}^{M}[A_{vs}]\{I_s\}\Big|_{t-\frac{R_{vs}}{c}}-\sum_{s=1}^{M}[A_{vs}^*]\{I_s\}\Big|_{t-\frac{R_{vs}^*}{c}}=[B_v]\{E_v\}\Big|_{t-\frac{|x-x'|}{c}} \\
&+\sum_{s=1}^{M}[C_{vs}]\left\{\sum_{n=0}^{\infty}I_s^n\right\}\Bigg|_{t-\frac{x-x_{0v}}{c}-\frac{2n}{c}L_v-\frac{R_{vs}^{(0)}}{c}}-\sum_{s=1}^{M}[C_{vs}^*]\left\{\sum_{n=0}^{\infty}I_s^n\right\}\Bigg|_{t-\frac{x-x_{0v}}{c}-\frac{2n}{c}L_v-\frac{R_{vs}^{*(0)}}{c}} \\
&-[D_v]\left\{\sum_{n=0}^{\infty}E_v^n\right\}\Bigg|_{t-\frac{x-x_{0v}}{c}-\frac{2n}{c}L_v-\frac{|x'-x_{0v}|}{c}}-\sum_{s=1}^{M}[E_{vs}]\left\{\sum_{n=0}^{\infty}I_s^n\right\}\Bigg|_{t-\frac{x-x_{0v}}{c}-\frac{2n+1}{c}L_v-\frac{R_{vs}^{(L)}}{c}} \\
&+\sum_{s=1}^{M}[E_{vs}^*]\left\{\sum_{n=0}^{\infty}I_s^n\right\}\Bigg|_{t-\frac{x-x_{0v}}{c}-\frac{2n+1}{c}L_v-\frac{R_{vs}^{*(L)}}{c}}+[D_v]\left\{\sum_{n=0}^{\infty}E_v^n\right\}\Bigg|_{t-\frac{x-x_{0v}}{c}-\frac{2n+1}{c}L_v-\frac{|x_{Lv}-x'|}{c}} \\
&+\sum_{s=1}^{M}[E_{vs}]\left\{\sum_{n=0}^{\infty}I_s^n\right\}\Bigg|_{t-\frac{x_{Lv}-x}{c}-\frac{2n}{c}L_v-\frac{R_{vs}^{(L)}}{c}}-\sum_{s=1}^{M}[E_{vs}^*]\left\{\sum_{n=0}^{\infty}I_s^n\right\}\Bigg|_{t-\frac{x_{Lv}-x}{c}-\frac{2n}{c}L_v-\frac{R_{vs}^{*(L)}}{c}} \\
&-[D_v]\left\{\sum_{n=0}^{\infty}E_v^n\right\}\Bigg|_{t-\frac{x_{Lv}-x}{c}-\frac{2n}{c}L_v-\frac{|x_{Lv}-x'|}{c}}-\sum_{s=1}^{M}[C_{vs}]\left\{\sum_{n=0}^{\infty}I_s^n\right\}\Bigg|_{t-\frac{x_{Lv}-x}{c}-\frac{2n+1}{c}L_v-\frac{R_{vs}^{(0)}}{c}} \\
&+\sum_{s=1}^{M}[C_{os}^*]\left\{\sum_{n=0}^{\infty}I_s^n\right\}\Bigg|_{t-\frac{x_{Lo}-x}{c}-\frac{2n+1}{c}L_o-\frac{R_{os}^{*(0)}}{c}}+[D_o]\left\{\sum_{n=0}^{\infty}E_o^n\right\}\Bigg|_{t-\frac{x_{Lo}-x}{c}-\frac{2n+1}{c}L_o-\frac{|x'-x_{0o}|}{c}}
\end{aligned}
\tag{9.211}
$$

where $\{E\}$ denotes excitation vector and the space-dependant matrices are given by:

$$[A_{vs}] = \int_{\Delta l_j}\int_{\Delta l_i} \frac{1}{4\pi R_{vs}} \{f\}_j \{f\}_i^T \mathrm{d}x' \mathrm{d}x \tag{9.212}$$

$$[A_{vs}^*] = \int_{\Delta l_j}\int_{\Delta l_i} \frac{r_{vs}(\theta)}{4\pi R_{vs}^*} \{f\}_j \{f\}_i^T \mathrm{d}x' \mathrm{d}x \tag{9.213}$$

$$[B_v] = \frac{1}{2Z_0}\int_{\Delta l_j}\int_{\Delta l_i} \{f\}_j \{f\}_i^T \mathrm{d}x' \mathrm{d}x \tag{9.214}$$

$$[C_{vs}] = \int_{\Delta l_j}\int_{\Delta l_i} \frac{1}{4\pi R_{vs}^{(0)}} \{f\}_j \{f\}_i^T \mathrm{d}x' \mathrm{d}x \tag{9.215}$$

$$[C_{vs}^*] = \int_{\Delta l_j}\int_{\Delta l_i} \frac{r_{vs}(\theta)}{4\pi R_{vs}^{*(0)}} \{f\}_j \{f\}_i^T \mathrm{d}x' \mathrm{d}x \tag{9.216}$$

$$[D_v] = \frac{1}{2Z_0}\int_{\Delta l_j}\int_{\Delta l_i} \{f\}_j \{f\}_i^T \mathrm{d}x' \mathrm{d}x \tag{9.217}$$

$$[E_{vs}] = \int_{\Delta l_j}\int_{\Delta l_i} \frac{1}{4\pi R_{vs}^{(L)}} \{f\}_j \{f\}_i^T \mathrm{d}x' \mathrm{d}x \tag{9.219}$$

Expressions with summation from $n = 0$ to infinity pertain to reflections from the wire ends. Since the considered time interval is finite, it is sufficient to take into account only those reflections that appear within the given interval, thus replacing infinity with finite integer value. Shorter observed intervals require smaller number of summands, and vice versa.

Having performed the space-domain discretization, the weighted residual approach is used for the time-domain discretization as well. The global matrix system can be generally written as:

$$[A]\{I\}\Big|_{t-\frac{R_{vs}}{c}} - [A^*]\{I\}\Big|_{t-\frac{R_{vs}^*}{c}} = \{g\} \tag{9.220}$$

while the time-domain solution on the ith finite element is given by:

$$I_i(t) = \sum_{k=1}^{N_t} I_i^k T^k(t') \tag{9.221}$$

where I_i^k are unknown coefficients, T_k are the linear time-domain shape functions and N_t is the total number of time samples.

Applying the weighted residual approach to eqn (9.220) leads to:

$$\int_{t_k}^{t_k+\Delta t} \left([A]\{I\}\Big|_{t-\frac{R_{VS}}{c}} - [A^*]\{I\}\Big|_{t-\frac{R^*_{VS}}{c}} - \{g\}\theta_k \right) \mathrm{d}t = 0; \; k = 1, 2, \ldots, N_t \qquad (9.222)$$

where θ_k denotes the set of time-domain weights.

Using Dirac impulses for the test functions, time sampling is provided and eqn (9.222) becomes:

$$[A]\{I\}\Big|_{t_k-\frac{R_{VS}}{c}} - [A^*]\{I\}\Big|_{t_k-\frac{R^*_{VS}}{c}} = \{g\}|_{all\,previous\,discrete\,instants} \qquad (9.223)$$

The transient current at the present instant can be expressed using currents and excitations at all previous instances and excitation at the present instant if the following inequality is satisfied.

$$\Delta x \geq c\Delta t \qquad (9.224)$$

If condition (9.224) is satisfied, the transient current for a jth space node and kth time node can be obtained via a recurrence formula. If the members related to the current at present instance t_k in eqn (9.223) are separated, eqn (9.223) can be written as follows:

$$A_{jj}I_j|_{t_k} + [\overline{A}]\{I\}\Big|_{t_k-\frac{R_{VS}}{c}} - [A^*]\{I\}\Big|_{t_k-\frac{R^*_{VS}}{c}} = \{g\}|_{all\,previous\,discrete\,instants} \qquad (9.225)$$

where overbar indicates the absence of diagonal members.

The first term in eqn (9.225) pertains to the current at the jth space node and kth time node, that is present instance. Other terms are related to all previous instances. The recurrence formula for the transient current at jth space node can be written as follows:

$$I_j|_{t_k} = \frac{-\sum_{i=1}^{N}\left(\overline{A}_{ji}I_i|_{t_k-\frac{R_{VS}}{c}} + A^*_{ji}I_i|_{t_k-\frac{R^*_{VS}}{c}} \right) + g_j|_{all\,previous\,discrete\,instants}}{A_{jj}} \qquad (9.226)$$

where N is total number of space elements, and $k = 1, 2, \ldots, N_t$ is the index of the time node.

9.2.7 Computational Examples

This section deals with a number of computational examples related to the single wire, two coupled wires, and straight wire array above perfectly conducting ground and dielectric half space, respectively.

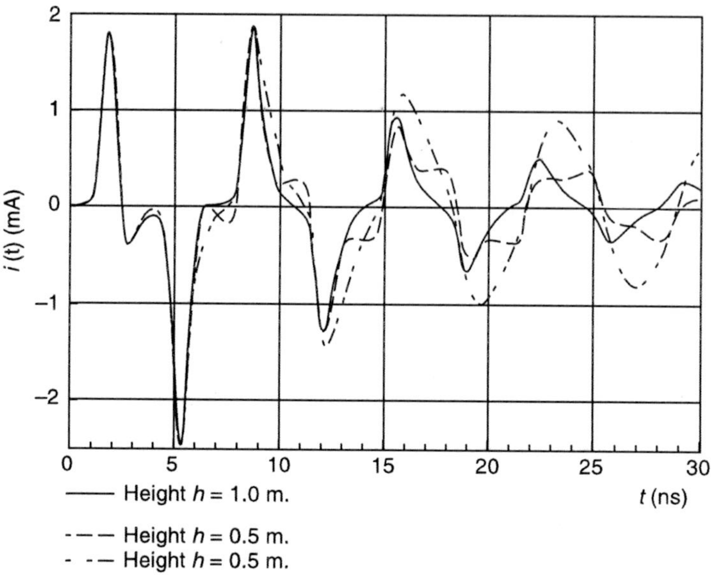

Figure 9.30 Transient response of the dipole above PEC ground for Gaussian voltage source.

9.2.7.1 Thin Wires Over a Perfect Ground A thin wire dipole located above a perfect ground plane, excited by a Gaussian voltage pulse and a unit step function voltage is considered. To study the effect of the PEC interface, the wire at various heights $h = 1$ m, 0.5 m, and 0.25 m above the ground plane is observed. The dimensions of the wire ($L = 1$ m, $a = 2$ mm) and the Gaussian pulse parameters are the same as in the case of a free space.

The transient response results for the Gaussian and step function excitation are shown in Figures 9.30 and 9.31, respectively.

At the height $h = 1$ m, there is a slight discrepancy between the actual transient response and the case of a free space discussed earlier in Figure 9.4. Essentially, there is no difference in the three current waveforms of the wire at different heights at the early time portion, which is reasonable as the scattered field from the wire takes time to propagate to the ground and return after being reflected. During this initial interval there is no effect of the ground.

On the contrary, in the highly resonant case where $h = 0.25$ m the current amplitude decays more slowly than when the wire is further away from the ground plane. In the later times, this field reflected from the ground plane acts on the induced current resulting in the diverging waveforms shown in Figures 9.30 and 9.31.

A thin wire scatterer located above a perfect ground plane, excited by a Gaussian voltage pulse and a unit step function voltage is considered. The scatterer is located horizontally at a height $h = 0.25$ m over the ground plane. The incident field has polarization parallel to the axis of the wire and also the ground plane. The corresponding transient responses are shown in Figures 9.32 and 9.33, respectively.

The effect of the PEC ground plane for the transient response is obvious.

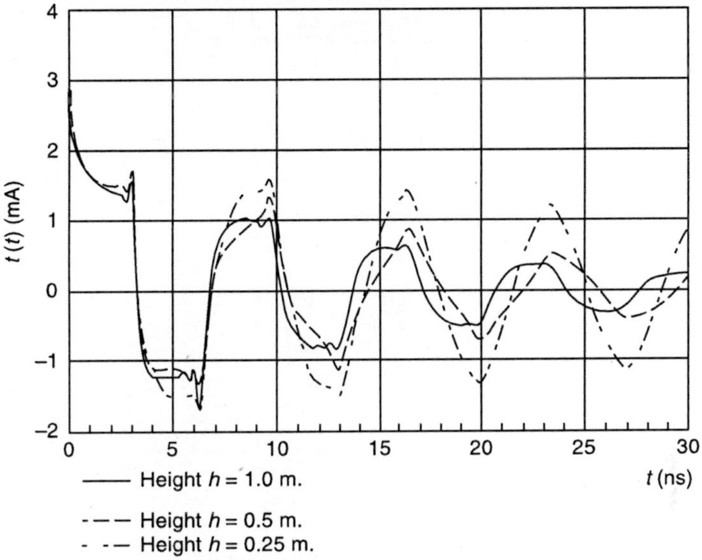

Figure 9.31 Transient response of the dipole above PEC ground for step function voltage source.

Figures 9.34 and 9.35 are related to the set of coupled wire scatterers in free space and above *PEC* ground with $L = 1$ m, $a = 2$ mm, $d = 0.5$ m, $h = 0.25$ m, $h = 0.5$ m, illuminated by the Gaussian plane wave with $g = 2 \times 10^9\ \mathrm{s}^{-1}$ and $t_0 = 2$ ns.

It is visible from Figures 9.34 and 9.35 that there is no effect of the *PEC* reflecting medium which does not manifest at very early times. The variations in the

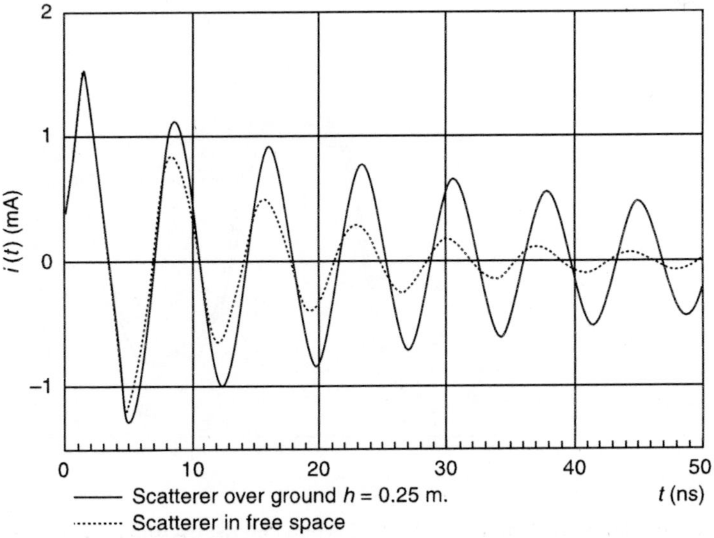

Figure 9.32 Scatterer over ground plane excited by broadside Gaussian plane wave.

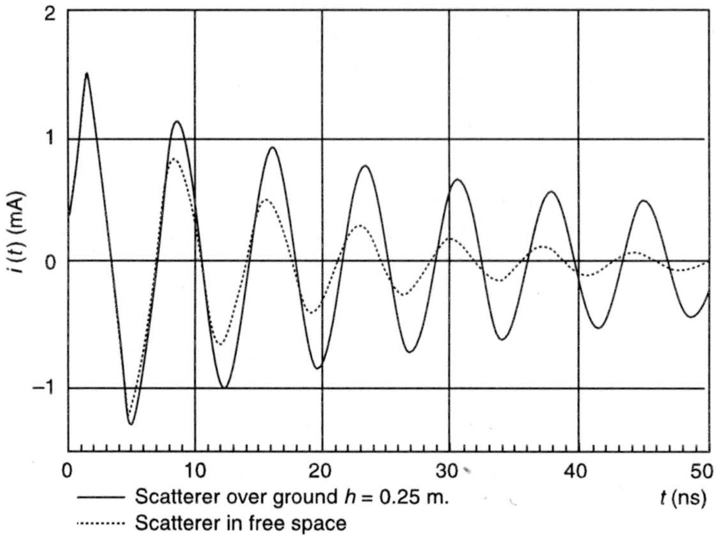

Figure 9.33 Scatterer over ground plane excited by broadside step plane wave.

transient response are evidently as a result of the effect of the field reflected from the *PEC* ground which would otherwise be lost in free space.

The associated energy measures of H-field, E-field and the total energy collected by the system of coupled wires is shown in Figures 9.36 and 9.37, respectively.

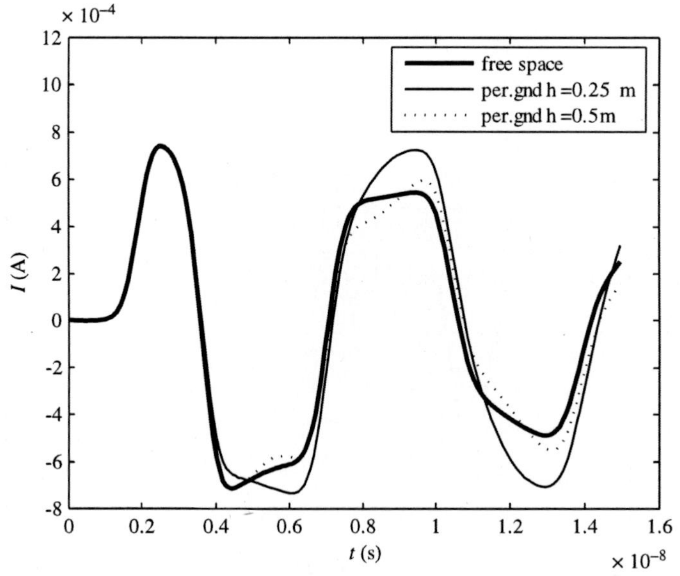

Figure 9.34 Influence of a presence of a PEC ground on the transient response at the center of the near side wire ($L = 1$ m, $a = 2$ mm, $d = 0.5$ m).

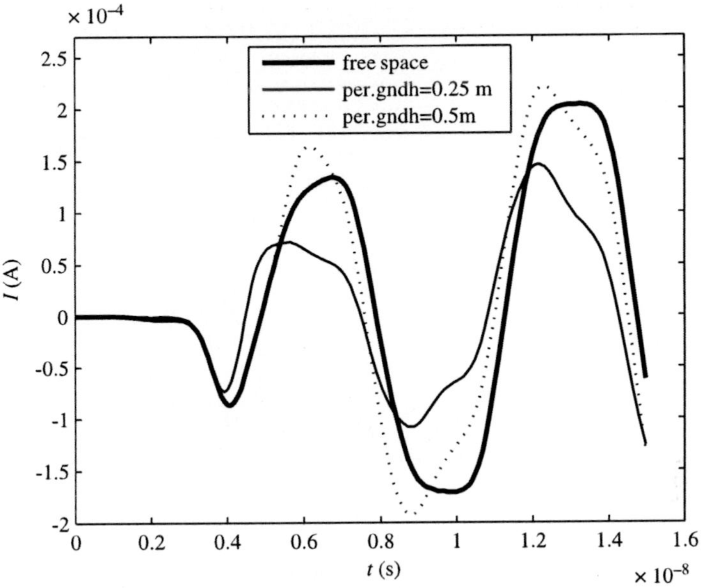

Figure 9.35 Influence of a presence of a PEC ground on the transient response at the center of the far side wire ($L = 1$ m, $a = 2$ mm, $d = 0.5$ m).

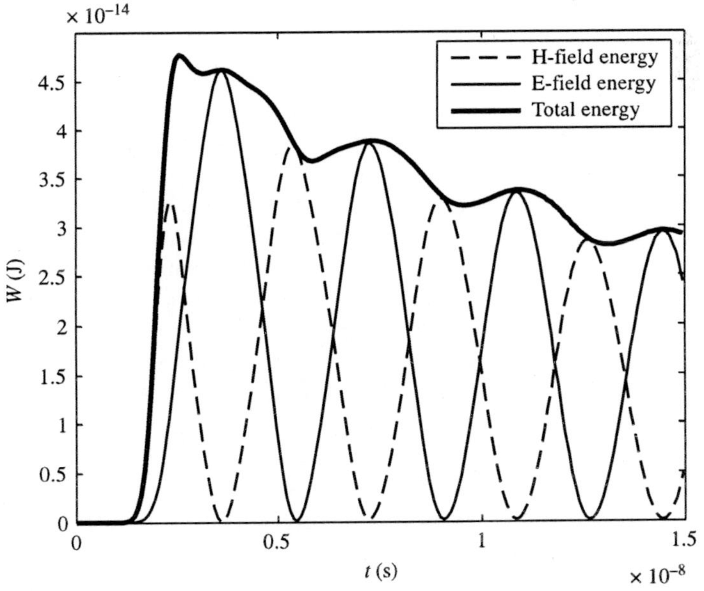

Figure 9.36 Measures of H-field, E-field, and the total energy collected by the near-side wire in the presence of a PEC ground ($L = 1$ m, $a = 2$ mm, $d = 0.5$ m, $h = 0.25$ m).

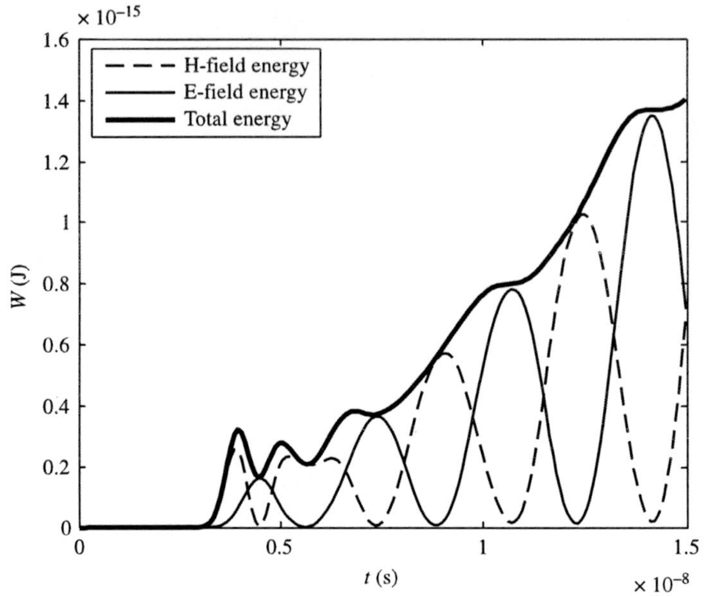

Figure 9.37 Measures of H-field, E-field, and the total energy collected by the far-side wire in the presence of a PEC ground ($L = 1$ m, $a = 2$ mm, $d = 0.5$ m, $h = 0.25$ m).

9.2.7.2 Thin Wires over a Dielectric Half-Space In Figures 9.38–9.41 the solutions for the free-space for the perfectly conducting half space and for the lossy half-space are compared. Figure 9.38 shows the source current of the antenna of length $L = 1$ m and radius $a = 6.74$ mm located at the height $h = 1$ m from the interface and excited by Gaussian pulse ($V_0 = 1$ V, $g = 1.5 \times 10^9\,\text{s}^{-1}$, $t_0 = 1.43$ ns). Figure 9.39 shows the electric field in the broadside direction, for the same wire dimensions at the distance $h = 0.5$ m from the interface at the observation point $T(x = L/2, y = 0, z = 20\,m)$. Figure 9.40 shows the induced current at the center of the scatterer of dimensions $L = 1$ m, $a = 2$ mm horizontally located at the height $h = 0.25$ m over the ground. Figure 9.41 shows the scattered field at the observation point $T(x = L/2, y = 0, z = 20\,m)$.

From Figures 9.38 and 9.40, it is obvious that the influence of the image wire is not noticed before the expiration of the time required for the signal to travel the distance between the real source and image source, respectively. It is also obvious from Figures 9.39 and 9.41 that the radiated field and the scattered field practically vanish when dealing with a perfectly conducting plane. This is plausible because the observation point is located symmetrically with respect to the real wire and to the image wire, whereas the distance from the observation point to the interface is much bigger than the distance between the source wire and the image wire (The currents along source wire and image wire flow in the opposite directions in accordance with the theory of images).

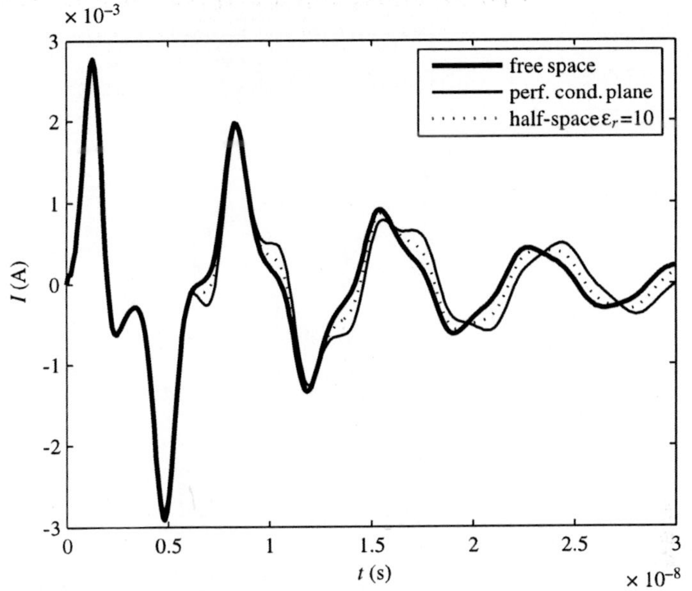

Figure 9.38 Source current dipole antenna—comparison of results for free space, perfectly conducting half-space, and dielectric half-space. (D. Poljak, V. Roje, Time-Domain Calculation of the Parameters of Thin Wire Antennas and Scatterers in Half-Space Configuration. // *IEE Proceedings : microwaves, antennas and propagation.* 145 (1998), 1; 57–63).

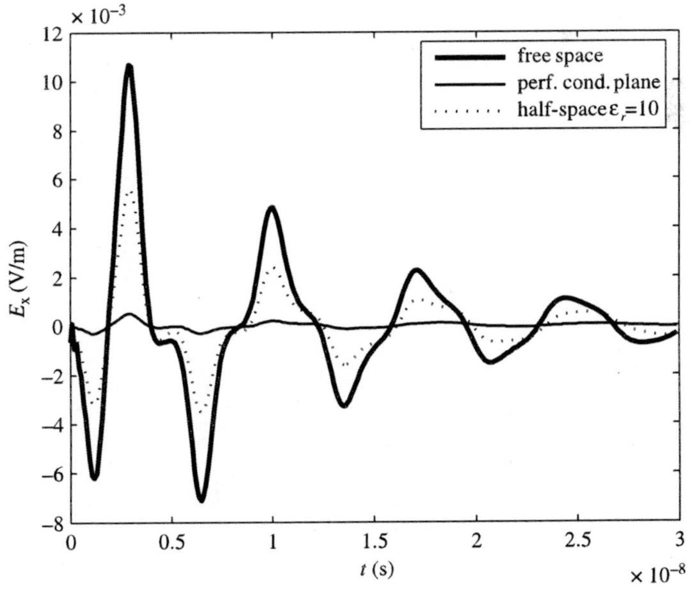

Figure 9.39 Broadside radiated electric field of linear antenna with Gaussian time dependence. (D. Poljak, V. Roje, Time-Domain Calculation of the Parameters of Thin Wire Antennas and Scatterers in Half-Space Configuration. // *IEE Proceedings : microwaves, antennas and propagation.* 145 (1998), 1; 57–63).

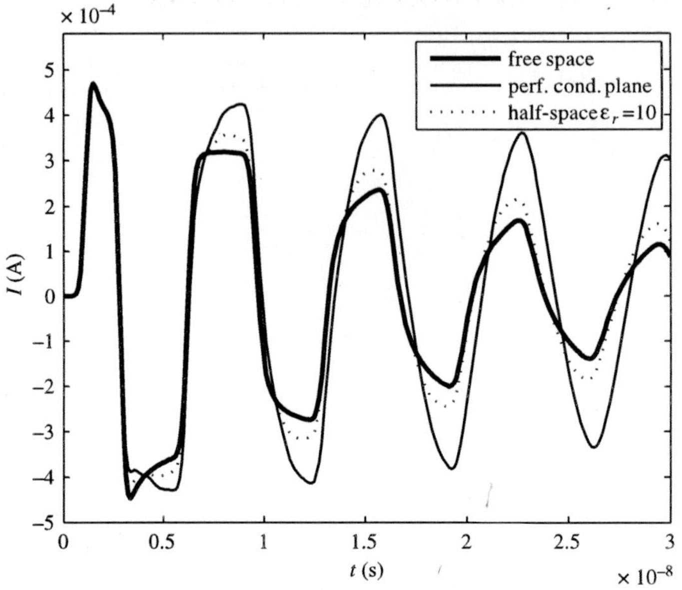

Figure 9.40 Induced current at the center of straight thin wire scatterer located above half-space. (D. Poljak, V. Roje, Time-Domain Calculation of the Parameters of Thin Wire Antennas and Scatterers in Half-Space Configuration. // *IEE Proceedings : microwaves, antennas and propagation.* 145 (1998), 1; 57–63).

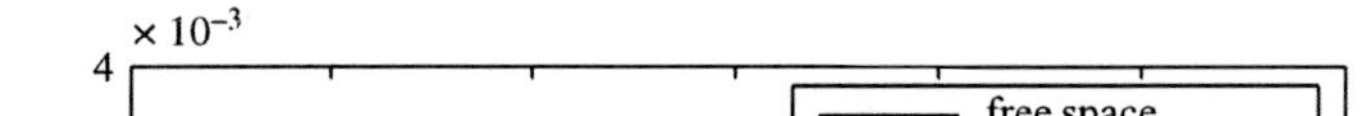

Figure 9.41 Bradside scattered electric field of linear antenna with Gaussian time dependence. (D. Poljak, V. Roje, Time-Domain Calculation of the Parameters of Thin Wire Antennas and Scatterers in Half-Space Configuration. // *IEE Proceedings : microwaves, antennas and propagation.* 145 (1998), 1; 57–63).

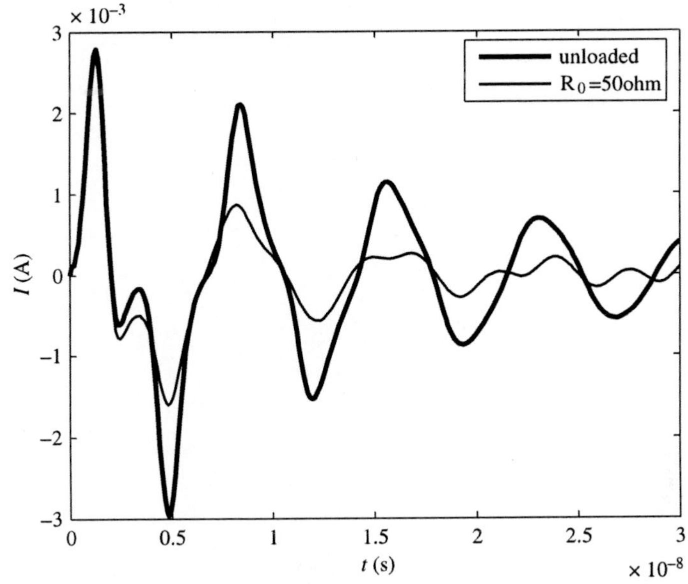

Figure 9.42 Transient response of a loaded dipole antenna radiating over dielectric half-space with $\varepsilon_r = 10$ at the height $h = 0.25\,\mathrm{m}$, $R_0 = 50\Omega$. (D. Poljak: Transient Response of Resistively Loaded Straight Thin Wire in Half-Space Configuration, *Journ. Elctromagn. Waves and Applic., No. 6, Vol. 12, 775–787, 1998*).

Figures 9.42–9.45 are related to the case of a loaded wire above a dielectric half-space, with $\varepsilon_r = 10$, driven with Gaussian pulse (9.110). The length of antenna is $L = 1$ m and radius is $a = 6.74$ mm. The Gaussian pulse parameters are $V_0 = 1\,\mathrm{V}$, $g = 1.5 \times 10^9$, and $t_0 = 1.43 \times 10^{-9}$ s. The loading function is of the form given in (9.35). The influence of a resistive loading is obvious with varying the values of parameter R_0. The end reflections are decreasing by the increase of the parameter R_0.

The end reflections are decreasing by increasing the values of parameter R_0, and by increasing the height h over ground.

Next two figures are related to the nonlinearly loaded horizontal wire located above a dielectric half-space. Figure 9.46 shows the driving point current of a dipole antenna with length $L = 1$ m, $a = 2$ mm located above a dielectric half-space ($\varepsilon_r = 10$) with the height as a parameter and center-fed by a Gaussian pulse voltage source of (9.110) with $V_0 = 1\,\mathrm{V}$, $g = 3.25 \times 10^{+9}\,\mathrm{s}^{-1}$, and $t_0 = 1.389$ ns. Figure 9.47 shows the induced current at the midpoint of the wire of the same dimensions but now operating in the scattering mode.

Numerical results clearly demonstrate that the height of the wire above the dielectric medium has negligible influence on the transient current flowing through the nonlinear element. The effects of the two-media configuration are negligible in comparison to the effects of the nonlinear loading.

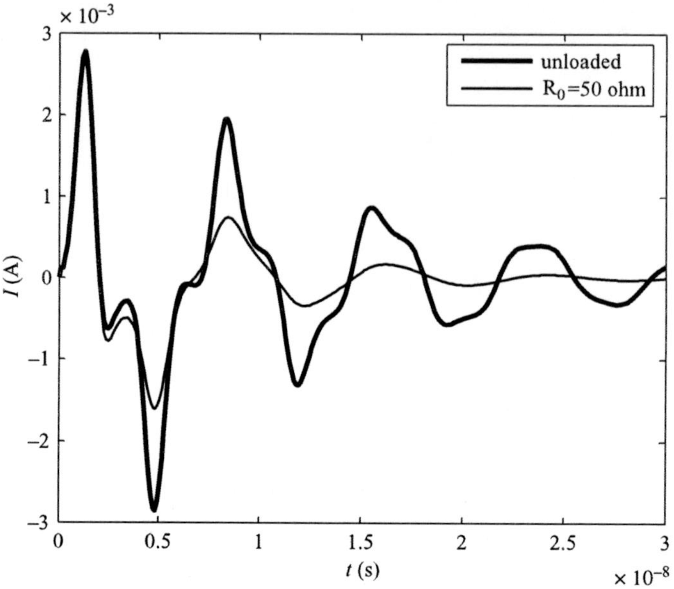

Figure 9.43 Transient response of a loaded dipole antenna radiating over dielectric half-space with $\varepsilon_r = 10$ at the height $h = 0.5\,\text{m}$, $R_0 = 50\,\Omega$. (D. Poljak: Transient Response of Resistively Loaded Straight Thin Wire in Half-Space Configuration, *Journ. Elctromagn. Waves and Applic., No. 6, Vol. 12, 775–787, 1998*).

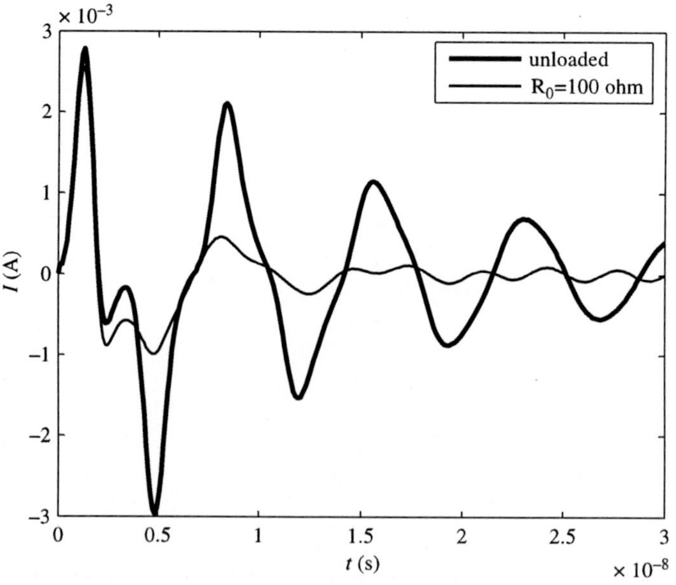

Figure 9.44 Transient response of a loaded dipole antenna radiating over dielectric half-space with $\varepsilon_r = 10$ at the height $h = 0.25$, $R_0 = 100\,\Omega$. (D. Poljak: Transient Response of Resistively Loaded Straight Thin Wire in Half-Space Configuration, *Journ. Elctromagn. Waves and Applic., No. 6, Vol. 12, 775–787, 1998*).

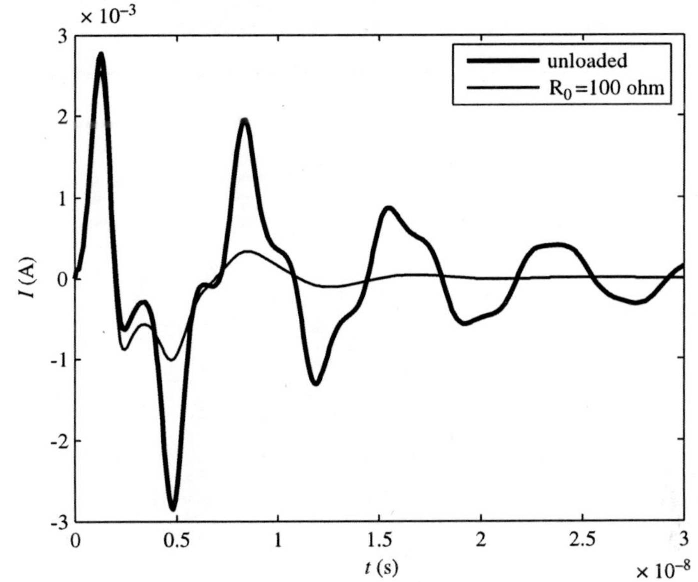

Figure 9.45 Transient response of a loaded dipole antenna radiating over dielectric half-space with $\varepsilon_r = 10$ at height $h = 0.5$ m, $R_0 = 100\Omega$. (D. Poljak: Transient Response of Resistively Loaded Straight Thin Wire in Half-Space Configuration, *Journ. Elctromagn. Waves and Applic., No. 6, Vol. 12, 775–787, 1998*).

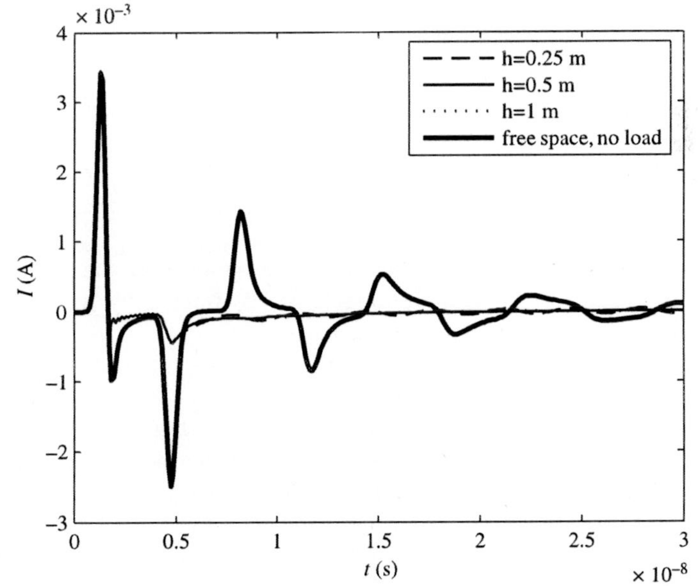

Figure 9.46 Driving point current of the nonlinearly loading dipole antenna ($R_1/R_2 = 0/700\Omega$) with $L = 1$m, $a = 2$ mm, placed horizontally above a dielectric ground and excited by a Gaussian pulse voltage source. (Poljak, D.; Yoong Tham, C.; McCowen, A. Transient Response of Nonlinearly Loaded Wires in a Two Media Configuration, *IEEE Transactions on Electromagnetic Compatibility.* 46 (2004), 1; 121–125).

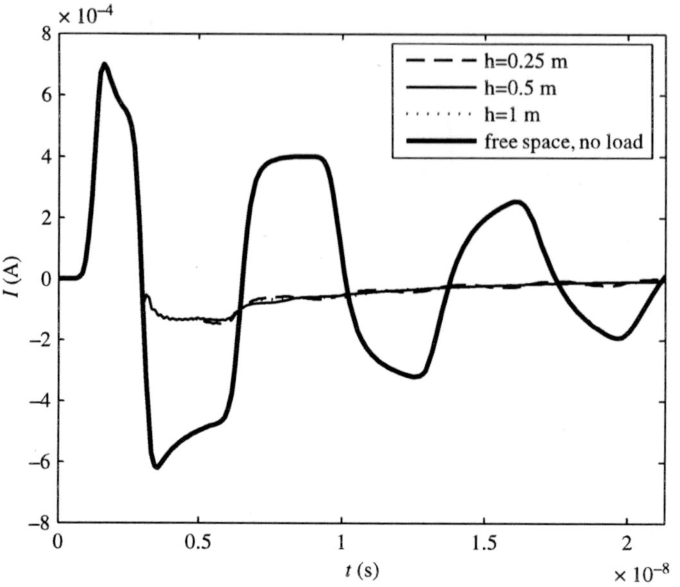

Figure 9.47 Current induced at the midpoint of the nonlinearly loading dipole antenna ($R_1/R_2 = 0/700\Omega$) with $L = 1$m, $a = 2$ mm, placed horizontally above a dielectric ground and excited by a Gaussian pulse plane wave. (Poljak, D.; Yoong Tham, C.; McCowen, A. Transient Response of Nonlinearly Loaded Wires in a Two Media Configuration, *IEEE Transactions on Electromagnetic Compatibility.* 46 (2004), 1; 121–125).

It is worth pointing out that the effects of the dielectric half-space can be neglected in comparison to the effects of the nonlinear load.

Next four Figures 9.48–9.51 show the influence of the height h above the lower medium and of separation d between wires to the transient current induced on both wires.

Analyzing the obtained numerical results, it is obvious that the transient current induced at the center of a parasitic wire is more affected by the presence of a reflecting half-space. The smaller is the distance d between the wires and the higher is the height h over interface, the more negligible is the influence of the dielectric half-space to the actual transient response.

Next example deals with the three-wire antenna array horizontally located over a dielectric half-space, when the central wire is excited by a time-dependent Gaussian pulse voltage source. The parameters of the Gaussian pulse are: $V_0 = 1$ V, $g = 2 \times 10^9\,\mathrm{s}^{-1}$, and $t_0 = 2$ ns, the entire length of the wires is $L = 1$ m, while the radius of all wires is $a = 2$ mm, and the separation between the neighboring wires is $d = 0.5$ m. The array is located above dielectric medium ($\varepsilon_r = 10$) at height $h = 1$ m.

The transient currents induced at wire centers are shown in Figures 9.52 and 9.53, while Figures 9.54 and 9.55 show the energy stored in the near field of the wires A, B, and C respectively.

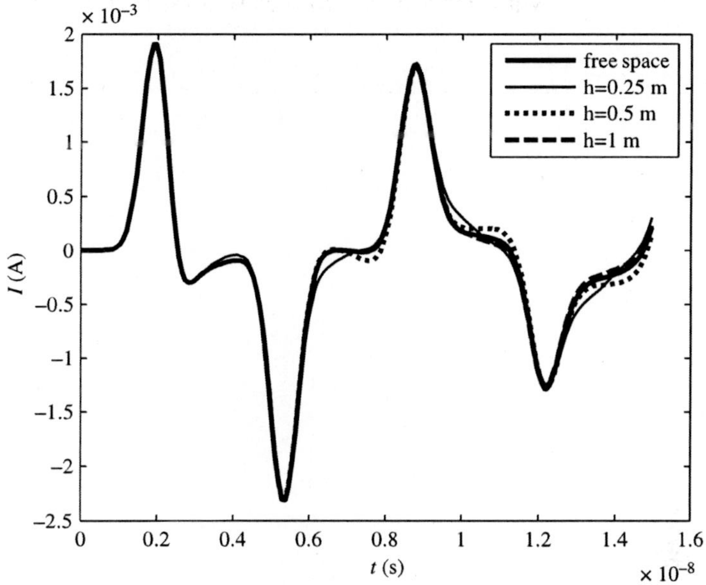

Figure 9.48 Current at the center of the wire A (active wire) versus time with height h over interface as parameter ($L = 1\text{m}$, $a = 2\,\text{mm}$, $\varepsilon_r = 9$, $h = 0.25\,\text{m}$). (D. Poljak, C. Y. Tham, A. Mc. Cowen, V. Roje: Transient Analysis of Two Coupled Horizontal Wires over a Real Ground, *IEE Proc. Microw. Antennas Propag., Vol. 147, No. 2, pp 87–94, April 2000*).

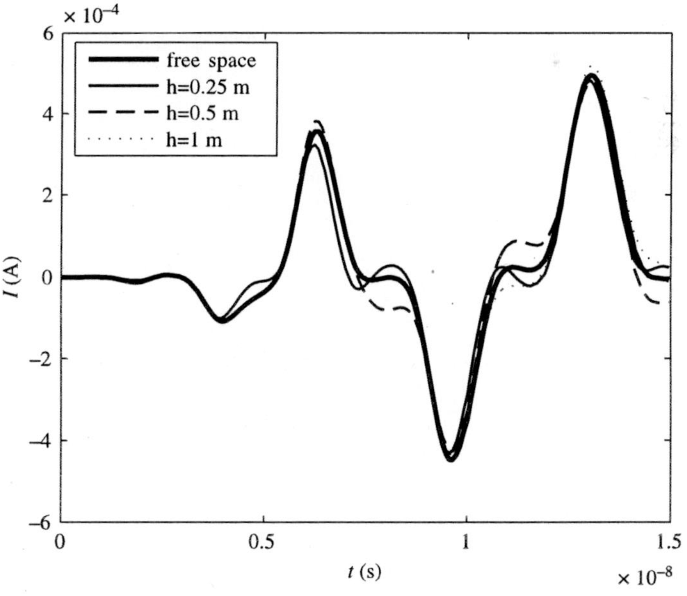

Figure 9.49 Current at the center of wired B (passive wire) versus time with height h over interface as parameter ($L = 1\text{m}$, $a = 2\,\text{mm}$, $\varepsilon_r = 9$, $h = 0.25\,\text{m}$). (D. Poljak, C. Y. Tham, A. Mc. Cowen, V. Roje: Transient Analysis of Two Coupled Horizontal Wires over a Real Ground, *IEE Proc. Microw. Antennas Propag., Vol. 147, No. 2, pp 87–94, April 2000*).

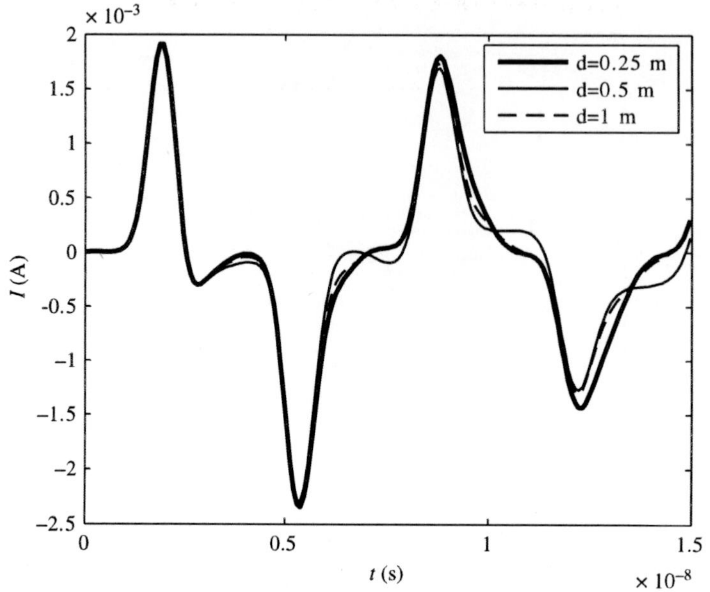

Figure 9.50 Current at the center of wire A (active wire) versus time with distance d between the wires as a parameter ($L = 1$m, $a = 2$ mm, $\varepsilon_r = 9$, $d = 0.5$ m). (D. Poljak, C. Y. Tham, A. Mc. Cowen, V. Roje: Transient Analysis of Two Coupled Horizontal Wires over a Real Ground, *IEE Proc. Microw. Antennas Propag., Vol. 147, No. 2, pp 87–94, April 2000*).

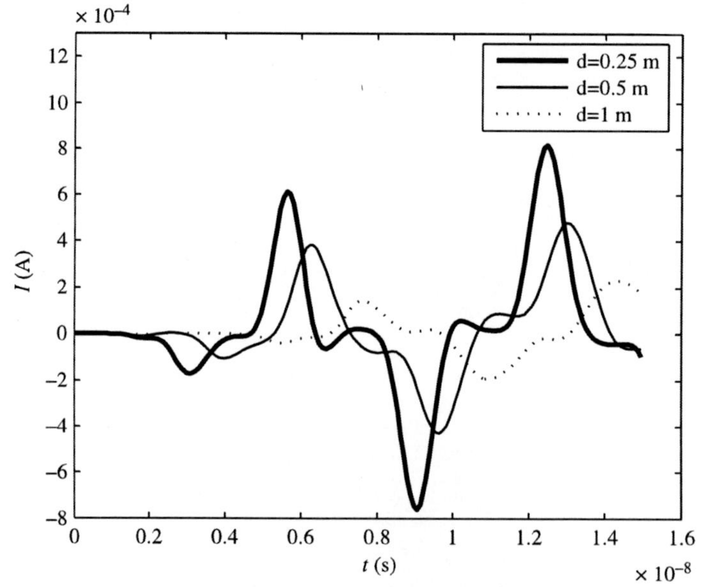

Figure 9.51 Current at the center of wire B (passive wire) versus time with distance d between the wires as a parameter ($L = 1$m, $a = 2$ mm, $\varepsilon_r = 9$, $d = 0.5$ m). (D. Poljak, C. Y. Tham, A. Mc. Cowen, V. Roje: Transient Analysis of Two Coupled Horizontal Wires over a Real Ground, *IEE Proc. Microw. Antennas Propag., Vol. 147, No. 2, pp 87–94, April 2000*).

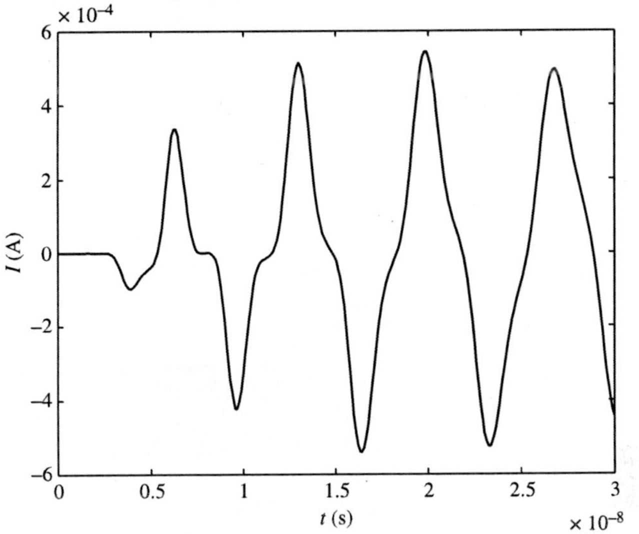

Figure 9.52 Transient current induced at the center of the passive wire A and C. (Poljak, D.; Miller, E. K.; Tham, C. Y.; Antonijević, S.; Doric, V., Time Domain Analysis of the Energy Stored in the Near Field of Multiple Straight-Wires Above Dielectric Half-Space // *Proceedings of Joint 9th International Conference on Electromagnetics in Advanced Applications ICEAA '05 and 11th European Electromagnetic Structures Conference EESC '05*.2005. 279–282).

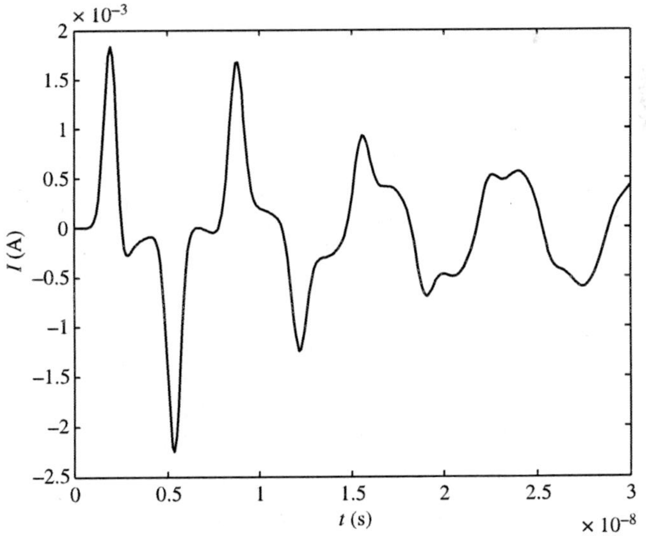

Figure 9.53 Transient current induced at the center of the active wire B. (Poljak, D.; Miller, E. K.; Tham, C. Y.; Antonijević, S.; Doric, V., Time Domain Analysis of the Energy Stored in the Near Field of Multiple Straight-Wires Above Dielectric Half-Space // *Proceedings of Joint 9th International Conference on Electromagnetics in Advanced Applications ICEAA '05 and 11th European Electromagnetic Structures Conference EESC '05*.2005. 279–282).

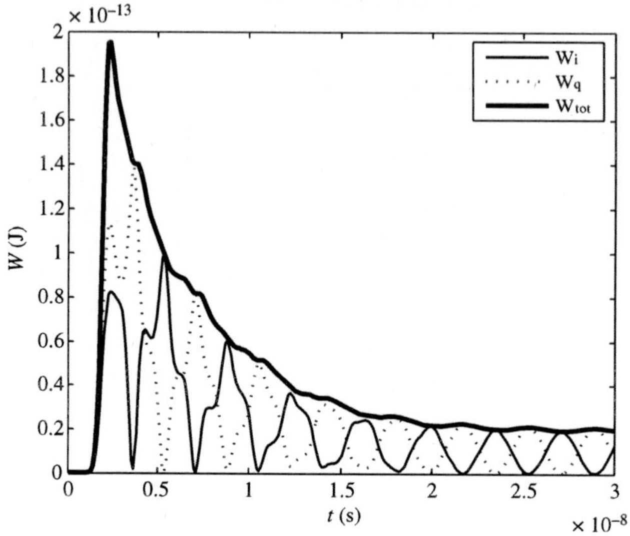

Figure 9.54 The H-field (W_I) E-field (W_q), and total energy (W_{tot}) energy measures as a function of time for the active wire of Figure 9.52. (Poljak, D.; Miller, E. K.; Tham, C. Y.; Antonijević, S.; Doric, V., Time Domain Analysis of the Energy Stored in the Near Field of Multiple Straight-Wires Above Dielectric Half-Space // *Proceedings of Joint 9th International Conference on Electromagnetics in Advanced Applications ICEAA '05 and 11th European Electromagnetic Structures Conference EESC '05*.2005. 279–282).

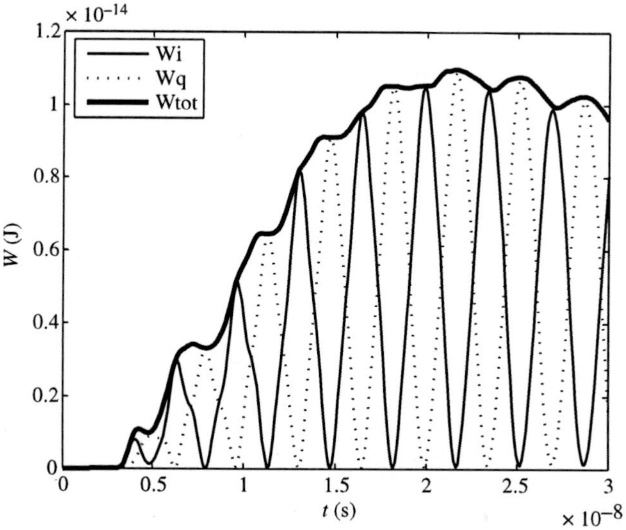

Figure 9.55 The H-field (W_I) E-field (W_q) and total energy (W_{tot}) energy measures as a function of time for the passive wire of Figure 9.53. (Poljak, D.; Miller, E. K.; Tham, C. Y.; Antonijević, S.; Doric, V., Time Domain Analysis of the Energy Stored in the Near Field of Multiple Straight-Wires Above Dielectric Half-Space // *Proceedings of Joint 9th International Conference on Electromagnetics in Advanced Applications ICEAA '05 and 11th European Electromagnetic Structures Conference EESC '05*.2005. 279–282).

9.3 REFERENCES

[1] E. K. Miller, J. A. Landt: Direct time-domain techniques for transient radiation and scattering from wires, *Proc. IEEE, 168(11)*, November 1980, pp. 1396–1423.

[2] E. K. Miller: PCs for AP and EM reflections, *IEEE Ant. Prop. Mag., 40(1)*, February 1998, pp. 96–100 and *41(2)*, April 1998, pp. 92–95.

[3] A. G. Tijhuis, Z. Q. Peng, A. R. Bretones: Transient excitation of a straight thin-wire segment: A New Look at an Old Problem, *IEEE Trans. AP 40(10)*, October 1992, pp. 1132–1146.

[4] T. K. Liu, K. K. Mei: A time domain integral equation solution for linear antennas and scatterers, *Radio Sci., Vol. 8*, August-September 1973, pp. 797–804.

[5] D. Poljak, V. Roje: Transient response of a thin wire in a two media configuration, *Proc. of EMC '97 Symposium*, Zurich, 1997, pp. 293–296.

[6] D. Poljak: Transient response of resistively loaded straight thin wire in half-space configuration, *J. Elctromagn. Waves Appl., 12(6)*, 1998, pp. 775–787.

[7] E. K. Miller, A. J. Poggio, G. J. Burke: An integro-differential equation technique for the time-domain analysis of thin wire structure, Part I, The numerical method, *J. Comput. Phys, 12(1)*, 1973, pp. 24–48, Part II, numerical results, *J. Comput Phys., 12(2)*, 1973, pp. 210–233.

[8] A. Salinas, R. G. Martin, A. R. Bretones, I. S. Garcia: Modeling of straight thin wires using time domain electric field integral equations, *IEE Proc. Microw. Antennas Prop. 141(2)*, April 1994, pp. 123–126.

[9] E. K. Miller: Time domain modeling in electromagnetics, *J Electro. Waves Appl., 8, No. 9–10*, 1994, pp. 1125–1172.

[10] D. Poljak, C. Y. Tham: Integral Equation techniques in Transient Electromagnetics, *WIT Press, Southampton-Boston*, 2003.

[11] D. Poljak: Time Domain Techniques in Computational Electromagnetics, *WIT Press, Southampton-Boston*, 2004.

[12] L. B. Felsen (Ed.): Transient Electromagnetic Waves, *Springer-Verlag, Berlin*, 1976.

[13] T. K. Liu, F. M. Tesche: Analysis of antennas and scatterers with nonlinear loads, *IEEE Trans. AP 24(2)*, March 1976, pp. 131–139.

[14] J. A. Landt, E. K. Miller, F. J. Deadrick: Time domain modeling of nonlinear loads, *IEEE Trans. AP 31(1)*, January 1983, pp. 121–126.

[15] H. Schuman: Time domain scattering from a nonlinearly loaded wire, *IEEE Trans. AP 22(4)*, April 1974, pp. 611–613.

[16] T. K. Liu, F. M. Tesche: Transient excitation of an antenna with a nonlinear load: Numerical and Experimental Results, *IEEE Trans. AP 25(7)*, March 1977, pp. 131–139.

[17] J. A. Landt: Network loading of thin wire antennas and scatterers in the time domain, *Radio Science, 16(6)*, November-December 1981, pp. 1241–1247.

[18] R. Leubbers, J. Beggs, K. Chamberlin: Finite difference time domain calculation of transients in antennas with nonlinear loads, *IEEE Trans. AP 41(5)*, May 1993, pp. 566–573.

[19] K. Aygun, B. Shanker, E. Michielson: Analysis of nonlinearly loaded antennas and circuits using the multilevel plane wave time domain algorithm, *Proc. ICEAA '99*, Turin, Italy, September 13–17, 1999, pp. 741–744.

[20] D. Poljak, E. K. Miller, C. Y. Tham: Root-mean-square measure of nonlinear effects to the transient response of thin wires, *IEEE Trans. Antennas Propagation 51(12)*, 2003, pp. 3280–3283.

[21] D. Poljak, C. Y. Tham, A. McCowen: Transient response of nonlinearly loaded wires in a two media configuration, *IEEE Trans. Electromagn. Compatibility 46(1)*, 2004, pp. 121–125.

[22] S. M. Rao, T. K. Sarkar, S. A. Dianat: A novel technique to the solution of transient electromagnetic scattering from thin wires, *IEEE Trans. AP 34(5)*, May 1986, pp. 630–634.

[23] D. Poljak, E. K. Miller, C. Y. Tham: Time domain energy measures for thin wire antennas and scatterers, *IEEE AP Magazine*, February 2002.

[24] D. Poljak, E. K. Miller, C. Y. Tham, S. Antonijević, V. Doric: Time domain analysis of the energy stored in the near field of multiple straight-wires above dielectric half-space, Proceedings of Joint 9th International Conference on Electromagnetics in Advanced Applications ICEAA '05 and 11th European Electromagnetic Structures Conference EESC '05, 2005, pp. 279–282.

[25] Y. Rahmat-Samii, P. Parhami, R. Mittra: Loaded horizontal antenna over an imperfect ground, *IEEE Trans., Antennas Propagation, AP 26(6)*, November 1978, pp. 789–796.

[26] D. Poljak, V. Roje: Time Domain Boundary Element Analysis of Resitively Loaded Wire, Boundary Elements XX, A. Kossab, C. A. Berbbia, M. Chopra (Eds), *Computational Mechanics Publications, Antony Rowe LTD., Southampton*, 1998.

[27] D. Poljak, V. Roje, N. Kovac; Time domain modeling of coupled wires by the finite element integral equation method, *Proc. SoftCOM '2000*, Split, 2000, pp. 299–308.

[28] D. Poljak: Electromagnetic Modeling of Wire Antenna Structures, *WIT Press, Southampton - Boston*, 2002.

[29] D. Poljak, V. Roje: Finite element technique for solving time-domain Hallen integral equation, *Proc. of ICAP '97*, Edinburgh, 1997, pp. 1225–1228.

[30] E. K. Miller, A. J. Poggio, G. J. Burke, E. S. Selden: Analysis of wire antennas in the presence of a conducting half-space, Part II. The horizontal antenna in free space, *Can. J. Phys., 50*, 1972, pp. 2614–2627.

[31] T. K. Sarkar: Analysis of arbitrarily oriented thin wire antennas over a plane imperfect ground, *Archiv fur Elektronik and Ubertragungstechnik, 31*, 1977, pp. 449–457.

[32] P. R. Barnes, F. M. Tesche: On the direct calculation of a transient plane wave reflected from a finitely conducting half-space, *IEEE Trans. EMC, 33(2)*, 1991, pp. 90–96.

[33] A. R. Bretones, A. G. Tijhuis: Transient excitation of two coupled wires over an interface between two dielectric half-spaces, *Radio Sci, 32(1)*, January–February 1997, pp. 25–41.

[34] A. G. Tijhuis, K. Belkebir, P. Zwamborn, A. R. Bretones: Marching on in anything: solving electromagnetic field equations with a varying physical parameter, *Proceedings of ICEAA '97*, September 15–18, Turin, Italy, 1997, pp. 175–178.

[35] D. Poljak, C. Y. Tham, A. McCowen, V. Roje: Transient analysis of two coupled horizontal wires over a real ground, *IEE Proc. Microw. Antennas Propag., 147(2)*, April 2000, pp. 87–94.

[36] D. Poljak (Ed): Time Domain Techniques in Computational Electromagnetics, *WIT-Press, Southampton, Boston, USA*, 2004.

PART III

COMPUTATIONAL MODELS IN ELECTROMAGNETIC COMPATIBILITY

The practical success of an idea, irrespective of its inherent merit is dependent on the attitude of the contemporaries.

Nikola Tesla

10

TRANSMISSION LINES OF FINITE LENGTH: GENERAL CONSIDERATIONS

The analysis of electromagnetic field coupling to overhead wires [1–16] or buried cables [17–21] can be performed by using the transmission line model or the thin-wire antenna in either frequency or time domain. Transient response of a transmission line of interest can be obtained directly by solving the time domain equations or by the indirect approach, that is, solving the frequency counterpart of the time domain equations. Using the indirect approach, the frequency spectrum has to be calculated and then the desired transient response is computed by means of the inverse Fourier transform.

Many cases of practical interest related to electromagnetic field coupling to transmission lines may be handled by means of the transmission line (TL) models. TL analysis includes two types of wave propagation:

- free space propagation, that is, incident electromagnetic field exciting the line;
- propagation of induced currents and voltages along the line.

The use of TL models require the line length to be significantly larger than the separation of the conductors and also larger than the actual height above ground or burial depth.

If these conditions are not satisfied, the line acts as a loop antenna and TL models are not appropriate, that is, the implementation of TL models provides only the computation of TL current components.

The approach, based on wire antenna theory, is more rigorous as the TL theory does not provide a complete solution for the excitation of victim equipment by an incident field, if the wavelength of the electromagnetic field coupling to a transmission line is comparable to or less than the transverse electrical dimensions of the line. The TL approximation is usually considered as a compromise between a quasistatic approximation and the receiving antenna model [1–16]. This approach, though sufficient approximation if long lines with electrically small cross sections are considered, fails if one deals with the lines of finite length and high frequency excitations.

As the frequency of the electromagnetic fields increases and the wavelength becomes comparable to the electrical dimensions of the source and victim equipment, the EMC models based on quasistatic approximation are found to be inaccurate [1]. In addition, though the simple transmission line models are often quite efficient, sometimes the behavior of the induced responses is to be evaluated using the antenna theory. An important aspect of this subject is to realize that transmission line theory does not provide a complete solution for the excitation of victim equipment by an incident field. Basically, it gives only an approximate solution.

Therefore, the TL model fails to predict resonances, fails to take into account properly the presence of a lossy ground, and the effects at the line ends also cannot be taken into account utilizing this approach [1–6]. So, when the above-ground or below-ground transmission lines of the finite length are of interest, the receiving antenna theory has to be used.

Namely, the wavelike behavior of the induced responses at higher frequencies requires a more general approach that is based on integral equations arising from wire antenna theory. On the contrary, the restrictions of the wire antenna model applied to transmission lines are often related to the long computational time required for the calculations pertaining to the long lines.

10.1 TRANSMISSION LINE THEORY METHOD

A transmission line is a class of wire structures where the radius a is small compared to the axial length L as well as the wavelength λ (i.e., $L \gg a$ and $\lambda \gg a$). For multiple lines in parallel separated by distance d, both the values for a and d are small compared to the wavelength ($\lambda \gg d$) and that ($d \gg a$). L is typically large and gets to be well over $\lambda/5$. The problems analyzed include lines in free space, effect of ground plane, coupling between lines, and coupling of external fields to the lines.

As the configuration being analyzed becomes electrically large, the wave propagating effect is significant and must be taken into account. In the transmission line approach, distributed circuit parameters are used in the analysis. However, all other assumptions and conditions stated in the lumped parameter circuit model still apply in the transmission line model. The set of governing equations of the source free transmission line model for a two parallel conductor system in the time domain are given by [1,11]

$$\frac{\partial v(x,t)}{\partial x} + R'i(x,t) + L'\frac{\partial i(x,t)}{\partial t} = 0 \tag{10.1}$$

$$\frac{\partial i(x,t)}{\partial x} + G'v(x,t) + C'\frac{\partial v(x,t)}{\partial t} = 0 \tag{10.2}$$

where $v(x, t)$ and $i(x, t)$ are the transient voltage and current, respectively, and R', L', G' and C' are the per unit length resistance, inductance, conductance, and capacitance, respectively.

The frequency domain counterpart of expressions (10.1) and (10.2) is [1,11]

$$\frac{\mathrm{d}V(x)}{\mathrm{d}x} + Z'I(x) = 0 \tag{10.3}$$

$$\frac{\mathrm{d}I(x)}{\mathrm{d}x} + Y'V(x) = 0 \tag{10.4}$$

where $Z' = R' + j\omega L'$ and $Y' = G' + j\omega C'$ are the per unit length impedance and admittance, respectively.

The general solution of the set of eqns (10.3) and (10.4) is given in the form of the positive and negative traveling voltage and current waves on the line [11]

$$V(x) = ae^{-\gamma x} + be^{\gamma x} \tag{10.5}$$

$$I(x) = \frac{1}{Z_C}(ae^{-\gamma x} - be^{\gamma x}) \tag{10.6}$$

where $\gamma = \sqrt{Z'Y'}$ is the propagation constant, $Z_C = \sqrt{Z'/Y'}$ is the characteristic impedance of the line, and a and b are complex constants.

To analyze the electromagnetic field coupling to transmission lines using the Telegrapher's equations (10.1) and (10.2), the excitation in the form of electric field has to be included into the set of eqns (10.1) and (10.2), which now take the following form:

$$\frac{\partial v(x,t)}{\partial x} + R'i(x,t) + L'\frac{\partial i(x,t)}{\partial t} = E_x^{exc}(x,t) \tag{10.7}$$

$$\frac{\partial i(x,t)}{\partial x} + G'v(x,t) + C'\frac{\partial v(x,t)}{\partial t} = 0 \tag{10.8}$$

where$E_x^{exc}(x,t)$ is the electric field tangential to the transmission line.

In the frequency domain, the set of eqns (10.7) and (10.8) become

$$\frac{\mathrm{d}V(x)}{\mathrm{d}x} + Z'I(x) = E_x^{exc}(x,t) \tag{10.9}$$

$$\frac{\mathrm{d}I(x)}{\mathrm{d}x} + Y'V(x) = 0 \tag{10.10}$$

This method is commonly used for the assessment of transient responses of single or parallel conductors with or without load. In those cases where the assumptions and conditions on which it is based are satisfied, good results will be obtained.

However, the transmission line method has many limitations and restrictions, particularly, on the line geometry. In Ref. [11], many cases are analyzed using the time domain antenna theory approach (featuring the GB-IBEM) and the frequency domain antenna theory approach (featuring the MoM). Highly resonant transmission lines also pose severe difficulty to the conventional frequency domain MM approach based on the FFT. This will be illustrated in the examples considered. The wide variety of problem configurations and geometry that are solved successfully with the methodologies illustrate the versatility of these methods.

For the case of short circuited lossless conductors with no load, the use of the fast Fourier transform (FFT) technique on frequency domain data from transmission line theory will result in totally unpredictable outcome [11]. The problem is due to the fact that current grows to infinity at resonant points as there is no loss and radiation resistance to limit its flow. The wide variety of problem configurations and geometry that are solved successfully with the methodologies illustrate the versatility of these methods.

This chapter outlines the basic principles of the analysis of electromagnetic field coupling to overhead and buried transmission lines via TL method and antenna theory approach, respectively, in the frequency and time domain, respectively.

The numerical solution has been carried out via the Galerkin-Bubnov indirect boundary element method (GB-IBEM) presented in Chapters 8 and 9.

10.2 ANTENNA MODELS OF THE TRANSMISSION LINES

The wavelike behavior of the fields at these higher frequencies requires a more general approach in modeling of EMC problems, which is based on integral equations arising from wire antenna theory. The above-ground and buried cables, respectively, are subjected to electromagnetic fields arriving from a distant source and inducing current to flow along the lines. The key to understand the behavior of induced fields is the knowledge of current distribution induced on the lines. The currents induced in victim equipment by the incident fields also generate scattered fields that propagate away from the equipment.

There are several possible approaches for determining the current distribution on a receiving wire antenna structure. The simplest procedure is to use an approximation to current distribution. However, these types of approximations are not always sufficient and a more accurate approach related to the solution of Maxwell's equations for the receiving antenna problem yielding an integral equation for current is to be used.

Essentially, there are two types of propagation to be analyzed: the free space propagation of an incident field that excites the victim wire structure and the propagation of the induced voltage and current distributions along the wires. Induced transient response, that is, the induced currents and voltages along the victim equipment can be obtained by solving the corresponding antenna integral equations in the frequency and time domain using various numerical techniques. An efficient numerical solution of the corresponding integral equations in the frequency and time domain discussed in details in Chapters 8 and 9 are applied to the analysis of above-ground and below-ground transmission lines in Chapters 11 and 12, respectively. Some general remarks regarding the antenna models of transmission lines are presented in Sections 10.2.3. and 10.2.4.

10.2.1 Above-Ground Transmission Lines

The analysis of transmission lines using the antenna models can be carried out in either frequency [1,7–9] or time domain [11–14].

An extension of the standard TL approach to the combined electromagnetic field to transmission line coupling equations valid for overhead wires has been presented in Refs. [2, 3]. The approach presented in Refs. [2, 3] is based on the perfectly conducting (PEC) ground approximation. The improvement of this important contribution can be done in two directions:

- to include the effect of a lossy ground in Tkatchenko et al. corrected form of telegrapher's equations;
- to provide a clear relationship between TL equations and antenna theory.

The above represent the objectives of the study available in Ref. [8], in which a generalized type of the Telegrapher's equations including the presence of a lossy ground is presented. The derivations in Ref. [8] provide a clear correlation between the rigorous antenna theory approach to the analysis of finite length lines based on the standard Pocklington integral equation formulation and the classic Telegrapher's equation approach arising from the transmission line theory.

In Ref. [8], the standard set of Telegrapher's equations is deduced from the generalized Telegrapher's equations. The influence of a lossy ground is taken into account via the Fresnel reflection coefficient appearing within the Green's function. The corresponding correlation between the antenna theory approach and TL approximation in the time domain is anticipated in the future work.

10.2.2 Below-Ground Transmission Lines

The electromagnetic field coupling to wires buried in a lossy medium is of great practical interest for many EMC applications [17–21], such as transient analysis of power and communications cables. Basically, the buried wire can represent a telephone cable, power cable, or a cylindrical antenna operating at a very low frequency (VLF). Some important applications are also related to submarine communication (long dipoles submerged in water), geophysical probing, and electromagnetic stimulation of biological tissue. In the past, the transient excitation of buried wires, as one of the major causes of malfunction of telecommunication and power lines, has been mostly related to the lightning discharge problems [1].

Generally, the electromagnetic field coupling to underground wires configurations [17,18] has been investigated to a lesser extent than coupling to above-ground lines.

The studies dealing with buried wires are usually based on an approximate transmission line (TL) approach [1]. The TL approach can be considered as a compromise between a quasistatic approximation and wire antenna model and it is mostly related to infinite or at least very long buried wires.

However, the effects at the line ends cannot be taken into account utilizing the TL approach [6]. Also, the effect of the earth-air interface has usually been neglected featuring the assumption that the wire is buried at a very large depth [17].

This approach, though sufficient approximation if long lines with electrically-small cross sections are considered, fails if one deals with the lines of finite length and high frequency excitations.

Namely, the TL model fails to predict resonances, fails to take into account properly the presence of a lossy ground, and the effects at the line ends also cannot be taken into account utilizing this approach [1, 9,11]. Consequently, when the lines of the finite length are of interest, the receiving antenna theory has to be used.

Thus, the wavelike behavior of the induced responses at higher frequencies requires a more general approach that is based on integral equations arising from wire antenna theory. On the contrary, the principal restriction of the wire antenna model applied to overhead lines is often related to the long computational time required for the calculations pertaining to the long lines.

The transient analysis of buried cables using the wire antenna theory can be carried out in either frequency or time domain. Essentially, there is no definitive advantage that could be gained using the indirect frequency domain approach or the direct time domain solution method.

Generally, if the solution for a large number of incident waves arriving from various directions is of interest then the frequency domain approach is appropriate.

On the contrary, for some EMC applications in which one requires accurate frequency data, the frequency samples obtained by the use of Fourier transform do not ensure accurate results. The time domain approach would be a better choice if the transient response is required only for the early time behavior, since the frequency domain approach requires computing of the frequency response up to the maximum effective frequency, and the entire range of frequency spectrum is to be transformed.

10.3 REFERENCES

[1] F. Tesche, M. Ianoz, F. Carlsson: EMC Analysis Methods and Computational Models, *John Wiley & Sons, New York*, 1997.

[2] S. Tkatchenko, F. Rachidi. M. Ianoz: Electromagnetic field coupling to a line of a finite length: theory and fast iterative solutions in frequency and time domains, *IEEE Trans, EMC, 37 (4)*, 1995, pp 509–518.

[3] S. Tkatchenko, F. Rachidi, M. Ianoz: High frequency electromagnetic field coupling to long terminated lines, *IEEE Trans. EMC, 43 (2)*, 2001, pp 117–129.

[4] M. Ianoz: Electromagnetic field coupling to lines, cables and networks, a review of problems and solutions, International Proceedings on Electromagnetics in Advanced Applications, ICEAA'95, *Turin, Italy*, September 12–15, 1999, pp 75–80.

[5] F. Arreghini, M. Ianoz, F. Rachidi: Frequency and time domain applications in EMP coupling; a comparison between different methods of calculations, electromagnetic field coupling to lines, cables and networks, a review of problems and solutions, International Proceedings on Electromagnetics in Aerospace Applications, ICEAA'91, *Turin, Italy*, September 17–20, 1991, pp 209–214.

[6] P. Degauque, A. Zeddam: Remarks on the transmission approach to determining the current induced on above-ground cables, *IEEE Trans. EMC, 30 (1)*, 1998, pp 77–80.

[7] D. Poljak: New numerical approach in analysis of thin wire radiating over lossy half-space, *Int. J. Numerical Methods Eng. 38 (22)*, 1995, pp 3803–3816.

[8] D. Poljak, F. Rachidi, S. Tkachenko: Generalized form of telegrapher's equations for the electromagnetic field coupling to finite length lines above a lossy ground, submitted to *IEEE Trans. EMC*.

[9] D. Poljak: Electromagnetic Modelling of Wire Antenna Structures, *WIT Press, Southampton, Boston*, 2001.

[10] A. K. Agrawal, H. J. Price, S. H. Gurbaxani: Transient response of a multiconductor transmission line excited by a nonuniform electromagnetic field, *IEEE Trans. EMC, 22 (2)*, May 1980, pp 119–129.

[11] D. Poljak, C. Y. Tham: Integral equation techniques in electromagnetics, *WIT Press, Southampton, Boston*, 2003.

[12] D. Poljak, V. Roje: Time domain modeling of electromagnetic field coupling to transmission lines, Proceedings 1988 IEEE EMC Symposium, *Denver, USA*, August 1998, pp 1010–1013.

[13] D. Poljak, C. Y. Tham, A. McCowen, V. Roje: Electromagnetic Pulse Excitation of Multiconductor Transmission Lines, ICEAA '99, *Turin, Italy*, September 13–17, 1999, pp 789–792.

[14] D. Poljak, E. K. Miller, C. Y. Tham, S. Antonijevic, V. Doric: Time domain analysis of the energy stored in the near field of multiple straight-wires above dielectric half-space, Proceedings of the ICEAA, *Turin*, September 2005.

[15] C. Y. Tham, A. McCowen, M. S. Towers, D. Poljak: Dynamic adaptive sampling technique in frequency-domain transient analysis, IEEE Trans. EMC, 44 (4), November 2002, pp 522–528.

[16] S. Tkatchenko, F. Rachidi, L. M. Ianoz: An asymptotic approach for the calculation of electromagnetic field coupling to long terminated lines, International Symposium on Electromagnetic Compatibility, EMC'98, *Roma*, September 14–18, 1998, pp 605–609.

[17] G. E. Bridges: Transient plane wave coupling to bare and insulated cables buried in a lossy half-space, *IEEE Trans. EMC, 37 (1)*, February 1995, pp 62–70.

[18] L. D. Grcev, F. E. Menter: Transient electromagnetic fields near large earthing systems, *IEEE Trans. Magn., 32*, May 1996, pp 1525–1528.

[19] D. Poljak, Electromagnetic modeling of finite length wires buried in a lossy half-space, Engineering Analysis with Boundary Elements, Vol. 26, January 2002, pp 81–86.

[20] V. Dorić, D. Poljak, V. Roje: Transient plane wave coupling to a finite length wire buried in a conductive ground, Boundary Elements XVII, 2005, pp 609–617.

[21] D. Poljak, V. Doric: Time domain modeling of electromagnetic field coupling to finite length wires embedded in a dielectric half-space, *IEEE Trans. EMC, 47 (2)*, May 2005, pp 247–253.

11

ELECTROMAGNETIC FIELD COUPLING TO OVERHEAD LINES: FREQUENCY DOMAIN AND TIME DOMAIN ANALYSIS

The analysis of finite length transmission lines carried out in this chapter is fully based on the thin wire antenna models presented in Chapters 9 and 10, respectively. Frequency domain approach uses the Pocklington integro-differential equation types, whereas the time domain formulation is based on the Hallen integral equation types. In both cases, the effect of a two-media configuration is taken into account by means of the reflection coefficient approximation. The corresponding equations, arising from antenna models, are handled numerically via the Galerkin–Bubnov indirect boundary element method (GB-IBEM). Numerical results are presented for frequency and transient response of considered transmission lines and somewhere compared to transmission line approximation. It is shown that the TL approximation can result in a significant underestimation of the induced currents.

11.1 FREQUENCY DOMAIN ANALYSIS: DERIVATION OF GENERALIZED TELEGRAPHER'S EQUATIONS

This section deals with a derivation of the generalized form of Telegrapher's equations for electromagnetic field coupling to finite length transmission lines.

The transmission line of a finite length L and radius a located at height h above an imperfect ground and illuminated by an incident electromagnetic field, shown in Figure 11.1, is considered.

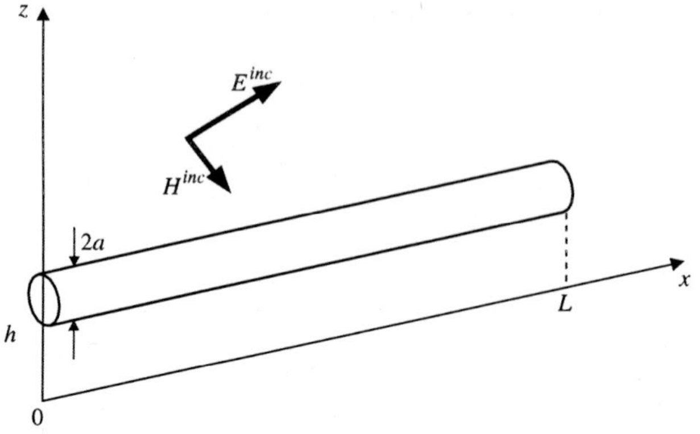

Figure 11.1 Finite length single wire transmission line above a lossy ground.

The line is subjected to electromagnetic fields arriving from a distant source and inducing current to flow along the line. The key to understanding the behavior of induced fields is the knowledge of current distribution induced on the line. This current also generates scattered fields that not only propagate away from the equipment but also generate the so-called scattered voltage along the line [1,2]. The governing equations for the current induced along the wire arising from the antenna model can be derived by enforcing the interface conditions for the tangential components of the electric field, as it is presented in Chapter 8.

The transmission line equations [1] can be deduced from thin wire antenna relationships, which is derived in this section.

The current distribution induced along the line is governed by the Pocklington integro-differential equation:

$$E_x^{exc} = j\omega \frac{\mu}{4\pi} \int_0^L I(x') g(x,x') \mathrm{d}x' - \\ - \frac{1}{j4\pi\omega\varepsilon} \frac{\partial}{\partial x} \int_0^L \frac{\partial I(x')}{\partial x'} g(x,x') \mathrm{d}x' \tag{11.1}$$

where $I(x')$ is the equivalent current distribution induced along the line, $\vec{E}^{exc}$ is the excitation field component representing the sum of the incident field $\vec{E}^{inc}$ and field reflected from the lossy ground $\vec{E}^{ref}$:

$$\vec{E}^{exc} = \vec{E}^{inc} + \vec{E}^{ref} \tag{11.2}$$

while $g(x,x')$ stands for the Green's function given by:

$$g(x,x') = g_0(x,x') - R_{\mathrm{TM}} g_i(x,x') \tag{11.3}$$

where $g_0(x,x')$ is the free space-Green function:

$$g_0(x,x') = \frac{e^{-jk_0 R_0}}{R_0} \tag{11.4}$$

and $g_i(x,x')$ arising from the image theory is given by:

$$g_i(x,x') = \frac{e^{-jk_0 R_i}}{R_i} \tag{11.5}$$

R_0 and R_i denote the corresponding distance from the source to the observation point, respectively.

R_{TM} is the reflection coefficient for the transverse polarization:

$$R_{\mathrm{TM}} = \frac{n\cos\Theta - \sqrt{n - \sin^2\Theta}}{n\cos\Theta + \sqrt{n - \sin^2\Theta}} \tag{11.6}$$

which accounts for the presence of a lossy half-space and the refraction index n is given by:

$$n = \varepsilon_r - j\frac{\sigma}{\omega\varepsilon_0} \tag{11.7}$$

where argument θ is defined as:

$$\Theta = \operatorname{arctg}\frac{|x - x'|}{2h} \tag{11.8}$$

To derive the telegrapher's type equations for the currents induced along the line, the so-called scattered voltage concept has to be included in the formulation [1,3].

The scattered voltage along the line above a lossy ground is defined as an integral of a scattered vertical electric field component from the point in the remote soil to the line surface:

$$V^{sct}(x) = -\int_{-\infty}^{h} E_z^{sct}(x,z)\,\mathrm{d}z \tag{11.9}$$

The vertical component of the electric field can be expressed in terms of the scalar potential gradient, that is:

$$E_z^{sct} = -\frac{\partial\varphi}{\partial z} \tag{11.10}$$

and the scattered voltage along the line can be written as follows:

$$V^{sct}(x) = \int_{-\infty}^{h} \frac{\partial \varphi}{\partial z} \mathrm{d}z = \frac{\mathrm{d}}{\mathrm{d}z} \int_{-\infty}^{h} \varphi(x, z) \mathrm{d}z \tag{11.11}$$

Performing the integration from the infinite soil to the electrode surface simply gives:

$$V^{sct}(x) = \varphi(x, z)\big|_{z=-\infty}^{z=h} \tag{11.12}$$

Assuming the scalar potential in the remote soil to be zero yields:

$$V^{sct}(x) = \varphi(x, h) \tag{11.13}$$

where the scalar potential is defined as:

$$\varphi(x) = \frac{1}{4\pi\varepsilon} \int_{0}^{L} q(x') g(x, x') \mathrm{d}x' \tag{11.14}$$

where $q(x)$ is the linear charge distribution along the line.

The linear charge density and the current distribution along the line are related through the equation of continuity [3,4]:

$$q = -\frac{1}{j\omega} \frac{\mathrm{d}I}{\mathrm{d}x} \tag{11.15}$$

Substituting eqn (11.15) into eqn (11.14) yields:

$$\varphi(x) = -\frac{1}{j4\pi\omega\varepsilon} \int_{0}^{L} \frac{\partial I(x')}{\partial x'} g(x, x') \mathrm{d}x' \tag{11.16}$$

Combining eqns (11.16) and (11.13), the final relationship for the scattered voltage along the transmission line of finite length above a lossy ground is obtained:

$$V^{sct}(x) = -\frac{1}{j4\pi\omega\varepsilon} \int_{0}^{L} \frac{\partial I(x')}{\partial x'} g(x, x') \mathrm{d}x' \tag{11.17}$$

Using expression (11.17), eqn (11.1) can be written in the form:

$$E_x^{exc} = j\omega \frac{\mu}{4\pi} \int_{0}^{L} I(x') g(x, x') \mathrm{d}x' + \frac{\partial V^{sct}(x)}{\partial x} \tag{11.18}$$

By rearranging eqns (11.17) and (11.18) the general form of the telegrapher's equations, valid for finite length lines, (antennas) is obtained:

$$\frac{\mathrm{d}V^{sct}(x)}{\mathrm{d}x} + j\omega\frac{\mu}{4\pi}\int_0^L I(x')g(x,x')\mathrm{d}x' = E_x^{exc} \tag{11.19}$$

$$\int_0^L \frac{\partial I(x')}{\partial x'} g(x,x')\mathrm{d}x' + j4\pi\omega\varepsilon V^{sct}(x) = 0 \tag{11.20}$$

The set of eqns (11.19–11.20) can also be written in the following form:

$$\frac{\mathrm{d}V^{sct}(x)}{\mathrm{d}x} + j\omega L''\int_0^L I(x')g(x,x')\mathrm{d}x' = E_x^{exc} \tag{11.21}$$

$$\int_0^L \frac{\partial I(x')}{\partial x'} g(x,x')\mathrm{d}x' + j\omega C''V^{sct}(x) = 0 \tag{11.22}$$

where the corresponding equivalent inductance and capacitance per unit length of the line are given by relations:

$$L'' = \frac{\mu}{4\pi} \tag{11.23}$$

$$C'' = 4\pi\varepsilon \tag{11.24}$$

The correlation of the generalized telegrapher's equation (11.21) and (11.22) with the classic telegrapher's equation [1] can be performed in a rather straightforward manner.

Assuming the case of perfectly conducting (PEC) ground with the following conditions satisfied:

$$\begin{aligned} k\cdot h &\ll 1 \\ L &\gg 2h \end{aligned} \tag{11.25}$$

and by taking into account the fact that the effective length for the integration along the line is approximately $2h$ [4] the quantities $I(x')$ and $\partial I(x')/\partial x'$ can be removed from the integral operator and the characteristic integral term over Green function is:

$$\int_0^L g(x,x')dx' = 2\ln\frac{2h}{a} \tag{11.26}$$

which simply gives the following equations:

$$V^{sct}(x) + \frac{1}{j4\pi\omega\varepsilon} \cdot 2\ln\frac{2h}{a} \cdot \frac{\partial I(x)}{\partial x} = 0 \tag{11.27}$$

$$E_x^{exc}(x) = j\omega\frac{\mu}{4\pi} \cdot 2\ln\frac{2h}{a} \cdot I(x) + \frac{\partial V^{sct}(x)}{\partial x} \tag{11.28}$$

Equations (11.27) and (11.28) can be rewritten in the standard form of the Agrawal et al. [5] field-to-transmission line coupling equations:

$$\frac{\mathrm{d}V^{sct}(x)}{\mathrm{d}x} + j\omega L' I(x) = E_x^{exc}(x) \tag{11.29}$$

$$\frac{\mathrm{d}I(x)}{\mathrm{d}x} + j\omega C' V^{sct}(x) = 0 \tag{11.30}$$

where the corresponding equivalent inductance and capacitance per unit length of the above-ground line are given by:

$$L' = \frac{\mu}{4\pi}\ln\frac{2h}{a} \tag{11.31}$$

$$C' = \frac{2\pi\varepsilon}{\ln\frac{2h}{a}} \tag{11.32}$$

The induced current distribution along the line are obtained by solving eqn (11.1) via the Galerkin Bubnov indirect boundary element method (GB-IBEM) as it is presented in Chapter 8.

Once the current distribution is determined, the scattered voltage defined by relation (11.17) can be readily computed using the boundary element formalism. Assuming a linear approximation over a boundary element the shape functions are given by:

$$f_i = \frac{x_{i+1} - x'}{\Delta x} \qquad f_{i+1} = \frac{x' - x_i}{\Delta x} \tag{11.33}$$

and the differentiation of the current distribution variation along the segment:

$$\frac{\partial I(x')}{\partial x'} = I_i \frac{\partial f_i}{\mathrm{d}x'} + I_{i+1}\frac{\partial f_{i+1}}{\mathrm{d}x'} \tag{11.34}$$

is simply:

$$\frac{\partial I(x')}{\partial x'} = \frac{I_{i+1} - I_i}{\Delta x} \tag{11.35}$$

The scattered voltage then can be written as:

$$V^{sct}(x) = -\frac{1}{j4\pi\omega\varepsilon}\sum_{i=1}^{M}\int_{\Delta l_i}\{D\}^T g(x,x')\mathrm{d}x' \cdot I_i \tag{11.36}$$

and furthermore, it follows:

$$V^{sct}(x) = -\frac{1}{j4\pi\omega\varepsilon}\sum_{i=1}^{M}\int_{x_i}^{x_{i+1}}\frac{I_{i+1}-I_i}{\Delta x}g(x,x')\mathrm{d}x' \tag{11.37}$$

where M is total number of elements along the entire line.

Therefore, the analysis of a single wire transmission line can be carried out approximately, by using Telegrapher's equations (11.29) and (11.30), or rigorously using the Pocklington integro-differential equation (11.1), that is, generalized Telegrapher's equations (11.21) and (11.22).

The accurate analysis of multiple transmission lines of finite length can be carried out by solving the set of coupled Pocklington integro-differential equations as it is presented in Chapter 8.

11.2 FREQUENCY DOMAIN COMPUTATIONAL RESULTS

This section presents some illustrative computational examples related to the single and multiple wire transmission lines, respectively.

11.2.1 Single Wire Above an Imperfect Ground

The excitation field at the conductor surface is the time-harmonic plane wave. The transmission line of interest is characterized by a variable length, with conductor radius $a = 1\,\mathrm{cm}$, whereas the height above ground is $h = 2.5\,\mathrm{m}$. The excitation field at the conductor surface is the time-harmonic plane wave for the case of the normal incidence:

$$E^{exc} = E_0(1 - R_{\mathrm{TM}}e^{-j2kh}) \tag{11.38}$$

The relative dielectric constant of the ground is $\varepsilon_r = 10$ and the ground conductivity is $\sigma = 0.01$ s/m. The magnitude of the plane wave is $E_0 = 1$ V/m and the frequency is $f = 50$ MHz.

Figures 11.2–11.7 present the real and imaginary part of the induced current along the line, for different line lengths, that is, $L = 5\,\mathrm{m}$, $L = 10$ m, and $L = 20$ m obtained by applying the transmission line model (TLM) (for a PEC ground and a lossy ground) and antenna theory. The calculations have been performed using the derived generalized telegrapher's equations (11.21) and (11.22), the classical transmission line theory for a perfectly conducting ground (PEC), and the classical transmission line theory for a lossy ground [1].

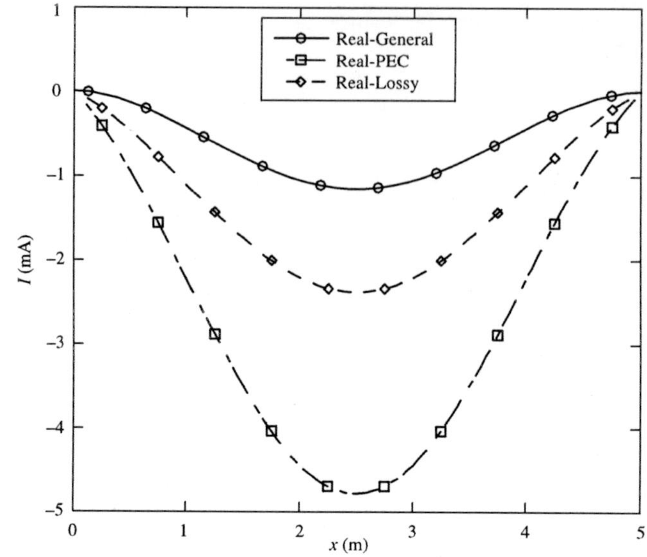

Figure 11.2 Real part of the induced current along the line $L = 5$ m. (D. Poljak, F. Rachidi, S. Tkachenko, Generalized Form of Telegrapher's Equations for the Electromagnetic Field Coupling to Finite Length Lines above a Lossy Ground, submitted to IEEE Trans. EMC).

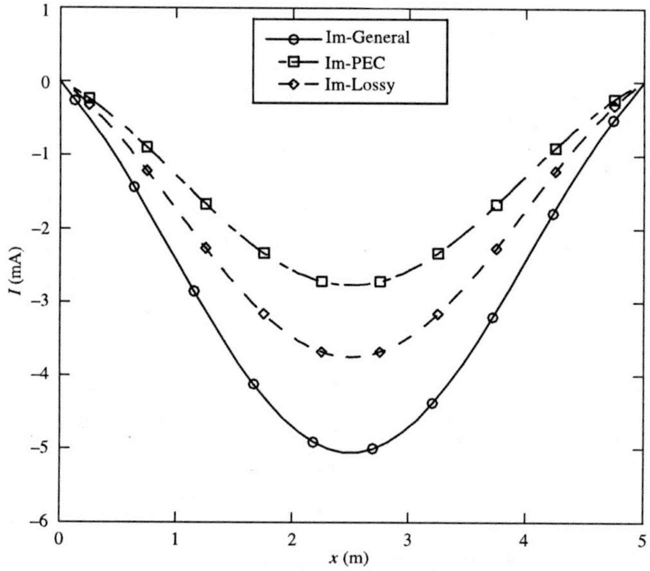

Figure 11.3 Imaginary part of the induced current along the line $L = 5$ m. (D. Poljak, F. Rachidi, S. Tkachenko, Generalized Form of Telegrapher's Equations for the Electromagnetic Field Coupling to Finite Length Lines above a Lossy Ground, submitted to IEEE Trans. EMC).

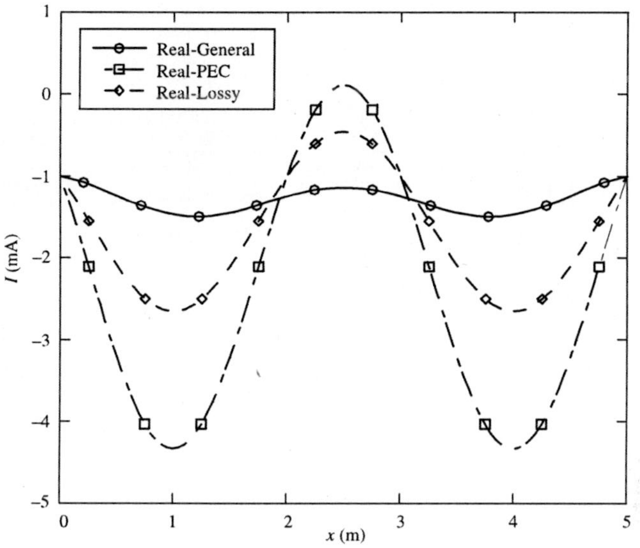

Figure 11.4 Real part of the induced current along the line $L = 10\,\text{m}$. (D. Poljak, F. Rachidi, S. Tkachenko, Generalized Form of Telegrapher's Equations for the Electromagnetic Field Coupling to Finite Length Lines above a Lossy Ground, submitted to IEEE Trans. EMC).

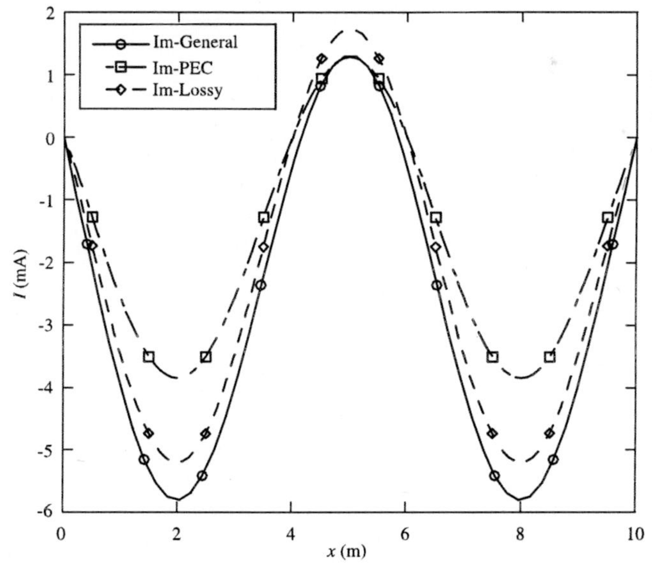

Figure 11.5 Imaginary part of the induced current along the line $L = 10$ m. (D. Poljak, F. Rachidi, S. Tkachenko, Generalized Form of Telegrapher's Equations for the Electromagnetic Field Coupling to Finite Length Lines above a Lossy Ground, submitted to IEEE Trans. EMC).

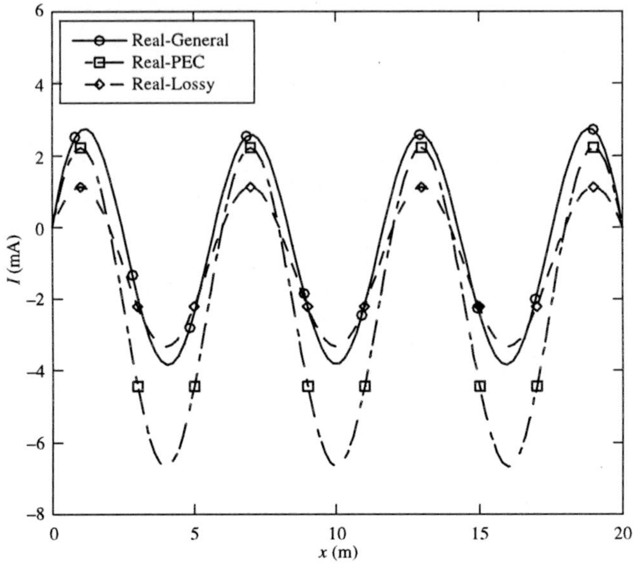

Figure 11.6 Real part of the induced current along the line $L = 20$ m. (D. Poljak, F. Rachidi, S. Tkachenko, Generalized Form of Telegrapher's Equations for the Electromagnetic Field Coupling to Finite Length Lines above a Lossy Ground, submitted to IEEE Trans. EMC).

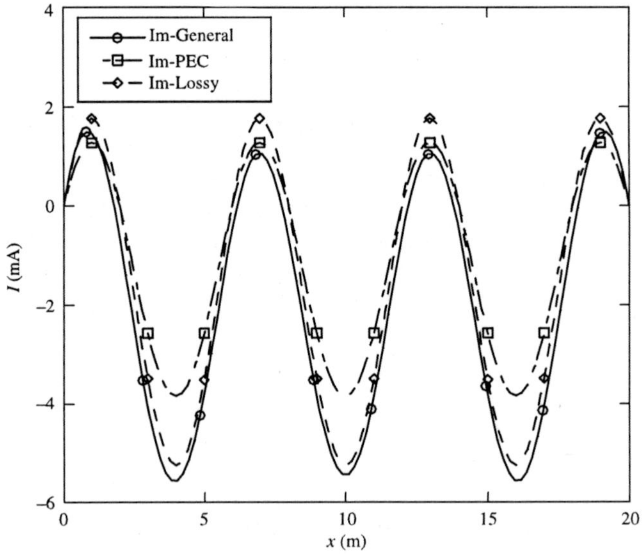

Figure 11.7 Imaginary part of the induced current along the line $L = 20$ m. (D. Poljak, F. Rachidi, S. Tkachenko, Generalized Form of Telegrapher's Equations for the Electromagnetic Field Coupling to Finite Length Lines above a Lossy Ground, submitted to IEEE Trans. EMC).

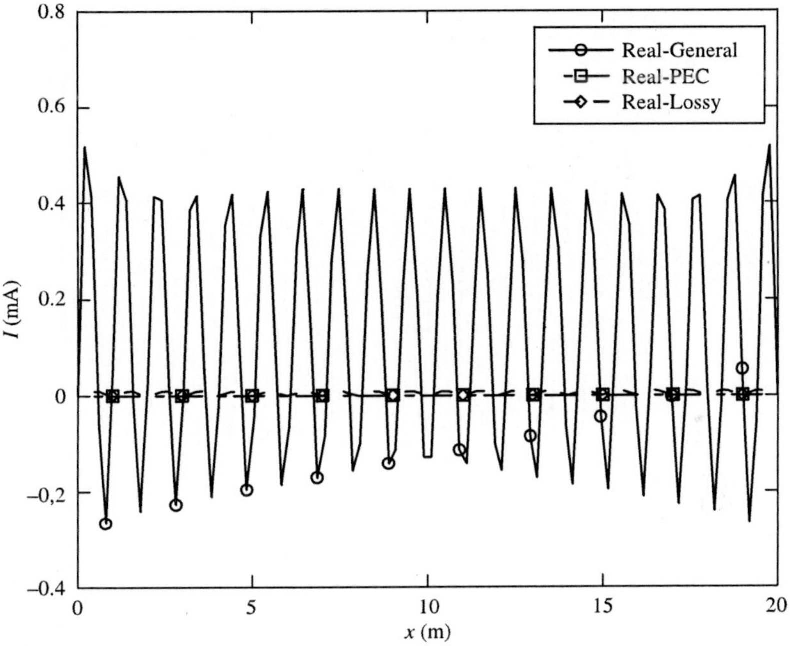

Figure 11.8 Real part of the induced current along the line $L = 20\,\text{m}$, $f = 300$ MHz. (D. Poljak, F. Rachidi, S. Tkachenko, Generalized Form of Telegrapher's Equations for the Electromagnetic Field Coupling to Finite Length Lines above a Lossy Ground, submitted to IEEE Trans. EMC).

Figures 11.2 to 11.5 clearly demonstrate that for relatively short line lengths ($L/h < 4$) the transmission line approximation fails to compute the current distribution along the line, accurately. The results obtained using the TL approximation become acceptable for longer lines ($L/h = 8$), as it is shown in Figures 11.6 and 11.7. It is worth noting that the simulations presented in Figures. 11.2–11.7 correspond to a 6-m wavelength incident field, about 2.5 times the height of the line.

Figures 11.8 and 11.9 present similar results for a 20-m long line, but for a frequency of the incident field equal to 300 MHz. This frequency is considered to be well beyond the validity of the TL approximation (corresponding to a wavelength of $h/2.5$) and the obtained results show a clear discrepancy (up to one order of magnitude) between the TL approximation and the generalized solutions.

11.2.2 Multiple Wire Transmission Line Above an Imperfect Ground

This section deals with numerical results obtained for currents induced on a multiple wire transmission line, as shown in Figure 11.10.

The computational example deals with a two-wire transmission line of wires length $L = 30$ m, radius $a = 0.15$ cm, with wire separation $D = 0.5$ m, and a height above ground $h = 1$ m. The properties of an imperfect ground are: $\varepsilon_r = 10$, $\sigma = 0.001$ mho/m.

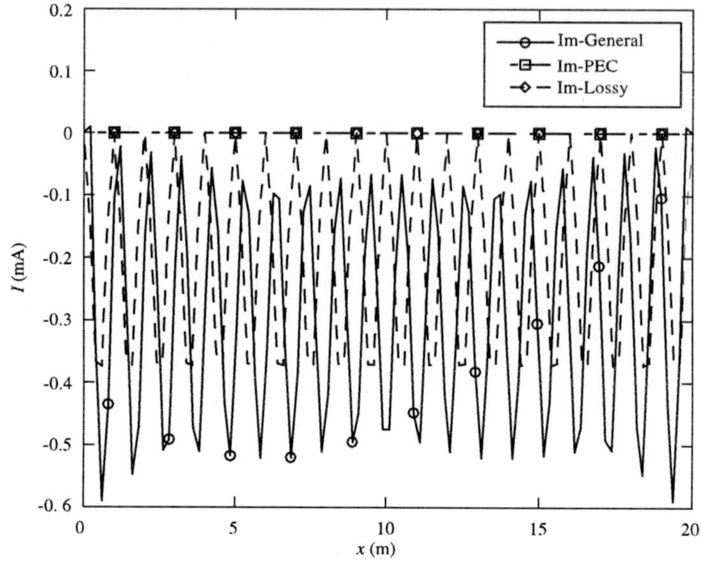

Figure 11.9 Imaginary part of the induced current along the line $L = 20\,\text{m}, f = 300\,\text{MHz}$. (D. Poljak, F. Rachidi, S. Tkachenko, Generalized Form of Telegrapher's Equations for the Electromagnetic Field Coupling to Finite Length Lines above a Lossy Ground, submitted to IEEE Trans. EMC).

The lines are excited by a plane wave:

$$\begin{aligned} E_x^{exc} &= E_x^i + E_x^r \\ &= E_0(\sin\alpha\sin\phi - \cos\alpha\cos\theta\cos\phi)e^{-jk_1\vec{n}_i\cdot\vec{r}} \\ &\quad + E_0(R_{\text{TE}}\sin\alpha\sin\phi \\ &\quad + R_{\text{TM}}\cos\alpha\cos\theta\cos\phi)e^{-jk_1\vec{n}_r\cdot\vec{r}} \end{aligned} \tag{11.39}$$

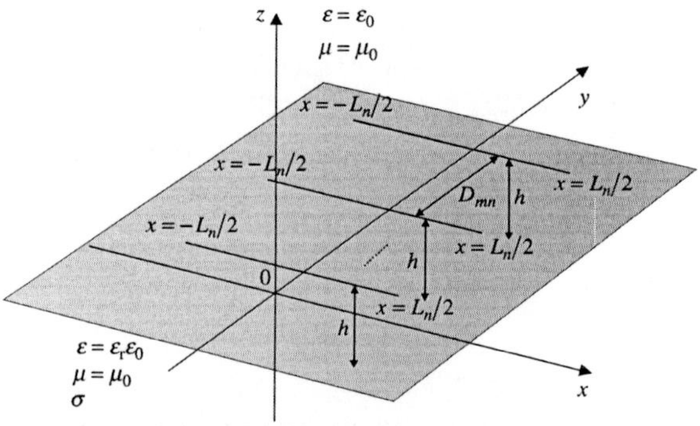

Figure 11.10 Multiple wire configuration.

where α is an angle between *E-field* vector and the plane of incidence, R_{TM} and R_{TE} are the vertical and horizontal Fresnel reflection coefficients at the air-earth interface given by:

$$R_{TM} = \frac{\underline{n}\cos\theta - \sqrt{\underline{n} - \sin^2\theta}}{\underline{n}\cos\theta + \sqrt{\underline{n} - \sin^2\theta}} \tag{11.40}$$

and

$$R_{TE} = \frac{\cos\theta - \sqrt{\underline{n} - \sin^2\theta}}{\cos\theta + \sqrt{\underline{n} - \sin^2\theta}} \tag{11.41}$$

$\vec{n}_i \cdot \vec{r}$ and $\vec{n}_r \cdot \vec{r}$ are distances from the origin point to the observation point for incident and reflected wave, respectively:

$$\begin{aligned} \vec{n}_i \cdot \vec{r} &= -x\sin\theta\cos\phi - y\sin\theta\sin\phi - z\cos\theta \\ \vec{n}_r \cdot \vec{r} &= -x\sin\theta\cos\phi - y\sin\theta\sin\phi + z\cos\theta \end{aligned} \tag{11.42}$$

The operating frequency is $f = 20\,\text{MHz}$, $E_0 = 1\,\text{V/m}$, and angle of incidence $\varphi = 180°$, $\theta = 30°$, $\alpha = 0°$. The results obtained for the real and imaginary part of the induced current distribution are shown in the Figure 11.11.

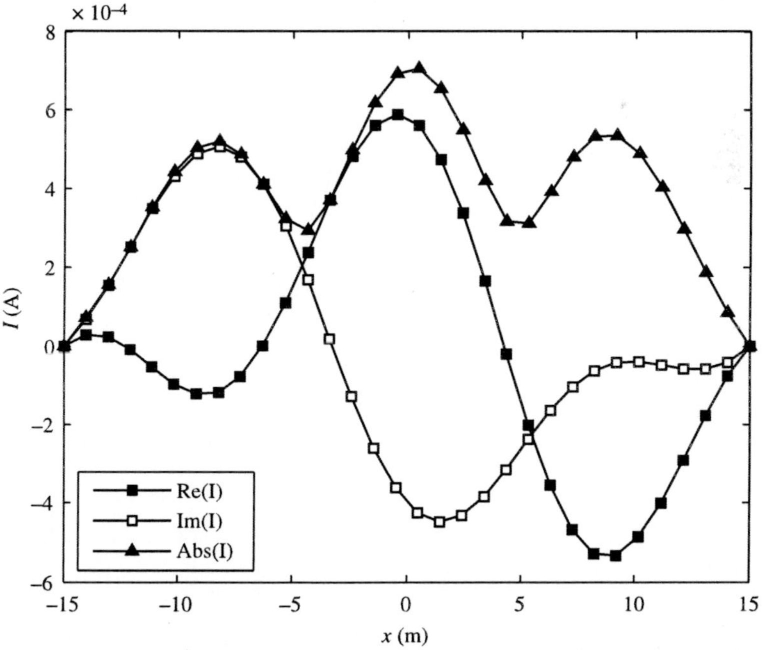

Figure 11.11 Current distribution along two coupled wires over imperfect ground with $f = 20\,\text{MHz}$, $\varepsilon_r = 10$, $\sigma = 0.001\text{mho/m}$, $h = 1\,\text{m}$, $D = 0.5\,\text{m}$, $L = 30\,\text{m}$, $a = 0.15\,\text{cm}$, $\varphi = 180°$, $\theta = 30°$, $\alpha = 0°$.

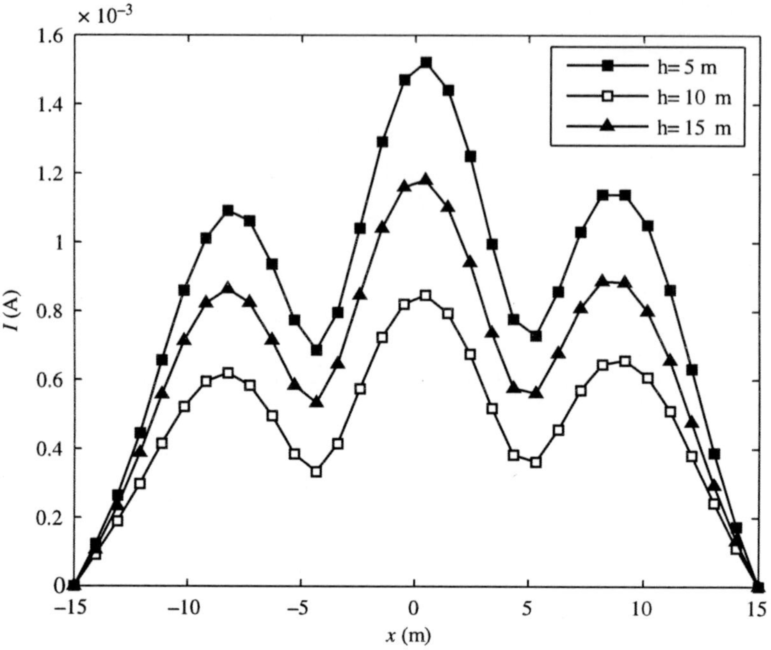

Figure 11.12 Current magnitudes for different values of the line height h.

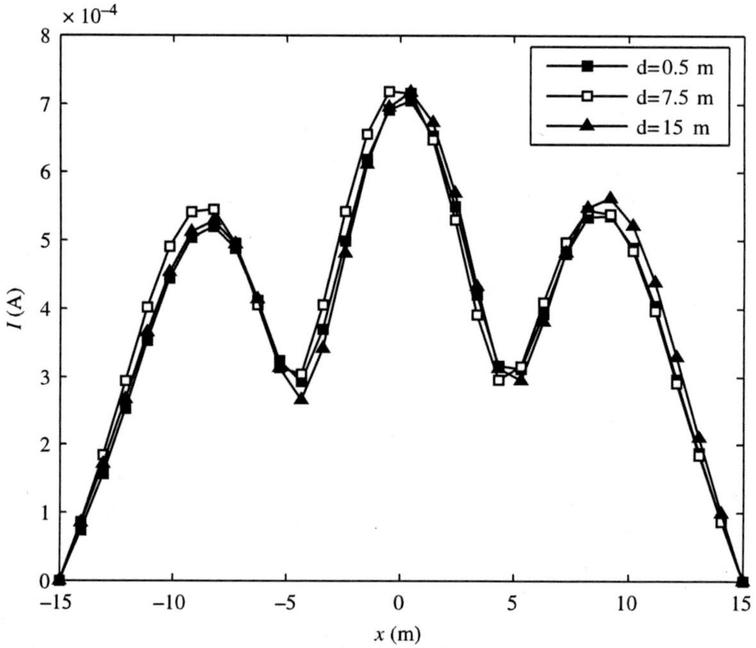

Figure 11.13 Current magnitudes for different values of the line height h.

Due to this specific angle of incidence, induced currents are the same on both wires. Figures 11.12 and 11.13 show the influence of variable height h above ground and separation D between the wires to the magnitudes of induced currents for the same wire configuration.

Figures 11.14 and 11.15 show the current distribution obtained for the same transmission line but with different angle of incidence $\varphi = 120°$ of the incident field.

11.3 TIME DOMAIN ANALYSIS

The time domain analysis of finite length transmission lines is carried out applying the receiving antenna theory (scattering theory) model, as presented in Chapter 9. The formulation is based on the corresponding Hallen integral equations for the case of a single wire transmission line, or on the set of Hallen integral equations for the case of multiple wire transmission line, respectively. The mathematical solution of these equations is carried out via time domain variant of the Galerkin Bubnov indirect boundary element method (GB-IBEM).

11.4 TIME DOMAIN COMPUTATIONAL EXAMPLES

This section deals with the computational examples related to single wire lines, two-wire and three wire lines, respectively. The cases of wires insulated in free

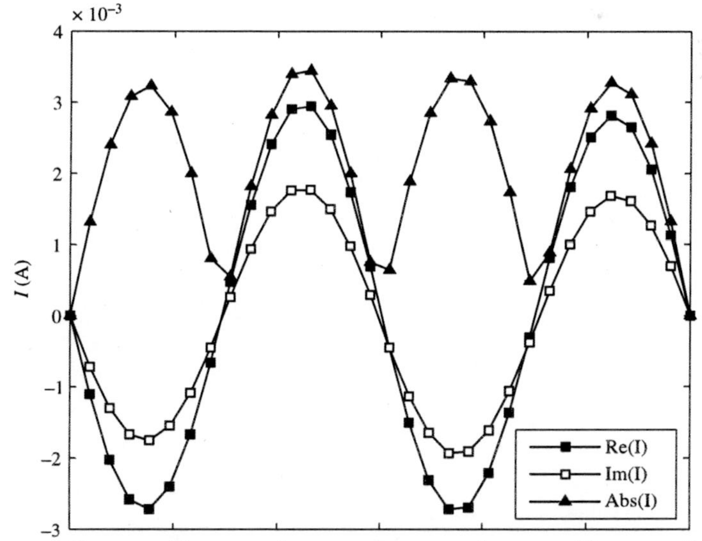

Figure 11.14 Current distribution along the first wire ($y = 0$) for angle of incidence $\Phi = 120°$, $\theta = 30°$, $\alpha = 0°$.

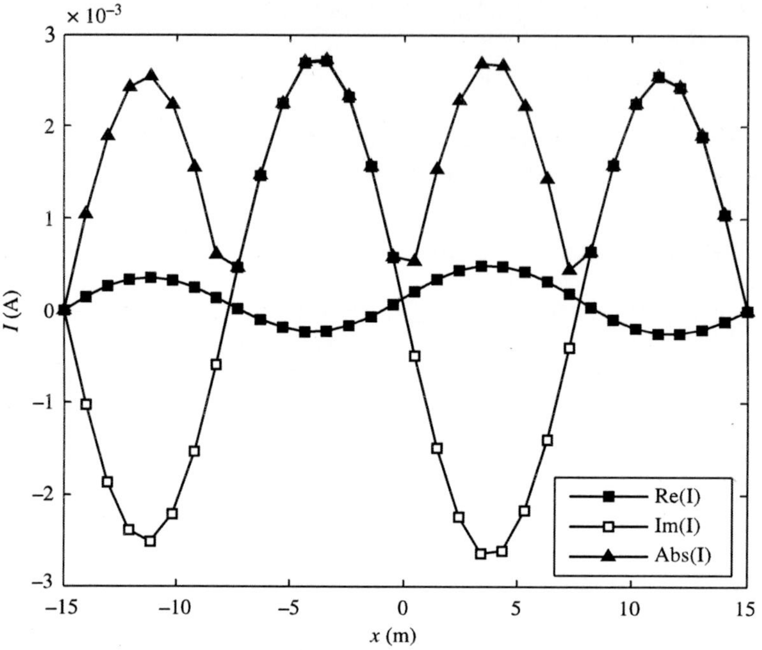

Figure 11.15 Current distribution along the second wire ($y = D$) for angle of incidence $\Phi = 120°$, $\theta = 30°$, $\alpha = 0°$.

space, above perfect ground and above a dielectric medium are considered. More illustrative numerical results on the subject can be found in Ref. [3].

11.4.1 Single Wire Transmission Line

First example in this subsection is related to the simplest transmission line—single wire in free space—Figure 11.16.

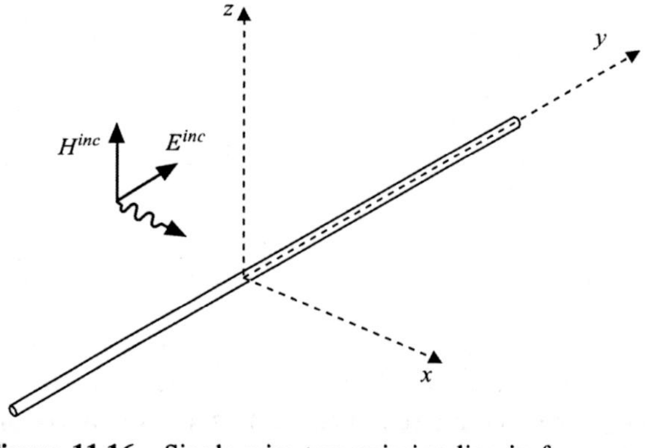

Figure 11.16 Single wire transmission line in free space.

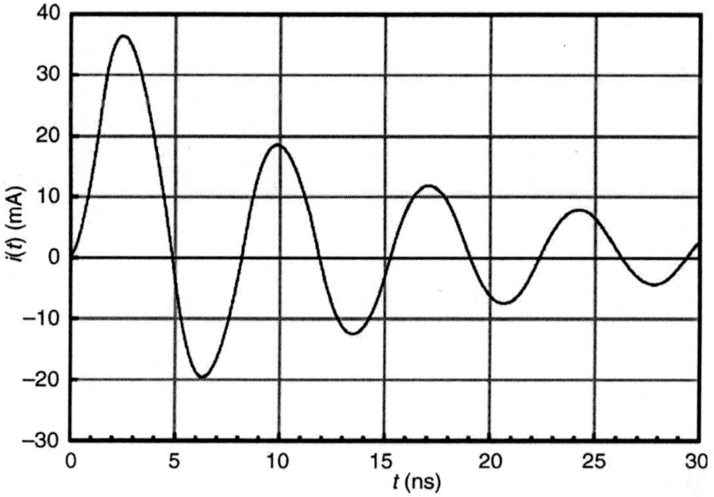

Figure 11.17 Transient current at the midpoint of the short line illuminated by the EMP plane wave.

Figure 11.17 shows the transient response at the midpoint of the very short line ($L = 1$ m, $a = 2$ mm) insulated in free space and illuminated by the broadside incident electromagnetic pulse (EMP) plane wave. EMP is often used to simulate the effect of lightning or nuclear explosion. The E-field of the wave is represented by the expression:

$$E^{inc}(t) = E_0(e^{-at} - e^{-bt}) \tag{11.43}$$

where $E_0 = 50.0\,\text{kV/m}$, $a = 4 \times 10^6\text{s}^{-1}$, $b = 4.76 \times 10^8\text{s}^{-1}$.

Figure 11.18 shows the results for the induced current at the center of the line of length $L = 130$ m and radius $a = 0.1$ m, insulated in free space and illuminated by a plane wave in the form of Gaussian pulse given by:

$$E^{inc}(t) = e^{g^2(t-t_0)^2} \tag{11.44}$$

with $g = 9.1 \times 10^7\text{s}^{-1}$ and $t_0 = 0.3334$ ns.

The results obtained by time domain scheme of GB-IBEM (or finite element integral equation method—FEIEM) seem to be in good agreement with results available from Ref. [6].

The transient response at the center of the wire of length $L = 10$ m and radius $a = 6.738$ cm. The wire is loaded at the center with concentrated resistive loading $R_L = 1\,\Omega$ as shown in Figure 11.19. The wire is insulated in free space and excited by the standard electromagnetic pulse (EMP) waveform (11.43) with $E_0 = 52.5$ V/m, $a = 4 \times 10^6\,\text{s}^{-1}$, $b = 4.76 \times 10^8\,\text{s}^{-1}$. The results calculated by the time domain FEIEM are compared with results published in Ref. [7]. Results obtained by different techniques are in good agreement.

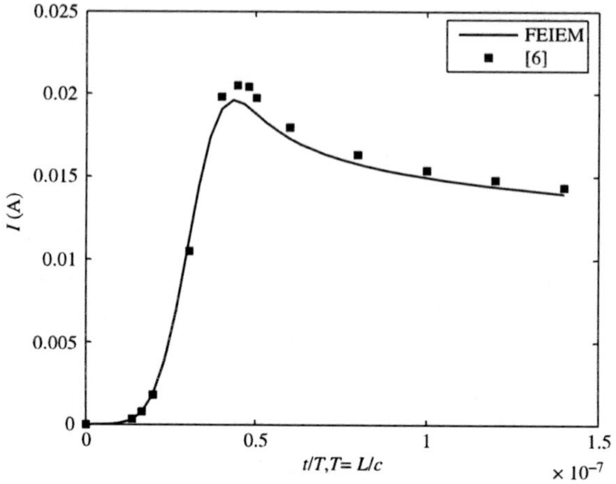

Figure 11.18 Induced current at the centre of unloaded line, $L = 130\,\text{m}$, $a = 0.1\,\text{m}$. (D. Poljak, V. Roje: "Time Domain Modeling of Electromagnetic Field Coupling to Transmission Lines", *Proc. 1988 IEEE EMC Symposium*, pp 1010–1013, Denver, USA, Aug. 1998).

Next set of figures is related to the single wire transmission line located over a perfect ground plane (Fig. 11.20). Figure 11.21 shows the transient response of a single wire line with length $L = 30\,\text{m}$ and radius $a = 1.5\,\text{mm}$ placed at height $h = 0.1\,\text{m}$ over a PEC ground. The line is illuminated by a broadside incident

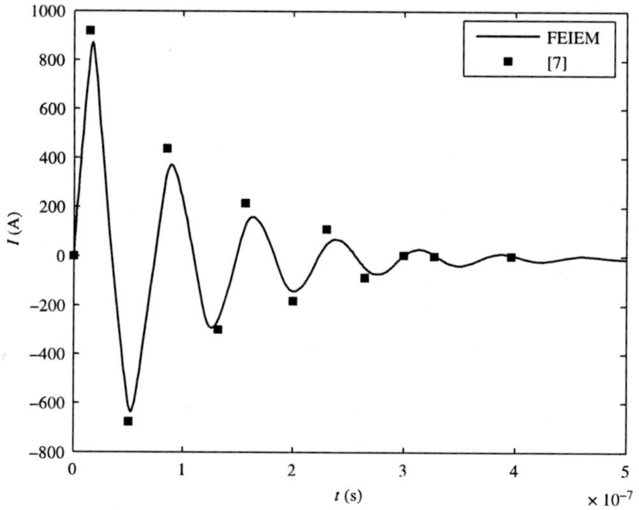

Figure 11.19 Induced current at the centre of the line, $L = 10\,\text{m}$, $a = 6.738\,\text{cm}$, $R_L = 1$ ohm. (D. Poljak, V. Roje: "Time Domain Modeling of Electromagnetic Field Coupling to Transmission Lines", *Proc. 1988 IEEE EMC Symposium*, pp 1010–1013, Denver, USA, Aug. 1998).

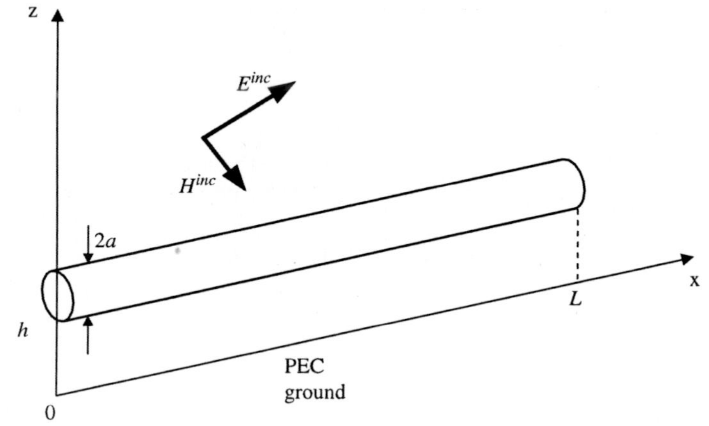

Figure 11.20 Single wire transmission line above a PEC ground.

EMP of the form given by eqn (11.43) with $E_0 = 1.05\,\text{V/m}$, $a = 4 \times 10^6\text{s}^{-1}$, $b = 4.76 \times 10^8\text{s}^{-1}$.

Next configuration of interest is a single wire of length $L = 20\,\text{m}$ and radius $a = 3\,\text{cm}$, placed above a perfectly conducting (PEC) ground at height $h = 5\,\text{m}$. The line is illuminated by the EMP with the following parameters: $E_0 = 65\,\text{kV/m}$, $a = 4 \times 10^7\text{s}^{-1}$, $b = 6 \times 10^8\text{s}^{-1}$.

Figure 11.22 shows the transient current induced at the point $x = 18\,\text{m}$ of the line. The results obtained via time domain GB-IBEM are compared with the results available from Ref. [8]. The results seem to agree satisfactorily.

Transient response of the 30 m long, loaded line, placed horizontally above a PEC ground plane at height $h = 0.1\,\text{m}$, illuminated by the EMP with the

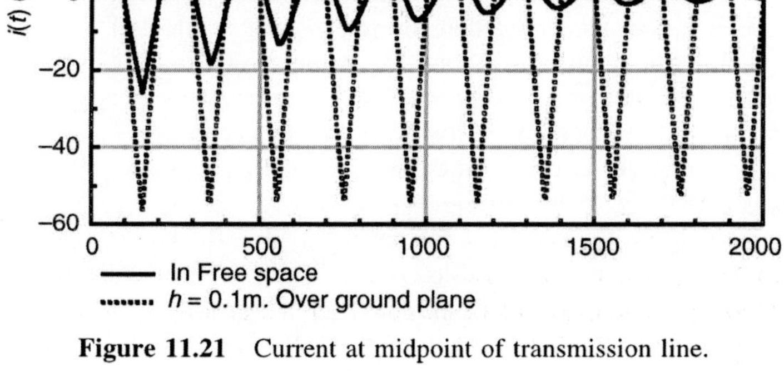

Figure 11.21 Current at midpoint of transmission line.

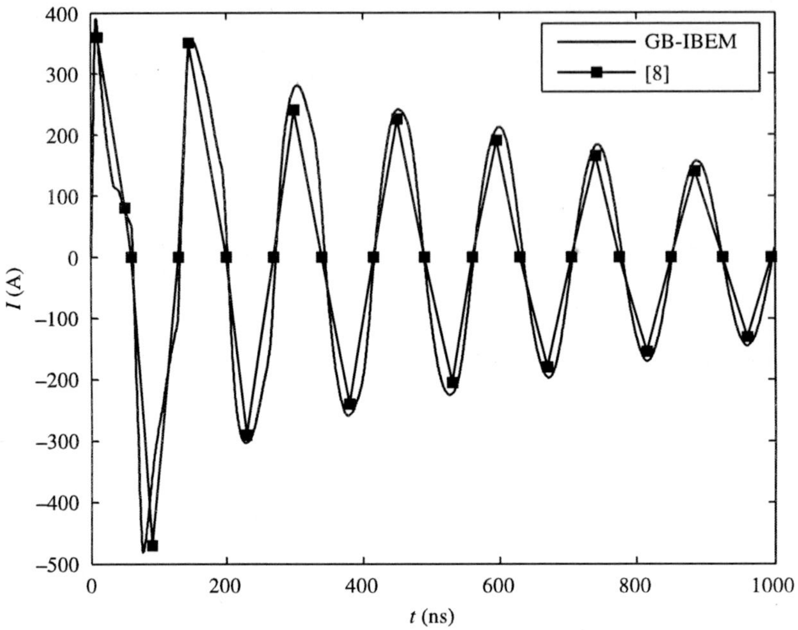

Figure 11.22 Transient current induced at the point $x = 18\,\text{m}$ of the wire of the configuration 2 compared with results published in Ref. [8].

parameters: $E_0 = 1.05\,\text{V/m}$, $a = 4 \times 10^6\text{s}^{-1}$, $b = 4.76 \times 10^8\text{s}^{-1}$, is shown in Figure 11.23. The line is loaded by a nonlinear resistive load at its midpoint. The transient load currents induced at the midpoint of the line for the cases of a 293 Ω linear resistor, a nonlinear load of $R_1/R_2 = 293/5000\,\Omega$, and when the load is removed, as shown in Figure 11.23.

The suppressing effect of the nonlinear loading is obvious.

Next example deals with a 30 m long line placed horizontally above a dielectric half-space and illuminated by the EMP excitation with $E_0 = 1.05\,\text{V/m}$, $a = 4 \times 10^6\text{s}^{-1}$, $b = 4.76 \times 10^8\text{s}^{-1}$. Figure 11.24 shows the transient load current induced at the center of the same line for the case of 293 Ω linear resistor, 5000 Ω linear load and for the case when the load is removed.

The corresponding total energy measure described in Chapter 9 is shown in Figure 11.25. It can be observed that the total energy collected by the wire from the incident EMP is reduced by about a factor of 3 for the 5000-Ω load, and declines nearly monotonically with the two smaller peaks separated by about 100 ns whereas for the two other load values the peaks are separated by about 200 ns. The latter effect is due to the 5000- Ω load essentially dividing a wire of length L into two wires of length $L/2$.

Figure 11.26 shows the transient current induced at the center of the horizontal line of length $L = 10\,\text{m}$ and radius $a = 6.74\,\text{cm}$ above dielectric half-space ($\varepsilon_r = 10$) at the height $h = 5\,\text{m}$ illuminated by a tangential electric field in the form of the EMP with $E_0 = 1\text{kV/m}$, $a = 4 \times 10^6\text{s}^{-1}$, $b = 4.76 \times 10^8\text{s}^{-1}$.

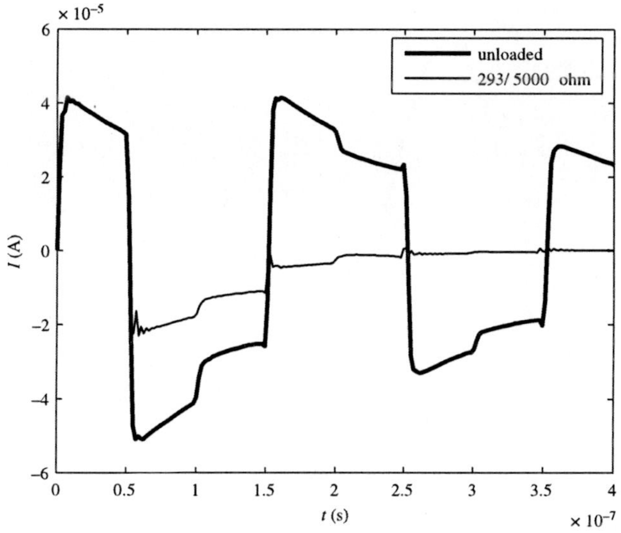

Figure 11.23 Transient response of a 30 m long line placed horizontally above a perfect ground at a height $h = 0.1\text{m}$ with resistive loadings at the midpoint ($R = 0$ and $R_1/R_2 = 293\,\Omega/5000\,\Omega$) illuminated by an EMP excitation.

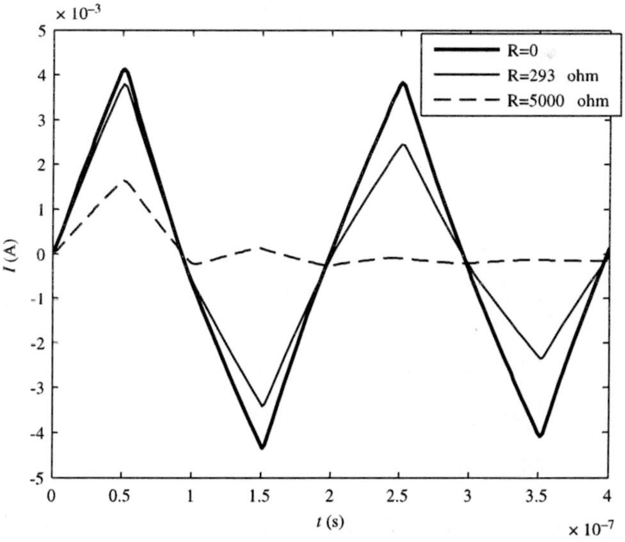

Figure 11.24 Transient response of a 30 m long line placed horizontally above ground at height $h = 0.1\text{m}$ with the various resistive loadings at the center ($R = 0$, $R = 293\,\Omega$ and $R = 5000\,\Omega$) illuminated by the EMP excitation. (D. Poljak, E. K. Miller, C. Y. Tham: "Time Domain Energy Measures for Thin-Wire Antennas and Scatterers", *IEEE Antennas & Propagation magazine*, Vol 44, No 1, pp 87–95, Feb. 2002).

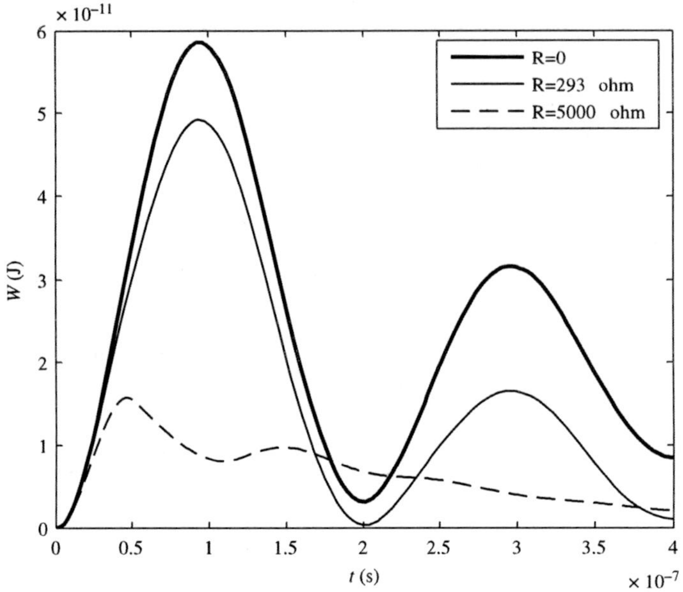

Figure 11.25 The total energy measure for resistively loaded 30 m long wire for load values $R = 0$, $R = 293\,\Omega$ and $R = 5000\,\Omega$. (D. Poljak, E. K. Miller, C. Y. Tham: "Time Domain Energy Measures for Thin-Wire Antennas and Scatterers", *IEEE Antennas & Propagation magazine*, Vol 44, No 1, pp 87–95, Feb. 2002).

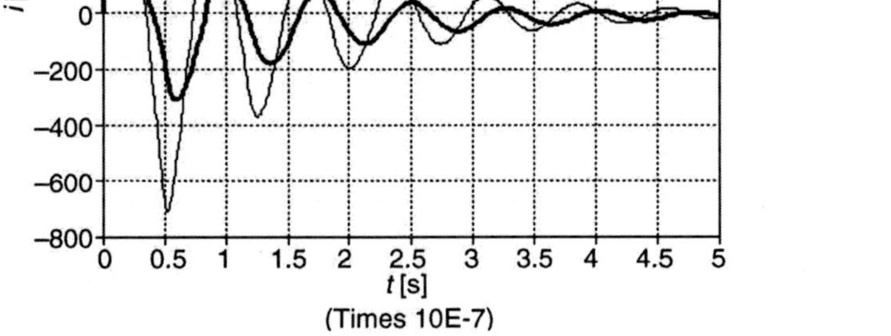

Figure 11.26 Transient current induced at the center of the single wire above dielectric half-space ($L = 1$ m, $a = 6.74$ cm, $h = 5$ m, $\varepsilon_r = 10$). (D. Poljak, S. Antonijevic, Time Domain Modeling of Electromagnetic Field Coupling to Multiple Transmission Lines Based on the Wire Antenna Theory//*Progress in Electromagnetics Research Symposium.* Pisa: Universita di Pisa, 2004. 149–152).

Taking into account the reflection from the dielectric medium, the total electric field tangential to the line is given by:

$$E_x^{exc}(t) = E_x^{inc}(t) - \frac{\varepsilon_r - \sqrt{\varepsilon_r}}{\varepsilon_r + \sqrt{\varepsilon_r}} E_x^{inc}\left(t - \frac{2h}{c}\right) \quad (11.45)$$

The results obtained for the given half-space problem are accompanied with the results obtained for the same wire insulated in free space and illuminated by the same excitation. Figure 11.27 shows the spatial distributions of the associated rms values.

The attenuating effect of the dielectric half-space is obvious.

11.4.2 Two Wire Transmission Line

Figure 11.28 shows the geometry of interest, that is, the transmission line consisting of two identical conductors placed horizontally over a PEC ground plane at height h.

The length of the conductors is $L = 20\,\text{m}$, the radius of both wires is $a = 6.74\,\text{cm}$, and the height above ground is $h = 5\,\text{m}$. The near side wire is illuminated by the EMP waveform (11.43) with $E_0 = 52.5\text{kV/m}$, $a = 4 \times 10^6\text{s}^{-1}$, $b = 4.76 \times 10^8\text{s}^{-1}$. Figures 11.29 and 11.30 show the transient response, obtained by the direct time domain approach (GB-IBEM) and by the indirect frequency domain method of moments (MOM)/inverse fourier transform (IFT) approach [5].

This conductor arrangement is used to show the effect of coupling by external incident fields. The far side conductor is assumed to lie in the shadow of the near side wire with respect to the incident EMP wave. The effect is that the far side conductor receives only the scattered radiation originating from the near side wire but not the external incident EMP. The current induced in the far side wire is therefore of lower amplitude since only the scattered field from the near side wire illuminates it.

11.4.3 Three Wire Transmission Line

In this section the configuration of interest is the multiple wire transmission line as shown in Figure 11.31.

First example is related to the transmission line consisting of a three identical wires with length of $L = 30\,\text{m}$ and radius $a = 3\,\text{cm}$, horizontally placed at height $h = 5\,\text{m}$ over a perfect ground. Wires are equally spaced with distance of $d = 3\,\text{m}$. The case of a normal incidence is analyzed (Fig. 11.31). The wire configuration is illuminated by the EMP (11.43) with $E_0 = 65\text{kV/m}$, $a = 4 \times 10^7\text{s}^{-1}$, $b = 6 \times 10^8\text{s}^{-1}$. Transient current induced at the midpoint of the central wire is shown in Figure 11.32, whereas Figure 11.33 presents the transient current induced at the midpoint of the side wires.

The last example is related to the multiple transmission line consisting of three identical wires with length $L = 30\,\text{m}$, $a = 3\,\text{cm}$, horizontally placed at height

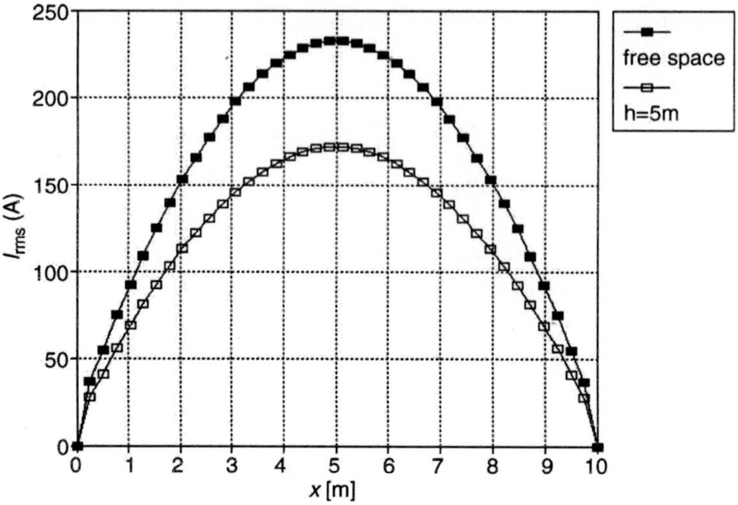

Figure 11.27 Spatial distribution of the rms values of the transient current along the single wire above dielectric half-space ($L = 1$ m, $a = 6.74$ cm, $h = 5$m, $\varepsilon_r = 10$). (D. Poljak, S. Antonijevic, Time Domain Modeling of Electromagnetic Field Coupling to Multiple Transmission Lines Based on the Wire Antenna Theory // *Progress in Electromagnetics Research symposium*. Pisa: Universita di Pisa, 2004. 149–152.

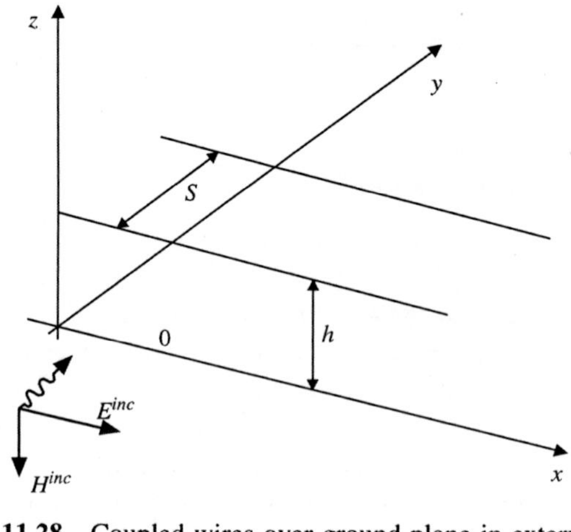

Figure 11.28 Coupled wires over ground plane in external field.

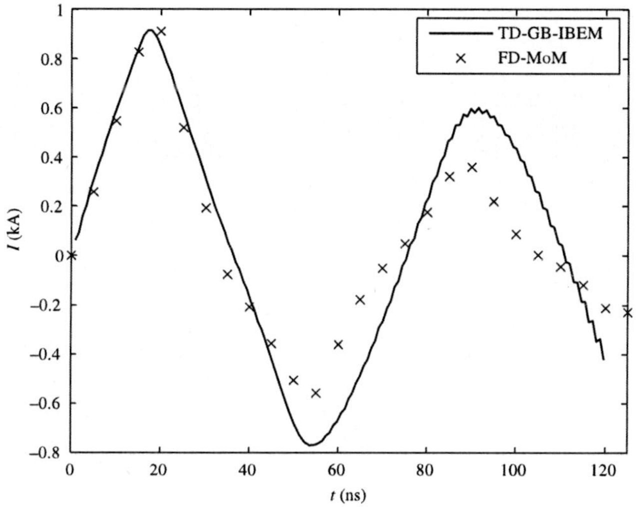

Figure 11.29 Coupled wires over perfect ground—current at the center of near-side wire ($L = 20\,\text{m}$, $a = 6.74\,\text{cm}$, $h = 5\,\text{m}$). (D. Poljak, C. Y. Tham, A.McCowen, V. Roje: "Electromagnetic Pulse Excitation of Multiconductor Transmission Lines", *ICEAA'99*, pp 789–792, Turin, Italy, Sept. 13–17, 1999).

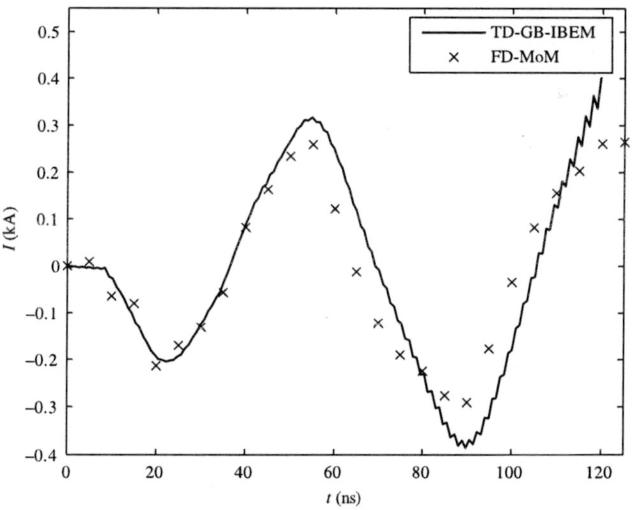

Figure 11.30 Coupled wires over perfect ground — current at the center of far-side wire ($L = 20\,\text{m}$, $a = 6.74\,\text{cm}$, $h = 5\,\text{m}$). (D. Poljak, C. Y. Tham, A.McCowen, V. Roje: "Electromagnetic Pulse Excitation of Multiconductor Transmission Lines", *ICEAA'99*, pp. 789–792, Turin, Italy, Sept. 13–17, 1999).

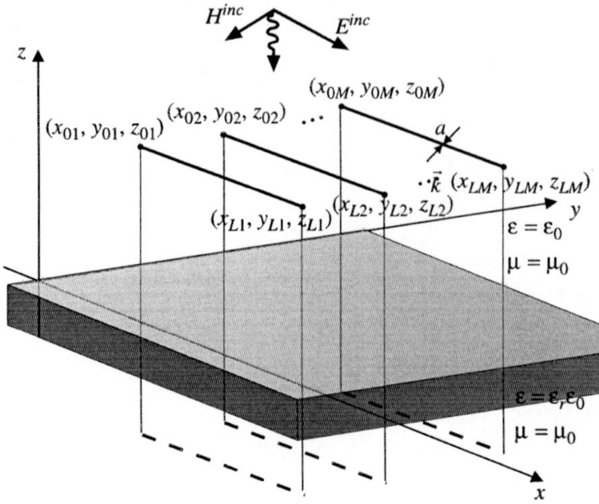

Figure 11.31 Multiple wires above a dielectric half-space.

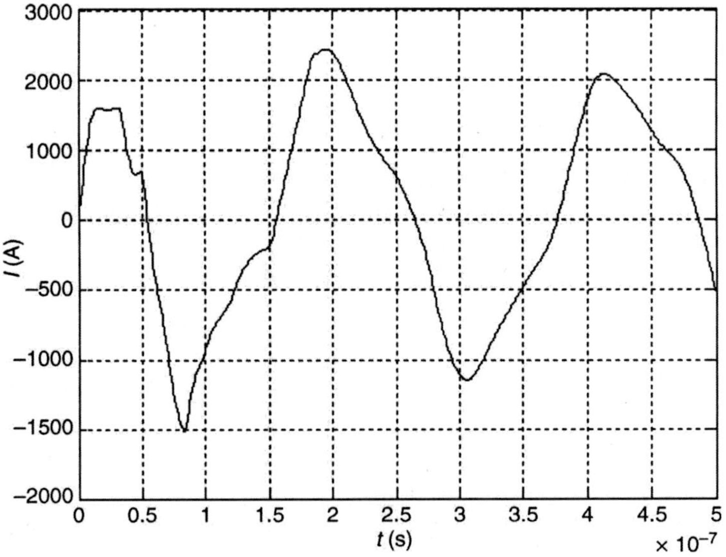

Figure 11.32 Transient current induced at the midpoint of the central wire ($L = 30$ m, $a = 3$ cm, $h = 5$ m, $d = 3$ m).

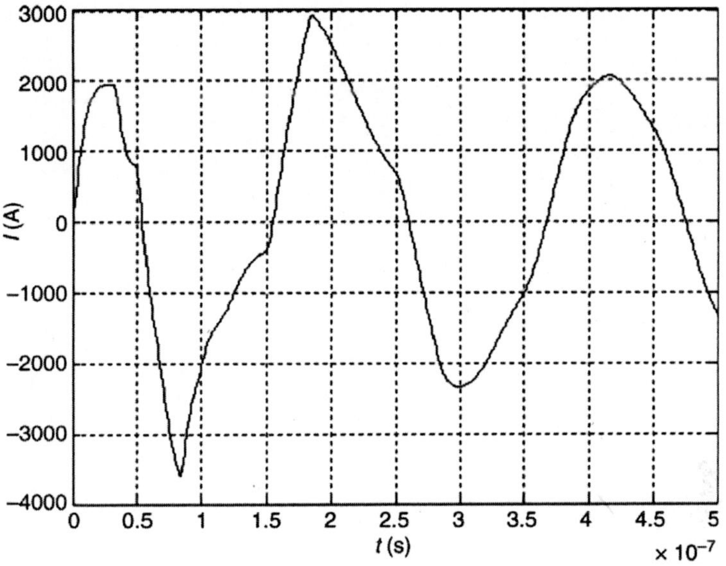

Figure 11.33 Transient current induced at the midpoint of the side wires ($L = 30$m, $a = 3$ cm, $h = 5$m, $d = 3$m).

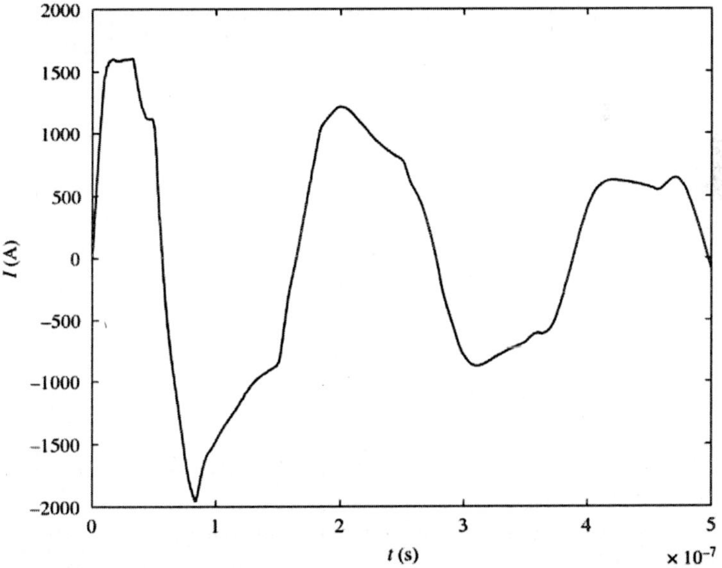

Figure 11.34 Transient current induced at the midpoint of the central wire ($L = 1$ m, $a = 6.74$ cm, $h = 5$ m, $d = 3$ m, $\varepsilon_r = 10$). (D. Poljak, S. Antonijevic, Time Domain Modeling of Electromagnetic Field coupling to Multiple Transmission Lines Based on the Wire Antenna Theory//*Progress in electromagnetics Research Symposium.* Pisa: Universita di Pisa, 2004. 149–152).

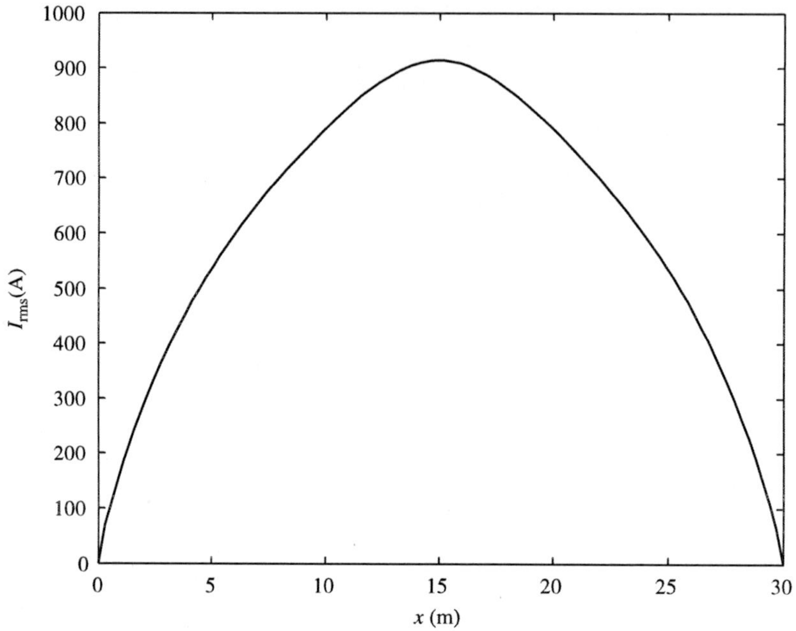

Figure 11.35 Spatial distribution of the rms values of the transient current along the central wire of the multiple wire configuration ($L = 1$ m, $a = 6.74$ cm, $h = 5$ m, $\varepsilon_r = 10$). (D. Poljak, S. Antonijevic, Time Domain Modeling of Electromagnetic Field coupling to Multiple Transmission Lines Based on the Wire Antenna Theory//*Progress in electromagnetics Research Symposium.* Pisa: Universita di Pisa, 2004. 149–152).

$h = 5$ m, over a dielectric half-space ($\varepsilon_r = 10$). The separation between wires is $d = 3$ m. This configuration is illuminated by the EMP (11.43) with $E_0 = 65$ kV/m, $a = 4 \times 10^7 \mathrm{s}^{-1}$, $b = 6 \times 10^8 \mathrm{s}^{-1}$. Figure 11.34 shows the transient response at the midpoint of the central wire. Figure 11.35 shows the spatial distributions of the associated rms values.

11.5 REFERENCES

[1] F. Tesche, M. Ianoz, F. Carlsson: EMC Analysis Methods and Computational Models, *John Wiley & Sons, New York*, 1997.

[2] D. Poljak: Electromagnetic Modelling of Wire Antenna Structures, *WIT Press, Southampton, Boston*, 2001.

[3] D. Poljak, F. Rachidi, S. Tkachenko: Generalized form of telegrapher's equations for the electromagnetic field coupling to finite length lines above a lossy ground, submitted to *IEEE Trans. EMC*

[4] S. Tkatchenko, F. Rachidi. M. Ianoz: Electromagnetic field coupling to a line of a finite length: theory and fast iterative solutions in frequency and time domains, *IEEE Trans. EMC 37 (4)*, 1995, pp 509–518.

[5] D. Poljak, C. Y. Tham: Integral Equation Techniques in Transient Electromagnetics, *WIT Press, Southampton-Boston*, 2003.

[6] J. A. Landt, E. K. Miller: Transient response of the infinite cylindrical antennas and scatterers, *IEEE Trans. AP 24 (3)*, March 1976, pp 246–250.

[7] E. K. Miller, J. A. Landt: Direct time-domain techniques for transient radiation and scattering from wires, *Proc. IEEE, 68 (11)*, November 1980, pp 1396–1423.

[8] F. Delfino, R. Procopio, M. Rossi, A. Andreotti, L. Verolino: HF electromagnetic field coupling to a transmission line of finite length over a perfectly conducting ground, Proceedings of EMCROMA, *Sorrento, Italy*, 2002, pp 1231–1236.

12

THE ELECTROMAGNETIC FIELD COUPLING TO BURIED CABLES: FREQUENCY- AND TIME-DOMAIN ANALYSIS

This chapter deals with frequency-and time-domain analysis of buried cables using antenna models. A frequency-domain model is based on Pocklington integral equation, whereas the time-domain formulation deals with the Hallen integral equation approach. The strengths and weaknesses of both approaches are explained subsequently.

12.1 THE FREQUENCY-DOMAIN APPROACH

The frequency-domain antenna theory approach provides one to take into account the earth-air interface effects via rigorous Sommerfeld integral formulation. However, the evaluation of the Sommerfeld integrals is not possible to perform analytically, whereas the corresponding numerical solution is rather demanding and very time consuming [1–5]. Consequently, a simplified approach based on the reflection coefficient (RC) approximation is sometimes used [6,7].

The wire antenna approach for the analysis of the plane wave coupling to the buried cable of finite length with the effect of the half space included via the RC approximation has been proposed in Ref. [8]. This approach has been extended to the more complex case of a plane wave excitation with an arbitrary angle of incidence that has been reported in Ref. [9].

The frequency-domain formulation presented in this chapter is based on the Pocklington integro-differential equation whose kernel contains the reflection coefficient to take into account the earth-air interface.

This integral equation is solved via the Galerkin-Bubnov scheme of the boundary element method (GB-IBEM) [10]. Furthermore, the transient response of the wire is computed using the inverse Fourier transform.

12.1.1 Formulation in the Frequency Domain

The horizontal line of length L and radius a, buried in a lossy ground at depth d, is shown in Figure 12.1.

The current distribution along the buried wire is governed by the corresponding Pocklington integro-differential equation. This integro-differential equation can be derived by enforcing the interface conditions for the tangential electric field components.

Assuming the wire to be perfectly conducting, the total field along the wire surface vanishes, that is,

$$\vec{e}_x \cdot \vec{E}^{tot} = 0 \tag{12.1}$$

here the total field $\vec{E}^{tot}$ is the sum of the excitation $\vec{E}^{exc}$ and scattered field $\vec{E}^{sct}$ field components

$$\vec{E}^{tot} = \vec{E}^{exc} + \vec{E}^{sct} \tag{12.2}$$

where the excitation field component represents the sum of the incident field $\vec{E}^{inc}$ and the field reflected from the lossy ground $\vec{E}^{ref}$,

$$\vec{E}^{exc} = \vec{E}^{inc} + \vec{E}^{ref} \tag{12.3}$$

The scattered field component is given by

$$\vec{E}^{sct} = -j\omega\vec{A} - \nabla\varphi \tag{12.4}$$

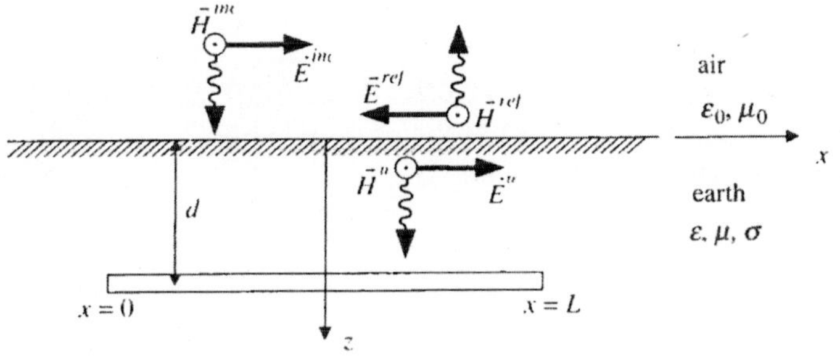

Figure 12.1 Geometry of the straight wire buried in a dissipative medium.

where $\vec{A}$ is the magnetic vector potential and φ is the electric scalar potential.

According to the widely used thin-wire approximation [7,11], only the axial component of the magnetic potential differs from zero and eqn (12.4) becomes

$$E_x^{sct} = -j\omega A_x - \frac{\partial \varphi}{\partial x} \tag{12.5}$$

where the magnetic vector potential and electric scalar potential are, respectively, defined as

$$A_x = \frac{\mu}{4\pi}\int_0^L I(x')g(x,x')\mathrm{d}x' \tag{12.6}$$

$$\varphi(x) = \frac{1}{4\pi\varepsilon_{eff}}\int_0^L q(x')g(x,x')\mathrm{d}x' \tag{12.7}$$

where ε_{eff} is the complex permittivity of the lossy ground that is given by

$$\varepsilon_{eff} = \varepsilon_r\varepsilon_0 - j\frac{\sigma}{\omega} \tag{12.8}$$

whereas $q(x)$ is the charge distribution along the line, $I(x')$ denotes the induced current along the line, and $g(x,x')$ stands for the Green's function given by

$$g(x,x') = g_0(x,x') - R_{TM}g_i(x,x') \tag{12.9}$$

and $g_0(x,x')$ denotes the lossy medium Green's function,

$$g_0(x,x') = \frac{e^{-jk_2R_1}}{R_1} \tag{12.10}$$

whereas $g_i(x,x')$ is, in accordance to the image theory, given by

$$g_i(x,x') = \frac{e^{-jk_2R_2}}{R_2} \tag{12.11}$$

where k_2 is the propagation constant of the lower medium and R_1 and R_2 are distances from the source point and from the corresponding image to the observation point, respectively, and are defined by

$$R_1 = \sqrt{(x-x')^2 + a^2}$$
$$R_2 = \sqrt{(x-x')^2 + 4d^2} \tag{12.12}$$

The influence of a nearby ground-air interface is taken into account by means of the Fresnel plane wave reflection coefficient,

$$R_{TM} = \frac{\frac{1}{n}\cos\theta - \sqrt{\frac{1}{n} - \sin^2\theta}}{\frac{1}{n}\cos\theta + \sqrt{\frac{1}{n} - \sin^2\theta}}; \ \theta = arctg\frac{|x - x'|}{2d}; \ n = \frac{\varepsilon_{eff}}{\varepsilon_0} \tag{12.13}$$

The linear charge density and the current distribution along the line are related through the equation of continuity [12]:

$$q = -\frac{1}{j\omega}\frac{dI}{dx} \tag{12.14}$$

Substituting eqn (12.14) into eqn (12.7) yields

$$\varphi(x) = -\frac{1}{j4\pi\omega\varepsilon_{eff}}\int_0^L \frac{\partial I(x')}{\partial x'} g(x, x')dx' \tag{12.15}$$

Combining eqns (12.5), (12.6), and (12.15) results in the following integral relationship for the scattered field:

$$E_x^{sct} = -j\omega\frac{\mu}{4\pi}\int_0^L I(x')g(x, x')dx' + \frac{1}{j4\pi\omega\varepsilon_{eff}}\frac{\partial}{\partial x}\int_0^L \frac{\partial I(x')}{\partial x'} g(x, x')dx' \tag{12.16}$$

Finally, eqns (12.3) and (12.16) result in the following integral equation for the unknown current distribution induced along the line:

$$E_x^{exc} = j\omega\frac{\mu}{4\pi}\int_0^L I(x')g(x, x')dx' - \frac{1}{j4\pi\omega\varepsilon_{eff}}\frac{\partial}{\partial x}\int_0^L \frac{\partial I(x')}{\partial x'} g(x, x')dx' \tag{12.17}$$

Integral eqn (12.17) is well known in antenna theory and represents one of the most commonly used variants of the Pocklington's integro-differential equation.

The electric field transmitted into the lossy ground and illuminating the buried line (Fig. 12.2) is given by

$$E_x^{exc} = E_x^{tr} = E_0(\Gamma_{TE}\sin\phi - \Gamma_{TM}\cos\theta_t\cos\phi)e^{-jk_2\vec{n}_t\cdot\vec{r}} \tag{12.18}$$

where α is the angle between *E-field* vector, whereas the plane of incidence and θ_t is defined by the Snell's law [1]:

$$k_1\sin\theta = k_2\sin\theta_t \tag{12.19}$$

where k_1 is free-space propagation constant.

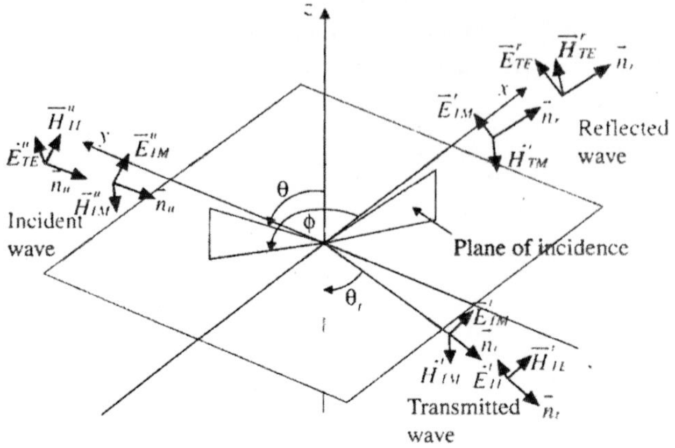

Figure 12.2 Incident, reflected, and transmitted wave from an air–earth interface.

Quantities Γ_{TM} and Γ_{TE} denote the vertical and horizontal Fresnel transmission coefficients, respectively, at the air-earth interface that are given by [6]

$$\Gamma_{TM} = \frac{2\sqrt{\underline{n}}\cos\theta}{\underline{n}\cos\theta + \sqrt{\underline{n} - \sin^2\theta}} \tag{12.20}$$

$$\Gamma_{TE} = \frac{2\cos\theta}{\cos\theta + \sqrt{\underline{n} - \sin^2\theta}} \tag{12.21}$$

and $\vec{n}_t \cdot \vec{r}$ is the distance from the origin point to the observation point at the wire surface.

$$\vec{n}_t \cdot \vec{r} = -x\sin\theta_t\cos\phi - y\sin\theta_t\sin\phi - z\cos\theta_t \tag{12.22}$$

Solving the Pocklington integro-differential eqn (12.17), the current distribution at the given frequency is obtained.

12.1.2 Numerical Solution of the Integral Equation

The numerical solution of (12.17) is obtained via Galerkin-Bubnov scheme of the indirect boundary element method that is outlined below. More detailed description of the method can be found in Ref. [1].

An operator form of the Pocklington integro-differential eqn (12.18) can be, for convenience, symbolically written as

$$KI = E \tag{12.23}$$

where K is a linear operator, I is the unknown function to be found for a given excitation E.

The unknown current is expanded into a finite sum of linearly independent basis function $\{f_i\}$ with unknown complex coefficient α_i,

$$I \cong I_n = \sum_{i=1}^{n} \alpha_i f_i \tag{12.24}$$

Substituting eqn (12.24) into eqn (12.23) yields

$$KI \cong KI_n = \sum_{i=1}^{n} \alpha_i K f_i = E_n = P_n(E) \tag{12.25}$$

where $P_n(E)$ is called a projection operator [1].

Now the residual R_n is formed as follows:

$$R_n = KI_n - E = P_n(E) - E \tag{12.26}$$

In accordance to the definition of the scalar product of functions in Hilbert function space, the error R_n is weighted to zero with respect to certain weight functions $\{W_j\}$, that is,

$$\langle R_n, W_j \rangle = 0; \quad j = 1, 2, \ldots, n \tag{12.27}$$

where the expression in bracket stands for a scalar product of functions given by:

$$\langle R_n, W_j \rangle = \int_{\Omega} R_n W_j^* \mathrm{d}\Omega \tag{12.28}$$

where denotes the actual calculation domain.

Since the operator K is linear, a system of linear equations is obtained by choosing $W_j = f_j$ that implies the Galerkin-Bubnov procedure. Thus, it can be written as

$$\sum_{i=1}^{n} \alpha_i \langle K f_i, f_j \rangle = \langle E, f_j \rangle; \quad j = 1, 2, \ldots, n \tag{12.29}$$

Equation (12.29) is the strong Galerkin-Bubnov formulation of the Pocklington integral eqn (12.18). Utilizing the integral equation kernel symmetry and taking into account the Dirichlet boundary conditions for the current at the free ends of the cylinder, after integration by parts eqn (12.29) becomes

$$\sum_{j=1}^{n} \alpha_i \frac{1}{j4\pi\omega\varepsilon_{eff}} \left[\int_0^L \frac{\partial f_j(x)}{\partial x} \int_0^L \frac{\partial f_i(x')}{\partial x'} g(x, x')\mathrm{d}x'\mathrm{d}x + k_2^2 \int_0^L f_j(x) \int_0^L f_i(x') g(x, x')\mathrm{d}x'\mathrm{d}x \right]$$

$$= \int_0^L E_x^{exc}(x) f_j(x)\mathrm{d}x; \quad \text{and} \quad j = 1, 2, \ldots, n \tag{12.30}$$

Equation (12.30) represents the weak Galerkin-Bubnov formulation of the integral eqn (12.23).

The resulting system of algebraic equations arising from the boundary element discretization of eqn (12.30) is given by

$$\sum_{j=1}^{M} [Z]_{ji}\{I\}_i = \{V\}_j; \quad \text{and} \quad j = 1, 2, \ldots, M \tag{12.31}$$

where $[Z]_{ji}$ is the local matrix representing the interaction of the ith source boundary element with the jth observation boundary element:

$$[z]_{ji} = \frac{1}{j4\pi\omega\varepsilon_{eff}}\left[\int_0^L \{D\}_j \int_0^L \{D\}^T g(x,x')\mathrm{d}x'\mathrm{d}x + k_2^2 \int_0^L \{f\}_j \int_0^L \{f'\}^T g(x,x')\mathrm{d}x'\mathrm{d}x\right] \tag{12.32}$$

The vector $\{I\}$ contains the unknown coefficients of the solution and represents the local voltage vector. Matrices $\{f\}$ and $\{f'\}$ contain the shape functions, whereas $\{D\}$ and $\{D'\}$ contain their derivatives; M is the total number of line segments, and Δl_i, Δl_j are the widths of ith and jth segments, respectively.

Functions $f_k(z)$ are the Lagrange's polynomials and $\{V\}_j$ is the local right-hand side vector for the jth observation segment,

$$\{V\}_j = \int_{\Delta l_j} E_x^{exc}\{f\}_j \mathrm{d}z \tag{12.33}$$

representing the local voltage vector.

Linear approximation over a boundary element is used as it has been shown that this choice provides accurate and stable results [1].

12.1.3 The Calculation of Transient Response

Calculating the current distribution along a buried wire in a wide frequency spectrum, one obtains the transfer function of the system $H(f)$.

To analyze the transient response of the buried wire, an incident field in the form of a single exponentially decaying function is used,

$$e_x^{tr}(t) = E_0 e^{-at} \tag{12.34}$$

in which Fourier transform is given by

$$E_x^{tr}(t) = \frac{E_0}{a + j2\pi f} \tag{12.35}$$

Now, the current distribution $I(f)$, that is, the frequency response of the wire to the particular excitation (12.35) is obtained by multiplying the frequency spectrum of the excitation function with the corresponding transfer function of the linear system $H(f)$,

$$I(f) = H(f)E_x^{tr}(f) \tag{12.36}$$

Applying the inverse Fourier transform (IFT) to the function $I(f)$ yields the transient current induced along the buried transmission line [13],

$$i(t) = \int_{-\infty}^{\infty} I(f)e^{j2\pi ft}\mathrm{d}\omega \tag{12.37}$$

As the system transfer function $H(f)$ is represented by a discrete set of values, and the actual frequency response $I(f)$ is also represented by the discrete set of values, the integral (12.37) thus cannot be solved analytically and one has to deal with the discrete Fourier transform (or in this case fast Fourier transform), that is,

$$i(t) = IFFT(I(f)) \tag{12.38}$$

Thus, the discrete set of the time-domain current values is defined by Ref. [21]:

$$i(n\Delta t) = F \cdot \sum_{k-0}^{N-1} I(k\Delta f)e^{jk\Delta fn\Delta t} \tag{12.39}$$

where F is the highest frequency taken into account, N is the total number of frequency samples, Δf is the sampling interval, and Δt is the time step.

12.1.4 Numerical Results

For comparison purposes, the current distribution induced along a conducting cylinder immersed in sea water is computed, assuming a unit incident field. The water parameters are $\varepsilon_r = 80$ and $\sigma = 4\,\mathrm{S/m}$, and the operating frequency is $f - 1\,\mathrm{MHz}$. The cylinder length is $L = 120\,\mathrm{m}$ and $L = 160\,\mathrm{m}$ with a radius of $a = 0.6\,\mathrm{m}$ and $a - 0.8\,\mathrm{m}$, respectively, while in Figure 12.3, the calculated current distributions are compared to the analytical results available from Ref. [14]. Agreement is found to be satisfactory.

Further numerical examples are related to the transmission lines buried in a lossy ground with permittivity $\varepsilon_r = 10$ and conductivity σ $0.001\,\mathrm{S/m}$. Conductor radius is $a = 1\,\mathrm{cm}$, while length L and burial depth d are varied. The wire is excited by the transmitted plane wave with a single exponential decaying form defined with eqn (12.35). The parameters of the exponential function are as follows: $E_0 = 1\mathrm{V/m}$, $a = (8.854 \times 10^{-8}\mathrm{s})^{-1}$.

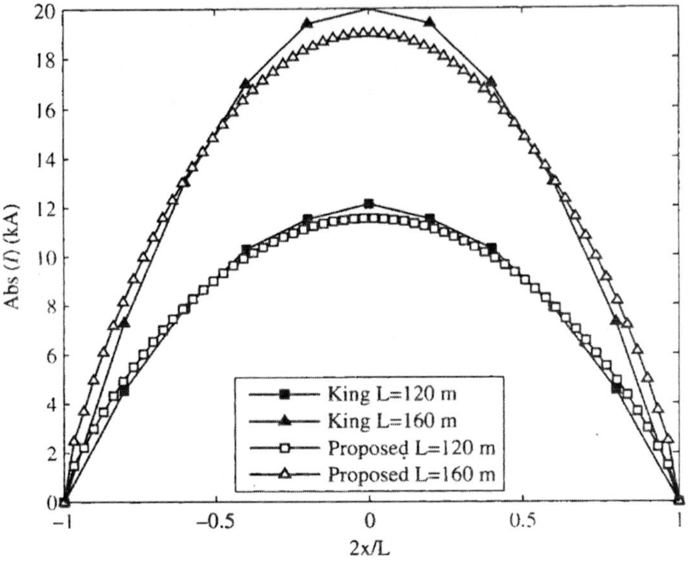

Figure 12.3 Comparison of numerical results obtained with different solution methods.

The transient response, that is, the current induced at the center of the wires having various lengths for the normal incidence is shown in Figure 12.4. The burial depth is $d = 1$m. The current wave reflections from the wire ends are clearly visible, particularly for the shorter wires. Figure 12.5 shows the influence of the burial depth on the induced current at the center of the 200 m long wire.

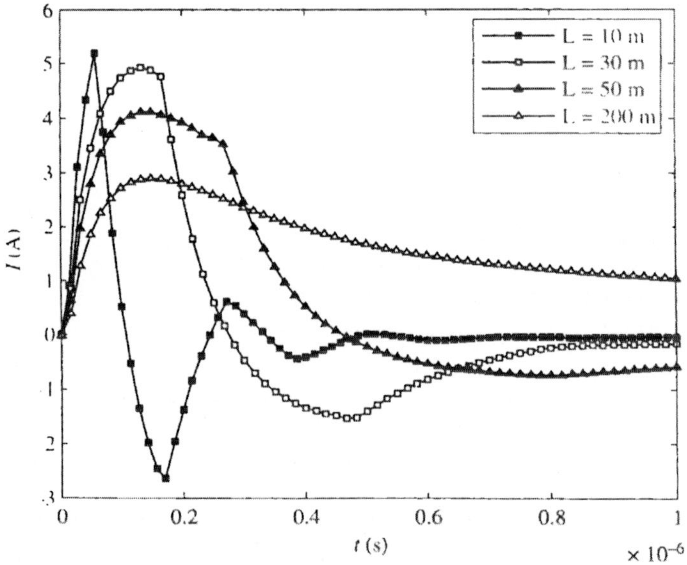

Figure 12.4 Transient response for different wire lengths.

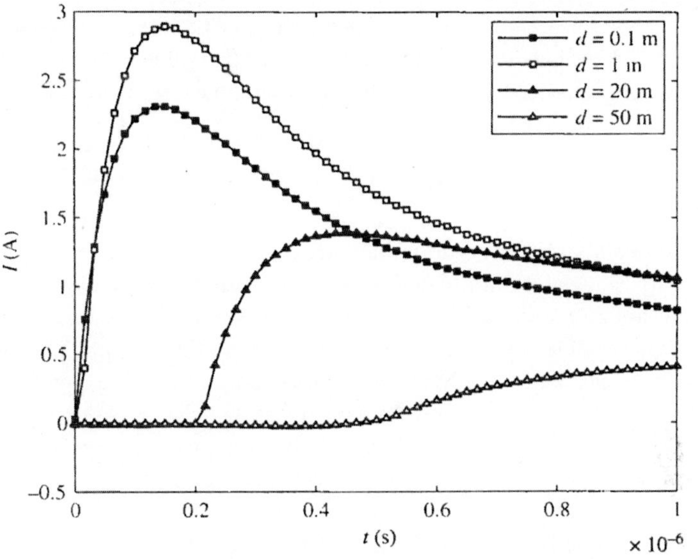

Figure 12.5 Transient response for different burial depth.

Both the time shift and the attenuation of the signal seem to increase with depth. The only exception occurs for the case of $d = 0.1$ m, where the induced current is lower than for the case of $d = 1$ m. This happens due to the vicinity of the nearby ground-air interface. The transient response of the wire of the transmission line of length $L = 30$ m buried at depth $d = 1$ m for various angles of incidence $\theta = 0°$, $60°$, $80°$ is shown in Figure 12.6.

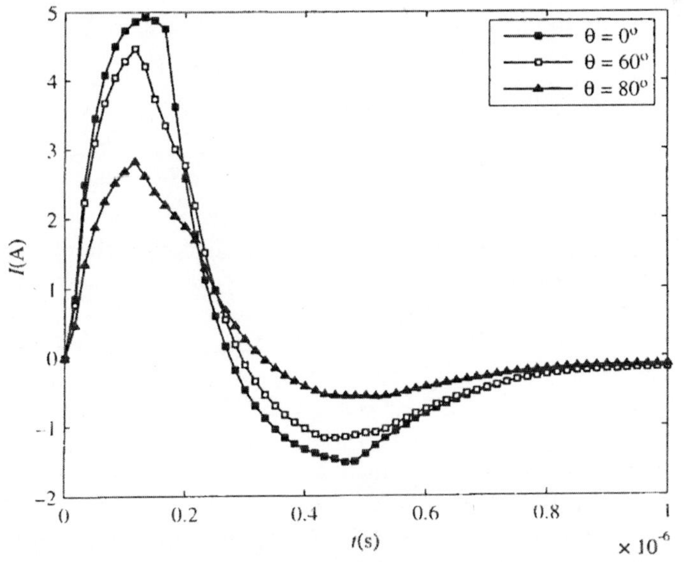

Figure 12.6 Transient response for different angle of incidence.

The transient response of the wire of the transmission line of length $L = 30$ m buried at depth $d = 1$m for various angles of incidence $\theta = 0^\circ, 60^\circ, 80^\circ$ is shown in Figure 12.6. For the higher angle θ, amplitude of the induced current at the wire center is smaller, since the tangential component of the transmitted electric field is also decreased.

12.2 TIME-DOMAIN APPROACH

This section deals with a time-domain study of a single straight wire embedded in a dielectric half-space and illuminated by a nonuniform transient electric field because this is important as a starting point in the transient analysis of wires buried in a lossy medium, which is of great practical interest for many EMC applications [6,7,11,15].

The analysis presented in this section is based on the wire antenna theory and is carried out directly in the time domain. Dealing with a rather simple geometry of a single wire embedded in a dielectric half-space, this section aims to introduce some basic ideas on the subject. The method can also be applied to the case of arbitrary wire configurations that are of much more practical importance in EMC applications.

The time-domain formulation presented in this section is based on the space-time Hallen integral equation for half-space problems [16]. The effects of the two-media configuration are taken into account via the corresponding reflection coefficient and the transmission coefficient. The transient current along the straight wire embedded in a dielectric half-space is obtained by solving the corresponding Hallen integral equation via the time-domain variant of the indirect Galerkin-Bubnov in Boundary Element Method (GB-BEM) [1].

After calculating the space-time current distribution along the wire, to further evaluate the obtained transient response, the time-domain energy measures are computed by spatially integrating the squared current and charge along the wire.

12.2.1 Formulation in the Time Domain

A perfectly conducting straight thin wire of length L and radius a immersed in a dielectric medium at depth d is considered as shown in Figure 12.7. The wire is illuminated by a horizontally polarized transient electric field tangential along its surface, that is, only the case of normal incidence is considered.

The mathematical framework of the problem is based on the wire antenna theory and thin wire approximation [7,11].

The transient induced current flowing along the straight wire embedded in a dielectric half-space is governed by the corresponding space-time integral equation. The time-domain formulation for the single wire problem in terms of the Pocklington or the Hallen integral equation type can be readily obtained as an extension of the wire in homogeneous dielectric medium.

The time-domain Pocklington type integral equations often suffer from numerical instabilities, that is, from nonphysical, rapidly growing oscillations at later instants of time [17].

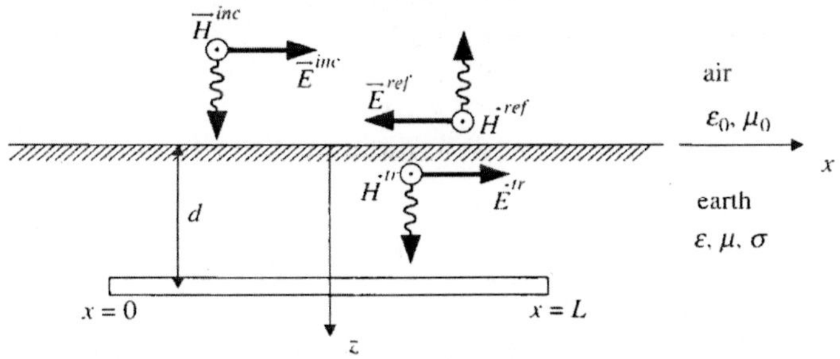

Figure 12.7 Geometry of the straight wire embedded in a dielectric half-space. (Poljak, D., Doric, V.: Time Domain Modeling of Electromagnetic Field Coupling to Finite Length Wires Embedded in a Dielectric Half-Space, *IEEE trans. EMC-47*, No 2., pp 247–253, May 2005).

On the contrary, the time-domain Hallen integral equation does not contain either space or time derivatives within its kernel, which are found to be the origin of numerical instabilities. Consequently, the Hallen equation has been proven to be attractive from the computational point of view.

The Hallen integral equation approach, already used for the transient analysis of straight wire configuration above ground [18–21], is applied to the problem of wires embedded in a dielectric half-space in this work. The Hallen integral equation for the straight wire in unbounded lossless medium can be readily derived from the corresponding Pocklington integral equation type.

Since the wire is perfectly conducting, the tangential component of the total field vanishes on the antenna surface, that is,

$$E_x^{inc} + E_x^{sct} = 0 \tag{12.40}$$

where E_x^{inc} is the incident field and E_x^{sct} is the scattered field on the metallic wire surface. From the first Maxwell equation

$$\nabla x\vec{E} = -\frac{\partial \vec{B}}{\partial t} \tag{12.41}$$

and using the vector magnetic potential $\vec{A}$,

$$\vec{B} = \nabla x\vec{A} \tag{12.42}$$

it follows

$$\vec{E}^{inc}\Big|_{tan} = \left(\frac{\partial \vec{A}}{\partial t} + \nabla\varphi\right)\Big|_{tan} \tag{12.43}$$

where $\vec{A}$ and φ are space-time dependent magnetic vector and electric scalar potential, respectively. These two potentials are to satisfy the Lorentz gauge,

$$\frac{\partial \varphi}{\partial t} + \frac{1}{\mu\varepsilon}\nabla\vec{A} = 0 \tag{12.44}$$

Differentiating eqn (12.43) and taking into account the Lorentz gauge (12.44), the wave equation for magnetic vector potential $\vec{A}$ is as follows:

$$\left(\frac{\partial^2 \vec{A}}{\partial t^2} - \frac{1}{\mu\varepsilon}\nabla(\nabla\vec{A})\right)\Bigg|_{tan} = \frac{\partial \vec{E}^{inc}}{\partial t}\Bigg|_{tan} \tag{12.45}$$

In accordance to the thin-wire approximation, only the axial component of the vector potential exists, so it can be written as

$$\frac{\partial^2 A_x}{\partial x^2} - \frac{1}{v^2}\frac{\partial^2 A_x}{\partial t^2} = -\frac{\partial \vec{E}^{inc}}{\partial t} \tag{12.46}$$

where v is the velocity of wave propagation in a homogeneous dielectric medium that is defined as follows:

$$v = \frac{1}{\sqrt{\mu\varepsilon_{rg}\varepsilon_0}} \tag{12.47}$$

where ε_{rg} is the lower medium relative permittivity where c denotes the velocity of light.

Equation (12.46) is valid on the surface of the perfect conductor and the solution can be represented by the sum of the homogeneous equation solution and particular solution of the inhomogeneous equation,

$$A_x(x,t) = A_x^h(x,t) + A_x^p(x,t) \tag{12.48}$$

In addition, the solution of the homogeneous wave equation is given as a superposition of incident and reflected wave [1],

$$A_x^h(x,t) = F_1\left(t - \frac{x}{v}\right) + F_2\left(t + \frac{x}{v}\right) \tag{12.49}$$

The particular solution is given by the integral [1]

$$\begin{aligned} A_x^p(x,t) &= \frac{\varepsilon v}{2}\int_{-\infty}^{x}\int_{-\infty}^{t-\frac{(x-x')}{c}} \frac{\partial E_x^{inc}(x',t')}{\partial t}\,\mathrm{d}t'\mathrm{d}x' \\ &+ \frac{\varepsilon v}{2}\int_{-\infty}^{x}\int_{-\infty}^{t-\frac{(x-x')}{c}} \frac{\partial E_x^{inc}(x',t')}{\partial t}\,\mathrm{d}t'\mathrm{d}x' \end{aligned} \tag{12.50}$$

Since the differential equation is related to the wire antenna surface, eqn (12.50) is simplified into

$$A_x^p(x,t) = \frac{1}{2Z_g}\int_0^L E_x^{inc}\left(x', t - \frac{|x-x'|}{v}\right)dx' \quad (12.51)$$

where Z_g is the corresponding impedance of a dielectric medium that is given by

$$Z_g = \sqrt{\frac{\mu}{\varepsilon_{rg}\varepsilon_0}} \quad (12.52)$$

On the contrary, the magnetic vector potential on the metallic wire surface on the left-hand side of eqn (12.48) may also be obtained as a solution of the wave equation

$$\nabla^2\vec{A} - \mu\varepsilon\frac{\partial^2\vec{A}}{\partial t^2} = -\mu\vec{J}(r,t) \quad (12.53)$$

where $\vec{J}(r,t)$ denotes the surface current density.

The solution of differential eqn (12.53) is usually obtained via the Green's function theory by introducing the auxiliary equation

$$\nabla^2 g - \mu\varepsilon\frac{\partial^2 g}{\partial t^2} = \delta(\vec{r}-\vec{r}', t-t') \quad (12.54)$$

solution of which is given in the form of retarded Dirac impulse

$$g(\vec{r}-\vec{r}', t-t') = \frac{\delta\left(t - t' - \frac{R}{v}\right)}{4\pi R} \quad (12.55)$$

where R is the distance from the source to the observation point.

Then the solution of eqn (12.53) using the Green's function approach may be written in the following form:

$$\vec{A}(r,t) = \frac{\mu}{4\pi}\int_{-\infty}^{t}\int_S \vec{J}_S(\vec{r}', t)\frac{\delta(t-t'-R/v)}{R}dS'dt' \quad (12.56)$$

Performing the time-domain integration, one obtains

$$\vec{A}(r,t) = \frac{\mu}{4\pi}\iint_S \frac{\vec{J}_s(r', t-R/v)}{R}dS' \quad (12.57)$$

According to the thin-wire approximation, the equivalent current along the wire is assumed to flow in the axis, whereas the observation points are located on the antenna surface, that is, it follows

$$I(x,t) = 2\pi a J_z(x,t) \quad (12.58)$$

and the axial component of the magnetic vector potential is given by

$$A_x(x,t) = \frac{\mu}{4\pi}\int_0^L \frac{I(x',t-R/v)}{R}\,\mathrm{d}x' \tag{12.59}$$

Combining relations (12.46) and (12.59) yields the Pocklington integral equation

$$\left[\frac{\partial^2}{\partial x^2} - \frac{1}{v^2}\frac{\partial^2}{\partial t^2}\right]\int_0^L \frac{I(x',t-R/v)}{4\pi R}\,\mathrm{d}x' = \frac{\partial E_x^{inc}}{\partial t} \tag{12.60}$$

whereas relations (12.48), (12.49), (12.51), and (12.59) lead to the space-time Hallen integral equation

$$\begin{aligned}\int_0^L \frac{I(x',t-R/v)}{4\pi R}\,\mathrm{d}x' &= F_0\left(t-\frac{x}{v}\right) + F_L\left(t-\frac{L-x}{v}\right)\\ &+ \frac{1}{2Z_g}\int_0^L E_x^{inc}\left(x',t-\frac{|x-x'|}{v}\right)\mathrm{d}x'\end{aligned} \tag{12.61}$$

but for the straight, finite length wire in an unbounded dielectric medium, the distance from the observation point is given by

$$R = \sqrt{(x-x')^2 + a^2} \tag{12.62}$$

The multiple reflections of the current at the free ends of the wire are taken into account by the unknown functions $F_0(t)$ and $F_L(t)$.

The Hallen integral equation for a homogeneous lossless medium can also be derived directly from the Pocklington equation by performing the straight-forward convolution [20].

The corresponding Hallen integral equation for the wire embedded in a dielectric half-space can be derived gradually.

As a first step, the Hallen integral equation for an unbounded medium (12.61) is transferred into the frequency domain

$$\begin{aligned}\int_0^L \frac{I(x',s)e^{-sR/v}}{4\pi R}\,\mathrm{d}x' &= F_0(s)e^{-sx/v} + F_L(s)e^{-s(L-x)/v}\\ &+ \frac{1}{2Z_g}\int_0^L E_x^{inc}(x',s)e^{-s|x-x'|/v}\,\mathrm{d}x'\end{aligned} \tag{12.63}$$

where $s = j\omega$ denotes the Laplace variable.

The frequency domain Hallen integral equation for a straight wire in a dielectric half-space is obtained by extending the integral eqn (12.63) with an additional term due to an image wire in the air located at height d above interface. This term contains the reflection coefficient Γ_{ref} for the TM polarization multiplied by the Green's function of the image wire in the air. In addition, the incident field E_x^{inc} appearing in the last term in eqn (12.61) has to be replaced by the corresponding transmitted field E_x^{tr}.

Thus, the resulting space frequency Hallen integral equation for the straight wire embedded in a dielectric medium becomes

$$\int_0^L \frac{I(x',s)e^{-\frac{sR}{v}}}{4\pi R}\,dx' - \int_0^L \Gamma_{ref}(\theta)\frac{I(x',s)e^{-\frac{sR^*}{v}}}{4\pi R^*}\,dx' = F_0(s)e^{-\frac{sx}{v}} + F_L(s)e^{-s\frac{(L-x)}{v}} + \frac{1}{2Z_g}\int_0^L E_x^{tr}(x',s)e^{-s\frac{|x-x'|}{v}}\,dx' \tag{12.64}$$

where R^* is the distance from the source point located at the image wire in the air to the observation point located at the wire immersed in a dielectric medium,

$$R^* = \sqrt{(x-x')^2 + 4d^2} \tag{12.65}$$

The space dependent reflection coefficient by which the ground-air interface effects are taken into account is defined by relation [6]

$$\Gamma_{ref}(\theta) = \frac{\frac{1}{\varepsilon_{rg}}\cos\theta - \sqrt{\frac{1}{\varepsilon_{rg}} - \sin^2\theta}}{\frac{1}{\varepsilon_{rg}}\cos\theta + \sqrt{\frac{1}{\varepsilon_{rg}} - \sin^2\theta}} \tag{12.66}$$

$$\theta = arctg\frac{|x-x'|}{2d}$$

Finally, the electric field transmitted into the dielectric medium E_x^{tr} is given as follows:

$$E_x^{tr} = \Gamma_{tr}(\theta_{tr})E_x^{inc} \tag{12.67}$$

where E_x^{inc} is the incident field in the air and $\Gamma_{tr}(\theta_{tr})$ is the corresponding transmission coefficient by which the air-ground interface effects are taken into account.

The space dependent transmission coefficient, for the case of normal incidence considered in this work, is given by [16]

$$\Gamma_{tr}(\theta_{tr}) = \left.\frac{2\sqrt{\varepsilon_{rg}}\cos\theta_{tr}}{\varepsilon_{rg}\cos\theta_{tr} + \sqrt{\varepsilon_{rg} - \sin^2\theta_{tr}}}\right|_{\theta_{tr}=0} = \frac{2}{\sqrt{\varepsilon_{rg}}+1} \tag{12.68}$$

where θ_{tr} is the angle of transmission

The time-domain Hallen equation for a straight thin wire in a dielectric half-space can now be obtained applying the inverse Laplace transform and the convolution theorem to the integral eqn (12.64). Thus, it follows

$$\int_0^L \frac{I(x', t - R/v)}{4\pi R} \mathrm{d}x' - \int_{-\infty}^{t} \int_0^L \Gamma_{ref}(\theta, \tau) \frac{I(x', t - R^*/v - \tau)}{4\pi R^*} \mathrm{d}x' \mathrm{d}\tau$$
$$= F_0\left(t - \frac{x}{v}\right) + F_L\left(t - \frac{L - x}{v}\right) + \frac{1}{2Z_g} \int_0^L E_x^{tr}\left(x', t - \frac{|x - x'|}{v}\right) \mathrm{d}x' \quad (12.69)$$

where the time-domain counterpart of the Fresnel reflection coefficient $\Gamma_{ref}(\theta, t)$ is

$$\Gamma_{ref}(\theta, t) = \Gamma_{ref}(\theta)\delta(t) \quad (12.70)$$

and $\delta(t)$ stands for the Dirac impulse.

The transmitted electric field in the dielectric medium E_x^{tr} represents a time-domain counterpart of relation (12.67) and can be obtained from the convolution integral,

$$E_x^{tr}(x, t) = \int_{x}^{t} E_x^{inc}(x, t - \tau)\Gamma_{tr}(\theta_{tr}, \tau)\mathrm{d}\tau \quad (12.71)$$

where $\Gamma_{tr}(\theta_{tr}, t)$ is the time-domain transmission coefficient counterpart of expression (12.68),

$$\Gamma_{tr}(\theta, t) = \Gamma_{tr}(\theta)\delta(t) \quad (12.72)$$

Substituting relations (12.72) and (12.68) into eqn (12.71), for the case of normal incidence ($\theta_{tr} = 0$), it follows

$$E_x^{tr} = \frac{2}{\sqrt{\varepsilon_{rg}} + 1} E_x^{inc}(t - t_0) \quad (12.73)$$

where $t_0 = d/v$ denotes the time delay.

It should be stated that a field reference point ($x = 0$, $z = 0$) has been used throughout this analysis.

Finally, combining relations (12.66–12.73), the resulting integral equation for the wire immersed in a dielectric medium becomes

$$\int_0^L \frac{I(x', t - R/v)}{4\pi R} \mathrm{d}x' - \int_0^L \frac{\frac{1}{\varepsilon_{rg}}\cos\theta - \sqrt{\frac{1}{\varepsilon_{rg}} - \sin^2\theta}}{\frac{1}{\varepsilon_{rg}}\cos\theta + \sqrt{\frac{1}{\varepsilon_{rg}} - \sin^2\theta}} \frac{I(x', t - R^*/v)}{4\pi R^*} \mathrm{d}x'$$
$$= F_0\left(t - \frac{x}{v}\right) + F_L\left(t - \frac{L - x}{v}\right)$$
$$+ \frac{1}{2Z_g} \int_0^L \frac{2}{\sqrt{\varepsilon_{rg}} + 1} E_x^{inc}\left(x', t - t_0 - \frac{|x - x'|}{v}\right) \mathrm{d}x' \quad (12.74)$$

The problem of a straight wire embedded in a dielectric medium governed by the space-time Hallen integral eqn (12.74) can be solved by prescribing, without any loss of generality, the zero edge $I(0,t) = I(L,t) = 0$ and initial conditions $I(x,0) = 0$.

The use of a convolution approach through eqns (12.69–12.72) to handle the reflection and transmission coefficients is not necessary for the case of dielectric half-space as they are only functions of the angle. This integral equation formulation is used as it can serve as a starting point in deriving the model for the wire buried in a medium with finite conductivity.

12.2.2 Time-Domain Energy Measures

The transient response of a straight thin-wire embedded in a dielectric half-space can be postprocessed by the time-domain energy measures based on spatial integrals of the squared current and charge induced along the wire.

The time-domain energy measures represented by the current and charge induced on an object yield insight into where and how much the object radiates as a function of time. These measures were originally proposed in Ref. [22] and reexamined in Refs. [20,23,24].

Upon solving the set of Hallen integral eqn (12.74) for the transient current along the wire located in a dielectric medium, the charge distribution along the wire can be determined from the continuity equation [20]

$$q = -\int_0^t \frac{\partial I(x', t)}{\partial x'} \mathrm{d}t \tag{12.75}$$

where q is the linear charge distribution along the embedded wire in a dielectric half-space.

Having found the current and charge, measures of the H-field (kinetic) and the E-field (static) energy densities are expressed as proposed in Ref. [22].

The H-field energy is represented by the relation

$$W_I = \frac{\mu_0}{4\pi} \int_0^L I^2(x', t)\mathrm{d}x' \tag{12.76}$$

while the E-field energy is measured by the following integral:

$$W_q = \frac{1}{4\pi\varepsilon_r\varepsilon_0} \int_0^L q^2(x', t)\mathrm{d}x' \tag{12.77}$$

The total energy stored in the near field is proportional to the sum of W_I and W_q [22].

Numerical procedures for the calculation of the current, charge, and time domain energy are outlined in the following section.

12.2.3 Time-Domain Numerical Solution Procedures

Time-domain modeling is a more demanding task than is the frequency-domain approach; however, the former provides not only a better insight into the physical transient phenomena but also some computational advantages [7,20,2]. The time-domain version of the indirect Galerkin-Bubnov boundary element method (GB-BEM) applied to the solution of various Hallen integral equation types provides the stable numerical results [18–21] and it is used for numerical handling of the straight thin wire in a dielectric half-space in Ref. [16].

As the time-domain solution procedure for the Hallen integral equation is stable for an arbitrary time interval, it does not require any smoothing procedure, contrary to most of the known techniques [17]. The space-time discretization is performed carefully, so that within one time increment the propagation on at least one space segment is considered, thus satisfying the inequality

$$\Delta t \leq \frac{\Delta x}{c} \tag{12.78}$$

Through the marching-on-in-time procedure, it is possible to obtain the solution for the current at a present time for each space node as a function of currents at previous instants, without requiring matrix inversion. It is also necessary to prescribe the initial values of current at the wire ends to start the stepping procedure.

According to the usual space-time discretization procedure, the local approximation for unknown current can be expressed in the form

$$I(x', t') = \{f\}^T \{I\} \tag{12.79}$$

where $\{f\}$ is a vector containing shape functions, and $\{I\}$ is the time-dependent solution vector. In addition, applying the weighted residual approach, the space boundary discretization of integral eqn (12.74) leads to the local equation system for ith source and jth observation boundary element:

$$\begin{aligned}
&\int_{\Delta l_j}\int_{\Delta l_i} \{f\}_j \{f\}_i^T \frac{1}{4\pi R} \mathrm{d}x' \mathrm{d}x \{I\}\Big|_{t-\frac{R}{v}} \\
&- \int_{\Delta l_j}\int_{\Delta l_i} \{f\}_j \{f\}_i^T \frac{\dfrac{1}{\varepsilon_{rg}}\cos\theta - \sqrt{\dfrac{1}{\varepsilon_{rg}} - \sin^2\theta}}{\dfrac{1}{\varepsilon_{rg}}\cos\theta + \sqrt{\dfrac{1}{\varepsilon_{rg}} - \sin^2\theta}} \frac{1}{4\pi R^*} \mathrm{d}x' \mathrm{d}x \{I\}\Big|_{t-\frac{R^*}{v}} \\
&= \int_{\Delta l_j} F_0\left(t - \frac{x}{v}\right) \{f\}_j \mathrm{d}x + \int_{\Delta l_j} F_L\left(t - \frac{L-x}{v}\right) \{f\}_j \mathrm{d}x \\
&+ \frac{1}{2Z_g} \int_{\Delta l_j}\int_{\Delta l_i} \frac{2}{\sqrt{\varepsilon_{rg}} + 1} E_x^{inc}\left(x', t - t_0 - \frac{|x - x'|}{c}\right) \{f\}_j \mathrm{d}x' \mathrm{d}x
\end{aligned} \tag{12.80}$$

Expression (12.80) can be written in the matrix form that is given by

$$
\begin{aligned}
&[A]\{I\}\Big|_{t-\frac{R}{c}} - [A_1^*]\{I\}\Big|_{t-\frac{R^*}{c}} \\
&= [B]\{E\}\Big|_{t-\frac{x-x'}{c}} + [C]\left\{\sum_{n=0}^{\infty} I^n\right\}\Bigg|_{t-\frac{x}{c}-\frac{2nL}{c}-\frac{R_0}{c}} \\
&- [C_1^*]\left\{\sum_{n=0}^{\infty} I^n\right\}\Bigg|_{t-\frac{x}{c}-\frac{2nL}{c}-\frac{R_0^*}{c}} - [B]\left\{\sum_{n=0}^{\infty} E^n\right\}\Bigg|_{t-\frac{x}{c}-\frac{2nL}{c}-\frac{x'}{c}} \\
&- [D]\left\{\sum_{n=0}^{\infty} I^n\right\}\Bigg|_{t-\frac{x}{c}-\frac{2n+1}{c}L-\frac{R_L}{c}} + [D_1^*]\left\{\sum_{n=0}^{\infty} I^n\right\}\Bigg|_{t-\frac{x}{c}-\frac{2n+1}{c}L-\frac{R_L^*}{c}} \\
&+ [B]\left\{\sum_{n=0}^{\infty} E^n\right\}\Bigg|_{t-\frac{x}{c}-\frac{2n+1}{c}L-\frac{L-x'}{c}} + [D]\left\{\sum_{n=0}^{\infty} I^n\right\}\Bigg|_{t-\frac{L-x}{c}-\frac{2nL}{c}-\frac{R_L}{c}} \\
&- [D_1^*]\left\{\sum_{n=0}^{\infty} I^n\right\}\Bigg|_{t-\frac{L-x}{c}-\frac{2nL}{c}-\frac{R_L^*}{c}} - [B]\left\{\sum_{n=0}^{\infty} E^n\right\}\Bigg|_{t-\frac{L-x}{c}-\frac{2nL}{c}-\frac{L-x'}{c}} \\
&- [C]\left\{\sum_{n=0}^{\infty} I^n\right\}\Bigg|_{t-\frac{L-x}{c}-\frac{2n+1}{c}L-\frac{R_0}{c}} + [C_1^*]\left\{\sum_{n=0}^{\infty} I^n\right\}\Bigg|_{t-\frac{L-x}{c}-\frac{2n+1}{c}L-\frac{R_0^*}{c}} \\
&+ [B]\left\{\sum_{n=0}^{\infty} E^n\right\}\Bigg|_{t-\frac{L-x}{c}-\frac{2n+1}{c}L-\frac{x'}{c}}
\end{aligned}
\tag{12.81}
$$

where $\{E\}$ vector denotes the excitation function and the space-time dependent matrices are given as follows:

$$[A] = \int_{\Delta l_j}\int_{\Delta l_i} \frac{1}{4\pi R}\{f\}_j\{f\}_i^T \mathrm{d}x'\mathrm{d}x \tag{12.82}$$

$$[A_1^*] = \int_{\Delta l_j}\int_{\Delta l_i} \frac{\frac{1}{\varepsilon_{rg}}\cos\theta - \sqrt{\frac{1}{\varepsilon_{rg}} - \sin^2\theta}}{\frac{1}{\varepsilon_{rg}}\cos\theta + \sqrt{\frac{1}{\varepsilon_{rg}} - \sin^2\theta}} \frac{1}{4\pi R^*}\{f\}_j\{f\}_i^T \mathrm{d}x'\mathrm{d}x \tag{12.83}$$

$$[B] = \frac{1}{2Z_0}\int_{\Delta l_j}\int_{\Delta l_i} \{f\}_j\{f\}_i^T \mathrm{d}x'\mathrm{d}x \tag{12.84}$$

$$[C] = \int_{\Delta l_j}\int_{\Delta l_i} \frac{1}{4\pi R_0}\{f\}_j\{f\}_i^T \mathrm{d}x'\mathrm{d}x \tag{12.85}$$

$$[C_1^*] = \int_{\Delta l_j}\int_{\Delta l_i} \frac{\frac{1}{\varepsilon_{rg}}\cos\theta - \sqrt{\frac{1}{\varepsilon_{rg}} - \sin^2\theta}}{\frac{1}{\varepsilon_{rg}}\cos\theta + \sqrt{\frac{1}{\varepsilon_{rg}} - \sin^2\theta}} \frac{1}{4\pi R_0^*}\{f\}_j\{f\}_i^T \mathrm{d}x'\mathrm{d}x \tag{12.86}$$

$$[D] = \int_{\Delta l_j} \int_{\Delta l_i} \frac{1}{4\pi R_L} \{f\}_j \{f\}_i^T \mathrm{d}x' \mathrm{d}x \tag{12.87}$$

$$[D_1^*] = \int_{\Delta l_j} \int_{\Delta l_i} \frac{\frac{1}{\varepsilon_{rg}} \cos\theta - \sqrt{\frac{1}{\varepsilon_{rg}} - \sin^2\theta}}{\frac{1}{\varepsilon_{rg}} \cos\theta + \sqrt{\frac{1}{\varepsilon_{rg}} - \sin^2\theta}} \frac{1}{4\pi R_L^*} \{f\}_j \{f\}_i^T \mathrm{d}x' \mathrm{d}x \tag{12.88}$$

Having completed the space discretization procedure, the weighted residual approach is used for the time discretization procedure as well.

Assuming that solution in time on the ith space segment can be expressed as

$$I_i(t') = \sum_{k=1}^{N_t} I_i^k T^k(t') \tag{12.89}$$

where I_i^k are the unknown coefficients and T^k are the time-domain shape functions. Choosing the Dirac impulses as test functions, the recurrence formula for the space-time varying current can be written as

$$I_j\Big|_{t_k} = \frac{-\sum_{i=1}^{N_g} \left(\bar{A}_{ji} I_i\Big|_{t_k - \frac{R}{c}} - A_{ji_1}^* I_{i_1}\Big|_{t_k - \frac{R^*}{c}} \right) + g_{j_1} * \Big|_{all\ retarded\ times}}{A_{jj}} \tag{12.90}$$

where N_g denotes the total number of global nodes, A_ji are the global matrix terms, $g_{jl}*$ is the whole right-hand side of expression (12.81) containing the excitation and the currents at previous instants, and the overbar line denotes that the self term is omitted.

Once the current distribution is obtained by solving the integral eqn (12.74) via the GB-IBEM, the energy-measure integrals (12.76) and (12.77) can be evaluated.

First, the charge distribution is determined by solving the integral

$$q = -\sum_{i=1}^{M} \sum_{k=1}^{N_t} \int_{\Delta t_k} \frac{\partial}{\partial x} \{f\}_i^T \{I\}_i \mathrm{d}t \tag{12.91}$$

where M denotes the total number of segments, while N_t stands for the total number of time steps.

The solution of integral in eqn (12.91) is carried out analytically and given in the form

$$q = -\frac{1}{2v} \sum_{i=1}^{M} \sum_{m=1}^{N_t} (I_{i+1}^m + I_{i+1}^{m+1} - I_i^m - I_i^{m+1}) \tag{12.92}$$

The H-field energy measure is obtained by evaluating the integral

$$W_I = \frac{\mu_0}{4\pi} \sum_{i=1}^{M} \int_{\Delta t_k} \frac{\partial}{\partial x'} (\{f\}_i^T \{I\}_i)^2 \mathrm{d}x' \tag{12.93}$$

The solution is available in the closed form and is given by

$$W_I = 10^{-7} \frac{\Delta x}{3} \sum_{i=1}^{M} [(I_i^k)^2 + I_i^k I_{i+1}^k + (I_{i+1}^k)^2]; \quad k = 1, 2, \ldots, N_t \tag{12.94}$$

The E-field energy is obtained from the integral

$$W_q = \frac{1}{4\pi\varepsilon_r\varepsilon_0} \sum_{i=1}^{M} \int_{\Delta t_k} \frac{\partial}{\partial x'} (\{f\}_i^T \{q\}_i)^2 \mathrm{d}x' \tag{12.95}$$

for which the solution is then

$$W_q = \frac{1}{4\pi\varepsilon_r\varepsilon_0} \frac{\Delta x}{3} \sum_{i=1}^{M} [(q_i^k)^2 + q_i^k q_{i+1}^k + (q_{i+1}^k)^2]; \quad k = 1, 2, \ldots, N_t \tag{12.96}$$

and the total energy measure is given by the sum of W_I and W_q.

12.2.4 Computational Examples

The first example is related to the straight wire of length $L = 5$ m and radius $a = 1$ cm embedded in the dielectric half-space with $\varepsilon_r = 10$ at the depth $d = 1$ m. The wire is illuminated by the transmitted part of the electromagnetic pulse (EMP) incident waveform

$$E^{inc}(t) = E_0(e^{-at} - e^{-bt}) \tag{12.97}$$

where the EMP parameters are $E_0 = 52.5\,\mathrm{kV/m}$, $a = 4 \times 10^6\mathrm{s}^{-1}$, $b = 4.78 \times 10^8\mathrm{s}^{-1}$.

The transient current induced at the wire center is shown in Figure 12.8.

The transient response obtained using the direct time-domain approach seems to be in a satisfactory agreement with the results computed via the indirect frequency domain approach. The multiple reflections of the transient current at the wire open ends are due to the reflected waves. This effect can also be observed in Figures 12.9–12.11.

The curves shown in Figure 12.9 for the transient current induced at the center of the wire embedded in the dielectric half-space at depth $d = 1$m, $d = 10$m, and $d = 20$ m represent time delayed waveforms due to the propagation delay necessary for the incident electric field to reach the wire. A slight amplitude and waveform variation as a function of depth indicate the minimal interface effect.

A slight proximity effect of the ground-air interface is visible for the case of the transient current induced along the wire placed at depth $d = 1$m below ground, while only the time delay is noticeable for the curves related to the wires located at depth $d = 10$ and $d = 20$ m, respectively.

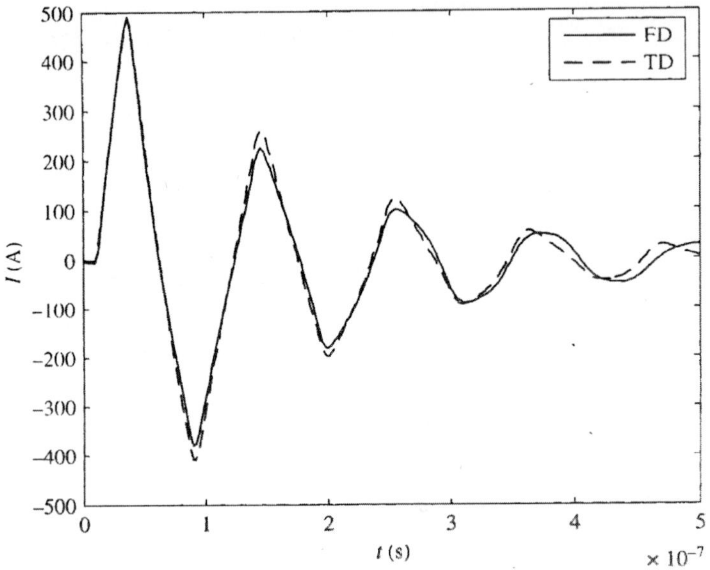

Figure 12.8 Transient current induced at the center of the straight wire ($L = 5$ m, $a = 1$ cm, $\varepsilon_r = 10$, $d = 1$m). (Poljak, D., Doric, V.: Time Domain Modeling of Electromagnetic Field Coupling to Finite Length Wires Embedded in a Dielectric Half-Space. *IEEE trans. EMC-47*, No 2., pp 247–253, May 2005).

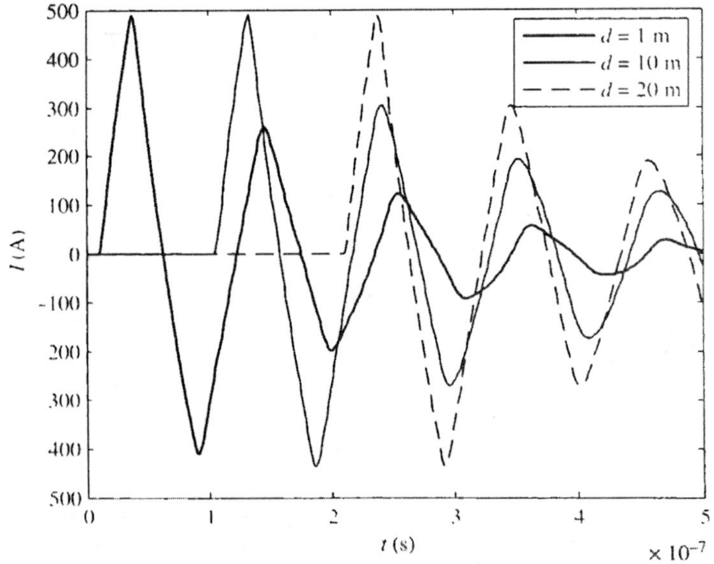

Figure 12.9 Transient current induced at the center of the straight wire ($L = 5$ m, $a = 1$ cm, $\varepsilon_r = 10$) for various depths. (Poljak, D., Doric, V.: Time Domain Modeling of Electromagnetic Field Coupling to Finite Length Wires Embedded in a Dielectric Half-Space. *IEEE trans. EMC-47*, No 2., pp 247–253, May 2005).

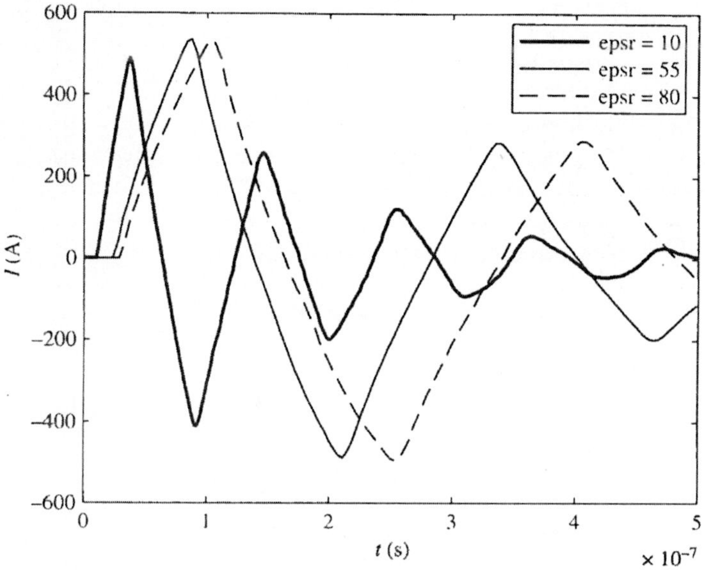

Figure 12.10 Transient response of the straight wires ($L = 5\,\text{m}$, $a = 1\,\text{cm}$) for various permittivity values. (Poljak, D., Doric, V.: Time Domain Modeling of Electromagnetic Field Coupling to Finite Length Wires Embedded in a Dielectric Half-Space, *IEEE trans. EMC-47*, No 2., pp 247–253, May 2005).

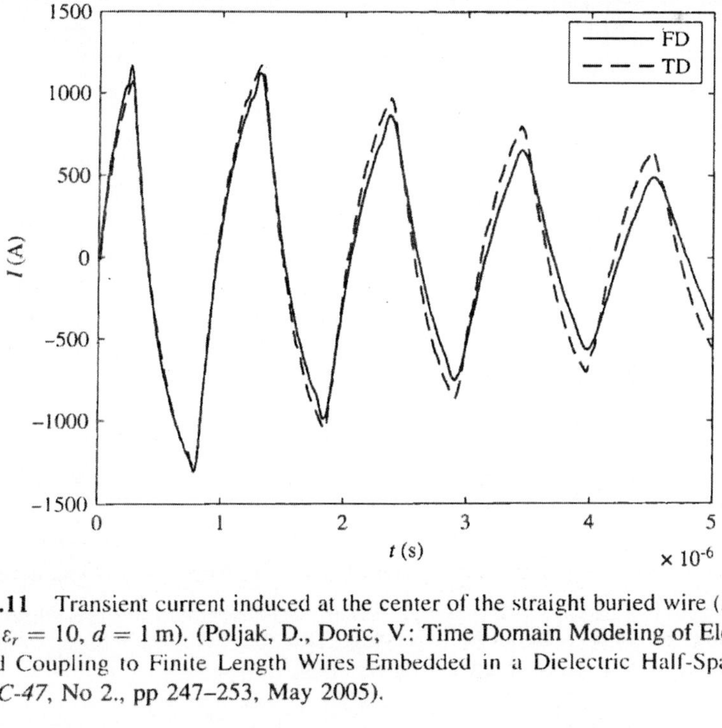

Figure 12.11 Transient current induced at the center of the straight buried wire ($L = 50\,\text{m}$, $a = 1\,\text{cm}$, $\varepsilon_r = 10$, $d = 1\,\text{m}$). (Poljak, D., Doric, V.: Time Domain Modeling of Electromagnetic Field Coupling to Finite Length Wires Embedded in a Dielectric Half-Space, *IEEE trans. EMC-47*, No 2., pp 247–253, May 2005).

The initial delay of the first peak is caused by the arrival time of the incident field and strongly depends on the depth d. After that time, the transient behavior of the incident current is the same.

Furthermore, Figure 12.10 shows the transient current induced at the center of $L = 5$ m long wire buried at $d = 1$ m below the interface for the various values of permittivity. Thus, three typical values of relative dielectric constant are chosen: $\varepsilon_r = 10$ (ground), $\varepsilon_r = 55$ (brain permittivity at GSM frequencies), $\varepsilon_r = 80$ (sea water). It can be observed from Figure 12.10 that permittivity of the medium strongly affects the transient response of the wire embedded in the dielectric medium.

As shown in Figure 12.10, the initial delay of the transient waveform first peak is significantly influenced by the relative permittivity of a dielectric semiinfinite medium ε_{rg}. Also, time delay of the reflections from the wire free ends is influenced by the velocity of the propagation in the dielectric medium. Smaller the value of ε_{rg} the greater is the velocity v while the oscillation period is decreased.

Figure 12.11 shows the transient response of the straight thin wire of length $L = 50$ m and radius $a = 1$ cm embedded in the dielectric half-space with $\varepsilon_r = 10$ at the depth $d = 1$m. The wire is illuminated by the EMP waveform (12.97). This transient response has been computed by means of the direct time-domain and the indirect frequency-domain methods, respectively. The numerical results obtained by the different approaches agree favorably again.

Comparison of the transient current from Figure 12.11 with the current waveform from Figure 12.8 clearly shows the influence of the wire length to the transient behavior.

It is also visible from Figure 12.5 that the dominant effect to the transient behavior of the induced current along the line is due to the multiple reflections of the current wave from the line open ends.

There is a slight frequency shift in Figures 12.8 and 12.11 between time-domain and frequency-domain results, particularly, for later time instants. Consequently, there are some points to be clarified regarding the frequency-domain modeling.

The transients of highly resonant structures have very long duration, while their related frequency spectra contain sharp peaks. Coarse frequency resolution cannot resolve the resonant points accurately, thus resulting in errors in transient waveforms.

The problem of analyzing transients of highly resonant structures in the frequency domain is the inability to know beforehand the frequency resolution required for sampling the spectrum. Dynamic adaptive sampling can be used to overcome this problem and more details can be found in Refs. [7,25]. An equivalent problem in analyzing transients in time domain is the inability to know the time duration of the waveform.

The related frequency spectrum of the impulse and transient response of a 50 m long line immersed in a dielectric medium with $\varepsilon_r = 10$ at depth $d = 1$ m as shown in Figure 12.12 and Figure 12.13, respectively.

As it is visible from Figures 12.12 and 12.13, the frequency response spectra contain a number of peaks that decrease with frequency. Although the frequency range of these spectra is infinite, signal amplitudes for the frequencies higher than 50 MHz are low enough to be neglected. On the contrary, the location and the amplitude values of the first two peaks strongly affect the obtained transient waveform.

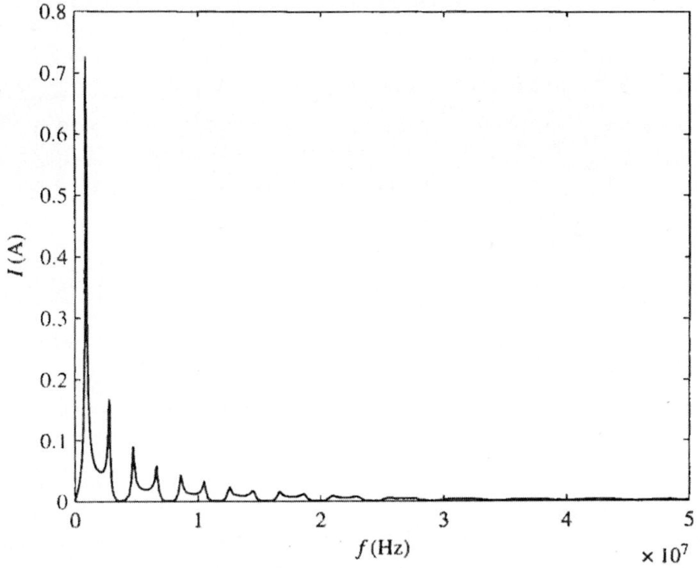

Figure 12.12 Frequency spectrum of the impulse response at the center of the straight wire ($L = 50\,\text{m}$, $a = 1\,\text{cm}$, $\varepsilon_r = 10$, $d = 1\,\text{m}$). (Poljak, D., Doric, V.: Time Domain Modeling of Electromagnetic Field Coupling to Finite Length Wires Embedded in a Dielectric Half-Space, *IEEE trans. EMC-47*, No 2., pp 247–253, May 2005).

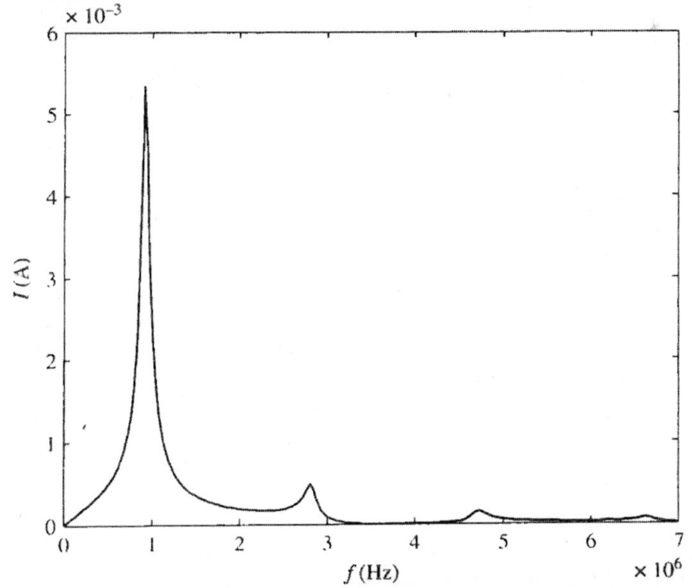

Figure 12.13 Frequency spectrum of the transient response at the center of the straight wire ($L = 50\,\text{m}$, $a = 1\,\text{cm}$, $\varepsilon_r = 10$, $d = 1\,\text{m}$). (Poljak, D., Doric, V.: Time Domain Modeling of Electromagnetic Field Coupling to Finite Length Wires Embedded in a Dielectric Half-Space, *IEEE trans. EMC-47*, No 2., pp 247–253, May 2005).

Therefore, it is very important for frequency samples to contain exact frequencies of the first two peaks. This could be achieved by using the sufficiently fine frequency step (resulting in large number of samples) or by means of the dynamic adaptive sampling technique presented in Ref. [25]. In this work, a number of $N = 2^{15}$ samples over the frequency range of 50 MHz have been used in order to obtain satisfactory convergence rate.

Since the current distribution evaluation for a single frequency takes around 30 (depending on the computer speed), the calculation of current amplitude for each sample is obviously impossible. Thus, about 150 samples over 50 MHz frequency range have been calculated, while the rest of the samples have been interpolated using cubic splines.

The effect of the interface and the validity of the Fresnel reflection coefficient approximation in modeling the interface should be discussed as well.

A relatively short wire ($L = 0.5$ m) is chosen to stress out the earth-air interface effect. Figure 12.14 shows the transient current induced at the center of 0.5 m long wire for various depths. As it is obvious from Figure 12.14, there is a slight variation in amplitude, waveform, and time shift for depths $d = 0.25$ m, $d = 1$ m, and $d = 2$ m. Only the curve for $d = 0.1$ m clearly demonstrates the influence of the earth-air interface in both the amplitude and waveform of the actual transient response.

Figure 12.15 shows the behavior of the energy stored in the near field of the wire, that is, it represents the decrease of the total energy of the wire with time once the exciting pulse vanishes.

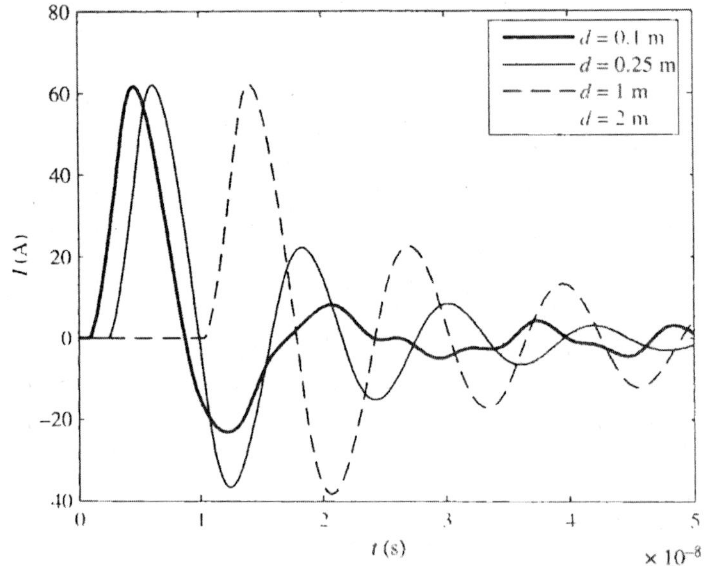

Figure 12.14 Transient current induced at the center of the straight wire ($L = 0.5$ m, $a = 1$cm, $\varepsilon_r = 10$) for various depths. (Poljak, D., Doric, V.: Time Domain Modeling of Electromagnetic Field Coupling to Finite Length Wires Embedded in a Dielectric Half-Space, *IEEE trans. EMC-47*, No 2., pp 247–253, May 2005).

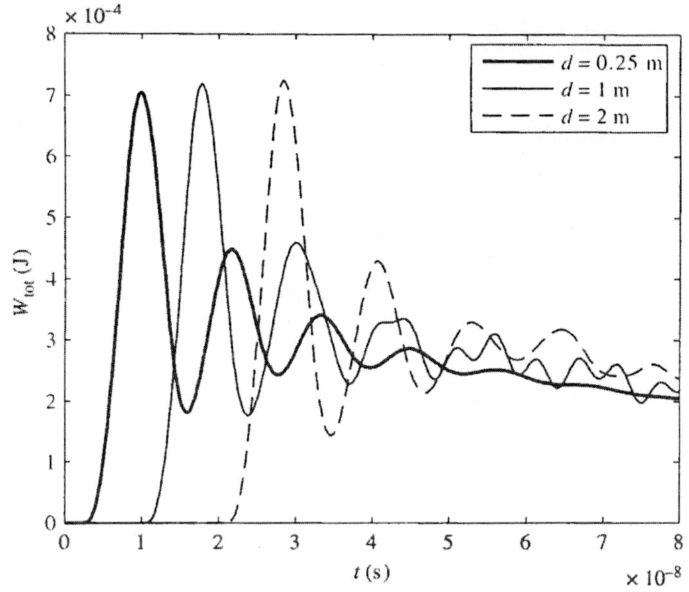

Figure 12.15 The measure of total energy (W_{tot}) energy as a function of time for various depths.

Figure 12.15 also demonstrates the absorbing effect of the dielectric half-space when the wire is brought closer to the interface.

The evaluation of the validity of the proposed model is not an easy task even in the frequency domain and even for the simplified case of a dielectric half-space. Generally, the validity of various approximations depends on the relationship between the spectral content and the direction of the incident field, the electrical properties of the earth, and to a lesser extent the burial depth of the line. The computational aspects in the frequency domain modeling of thin-wire antennas in the presence of a lossy half-space have been reported in a number of papers and among the most cited ones are Refs. [3,4].

The frequency-domain analysis of the buried wire scatterer has been carried out in the previous work of the first author via the rigorous Sommerfeld integral [5] and via the approximate reflection coefficient (RC) approach [8] to account for the influence of the earth-air interface reflected field upon the straight wire scatterer current distribution.

The Sommerfeld integral approach has been found to be numerically stable and reliable for straight horizontal line brought to within 10^{-6} wavelengths of the interface [4,5]. The rigorous and approximate results are in a very good agreement for depths greater or at least equal to

$$d \geq \frac{\lambda_0}{4\sqrt{\varepsilon_r}}, \quad \text{where } \lambda_0 = \frac{c}{f} \tag{12.98}$$

As a rough guideline, the reflection coefficient approach to account for the reflection from the earth-air interface has been found to produce results generally within 10% of those obtained using rigorous, but computationally very expensive, Sommerfeld integral approach.

Furthermore, for wire depths

$$d \leq \frac{\lambda_0}{10\sqrt{\varepsilon_r}} \tag{12.99}$$

the qualitative dependence of the input admittance (obtained by using RC approach) upon depth is found to be generally correct, but these results primarily differ from the rigorous results due to a slight shift in the maxima with respect to depth.

Essentially, from the time-domain point of view, this condition should be satisfied for each component of the considered frequency spectrum.

It is also necessary to make at least a general trade-off between a dielectric half-space and an imperfectly conducting half-space. The absence of the ground conductivity (or at least low values of ground conductivity) causes strong resonance effect that is particularly noticeable in Figures 12.11–12.13.

From the findings in the frequency domain, the increasing ground conductivity is expected to decrease the response rapidly. In the realistic problems such as wire immersed inside sea water, ground probing, or ground penetrating radar conductivity should not be neglected. The influence of the finite conductivity reduces the external electric field and influences the propagation on the induced current. The finite conductivity can also delay the initial field. However, the corresponding Green's function, which is responsible for the second effect, is much more complicated and has to be calculated by the Sommerfeld integral. In this case, one would need to perform additional integration in the Hallen equation and the problem would become tremendously time consuming.

To roughly estimate whether a dielectric half-space approximation of a dissipative half space can be used, one should examine the absolute value of the refractive index of the earth that is given by

$$|n| = \sqrt{\varepsilon_{rg}^2 + \left(\frac{\sigma_g}{\omega\varepsilon_0}\right)^2} \tag{12.100}$$

Conductivity σ should be at least an order of magnitude lesser than $\omega\varepsilon_0$ to be neglected.

For example, a reasonably dry earth, $\varepsilon_{rg} = 10$, in the frequency band from around 1 kHz to 1 MHz requires the conductivity σ to vary from 10^{-6} to 10^{-4} S/m.

To sum up, the numerical results obtained via the different approaches agree satisfactorily, that is, the maximum deviation between the results is around 6%.

The transient response of the single straight wire immersed in a dielectric half-space has been found to be influenced to a greater extent by the line length and the permittivity of a dielectric medium, and to a lesser extent by the burial depth.

12.3 REFERENCES

[1] D. Poljak: Electromagnetic Modelling of Wire Antenna Structures, *WIT Press, Southampton, Boston*, 2001.

[2] D. Poljak: New numerical approach in analysis of thin wire radiating over lossy half-space, *Int. J. Num. Methods Eng., 38 (22)*, 1995, pp 3803–3816.

[3] E. K. Miller, A. J. Poggio, G. J. Burke, E. S. Selden: Analysis of wire antennas in the presence of a conducting half-space, part II. the horizontal antenna in free space, *Can. J. Phys.*, 50, 1972, pp 2614–2627.

[4] G. J. Burke, E. K. Miller: Modelling antennas near to and penetrating a lossy interface, *IEEE Trans. AP, 32 (10)*, October 1984, pp 1040–1049.

[5] D. Poljak, V. Roje: The integral equation method for ground wire input impedance, in: C. Constanda, J. Saranen, S. Seikkala (Ed.), Integral Methods in Science and Engineering, Vol. I Analytic Methods, *Longman*, 1997, pp 139–143.

[6] G. E. Bridges: Transient plane wave coupling to bare and insulated cables buried in a lossy half-space, *IEEE Trans. EMC, 37 (1)*, February 1995, pp 62–70.

[7] D. Poljak, C. Y. Tham: Integral equation techniques in electromagnetics, *WIT Press, Southampton, Boston*, 2003.

[8] D. Poljak: Electromagnetic modeling of finite length wires buried in a lossy half-space, Engineering Analysis with Boundary Elements, Vol. 26, January 2002, pp 81–86.

[9] V. Doric, D. Poljak, V. Roje: Transient plane wave coupling to a finite length wire buried in a conductive ground, Boundary Elements XVII, 2005, pp 609–617.

[10] D. Poljak, C. A. Brebbia: Indirect Galerkin-Bubnov boundary element method for solving integral equations in electromagnetics, *Eng. Anal. Boundary Elements, 28 (7)*, 2004, pp 771–777.

[11] F. Tesche, M. Ianoz, F. Carlsson: EMC Analysis Methods and Computational Models, *John Wiley & Sons, New York*, 1997.

[12] S. Tkatchenko, F. Rachidi, M. Ianoz: Electromagnetic field coupling to a line of a finite length: theory and fast iterative solutions in frequency and time domains, *IEEE Trans. EMC, 37 (4)*, 1995, pp 509–518.

[13] R. E. Ziemer, W. H. Tranter: Principles of Communications, *Houghton Mifflin company, Boston, Toronto*, 1995.

[14] R. W. P. King, P. F. Sforza, T. I. S. Boak: The current in a parasitic antenna in a dissipative medium, *IEEE Trans. Antennas Propagat.*, 1974, pp 809–814.

[15] L. D. Grcev, F. E. Menter: Transient electromagnetic fields near large earthing systems", *IEEE Trans. Magn.*, 32, May 1996, pp 1525–1528.

[16] D. Poljak, V. Doric: Time domain modeling of electromagnetic field coupling to finite length wires embedded in a dielectric half-space, *IEEE Trans. EMC, 47 (2)*, May 2005, pp 247–253.

[17] S. M. Rao, T. K. Sarkar, S. A. Dianat: A novel technique to the solution of transient electromagnetic scattering from thin wires, *IEEE Trans. AP*, 34, May 1986, pp 630–634.

[18] D. Poljak, V. Roje: Time domain modeling of electromagnetic field coupling to transmission lines, Proceeding 1988. IEEE EMC Symposium, *Denver, USA*, August 1998, pp 1010–1013.

[19] D. Poljak, C. Y. Tham, A. McCowen, V. Roje: Electromagnetic pulse excitation of multiconductor transmission Lines, ICEAA '99, *Turin, Italy*, September 13–17, 1999, pp 789–792.

[20] D. Poljak, E. K. Miller, C. Y. Tham: Time domain energy measures for thin-wire antennas and scatterers, *IEEE Antennas Prop. Mag., 44 (1)*, February 2002, pp 87–95.

[21] D. Poljak, C. Y. Tham, A. McCowen: Transient response of nonlinearly loaded wires in a two media configuration, *IEEE Trans. EMC 46 (1)*, 2004.

[22] E. K. Miller, J. A. Landt: Direct time-domain techniques for transient radiation and scattering from wires, *Proc. IEEE, 168 (11)*, November 1980, pp 1396–1423.

[23] E. K. Miller PCs for AP and EM Reflections, *IEEE Ant. & Prop. Mag., 40 (1)*, February 1998, pp 96–100, and *41 (2)*, April 1998, pp 92–95.

[24] D. Poljak, E. K. Miller, C. Y. Tham, S. Antonijevic, V. Doric: Time domain analysis of the energy stored in the near field of multiple straight-wires above dielectric half-space, Proceedings of the ICEAA, *Turin, Italy*, September 2005.

[25] C. Y. Tham, A. McCowen, M. S. Towers, D. Poljak: Dynamic adaptive sampling technique in frequency-domain transient analysis, *IEEE Trans. EMC, 44 (4)*, November 2002, pp 522–528.

13

SIMPLE GROUNDING SYSTEMS

Grounding systems, such as buried vertical or horizontal electrodes and large grounding grids, are an important part of safety and electrical equipment protection in industrial and power plants.

The principal task of such grounding systems is to ensure the safety of personnel and prevent damage of installations and equipment, that is, the configuration of grounding systems should avoid the values of transient step and touch voltages which determines the health hazard. The secondary purpose of grounding systems is to provide common reference voltage for all interconnected electrical and electronic systems.

Local inequalities of this reference potential and disturbances distributed along the grounding systems are a source of malfunction or even destruction of various components electrically connected to grounding systems. In particular, the induced transient voltages may cause serious damage of electronic equipment with low-signal levels that are sensitive to various types of electromagnetic interfaces.

Consequently, the analysis of the space-time distribution of transients along the grounding systems is necessary. One of the most important parameters arising from the transient analysis of a grounding system is the transient impedance.

In general, the transient response of any linear system could be obtained directly, by solving the time-domain equations, or by frequency-domain approach and inverse Fourier transform. The former approach is valid as far as the soil ionization effect can be neglected.

While the stationary behavior of grounding systems is well investigated, the transient analysis has been performed to a significantly lesser extent. Until now, there have been some studies dealing with the transient analysis of grounding systems based on analytical approaches, transmission line models, and electromagnetic models [1–5].

Analytical approaches can be referred to as the network concept, being based on the empirical approach and the lumped circuit theory. However, these approaches definitely suffer from too many approximations. The calculation of the equivalent electrical network parameters, due to conductive or inductive coupling, is rather difficult, particularly for the grounding systems of complex geometry.

Many conventional approaches for transient analysis of grounding systems are usually related to the transmission line method (TLM) [2,3]. TLM approach is, however, valid for long horizontal conductors, but not for vertical and especially interconnected conductors. Moreover, the effect of the air-earth interface has often been neglected assuming the conductors to be buried at very large depths [6]. Finally, TLM approach is not valid for low values of ground conductivity and it also neglects the effect of mutual coupling among the parts of the grounding system, thereby generating some errors.

The rigorous electromagnetic field approach to the analysis of the grounding system is based on wire antenna theory and is considered to be most accurate approach [4,5,7]. This model is based on solving the corresponding thin electric field integral equation (EFIE) for half-space problems.

According to this approach, the grounding system of interest is represented by a corresponding configuration of straight thin wire antennas buried in the surrounding soil. The soil is represented as a linear and homogeneous half-space characterized by its electrical parameters.

Through this rigorous approach, the effect of an imperfectly conducting half-space is taken into account by the Sommerfeld integrals, appearing in the integral equation kernel. This approach, however, suffers from too long computational time for the evaluation of broadband frequency spectrum. Consequently, the rigorous approach is often regarded as too complex for practical applications, especially for large grounding grids and it should be avoided wherever possible [8]. A useful study on the validity of quasistatic theory compared with full wave theory has been presented in Ref. [8]. One possible way of avoiding the computation of Sommerfeld integral is the use of modified image theory (MIT) [5,9]. The MIT approach takes into account only the electrical properties of the soil but not the burial depth as a parameter, and it is stated to be valid up to 1 MHz.

This chapter deals with the transient analysis of simple grounding systems, that is, with the transient impedance calculation of the vertical and horizontal grounding electrodes. The approach presented can be readily extended to more complex geometries related to more realistic grounding systems.

The vertical grounding electrode is analyzed in Section 13.1, while a more demanding case of horizontal electrode is considered in Section 13.2.

13.1 VERTICAL GROUNDING ELECTRODE

The transient analysis of grounding electrode presented in this section is based on the Pocklington integro-differential equation formulation [4,7] arising from the antenna theory. The influence of an earth-air interface has been taken into account via the simplified reflection coefficient approach (instead of analytically demanding and numerically time-consuming Sommerfeld integral approach [4,5,7–12]).

The current along the vertical grounding electrode is obtained by solving the corresponding integro-differential equation via the indirect scheme of the Galerkin–Bubnov boundary element method (GB-BEM) [7,9–11] and some illustrative numerical results are presented.

The feed-point voltage is evaluated using the procedure described in Ref. [13]. Firstly, the near electric field is expressed in terms of the current induced along the vertical electrode. The feed-point voltage can be obtained by integrating the normal electric field from the electrode surface to infinity. The integration is avoided by utilizing the weak formulation of the problem.

The input impedance of the grounding electrode, that is, the transfer function of the linear system, is defined as a ratio of evaluated voltage and injected lightning current at the feed point.

The frequency response of the grounding electrode is obtained by multiplying the input impedance spectrum with Fourier transform of the lightning current waveform. Finally, the transient impedance of the grounding electrode is computed by means of the inverse Fourier transform.

13.1.1 Integral Equation Formulation for the Vertical Grounding Electrode

The geometry of interest is a vertical straight wire of length L and radius a, buried in a lossy medium at depth d as shown in Figure 13.1. The wire is assumed to be

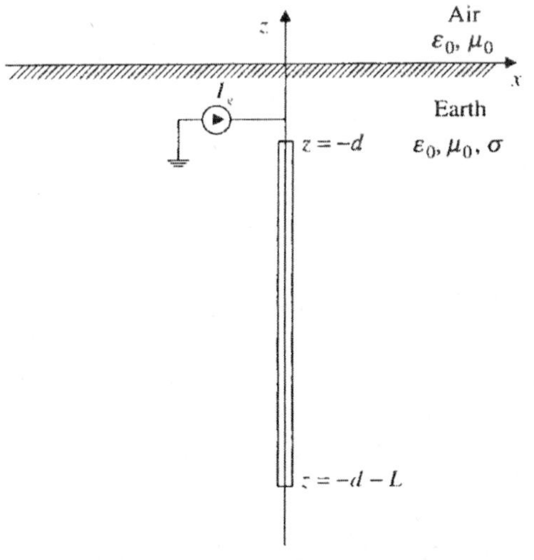

Figure 13.1 Vertical grounding electrode energized by a current source I_g. (D. Poljak and V. Doric, Wire Antenna Model for Transient, Analysis of Simple Groundingsystems, Part I: The Vertical Grounding Electrode, *Progress in Electromagnetics Research, PIER 64, 149–166, 2006*).

perfectly conducting and the wire dimensions satisfy the well-known thin wire approximation [4,7,8,11], so the current along the wire is z-directed only.

The starting point in the mathematical model is the assessment of the current distribution induced along the vertical electrode due to a time-harmonic excitation and for a number of frequencies within a frequency band of interest. This current distribution is governed by the Pocklington integro-differential equation. This integro-differential equation can be derived by expressing the electric field in terms of Hertz vector potential and by satisfying the boundary conditions for the tangential field components at the electrode surface.

The complete electric field induced in the vicinity of the straight thin wire of finite length buried in an imperfectly conducting half-space can be expressed in terms of Hertz vector potential Π [7]:

$$\vec{E} = \nabla(\nabla\vec{\Pi}) + k_1^2\vec{\Pi} \tag{13.1}$$

where k_1 is the phase constant of a lossy ground:

$$k_1^2 = -\omega^2\mu\varepsilon_{\text{eff}} \tag{13.2}$$

and ε_{eff} denotes the complex permittivity of the lossy ground:

$$\varepsilon_{\text{eff}} = \varepsilon_r\varepsilon_0 - j\frac{\sigma}{\omega} \tag{13.3}$$

where ε_r and σ are the relative permittivity and conductivity of the ground respectively, and ω denotes the operating frequency.

A corresponding integral equation formulation for the vertical grounding electrode using both the Sommerfeld and RC approach is presented in Sections 13.1.1.1 and 13.1.1.2, respectively.

13.1.1.1 Sommerfeld Integral Approach For the case of vertical grounding electrode energized by a current source I_g, the vector eqn (1) can be written as a set of two scalar equations for the normal x-component and tangential z-component of the electric field due to the vertical electrode:

$$E_x^V(x,z) = \frac{\partial^2\Pi_z^V}{\partial x\partial z} \tag{13.4}$$

$$E_z^V(x,z) = \left[\frac{\partial^2}{\partial z^2} + k_1^2\right]\Pi_z^V \tag{13.5}$$

where superscript V denotes the vertical electrode.

The vector potential z-component is given by Ref. [7]:

$$\Pi_z^V = \frac{1}{j4\pi\omega\mu\varepsilon_{\text{eff}}}\int_{-d-L}^{d}\left[g_0^V(x,z,z') - g_i^V(x,z,z') + k_0^2V_{11}\right]I(z')\mathrm{d}z' \tag{13.6}$$

where $I(z')$ is the unknown current distribution along the vertical straight wire, $g_0(x, z, z')$ denotes the free space Green's function of the form:

$$g_0^V(x, z, z') = \frac{e^{-jkR_{1v}}}{R_{1v}} \tag{13.7}$$

while $g_i(x,z,z')$ arises from the image theory and is given by

$$g_i^V(x, z, z') = \frac{e^{-jk_2 R_{2v}}}{R_{2v}} \tag{13.8}$$

where k_2 is the phase constant of free space:

$$k_2^2 = \omega^2 \mu \varepsilon_0 \tag{13.9}$$

and R_{1v} and R_{2v} are the distances from the wire in the ground and from its image in the air to the observation point in the lower medium, respectively.

The attenuation effect of the lossy ground is taken into account by the Sommerfeld integral term V_{11} [4]:

$$V_{11} = 2 \int_0^{\infty} \frac{e^{-\mu_1 (h-z)}}{k_2^2 \mu_1 + k_1^2 \mu_2} J_0(\lambda \rho) \lambda \mathrm{d}\lambda \tag{13.10}$$

where $J_0(\lambda\rho)$ is zero-order Bessel function of the first kind, while μ_1, μ_2, and ρ are given by

$$\mu_1 = (\lambda^2 - k_1^2)^{1/2}, \mu_2 = (\lambda^2 - k_2^2)^{1/2}, \rho = |z - z'| \tag{13.11}$$

The Pocklington integro-differential equation for the vertical straight wire buried in a lossy ground can now be obtained by enforcing the boundary conditions for the tangential electric field components on the perfectly conducting (PEC) wire surface. The total tangential electric field on the PEC wire surface at $x = a$ vanishes, that is:

$$E_z^{exc,V}(a, z) + E_z^{sct,V}(a, z) = 0 \tag{13.12}$$

where $E_z^{exc,V}$ denotes the excitation function and $E_z^{sct,V}$ is the scattered field along the electrode surface.

Combining the relations (13.5)–(13.12) leads to the Pocklington integro-differential equation for the vertical grounding electrode:

$$E_z^{exc,V} = -\frac{1}{j4\pi\omega\varepsilon_{\mathrm{eff}}} \int_{-d-L}^{-d} \left[\frac{\partial^2}{\partial z^2} + k_1^2\right] \left[g_0(z, z') - g_i(z, z') + k_0^2 V_{11}\right] I(z') \mathrm{d}z' \tag{13.13}$$

Solving the integral eqn (13.13), the current distribution along the electrode is obtained.

13.1.1.2 Reflection Coefficient Approach The repeated evaluation of the Sommerfeld integral (13.10) at several frequencies, by which the earth–air attenuation effect is taken into account, is rather a difficult and time consuming task [4,7–12].

Therefore, this work deals with a reflection coefficient (RC) approach of which the principal advantage versus rigorous Sommerfeld integral approach is simplicity of the formulation and significantly less computational cost.

For convenience, the integro-differential eqn (13.13) can be written in the form:

$$E_z^{exc} = -\frac{1}{j4\pi\omega\varepsilon_{\text{eff}}} \int_{-d-L}^{-d} G^V(z,z')I(z,z')\mathrm{d}z' \tag{13.14}$$

where $G(z,z')$ is the total Green's function given by:

$$G^V(z,z') = \left[\frac{\partial^2}{\partial z^2} + k_1^2\right]\left[g_0^V(z,z') - g_i^V(z,z') + k_0^2 V_{11}\right] \tag{13.15}$$

According to the RC approximation [7], the rigorous Green's function simplifies into

$$G^V(z,z') = \left[\frac{\partial^2}{\partial z^2} + k_1^2\right][g_0(z,z') + \Gamma g_i(z,z')] = \left[\frac{\partial^2}{\partial z^2} + k_1^2\right] g^V(z,z') \tag{13.16}$$

where Γ is the corresponding transmission coefficient [6] for the TM polarization:

$$\Gamma = \frac{\frac{1}{\underline{n}} - \sqrt{\frac{1}{\underline{n}}}}{\frac{1}{\underline{n}} + \sqrt{\frac{1}{\underline{n}}}} \tag{13.17}$$

and $\underline{n}$ is given by

$$\underline{n} = \frac{\varepsilon_{\text{eff}}}{\varepsilon_0} \tag{13.18}$$

Finally, the resulting Pocklington integro-differential equation for the vertical straight wire buried in a lossy half-space is given by

$$E_z^{exc,V} = -\frac{1}{j4\pi\omega\varepsilon_{\text{eff}}} \int_{-d-L}^{-d} \left[\frac{\partial^2}{\partial z^2} + k^2\right]\left[g_0^V(z,z') + \Gamma g_i^V(z,z')\right] I(z')\mathrm{d}z' \tag{13.19}$$

In this paper, the integro-differential eqn (13.19) is solved by means of the indirect Galerkin–Bubnov scheme of the boundary element method (GB-BEM). Solving the integral eqn (13.19), the equivalent current distribution is obtained.

13.1.1.3 Imposed Boundary Conditions In the analysis of the grounding electrodes, the excitation function is not given in the form of electric field, as the wire is not illuminated by the plane wave [4]. Hence, the left-hand side of the eqn (13.19) vanishes along the perfectly conducting (PEC) wire surface, that is,

$$E_x^{exc} = 0 \tag{13.20}$$

and the integro-differential eqn (13.19) becomes homogeneous.

The vertical electrode is energized by the injection of an arbitrary waveform current pulse produced by an ideal current generator with one terminal connected to the grounding electrode and the other one in the remote soil.

Thus, the excitation is given in the form of the current flowing into the electrode. This current source is included into the integral equation scheme through the simple boundary condition applied at the top of the electrode:

$$I(-d) = I_g \tag{13.21}$$

where I_g denotes the actual current generator.

In the frequency domain the unit current generator is always chosen, as its time domain counterpart is Dirac impulse.

13.1.2 The Evaluation of the Input Impedance Spectrum

The calculation of the vertical electrode input impedance is more demanding than in the antenna case, primarily because the input terminals are placed between electrode point and remote soil, and the tedious integration on the infinite integral cannot be avoided.

By solving the integral equations (13.19), the equivalent current distribution along the vertical electrode is obtained and the input impedance can be computed. It is worth noting that this input impedance depends only on the grounding system geometry and on the electrical properties of the surrounding soil.

The input impedance is simply defined by ratio

$$Z_{in} = \frac{V_g}{I_g} \tag{13.22}$$

where V_g and I_g are the values of the voltage and the current at the driving point, respectively.

The feed-point voltage is obtained by integrating the normal electric field component from the remote soil to the electrode surface, that is,

$$V_g = -\int_{\infty}^{a} \vec{E}d\vec{s} \tag{13.23}$$

Thus, the problem of obtaining the input impedance is related to the calculation of the feed-point voltage. The spectrum of input impedance is obtained by repeating this procedure in the wide frequency band.

For the given geometry of the vertical electrode, integral (13.23) becomes

$$V_g = -\int_{\infty}^{a} E_x^V(x, z)dx \tag{13.24}$$

Since only the Π_z component of the Hertz vector potential exists, the E_z component of the electric field is given by

$$E_x^V(x, z) = \frac{\partial^2 \Pi_z^V}{\partial x \partial z} = \frac{1}{j4\pi\omega\varepsilon_{\text{eff}}} \int_{-d-L}^{-d} I(z') \frac{\partial^2 g^V(x, z, z')}{\partial x \partial z} dz' \tag{13.25}$$

Numerical treatment of expression (13.25) can be simplified by featuring the weak formulation of the problem [7,9–11]. Namely, utilizing the property of Green's functions for the source and image wire, respectively:

$$\frac{\partial g_0^V(x, z, z')}{\partial z} = -\frac{\partial g_0^V(x, z, z')}{\partial z'} \tag{13.26}$$

$$\frac{\partial g_i^V(x, z, z')}{\partial z} = \frac{\partial g_i^V(x, z, z')}{\partial z'} \tag{13.27}$$

and performing the integration by parts it follows:

$$E_x^V(x, z) = -\frac{1}{j4\pi\omega\varepsilon_{\text{eff}}} \frac{d}{dx} \Bigg[I(z')g^V(x, z, z')\Big|_{z'=-d-L}^{z'=-d} - \int_{-d-L}^{-d} \frac{\partial I(z')}{\partial z'} g_0^V(x, z, z')dz' + \int_{-d-L}^{-d} \frac{\partial I(z')}{\partial z'} \Gamma g_i^V(x, z, z')dz' \Bigg] \tag{13.28}$$

Obviously, the mixed second-order differential operator is removed from the integral equation kernel G^V.

Furthermore, substituting eqn (13.25) into (13.24), the following expression for the feed-point voltage is obtained:

$$V_g = -\frac{1}{j4\pi\omega\varepsilon_{\rm eff}} \int_a^{\infty} \frac{\rm d}{{\rm d}x} \left[I(z')g^V(x,z,z') \Big|_{z'=-d-L}^{z'=-d} - \int_{-d-L}^{-d} \frac{\partial I(z')}{\partial z'} g_0^V(x,z,z'){\rm d}z' + \int_{-d-L}^{-d} \frac{\partial I(z')}{\partial z'} \Gamma g_i^V(x,z,z'){\rm d}z' \right] {\rm d}x \tag{13.29}$$

which simply leads to the following expression:

$$V_g = \frac{1}{j4\pi\omega\varepsilon_{\rm eff}} [I(-d)g^V(x,-d,z) - \int_{-d-L}^{-d} \frac{\partial I(z')}{\partial z'} g_0^V(x,z,z'){\rm d}z' + \int_{-d-L}^{-d} \frac{\partial I(z')}{\partial z'} \Gamma g_i^V(x,z,z'){\rm d}z' \Big|_{x=a}^{x=\infty} \tag{13.30}$$

in which the tedious numerical integration over infinite domain is avoided.

The desired input impedance of the vertical grounding electrode is finally defined by the following relation:

$$Z_{in} = \frac{1}{j4\pi\omega\varepsilon_{\rm eff} I_g} \left[I(-d)g^V(x,-d,z) - \int_{d-L}^{-d} \frac{\partial I(z')}{\partial z'} g_0^V(x,z,z'){\rm d}z' \Big|_{x=a}^{x=\infty} + \int_{-d-L}^{-d} \frac{\partial I(z')}{\partial z'} \Gamma g_i^V(x,z,z'){\rm d}z' \Big|_{x=a}^{x=\infty} \right. \tag{13.31}$$

Calculating this relation in a wide frequency range gives the frequency spectrum of the input impedance.

13.1.3 Numerical Procedures for Vertical Grounding Electrode

The Pocklington integro-differential eqn (13.19) is numerically solved via GB-IBEM which has been described in Chapter 8. The unknown current $I^e(x)$ along the electrode segment is given by

$$I^e(x') = \{f\}^T\{I\} \tag{13.32}$$

where $\{f'\}$ denotes the shape function while $\{I\}$ represents the unknown values of current along the electrode. Assembling the contributions from each segment yields the global equation system:

$$\sum_{i=1}^{M} [Z]_{ji}\{I\}_i = 0; \quad \text{and} \quad j = 1, 2, \ldots, M \tag{13.33}$$

where M is the total number of wire segments and $[Z]_{ji}$ is the mutual impedance matrix representing the interaction of the ith source boundary element with the jth observation boundary element:

$$[Z]_{ji} = -\frac{1}{j4\pi\omega\varepsilon_{\mathrm{eff}}}\left\{\int_{\Delta l_j}\int_{\Delta l_i}\{D\}_j\{D'\}_i^T g^V(z,z')\mathrm{d}z'\mathrm{d}z + k^2\int_{\Delta l_j}\int_{\Delta l_i}\{f\}_j\{f\}_i^T g^V(z,z')\mathrm{d}z'\mathrm{d}z\right\} \tag{13.34}$$

Throughout this work, the linear approximation over a boundary element is used as it has been shown that this choice provides accurate and stable results for various wire configurations [7,10–12,15]. Since the functions $f(x)$ are rèquired to be of class C^1 (once differentiable), a convenient choice for the shape functions over the finite elements is the family of Lagrange's polynomials given by:

$$L_i(x) = \prod_{j=1}^{m}\frac{x - x_j}{x_i - x_j};\ j \neq i \tag{13.35}$$

In this case, the linear interpolation over an element is used.

The excitation function in the form of the current generator I_g is taken into account as a forced condition in the first node of the solution vector:

$$I_1 = I_g \text{ and } I_g = 1e^{j0} \tag{13.36}$$

and the linear equation system can be properly solved.

Once determining the current distribution along the vertical electrode, the input impedance can be computed using the boundary element formalism as follows:

$$Z_{in} = \frac{1}{j4\pi\omega\varepsilon_{\mathrm{eff}} I_g}\Bigg[I(-d)g^V(x,-d,z) - \sum_{i=1}^{M}\left(\int_{\Delta l_i}\{D'\}g_0^V(x,z,z')\mathrm{d}z'\Big|_{x=a}^{x=\infty} - \int_{\Delta l_i}\{D'\}\Gamma g_i^V(x,z,z')\mathrm{d}z'\Big|_{x=a}^{x=\infty}\right)I_i \tag{13.37}$$

The input impedance spectrum is obtained by repeating this procedure within the efficient frequency bandwidth.

13.1.4 Calculation of the Transient Impedance

The transient impedance, an essential parameter in grounding system design, is defined as a ratio of time varying voltage and current at the driving point [4]:

$$z(0,t) = \frac{v(0,t)}{i(0,t)} \tag{13.38}$$

where $i(t)$ represents the excitation function, that is, the injected current at the top of the vertical electrode, as shown in Figure 13.1.

This injected current represents the lightning channel current usually expressed by the double exponential function:

$$i(t) = I_0 \cdot (e^{-\alpha t} - e^{-\beta t}); \quad t \geq 0 \tag{13.39}$$

where pulse rise time is determined by constants α and β, while I_0 denotes the amplitude of the current waveform.

The Fourier transform of the excitation function is defined by integral [14]

$$I(f) = \int_{\infty}^{\infty} i(t)e^{-j2\pi ft}dt \tag{13.40}$$

Integral (13.39) can be evaluated analytically [11]:

$$I(f) = I_0 \cdot \left(\frac{1}{\alpha + j2\pi f} - \frac{1}{\beta + j2\pi f}\right) \tag{13.41}$$

The frequency components up to few megahertzs are meaningfully present in the lightning current Fourier spectrum with strong decreasing importance from very low to the highest frequencies.

Multiplying the excitation function $I(f)$ with the input impedance spectrum $Z_{in}(fj)$ provides the frequency response of the grounding system:

$$V(f) = I(f)Z_{in}(f) \tag{13.42}$$

Applying the inverse fourier transform (IFT), a time-domain voltage counterpart is obtained. IFT of the function $V(f)$ is defined by the integral [14]

$$v(t) = \int_{\infty}^{\infty} V(f)e^{j2\pi ft}d\omega \tag{13.43}$$

As the frequency response $V(f)$ is represented by a discrete set of values, the integral (13.43) cannot be evaluated analytically and the discrete fourier transform, in this case the fast fourier transform algorithm, is used, that is,

$$v(t) = \text{IFFT}(V(f)) \tag{13.44}$$

Implementation of this algorithm inevitably causes an error due to discretization and truncation of essentially unlimited frequency spectrum. The discrete set of the time-domain voltage values is defined as [13]

$$v(n\Delta t) = F \cdot \sum_{k=0}^{N-1} V(k\Delta f)e^{jk\Delta fn\Delta t} \tag{13.45}$$

where F is the highest frequency taken into account, N is the total number of frequency samples, Δf is the sampling interval, and Δt is the time step.

Finally, the transient impedance of the vertical grounding electrode is computed from relation (13.37). The transient impedance should be recalculated for each excitation function while the input impedance spectrum depends only on the geometry of the grounding system and on the characteristics of the surrounding soil.

13.1.5 Numerical Results

Figure 13.2 shows the current distribution induced along the vertical grounding electrode of length $L = 1$m and radius $a = 5$ mm for two different burial depths. As results clearly demonstrate, the influence of the burial depth to the induced current is negligible.

Figures 13.3 and 13.4 show the frequency spectrum of the input impedance for the grounding electrodes of length $L = 1$ m and $L = 2$m, respectively, and radius $a = 5$ mm, buried in the ground at depth $d = 0.5$ m with $\varepsilon_r = 10$. The specific resistance of the ground is $\rho = 5400\,\Omega m$.

The results clearly demonstrate the influence of the electrode length to the input impedance.

The transient response of the vertical grounding electrode is evaluated using the relation (13.44). To avoid the discretization error, both the frequency range and the number of frequency samples have to be increased until satisfactory convergence is achieved. Although the amplitude of the signal almost vanishes for the frequencies higher then 1 MHz, satisfactory results are obtained for frequency range

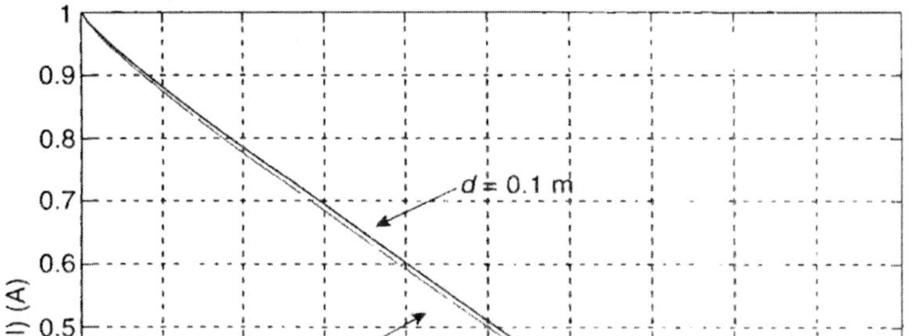

Figure 13.2 Current distribution along a grounding electrode ($L = 1$ m, $f = 1$ MHz, $\rho = 5400\,\Omega$ m). (D. Poljak and V. Doric, Wire Antenna Model for Transient, Analysis of Simple Groundingsystems, Part I: The Vertical Grounding Electrode, *Progress in Electromagnetics Research, PIER 64, 149–166, 2006*).

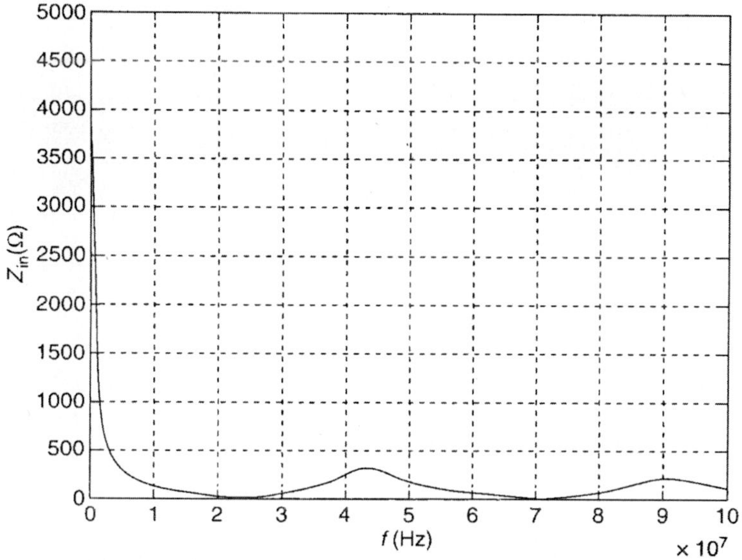

Figure 13.3 Input impedance spectrum ($L = 1$ m, d 0.5 m, $\rho = 5400\,\Omega$m). (D. Poljak and V. Doric, Wire Antenna Model for Transient, Analysis of Simple Groundingsystems, Part I: The Vertical Grounding Electrode, *Progress in Electromagnetics Research, PIER 64, 149–166, 2006*).

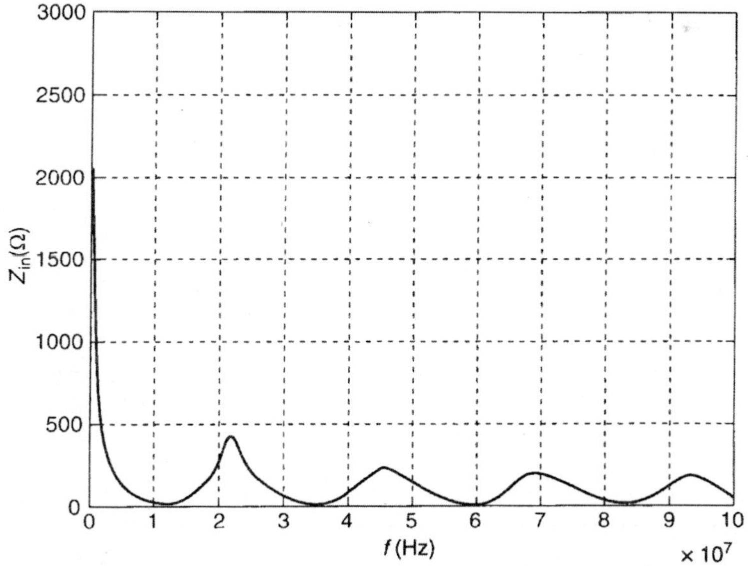

Figure 13.4 Input impedance spectrum ($L = 2$ m, $d - 0.5$ m, $\rho - 5400\,\Omega$m). (D. Poljak and V. Doric, Wire Antenna Model for Transient, Analysis of Simple Groundingsystems, Part I: The Vertical Grounding Electrode, *Progress in Electromagnetics Research, PIER 64, 149–166, 2006*).

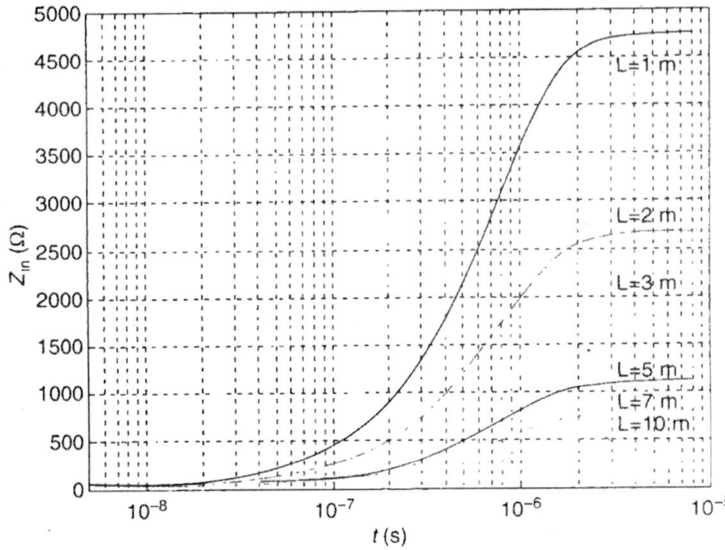

Figure 13.5 Transient impedance of the vertical grounding electrode with $d = 0.5$ m. $\rho = 5400\,\Omega$m computed for various wire lengths. (D. Poljak and V. Doric, Wire Antenna Model for Transient, Analysis of Simple Groundingsystems, Part I: The Vertical Grounding Electrode, *Progress in Electromagnetics Research, PIER 64, 149–166, 2006*).

of $F = 100\,\text{MHz}$ and $N = 2^{16}$ samples as higher frequencies contain some information that is significant for the early time response. Grounding electrode is excited by the double exponential current pulse with parameters $I_0 = 1.1043\text{A}$, $\alpha = 0.07924 \times 10^6\text{s}^{-1}$, $\beta = 4.0011 \times 10^6\text{s}^{-1}$. Figures 13.5–13.7 show different curves of the transient impedance of the vertical electrode for the case of variable wire length, burial depth, and ground specific resistance. Radius of all electrodes is $a = 5$ mm.

It can be observed that the transient impedance values vary from zero toward the certain steady state value. The transient impedance values are highly sensitive to the variations of electrode length, burial depth, and properties of the ground resistance.

This procedure shows some advantages over rigorous approaches based on Sommerfeld integrals, primarily in simplicity and computational efficiency. The method presented in this work for the case of vertical grounding electrode is readily applicable to the horizontal electrodes and complex grounding systems consisting of interconnected conductors. An extension of the reflection coefficient approach to the analysis of the horizontal grounding electrode is presented in Part II of this work.

13.2 HORIZONTAL GROUNDING ELECTRODE

Transient electromagnetic behavior of horizontal grounding electrode is of widespread interest in the protection of electrical and electronic systems [2,4,5,16]. This section deals with the transient impedance calculation of the horizontal grounding electrode.

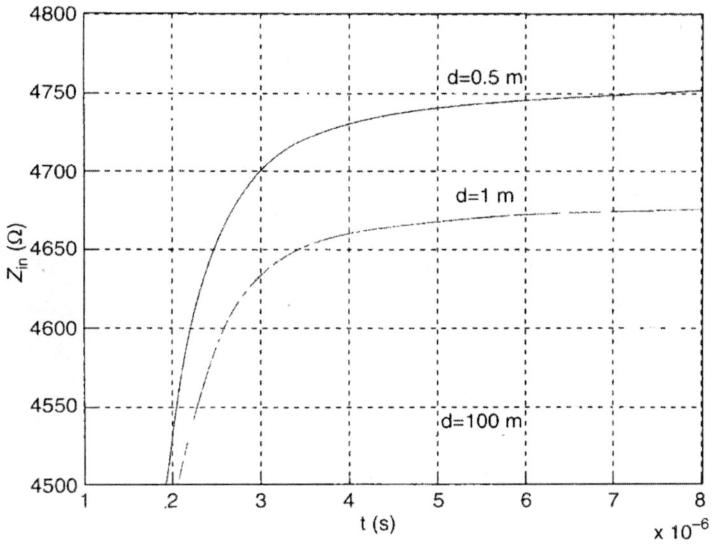

Figure 13.6 Transient impedance of the vertical electrode with $L = 1$ m and $\rho - 5400\,\Omega$m, computed for various burial depths. (D. Poljak and V. Doric, Wire Antenna Model for Transient. Analysis of Simple Groundingsystems, Part I: The Vertical Grounding Electrode, *Progress in Electromagnetics Research, PIER 64, 149–166, 2006*).

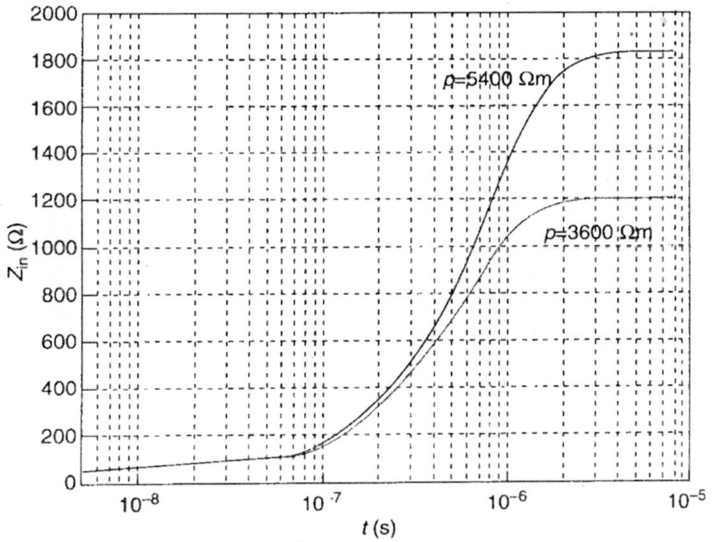

Figure 13.7 Transient impedance of the vertical grounding electrode with $L = 3$m and $d = 0.5$m computed for various values of the ground resistance. (D. Poljak and V. Doric, Wire Antenna Model for Transient. Analysis of Simple Groundingsystems, Part I: The Vertical Grounding Electrode, *Progress in Electromagnetics Research, PIER 64, 149–166, 2006*).

In principle, the procedure presented in this section is quite similar to the one for the vertical electrode, though the parts of formulation regarding the current distribution and input impedance are more demanding in the case of horizontal wire.

This model also poses problem in the frequency domain, and transient analysis of the grounding systems is obtained by applying the inverse Fourier transform, while the linear surge-protection components are modeled in the time domain. In general, it is not easy to collect these models [16].

Recently, a direct time-domain analysis of a straight wire antenna embedded in a dielectric half-space has been presented in Ref. [17] and can be regarded as an opener to the subject. Further work on the subject will deal with the full time-domain model including the ground conductivity.

The method presented in this paper is based on the Pocklington integro-differential equation formulation [4,17] and related reflection coefficient approximation to account for the presence of a lossy half-space (instead of analytically demanding and numerically time-consuming Sommerfeld integral approach [4,7,15]).

In principle, the procedure for the assessment of the horizontal electrode transient impedance is the same as in the previous case of the vertical electrode.

The current along the horizontal grounding electrode is obtained by solving the corresponding integro-differential equation via the indirect scheme of the Galerkin–Bubnov boundary element method (GB-BEM) [10,11,15].

The feed-point voltage can be obtained by analytically integrating the normal electric field from the electrode surface to infinity.

The input impedance of the horizontal electrode is obtained as a ratio of evaluated voltage and injected current at the feed point.

The frequency response of the electrode is obtained by multiplying the input impedance spectrum with Fourier transform of the given lightning current waveform.

Finally, the transient impedance of the horizontal grounding electrode is computed by means of the inverse Fourier transform.

13.2.1 Integral Equation Formulation for the Horizontal Electrode

The horizontal grounding electrode is represented by the straight end-fed wire of length L and radius a, buried in a lossy medium at depth d as shown in Figure 13.8 The

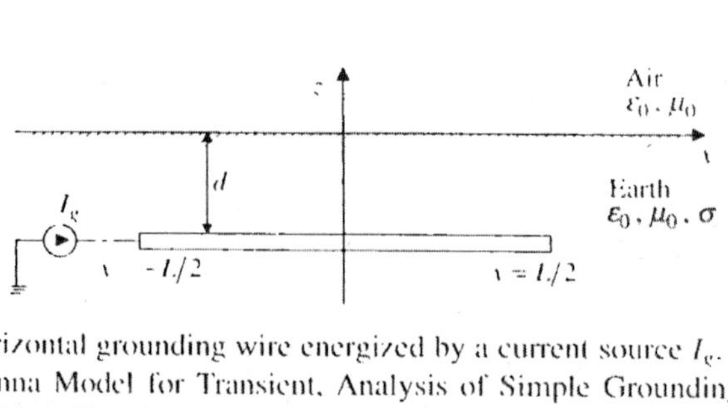

Figure 13.8 Horizontal grounding wire energized by a current source I_g. (D. Poljak and V. Doric, Wire Antenna Model for Transient, Analysis of Simple Groundingsystems, Part II: The Horizontal Grounding Electrode, *Progress in Electromagnetics Research, PIER 64, 167–189, 2006*).

electrode is assumed to be perfectly conducting and its dimensions satisfy the well-known thin wire approximation [4,7,10,11,15].

The Pocklington integro-differential equation for the horizontal electrode can be derived by expressing the electric field in terms of Hertz vector potential and by satisfying the appropriate boundary conditions for the tangential field components at the electrode surface, in the same manner as it was done for vertical electrode.

The complete electric field induced in the vicinity of the straight thin wire of finite length buried in an imperfectly conducting half-space can be expressed in terms of Hertz vector potential Π [7,11]:

$$\vec{E} = \nabla(\nabla\vec{\Pi}) + k_1^2\vec{\Pi} \tag{13.46}$$

where k_1 is the phase constant of a lossy ground:

$$k_1^2 = -\omega^2\mu\varepsilon_{\text{eff}} \tag{13.47}$$

and ε_{eff} denotes the complex permittivity of the lossy ground:

$$\varepsilon_{\text{eff}} = \varepsilon_r\varepsilon_0 - j\frac{\sigma}{\omega} \tag{13.48}$$

where ε_r and σ are relative permittivity and conductivity of the ground, respectively, and ω denotes the operating frequency.

A corresponding integral equation formulation for the horizontal grounding electrode using Sommerfeld and RC approach is presented in Sections 13.2.1.1 and 13.2.1.2.

13.2.1.1 Sommerfeld Integral Approach The scattered tangential electric field in the vicinity of the straight horizontal thin wire buried at depth d in the earth and expressed by vector eqn (13.46), can be written in terms of Hertz vector potential components Π_x and Π_z [7]:

$$E_x^H(x,z) = \left[\frac{\partial^2}{\partial x^2} + k_1^2\right]\Pi_x^H + \frac{\partial^2\Pi_z^H}{\partial x\partial z} \tag{13.49}$$

$$E_z^H(x,z) = \frac{\partial^2\Pi_x^H}{\partial x\partial z} + \left[\frac{\partial^2}{\partial z^{2^2}} + k_1^2\right]\Pi_z^H \tag{13.50}$$

where superscript H emphasizes the horizontal grounding electrode.

The vector potential components for the straight horizontal buried wire are given by [7]

$$\Pi_x^H = \frac{1}{j4\pi\omega\mu\varepsilon_{\text{eff}}}\int_{-L/2}^{L/2}\left[g_0^H(x,x',z) - g_i^H(x,x',z) + U_{11}\right]I(x')\mathrm{d}x' \tag{13.51}$$

$$\Pi_z^H = \frac{1}{j4\pi\omega\varepsilon_{\text{eff}}} \int_{-L/2}^{L/2} \frac{\partial W_{11}}{\partial x} I(x')dx' \tag{13.52}$$

where $I(x')$ is current distribution along the wire, $g_0^H(x, x', z)$ denotes the free-space Green's function of the form

$$g_0^H(x, x', z) = \frac{e^{-jk_1 R_{1h}}}{R_{1h}} \tag{13.53}$$

while $g_i^H(x, x', z)$ arises from the image theory and is given by

$$g_i^H(x, x', z) = \frac{e^{-k_2 R_{2h}}}{R_{2h}} \tag{13.54}$$

where R_{1h} and R_{2h} are the distances from the horizontal wire in the lossy ground and from its image in the air to the observation point in the lower medium, respectively.

The influence of the imperfectly conducting ground is taken into account by the attenuation terms in the form of Sommerfeld integrals U_{11} and W_{11}, defined as follows [4]:

$$U_{11} = 2\int_0^\infty \frac{e^{-\mu_1(d-z)}}{\mu_1 + \mu_2} J_0(\lambda\rho)\lambda d\lambda \tag{13.55}$$

$$W_{11} = 2\int_0^\infty \frac{(\mu_1 - \mu_2)e^{-\mu_1(d-z)}}{k_2^2\mu_1 + k_1^2\mu_2} J_0(\lambda\rho)\lambda d\lambda \tag{13.56}$$

where $J_0(\lambda\rho)$ is zero-order Bessel function of the first kind, while μ_1, μ_2, and ρ are given by

$$\mu_1 = (\lambda^2 - k_1^2)^{1/2}, \mu_2 = (\lambda^2 - k_2^2)^{1/2}, \rho = |x - x'| \tag{13.57}$$

Assuming the perfectly conducting wire [4,7], the total tangential electric field vanishes along the perfectly conducting electrode surface, that is,

$$E_x^{exc,H}(x, a) + E_x^{sct,H}(x, a) = 0 \tag{13.58}$$

where $E_x^{exc,H}$ is the excitation function and $E_x^{sct,H}$ is the corresponding scattered field along the electrode surface.

Taking into account the Sommerfeld integrals property:

$$\frac{\partial W_{11}}{\partial z} = k_1^2 V_{11} - U_{11} \tag{13.59}$$

Combining the relations (13.59) and (13.49) yields:

$$E_x^{exc,H} = -\frac{1}{j4\pi\omega\varepsilon_{\text{eff}}} \left\{ \int_{-L/2}^{L/2} \frac{\partial^2}{\partial x^2} \left[g_0^H(x,x') - g_i^H(x,x') + k_1^2 V_{11} \right] I(x')\mathrm{d}x' \right. \\ \left. + k_1^2 \int_{-L/2}^{L/2} \left[g_0^H(x,x') - g_i^H(x,x') + U_{11} \right] I(x')\mathrm{d}x' \right\} \tag{13.60}$$

Solving the integral eqn (15), the current distribution along the horizontal electrode is obtained.

13.2.1.2 Simplified Reflection Coefficient Approach The evaluation of the Sommerfeld integrals (13.55), (13.56) is rather a difficult task [4,7,15]. This approach requires repeated evaluation of Sommerfeld integrals and matrix inversion on several frequencies, for spectrum calculation purposes which is rather time consuming.

A simplified approach based on the modified image theory was proposed in Ref. [5] for handling the grounding systems and in Ref. [18] for the treatment of the plane wave coupling to the horizontal buried cables. However, the modified image theory is based on a quasi-static approximation of the phenomena and takes into account only the electrical properties of the soil but not the burial depth as a parameter.

This work deals with a reflection coefficient (RC) approach by which the angle of electric field reflection from the earth–air interface and the burial depth of the grounding electrode are both taken into account. The principal advantage of the RC approach versus rigorous Sommerfeld integral approach is the simplicity of the formulation and a significantly less computational cost.

For convenience, the integro-differential eqn (13.60) can be written in the form:

$$E_x^{exc} = -\frac{1}{j4\pi\omega\varepsilon_{\text{eff}}} \int_{-L/2}^{L/2} G^H(x,x')I(x')\mathrm{d}x' \tag{13.61}$$

where $G(x,x')$ is the total Green's function given by

$$G^H(x,x') = \frac{\partial^2}{\partial x^2} \left[g_0^H(x,x') - g_i^H(x,x') + k_1^2 V_{11} \right] \\ + k_1^2 \left[g_0^H(x,x') - g_i^H(x,x') + U_{11} \right] \tag{13.62}$$

According to the RC approximation, the simplified Green's function for the horizontal grounding electrode is given by

$$G^H(x,x') = \left[\frac{\partial^2}{\partial x^2} + k_1^2\right]\left[g_0^H(x,x') - \Gamma g_i^H(x,x')\right] = \left[\frac{\partial^2}{\partial x^2} + k_1^2\right] g^H(x,x') \quad (13.63)$$

while Γ is the corresponding reflection coefficient for the TM polarization [11]:

$$\Gamma = \frac{\frac{1}{\underline{n}}\cos\theta - \sqrt{\frac{1}{\underline{n}} - \sin^2\theta}}{\frac{1}{\underline{n}}\cos\theta + \sqrt{\frac{1}{\underline{n}} - \sin^2\theta}} \quad (13.64)$$

where θ and $\underline{n}$ are given by

$$\theta = arctg\frac{|x - x'|}{2d}; \quad \underline{n} = \frac{\varepsilon_{eff}}{\varepsilon_0} \quad (13.65)$$

Also, the Pocklington integro-differential equation for the straight horizontal wire buried in the lossy ground is

$$E_x^{exc,H} = \frac{1}{j4\pi\omega\varepsilon_{eff}} \int_{-L/2}^{L/2} \left[\frac{\partial^2}{\partial x^2} + k_1^2\right]\left[g_0^H(x,x') + \Gamma g_i^H(x,x')\right] I(x')\mathrm{d}x' \quad (13.66)$$

Solving the integral eqn (13.66) via Galerkin–Bubnov indirect boundary element method, the equivalent current distribution along the horizontal electrode is obtained.

Though the RC approximation is found to be an asymptotically correct form of the Sommerfeld integrals in the far-field region, the relationship between the rigorous and approximate field expressions is not in general so simple [12].

On the contrary, the RC approach provides a more efficient algorithm compared with the rigorous method. One possible way of doing a trade-off between the approximate and rigorous approach could be the comparison of numerical results, rather than analytical considerations [12,13].

13.2.1.3 ***Imposed Boundary Conditions*** In the analysis of the grounding electrodes, the excitation function cannot be defined in the form of an incident electric field, as this field vanishes along the perfectly conducting (PEC) wire surface, that is,

$$E_x^{exc} = 0 \quad (13.67)$$

so the integral equations (13.59) and (13.65) have become homogeneous [4,7].

The horizontal grounding electrode is energized by the injection of a current pulse of an arbitrary waveform produced by an ideal current source with one terminal connected to the grounding electrode and the other one grounded at infinity.

The excitation can be defined by the current flowing into the wire. This current source is incorporated into the integral equation scheme through the boundary condition:

$$I(-L/2) = I_g \tag{13.68}$$

where I_g denotes the current generator.

This unit current source is often used in the frequency-domain calculations as it is the counterpart of the time-domain current pulse.

13.2.2 The Evaluation of the Input Impedance Spectrum

As the horizontal grounding electrode is represented by the wire antenna model, the calculation of antenna input impedance against the grounding system input impedance presented in this work should be discussed and a corresponding trade-off should be made.

The calculation of the wire antenna input impedance is relatively a simple task, because the terminal points of the voltage generator are very close, that is, one deals with the so-called delta-function generator [4,7].

The calculation of grounding system input impedance is more demanding than in the antenna case, primarily because the input terminals are placed between ground electrode point and remote soil, and the integration on the infinite integral cannot be avoided.

The input impedance is given by the ratio

$$Z_{in} = \frac{V_g}{I_g} \tag{13.69}$$

where V_g and I_g are the values of the voltage and the current at the driving point, respectively.

The feed-point voltage can be calculated by integrating the normal electric field component from the electrode surface to infinity, that is,

$$V_g = -\int_{\infty}^{a} \vec{E} d\vec{s} \tag{13.70}$$

Hence, as in the case of vertical electrode, the problem of obtaining the input impedance is referred to the calculation of feed-point voltage. The input impedance spectrum is obtained by repeating this procedure in the wide frequency band.

For the case of horizontal grounding electrode, the integral (13.71) becomes

$$V_g = -\int_{\infty}^{a} E_z^H(x,z) dz \tag{13.71}$$

The radial electric field component, normal to the electrode, is now given by

$$E_z^H(x,z) = \frac{\partial^2 \Pi_x^H}{\partial x \partial z} = \frac{1}{j4\pi\omega\varepsilon_{\text{eff}}} \int_{-L/2}^{L/2} I(x') \frac{\partial^2 g^H(x,x',z)}{\partial x \partial z} \mathrm{d}x' \tag{13.72}$$

The kernel g^H has the following symmetry property:

$$\frac{\partial g^H(x,x',z')}{\partial x} = -\frac{\partial g^H(x,x',z)}{\partial x'} \tag{13.73}$$

and performing the integration by parts, expression (13.73) becomes

$$E_z(x,z) = -\frac{1}{j4\pi\omega\varepsilon_{\text{eff}}} \frac{d}{\mathrm{d}z}\left[I(x')g^H(x,x',z) \Big|_{x'=-L/2}^{x'=L/2} - \int_{-L/2}^{L/2} \frac{\partial I(x')}{\partial x'} g^H(x,x',z)\mathrm{d}x' \right] \tag{13.74}$$

It can be observed that the differentiation over x variable is avoided while a differentiation over z variable is replaced outside the integral sign.

Substituting eqn (13.74) into (13.72), it follows

$$V_g = -\frac{1}{j4\pi\omega\varepsilon_{\text{eff}}} \int_a^{\infty} \frac{d}{\mathrm{d}z}\left[I(x')g^H(x,x',z) \Big|_{x'=-L/2}^{x'=L/2} - \int_{-L/2}^{L/2} \frac{\partial I(x')}{\partial x'} g^H(x,x',z)\mathrm{d}x' \right] \mathrm{d}z \tag{13.75}$$

that simply leads to:

$$V_g = \frac{1}{j4\pi\omega\varepsilon_{\text{eff}}}\left[I(-L/2)g^H(x,-L/2,z) - \int_{-L/2}^{L/2} \frac{\partial I(x')}{\partial x'} g^H(x,x',z)\mathrm{d}x' \right]\Bigg|_{z=a}^{z=\infty} \tag{13.76}$$

The tedious numerical integration over infinite domain in (13.76) is avoided and the desired input impedance of the buried wire is determined by the relation

$$Z_{in} = \frac{1}{j4\pi\omega\varepsilon_{\text{eff}} I_g}\left[I(-L/2)g^H(x,-L/2,z) - \int_{-L/2}^{L/2} \frac{\partial I(x')}{\partial x'} g^H(x,x',z)\mathrm{d}x' \right]\Bigg|_{z=a}^{z=\infty} \tag{13.77}$$

However, the approximation of Sommerfeld integrals with Fresnel reflection coefficient (RC) should be performed very carefully, and validity limits of this approach should be clarified as well.

The Sommerfeld integral approach has been found to be numerically stable for buried horizontal wire brought to within 10^{-6} wavelengths of the interface [7,12,17]. In addition, the RC approach produces results roughly within 10% of those obtained via rigorous Sommerfeld integral approach.

13.2.3 Numerical Procedures for Horizontal Electrode

The Pocklington integro-differential equation for the horizontal grounding electrode is solved via GB-IBEM. Performing the boundary element discretization, the solution for the unknown current $I^e(x)$ along the boundary element can be written as

$$I^e(x') = \{f\}^T\{I\} \tag{13.78}$$

Assembling the contributions from each element, the resulting system of equations is given by

$$\sum_{i=1}^{M}[Z]_{ji}\{I\}_i = 0, \quad \text{and} \quad j = 1,2,\ldots,M \tag{13.79}$$

where M is the total number of wire segments and $[Z]_{ji}$ is the mutual impedance matrix representing the interaction of the ith source boundary element with the jth observation boundary element:

$$[Z]_{ji} = -\frac{1}{j4\pi\omega\varepsilon_{\text{eff}}}\left\{\int_{\Delta l_j}\int_{\Delta l_i}\{D\}_j\{D'\}_i^T g^H(x,x')\mathrm{d}x'\mathrm{d}x + k^2\int_{\Delta l_j}\int_{\Delta l_i}\{f\}_j\{f\}_i^T g^H(x,x')\mathrm{d}x'\mathrm{d}x\right\} \tag{13.80}$$

Matrices $\{f\}$ and $\{f'\}$ contain the shape functions while $\{D\}$ and $\{D'\}$ contain their derivatives, M is the total number of finite elements, and Δl_i, Δl_j are the widths of ith and jth boundary elements.

Throughout this work the linear approximation over a boundary element is used, as it has been shown that this choice provides accurate and stable results for various wire configurations [7,10–12,15]. Since the functions $f(x)$ are required to be of class C^1 (once differentiable), a convenient choice for the shape functions over the finite elements is the family of Lagrange's polynomials given by

$$L_i(x) = \prod_{j=1}^{m}\frac{x - x_j}{x_i - x_j}; j \neq i \tag{13.81}$$

The excitation function in the form of the current generator I_g is taken into account as a forced condition in the first node of the solution vector:

$$I_1 = I_g \text{ and } I_g = 1e^{j0} \tag{13.82}$$

and the linear equation system can be properly solved.

Once obtaining the current distribution along the grounding wire, the corresponding input impedance can be computed from the integral (13.77). Using the boundary element formalism integral, (13.77) becomes

$$Z_{in} = \frac{1}{j4\pi\omega\varepsilon_{\text{eff}} I_g} \left[I(-L/2) g^H(x, \; L/2, z) - \sum_{i=1}^{M} \int_{\Delta l_i} \{D'\} g^H(x, x', z) \mathrm{d}x' \right] \Bigg|_{z=a}^{z=\infty} I_i \tag{13.83}$$

The input impedance spectrum can be obtained by repeating this procedure within the efficient frequency bandwidth.

13.2.4 The Transient Impedance Calculation

The transient impedance of the horizontal grounding electrode is defined as a ratio of time varying voltage and current at the input terminals [1]:

$$z(0, t) = \frac{v(0, t)}{i(0, t)} \tag{13.84}$$

where $i(t)$ is the current injected at the end of the horizontal electrode (Fig. 13.8).

This injected current is related to the lightning channel current given by

$$i(t) = I_0 \cdot (e^{-\alpha t} - e^{-\beta t}); \quad t \geq 0 \tag{13.85}$$

where pulse rise time is determined by constants α and β, while I_0 denotes the amplitude of the current waveform. The calculation procedure for $v(t)$ is the same as in the case of the vertical electrode presented in Section 13.1.4.

13.2.5 Numerical Results

Figures 13.9 and 13.10 show the frequency spectrum of the input impedance for the grounding electrode of length $L = 1$ m, and radius $a = 5$ mm, buried at depth $d = 1$ m in the ground with $\varepsilon_r = 10$. Ground conductivity is $\sigma = 0.001$ s/m (dry sand, gravel) and $\sigma = 0.01$ s/m (loam, clay), respectively. The obtained results are compared with results available from Ref. [19]. The agreement between results calculated via different approaches is found to be satisfactory.

Figure 13.11 shows the transient impedance for the grounding electrode of same dimensions buried at depth $d = 1$ m compared with the results from Ref. [19].

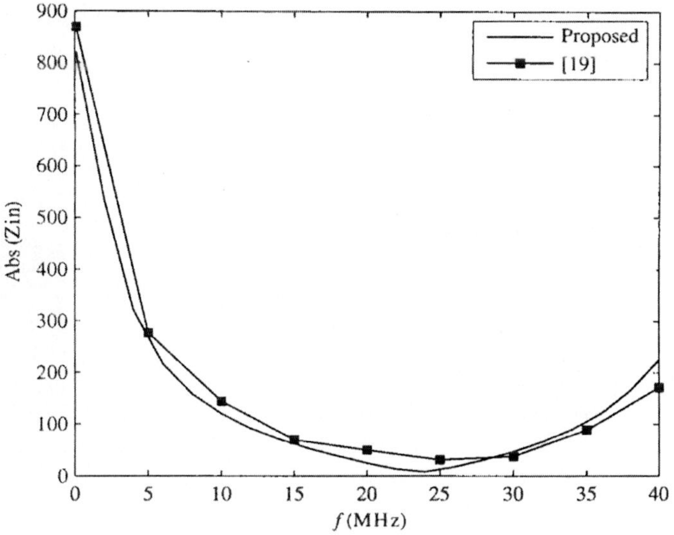

Figure 13.9 Input impedance spectrum of the grounding electrode: $L = 1$ m, $a = 5$ mm, $d = 1$ m, $\varepsilon_r = 10$, $\sigma = 0.001$s/m. (D. Poljak and V. Doric, Wire Antenna Model for Transient, Analysis of Simple Groundingsystems, Part II: The Horizontal Grounding Electrode, *Progress in Electromagnetics Research, PIER 64, 167–189, 2006*).

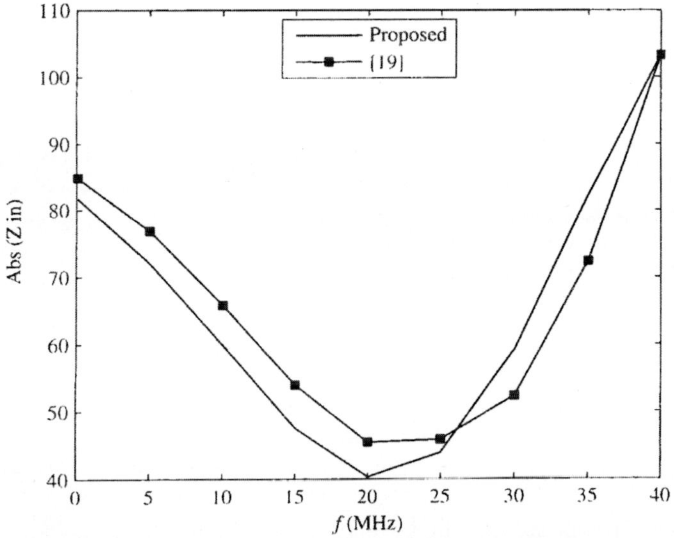

Figure 13.10 Input impedance spectrum of the grounding electrode: $L = 1$ m, $a = 5$ mm, $d = 1$ m, $\varepsilon_r = 10$, $\sigma = 0.01$s/m. (D. Poljak and V. Doric, Wire Antenna Model for Transient, Analysis of Simple Groundingsystems, Part II: The Horizontal Grounding Electrode, *Progress in Electromagnetics Research, PIER 64, 167–189, 2006*).

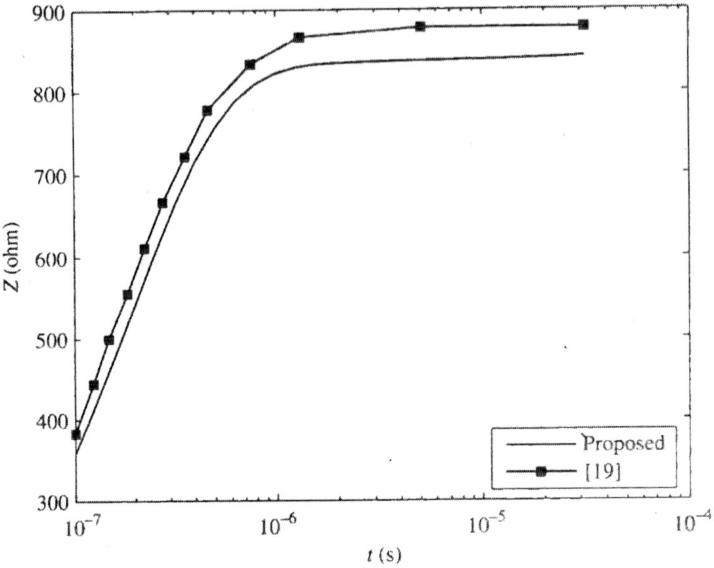

Figure 13.11 Transient impedance of the grounding electrode buried in the lossy ground. $L = 1$ m, $a = 5$ mm, $d = 1$ m, $\varepsilon_r = 10$, $\sigma = 0.001$s/m. (D. Poljak and V. Doric, Wire Antenna Model for Transient. Analysis of Simple Groundingsystems. Part II: The Horizontal Grounding Electrode, *Progress in Electromagnetics Research, PIER 64, 167–189, 2006*).

Grounding electrode is excited by the double exponential current pulse with parameters $I_0 = 1.1043$A, $\alpha = 0.07924 \times 10^6\,\text{s}^{-1}$, $\beta = 4.0011 \times 10^6\,\text{s}^{-1}$. The agreement is judged to be satisfactory again.

At this point, the choice of the parameters α and β are determined by rise time (time-to-maximum value of the current waveform) and duration time (time-to-half maximum). Thus, values $\alpha = 0.07924 \times 10^6\text{s}^{-1}$, $\beta = 4.0011 \times 10^6\text{s}^{-1}$ correspond to time-to-maximum 1 μs and time-to-half maximum 10 μs [18,19].

This excitation current pulse is chosen in a way to be as close as possible to the realistic lightning pulse. In realistic scenarios, the rise time of the current pulse varies from 0.1 μs to a few μs, while the duration time, defined as a time required for the current magnitude, decreases to the 50% of maximal value that ranges from 5 to 2000 μs.

The computational details regarding the use of inverse Fourier transform have been already discussed in Section 13.1.

Figures 13.12 and 13.13 show the input impedance spectra for horizontal grounding electrode of length $L = 1$ m and $L = 10$ m, respectively, buried at depth $d = 0.5$m in the lossy ground with $\varepsilon_r = 10$ and conductivity $\sigma = 0.00185$s/m (specific resistivity $\rho = 5400\Omega$m).

Figure 13.14 shows the frequency spectrum of the lightning strike current pulse. It is visible that this pulse practically vanishes for frequencies higher than 2 MHz. However, to obtain the proper information about the early time behavior of the transient response, higher harmonics need to be taken into

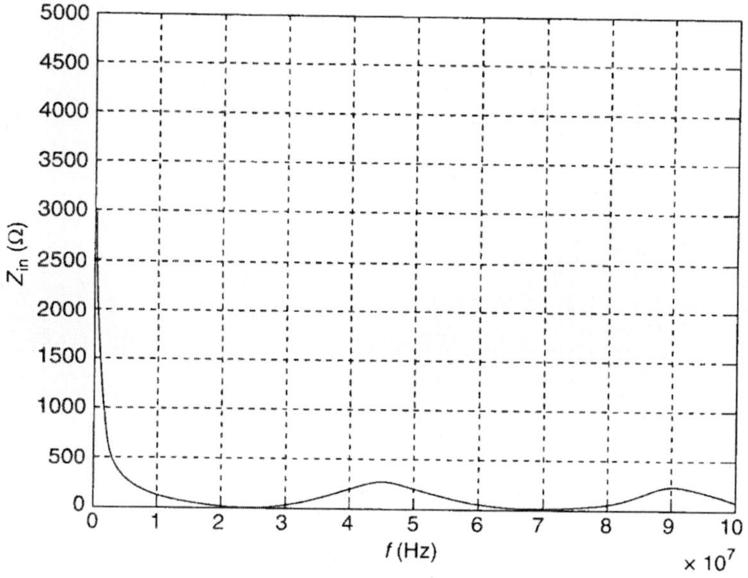

Figure 13.12 Input impedance spectrum ($L = 1$ m, $d = 0.5$ m, $\rho = 5400\,\Omega$m). (D. Poljak and V. Doric, Wire Antenna Model for Transient, Analysis of Simple Groundingsystems, Part II: The Horizontal Grounding Electrode, *Progress in Electromagnetics Research, PIER 64, 167–189, 2006*).

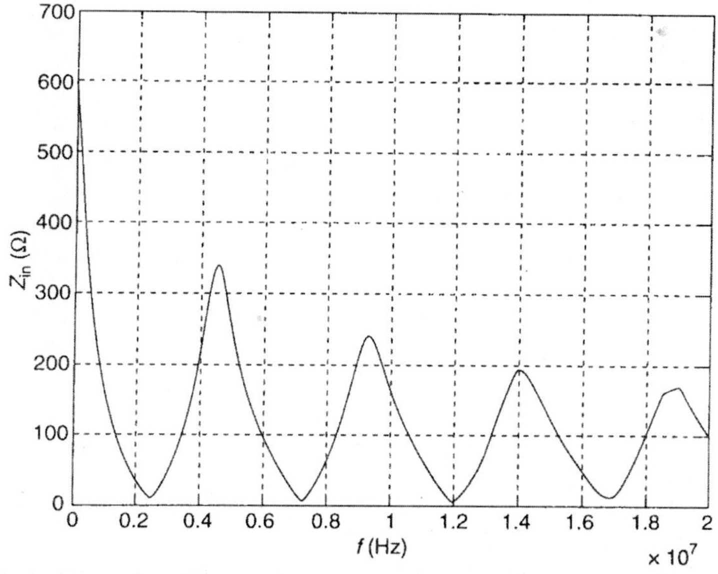

Figure 13.13 Input impedance spectrum ($L = 10$ m, $d = 0.5$ m, $\rho = 5400\,\Omega$m). (D. Poljak and V. Doric, Wire Antenna Model for Transient, Analysis of Simple Groundingsystems, Part II: The Horizontal Grounding Electrode, *Progress in Electromagnetics Research, PIER 64, 167–189, 2006*).

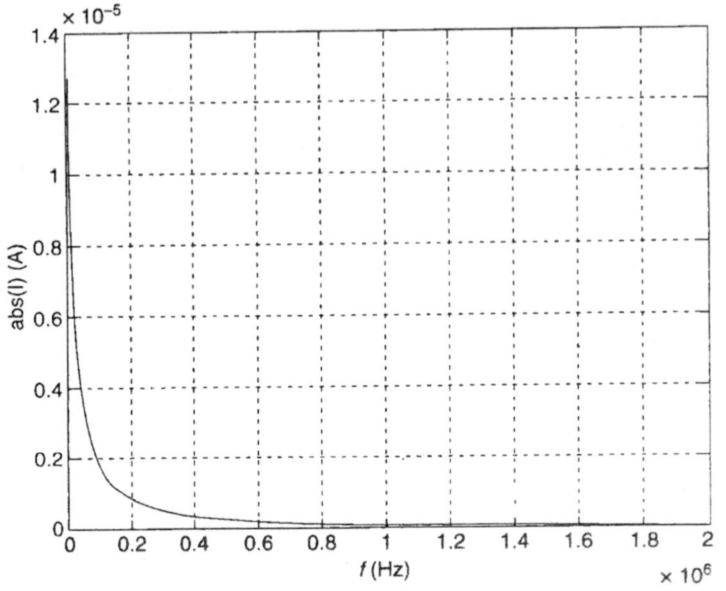

Figure 13.14 Frequency spectrum of the current pulse. (D. Poljak and V. Doric, Wire Antenna Model for Transient, Analysis of Simple Groundingsystems, Part II: The Horizontal Grounding Electrode, *Progress in Electromagnetics Research, PIER 64, 167–189, 2006*).

account. As the inverse fast fourier transform (IFFT) requires $\Delta t = 1/2f_{\max}$, it follows for $f_{\max} = 100\,\text{MHz}$ $\Delta t = 5\,\text{ns}$.

The frequency spectrum of the voltage induced at the grounding electrode input is obtained by multiplying the input impedance spectrum and the spectrum of the current pulse. Applying the IFFT, the transient voltage at the input of the grounding electrode is obtained. Figure 13.15 shows the normalized injected current and induced voltage waveform for the horizontal grounding electrode of length $L = 1$ m. A shift between the current excitation function and induced voltage clearly demonstrates the reactive character of the grounding impedance, contrary to the widely adopted assumption of the resistive grounding electrode behavior.

The ratio of time-dependent voltage and current gives the transient impedance of the grounding electrode. Figures 13.16–13.18 show the transient impedance waveform for various electrode lengths. It is obvious that the greater the conductor length is, the lower is the value of the transient impedance. Generally, the transient impedance at instant $t = 0$ has the zero value and reaches the stationary value after some time. The waveform presented in Figure 11 is due to the signal reflection from the wire ends. The physical explanation of the differences appearing in the values of the transient impedance visible in Figures 13.16–13.18 is related to the energy loss of the grounding electrode due to the radiation mechanism. In accordance to the inplemented model arising from the thin wire antenna theory,

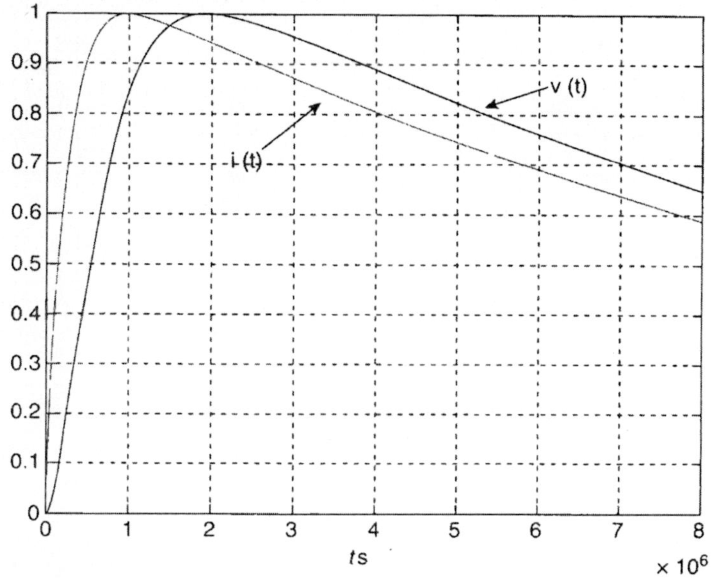

Figure 13.15 Normalized values of the current pulse and induced voltage. (D. Poljak and V. Doric, Wire Antenna Model for Transient, Analysis of Simple Groundingsystems, Part II: The Horizontal Grounding Electrode, *Progress in Electromagnetics Research, PIER 64, 167–189, 2006*).

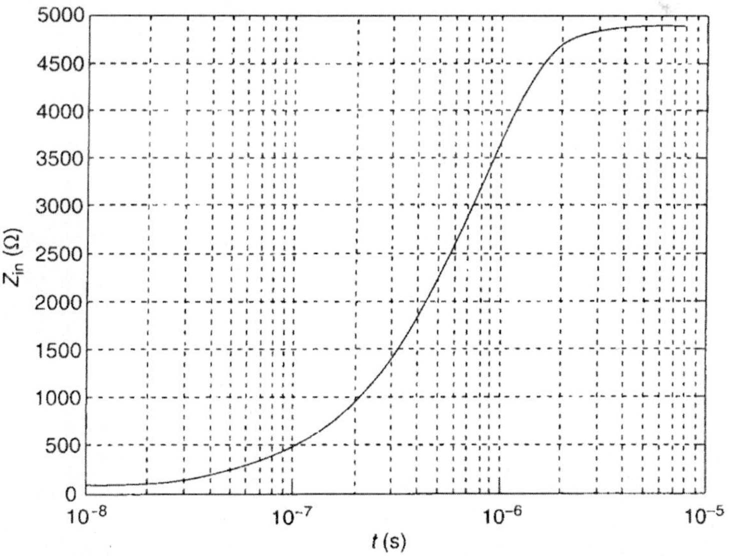

Figure 13.16 Transient impedance of the horizontal grounding electrode: $L = 1$ m, $d = 0.5$ m, $\rho = 5400\,\Omega$m. (D. Poljak and V. Doric, Wire Antenna Model for Transient, Analysis of Simple Groundingsystems, Part II: The Horizontal Grounding Electrode, *Progress in Electromagnetics Research, PIER 64, 167–189, 2006*).

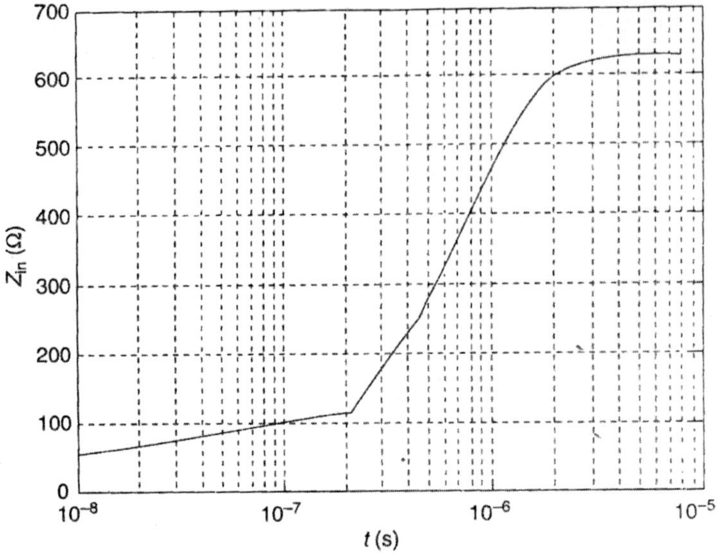

Figure 13.17 Transient impedance of the horizontal grounding electrode: $L = 10\,\text{m}$, $d = 0.5\,\text{m}$, $\rho = 5400\,\Omega\text{m}$. (D. Poljak and V. Doric, Wire Antenna Model for Transient. Analysis of Simple Groundingsystems, Part II: The Horizontal Grounding Electrode. *Progress in Electromagnetics Research, PIER 64, 167–189, 2006*).

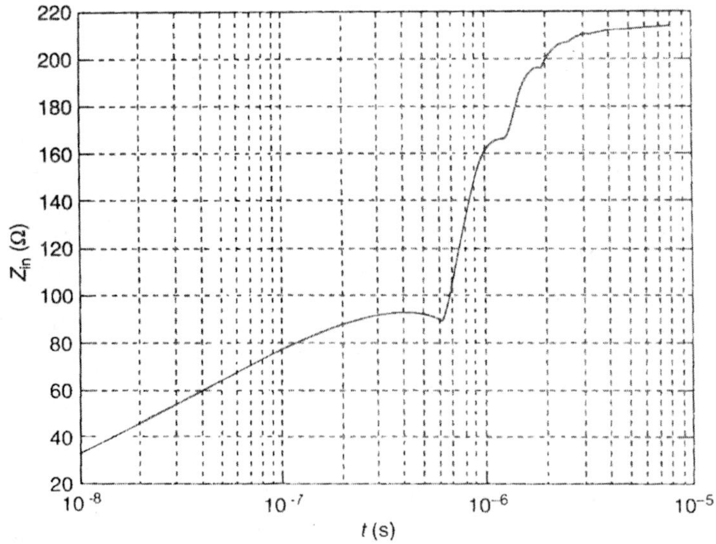

Figure 13.18 Transient impedance of the horizontal grounding electrode: $L = 30\,\text{m}$, $d = 0.5\,\text{m}$, $\rho = 5400\,\Omega\text{m}$. (D. Poljak and V. Doric, Wire Antenna Model for Transient. Analysis of Simple Groundingsystems, Part II: The Horizontal Grounding Electrode. *Progress in Electromagnetics Research, PIER 64, 167–189, 2006*).

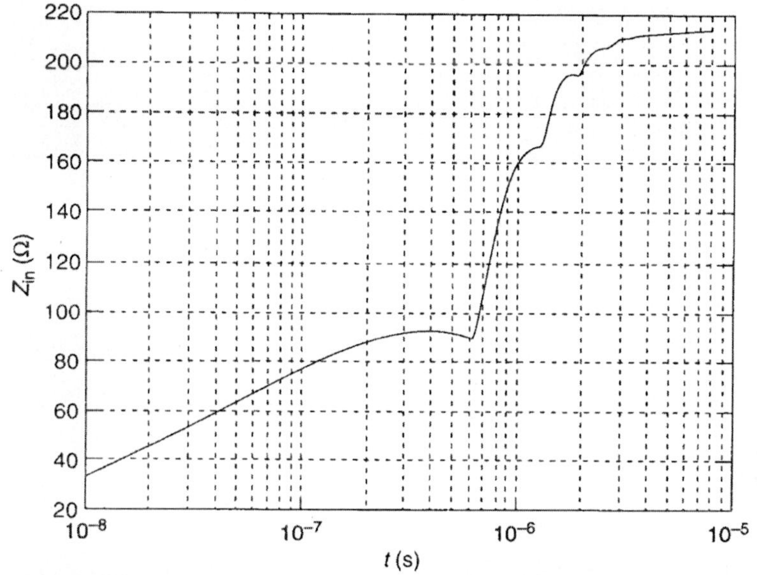

Figure 13.19 Transient impedance of the horizontal grounding electrode: $L = 10$ m, ρ · 5400 Ωm. (D. Poljak and V. Doric, Wire Antenna Model for Transient, Analysis of Simple Groundingsystems, Part II: The Horizontal Grounding Electrode, *Progress in Electromagnetics Research, PIER 64, 167–189, 2006*).

the horizontal grounding electrode behaves as the end-driven antenna immersed in a lossy medium.

Figure 13.19 shows the results for the various burial depths. It is visible that the impedance decreases as the burial depth increases. In particular, the impedance reaches the lowest value at $d = 10$ m where the influence of the earth–air interface is negligible. In realistic case, the soil at greater burial depth may have a higher humidity which would additionally decrease the input impedance.

In Figure 13.20, results for two different values of soil resistivity are compared. As expected, a lower value of soil resistivity implies a lower value of transient impedance.

In all examples presented so far, the transient impedance continuously grows from zero value to the stationary value (maximum value).

Figure 13.21 shows a transient impedance waveform for the case of low soil resistivity. Reached grounding resistance equals 9 Ω. However, observing the transient impedance it is obvious that the maximal value of the transient impedance is around 37 Ω, that is, four times higher. From this example it is visible that transient impedance calculation is very important, especially the early time transient response. The low value of the grounding resistance does not ensure lightning protection in the case of lightning strike.

Figure 13.22 shows the frequency spectrum of the input impedance. Comparing to spectra shown in Figures 13.9 and 13.10, it is visible that this spectrum is not monotonically decreasing.

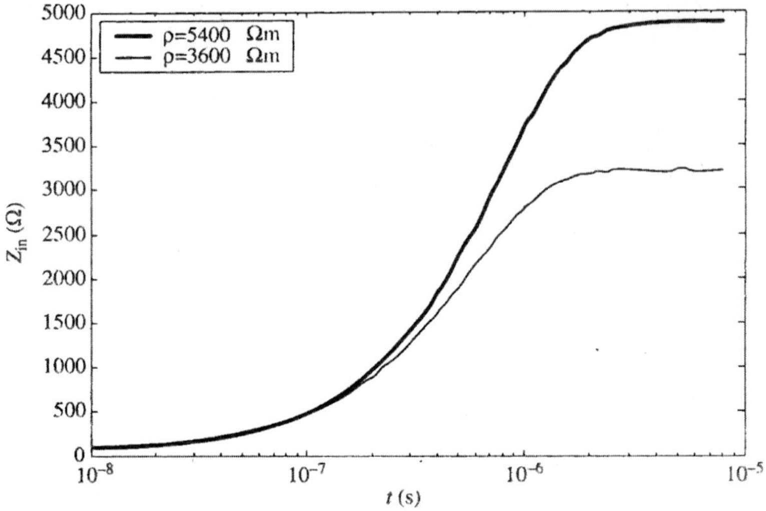

Figure 13.20 Transient impedance of the horizontal grounding electrode: $L - 1$ m. $d = 0.5$ m. (D. Poljak and V. Doric, Wire Antenna Model for Transient, Analysis of Simple Groundingsystems, Part II: The Horizontal Grounding Electrode, *Progress in Electromagnetics Research, PIER 64, 167–189, 2006*).

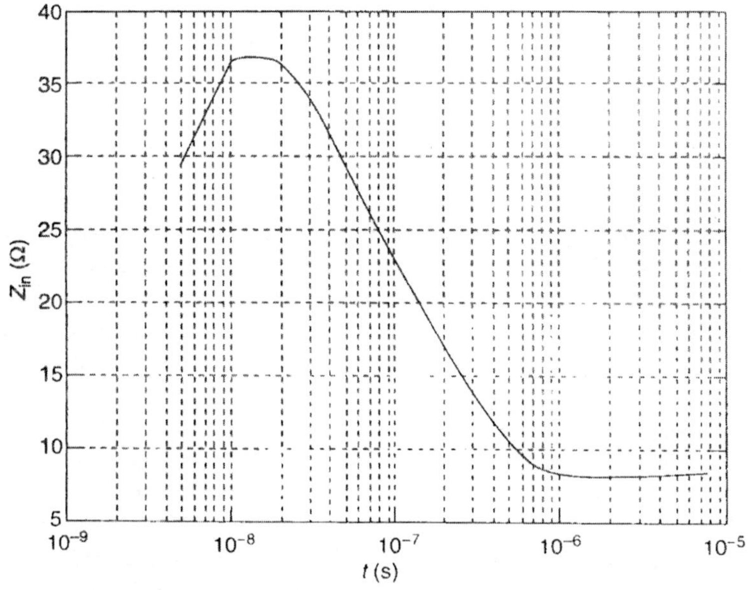

Figure 13.21 Transient impedance of the horizontal grounding electrode: $L = 10$ m, $d = 10$ m, ρ · 100 Ωm. (D. Poljak and V. Doric, Wire Antenna Model for Transient, Analysis of Simple Groundingsystems, Part II: The Horizontal Grounding Electrode, *Progress in Electromagnetics Research, PIER 64, 167–189, 2006*).

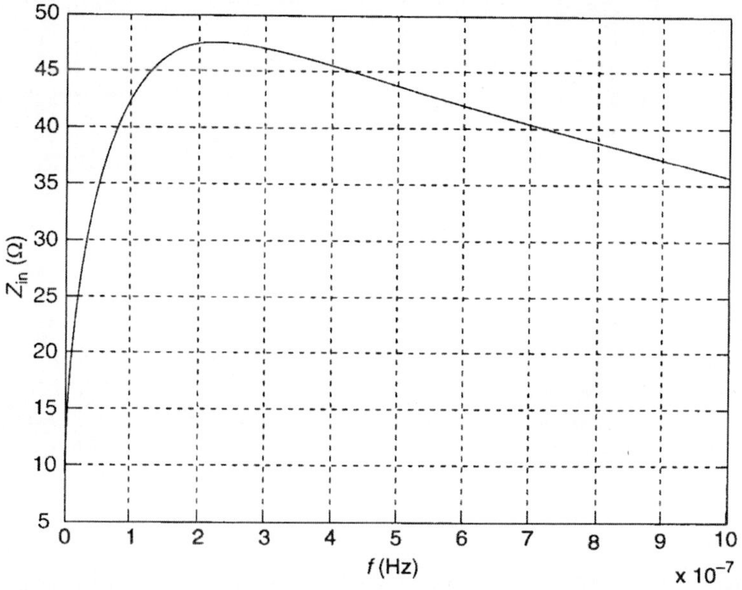

Figure 13.22 Input impedance spectrum ($L = 10$ m, $d = 10$ m, $\rho = 100\,\Omega$m). (D. Poljak and V. Doric, Wire Antenna Model for Transient, Analysis of Simple Groundingsystems, Part II: The Horizontal Grounding Electrode, *Progress in Electromagnetics Research, PIER 64, 167–189, 2006*).

13.3 TRANSMISSION LINE METHOD VERSUS ANTENNA THEORY APPROACH

As already mentioned, the transient modeling of grounding systems can be carried out featuring the transmission line model (TLM) [2,16,20] or antenna (electromagnetic) model (AM) [4,5,7]. Strengths and weaknesses of the approximate transmission line and rigorous antenna model, respectively, are as follows:

The principal feature of the TLM approach is simplicity and relatively low computational cost. However, though valid for long horizontal conductors, it is not convenient for modeling vertical and interconnected conductors. Furthermore, the effect of the air–earth interface has often been neglected, assuming the conductors to be buried at very large depths [6]. TLM approach is not only valid for low values of ground conductivity, it also neglects the effect of mutual coupling among the parts of the grounding system generating inevitable errors. To summarize, the TLM solutions are limited to a certain upper frequency depending on the dimensions of grounding electrodes and the electrical properties of the ground. The TLM approach is stated to be valid if the electric size is significantly smaller than a wavelength of the considered physical quantity in the earth. For typical grounding electrodes, TLM approach is considered to be valid for frequencies below MHz range. It is worth noting that at high frequencies the risk for equipment failure is greater than the possible hazard for human beings [8].

The rigorous electromagnetic field approach to the modeling of the grounding system is based on antenna theory [4,5,7]. The analysis based on AM involves the solution of corresponding electric field integral eqn (EFIE) for half-space problems. According to this approach, the grounding system of interest is represented by a configuration of buried straight thin wires. The ground is represented as a linear and homogeneous half-space characterized by its electrical properties. Using this rigorous approach, the effect of a dissipative half-space is taken into account by the Sommerfeld integrals, appearing in the integral equation kernel. This approach could be, in principle, useful for all cases [8]. However, it suffers from too long computational time for the evaluation of broadband frequency spectrum. Consequently, the rigorous approach is often regarded as too complex for practical applications, especially for large grounding grids and it has been avoided wherever possible [8]. A useful study on the validity of quasistatic theory compared with full wave theory has been presented in Ref. [8]. Also, the use of the modified image theory (MIT) [5,9] represents a possible way of avoiding the tedious calculation of Sommerfeld integrals. The MIT approach takes into account only the electrical properties of the soil, and not the burial depth as a parameter and it is stated to be valid up to 1 MHz.

This section deals with the transient impedance calculation of a horizontal grounding electrode using both antenna and transmission line approach, respectively. While the antenna model (AM) is formulated in a frequency domain (FD), the transmission line model (TLM) is carried out directly in the time domain.

The comparison of the direct time-domain approach based on the antenna model and transmission-line model has been presented in Ref. [21], while a trade-off between the direct TLM approach and the indirect frequency domain/IFT approach based on the antenna theory model for the analysis of transmission lines has been discussed in Ref. [22].

The full wave approach used in Ref. [22] has been based on the Pocklington integro-differential equation formulation [4,5,7] and related reflection coefficient approximation [23] to account for the presence of a dissipative half-space, thus avoiding the solution of analytically demanding and numerically time-consuming Sommerfeld integrals, as presented in Section 13.2.

The transmission line approach used in Refs. [21] and [22] is related to the direct TD solution of the Telegrapher's equations using the time domain finite element method (TD-FEM). The TD-FEM procedures used in Refs. [21] and [22] are outlined in Section 13.3.1.

13.3.1 Transmission Line Method (TLM) Approach to Modeling of Horizontal Grounding Electrode

The transmission line equations for a horizontal grounding wire excited by the current generator can be derived from the Maxwell's equations and can be expressed in terms of voltage and current induced along the wire. The geometry of interest is the

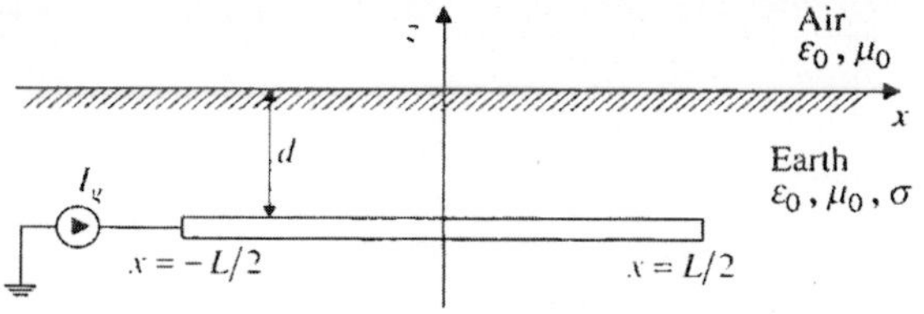

Figure 13.23 Horizontal grounding wire excited by a current generator I_g.

horizontal grounding electrode represented by the straight wire of length L and radius a, buried in a lossy medium at depth d, as shown in Figure 13.23.

The electrode is assumed to be perfectly conducting and driven at its end by a current source.

The set of time-domain Telegrapher's equations is given as follows [21,22]:

$$\frac{\partial u(x,t)}{\partial x} + L\frac{\partial i(x,t)}{\partial t} + Ri(x,t) = 0 \tag{13.86}$$

$$\frac{\partial i(x,t)}{\partial x} + C\frac{\partial u(x,t)}{\partial t} + Gu(x,t) = 0 \tag{13.87}$$

where u and i are the induced voltage and current on the wire, respectively.

The distributed parameters R, L, C, G for the geometry shown in Figure 13.23 have been computed in accordance to the procedure available in Ref. [24]. Using weighted residual approach to TD-FEM it follows

$$\int_{\Delta l_i} \left[\frac{\partial u^s(x,t)}{\partial x} + L\frac{\partial i(x,t)}{\partial x} + Ri(x,t)\right] W_j \mathrm{d}x = 0 \tag{13.88}$$

$$\int_{\Delta l} \left[\frac{\partial i(x,t)}{\partial x} + C\frac{\partial u^s(x,t)}{\partial x}\right] W_j \mathrm{d}x = 0 \tag{13.89}$$

Furthermore, the voltage and current along the line are expressed by a linear combination of shape functions:

$$uW_j \mathrm{d}x = \{f\}^T\{u\} \tag{13.90}$$

$$i(x,t) = \{f\}^T\{I\} \tag{13.91}$$

Using the explicit/implicit time-stepping procedure (θ-method) [22] transient voltage and current distribution along the wire is obtained. Within TLM, transient impedance is computed using relation (13.84), as well as AM.

13.3.2 Computational Examples

Figures 13.24–13.29 show the transient impedance waveform calculated via AM and TLM, respectively. Three electrode lengths for two values of ground conductivity with constant buried depth are analyzed.

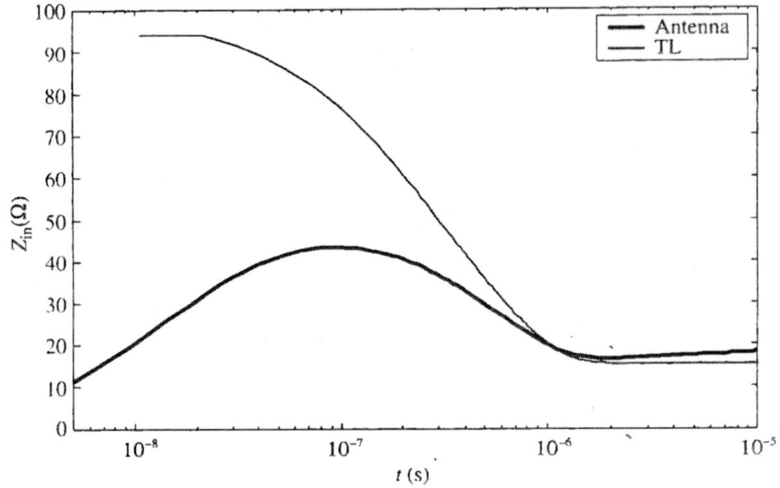

Figure 13.24 A comparison of the transient impedances obtained via different approaches ($L = 50\,\text{m}$, $d = 5\,\text{m}$, $\sigma = 0.001\text{s/m}$). (D. Poljak et al. ICLP 2006 D. Poljak, V. Doric, R. Lucic, S. Antonijevic, *V. Murko*, Comparison of Wire Antenna and Transmission Line Model for Transient Analysis of Grounding Electrodes, ICLP 2006).

General observation is related to the early time behavior of the grounding wire. What can be noticed from the results is that the TLM fails to predict accurate results for the early time behavior, up to around microsecond and more. (Of course, this is related to the high frequency content of the signal spectrum, which cannot be predicted by the TLM.)

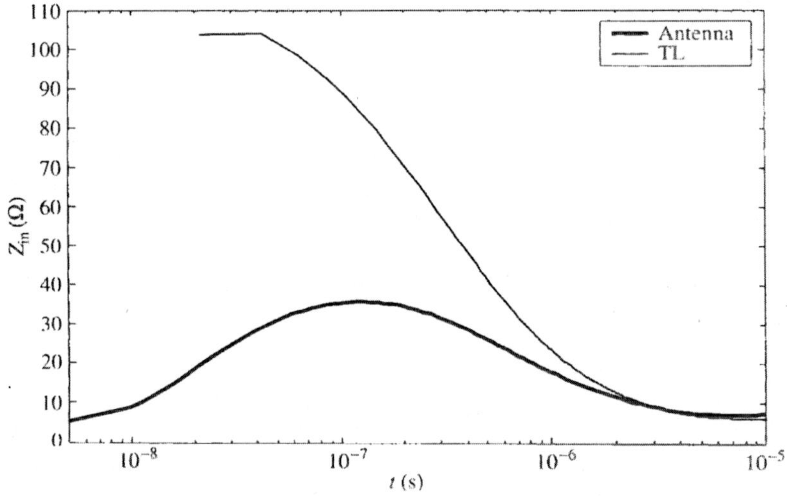

Figure 13.25 A comparison of the transient impedances obtained via different approaches ($L = 100\,\text{m}$, $d = 5\,\text{m}$, $\sigma = 0.001\text{s/m}$). (D. Poljak et al. ICLP 2006 D. Poljak, V. Doric, R. Lucic, S. Antonijevic, *V. Murko*, Comparison of Wire Antenna and Transmission Line Model for Transient Analysis of Grounding Electrodes, ICLP 2006).

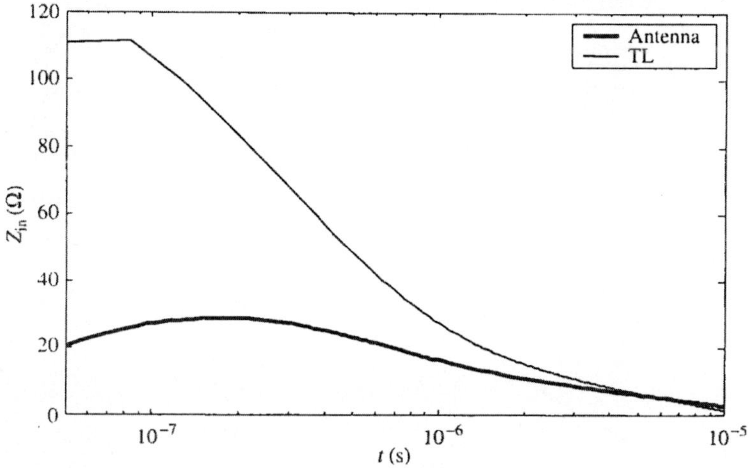

Figure 13.26 A comparison of the transient impedances obtained via different approaches ($L = 200$ m, $d = 5$ m, $\sigma = 0.001$s/m). (D. Poljak et al. ICLP 2006 D. Poljak, V. Doric, R. Lucic, S. Antonijevic, *V. Murko*, Comparison of Wire Antenna and Transmission Line Model for Transient Analysis of Grounding Electrodes, ICLP 2006).

For this particular case of rather low conductivity (σ = 1ms/m and 0.1 ms/m), TLM overestimates the values of transient impedance and is practically useless. For later time instants, there is a good agreement between the different approaches. The greater is the electrode length, the better is the agreement achieved between the curves.

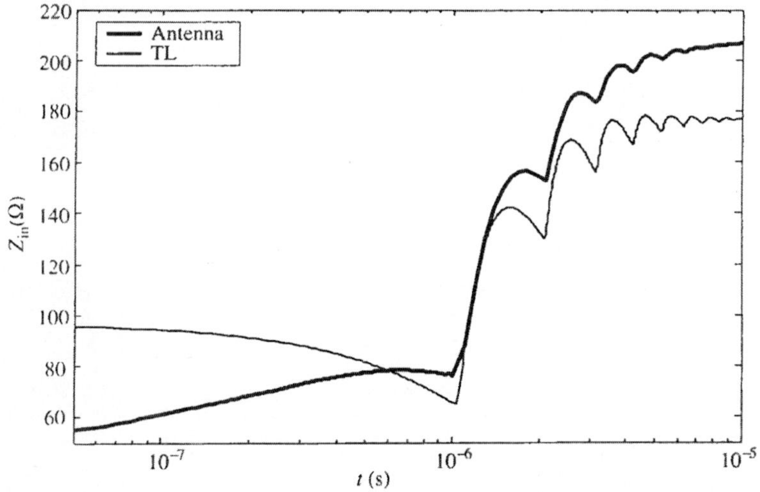

Figure 13.27 A comparison of the transient impedances obtained via different approaches ($L = 50$ m, $d = 5$ m, $\sigma = 0.0001$s/m). (D. Poljak et al. ICLP 2006 D. Poljak, V. Doric, R. Lucic, S. Antonijevic, *V. Murko*, Comparison of Wire Antenna and Transmission Line Model for Transient Analysis of Grounding Electrodes, ICLP 2006).

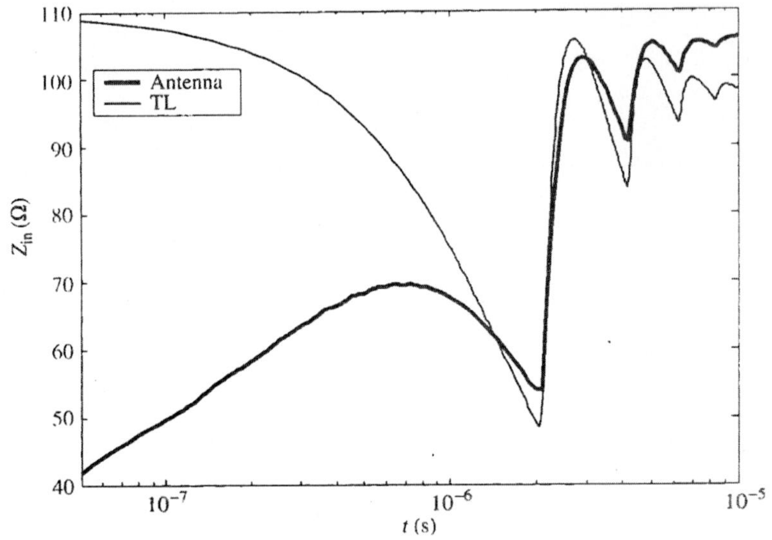

Figure 13.28 A comparison of the transient impedances obtained via different approaches ($L = 100\,\mathrm{m}$, $d = 5\,\mathrm{m}$, $\sigma = 0.0001\,\mathrm{s/m}$). (D. Poljak et al. ICLP 2006 D. Poljak, V. Doric, R. Lucic, S. Antonijevic, *V. Murko*, Comparison of Wire Antenna and Transmission Line Model for Transient Analysis of Grounding Electrodes, ICLP 2006).

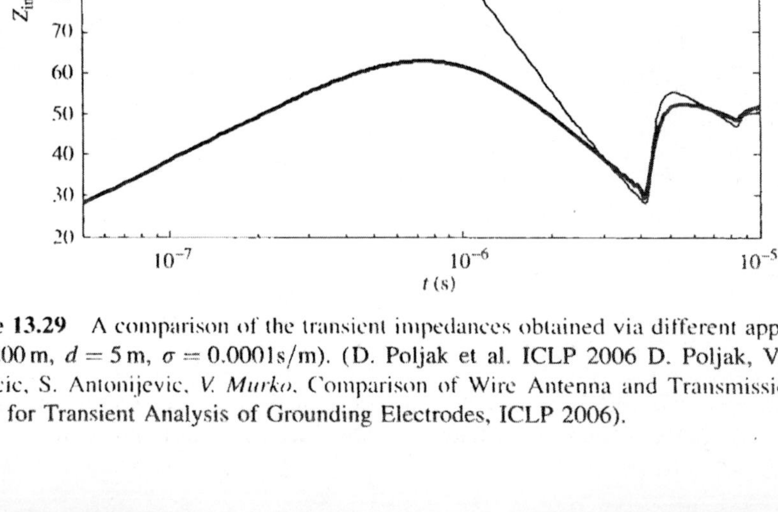

Figure 13.29 A comparison of the transient impedances obtained via different approaches ($L = 200\,\mathrm{m}$, $d = 5\,\mathrm{m}$, $\sigma = 0.0001\,\mathrm{s/m}$). (D. Poljak et al. ICLP 2006 D. Poljak, V. Doric, R. Lucic, S. Antonijevic, *V. Murko*, Comparison of Wire Antenna and Transmission Line Model for Transient Analysis of Grounding Electrodes, ICLP 2006).

Analyzing the obtained numerical results the conclusions drawn are summarized as follows:

- The TLM fails to predict accurate results for the early time behavior (up to the order of microsecond)
- The greater is the electrode length, the closer agreement between TLM and AM results is achieved.

Further analysis including the careful study of the related spectra is required for a future work.

Further extension of this analysis not only for the treatment of complex grounding systems consisting of interconnected conductors, but also for careful study of the related spectra will be reported in a future work.

13.4 MEASURES FOR QUANTIFYING THE TRANSIENT RESPONSE OF GROUNDING ELECTRODES

Further to the transient impedance concept for quantifying transient responses, widely adopted within the community, some additional measures of a horizontal electrode transient response have been proposed in Ref. [25].

These measures are instantaneous power, average power, and the total energy accumulated in the near field of the electrode.

As presented in Section 13.2, the frequency response of the horizontal electrode to a particular current source excitation is obtained by multiplying the input impedance spectrum with Fourier transform of the actual lightning current waveform. The corresponding transient response of the horizontal grounding electrode is calculated by applying the inverse Fourier transform.

Once determining the transient response of the electrode, the transient behavior can be quantified not only by using the standard concept of transient impedance, but also by using the additional circuit theory measures in terms of instantaneous power, average power, and the total energy stored in the electrode near field.

13.4.1 Transient Response Assessment

Transient voltage at the feed point ($x = -L/2$) (Fig. 13.8) can be obtained from the convolution integral:

$$v(x,t)\big|_{x=-L/2} = \int_0^t z_{in}(x,t)\big|_{x=-L/2}\, i(x,t-\tau)\big|_{x=-L/2}\, d\tau \qquad (13.92)$$

The frequency domain equivalent to the transient voltage $v(t)$, that is, the frequency response of the grounding system, is obtained by multiplying the frequency spectrum of the excitation function $I(f)$ with the frequency domain counterpart of the impulse response, that is, the input impedance spectrum $Z_{in}fj$:

$$V(f) = I(f)Z_{in}(f) \tag{13.93}$$

This injected current waveform is related to the lightning channel current:

$$i(t) = I_0 \cdot (e^{-\alpha t} - e^{-\beta t}); \quad t \geq 0 \tag{13.94}$$

where pulse rise time is defined by constants α and β, while I_0 is the amplitude of the current waveform.

The Fourier transform of the excitation function (13.94) is given by Ref. [14]:

$$I(f) = I_0 \cdot \left(\frac{1}{\alpha + j2\pi f} - \frac{1}{\beta + j2\pi f} \right) \tag{13.95}$$

Instead of solving convolution integral, as the time domain waveform of the input impedance (the impulse response of the system) is not known, the transient voltage is computed by applying the inverse fourier transform (IFT), that is, the inverse fast fourier transform algorithm:

$$v(n\Delta t) = F \cdot \sum_{k=0}^{N-1} V(k\Delta f) e^{jk\Delta f n\Delta t} \tag{13.96}$$

where F denotes the highest frequency taken into account, N is the total number of frequency samples, Δf is the sampling interval, and Δt is the time step.

13.4.2 Measures for Quantifying the Transient Response

The standard measure for analyzing the transient behavior of the horizontal grounding electrode is the transient impedance which is defined as a ratio of transient voltage and current at the input terminals:

$$z(x,t)\Big|_{x=-L/2} = \frac{v(x,t)}{i(x,t)}\Big|_{x=-L/2} \tag{13.97}$$

where $i(t)$ is the current injected at the end of the horizontal electrode.

Once obtaining the transient voltage flowing through the horizontal electrode, it is possible to calculate certain parameters by providing additional measures of this transient response. Such parameters can be found in the basic theory of electric circuits. The convenient parameters evaluating the horizontal electrode transient response, suggested in this work, are instantaneous power and the total energy stored in the near field of the electrode.

According to the basic theory of electric circuits, a time varying current delivers an average power to a grounding electrode. The amount of delivered power strongly depends on the particular waveform and it is given by the product:

$$p(x,t)\big|_{x=-L/2} = u(x,t)\cdot i(x,t)\big|_{x=-L/2} \tag{13.98}$$

The corresponding average power P_{av} is determined by the following integral relation:

$$P_{av} = \frac{1}{T_0}\int_0^{T_0} p(x,t)\big|_{x=-L/2}\,dt = \frac{1}{T_0}\int_0^{T_0} u(x,t)\cdot i(x,t)\big|_{x=-L/2}\,dt \tag{13.99}$$

In accordance to the circuit theory, the energy in the grounding electrode near field can be obtained by temporally integrating the instantaneous power:

$$W_{tot}(x,t)\big|_{x=-L/2} = \int_0^t p(x,t)\big|_{x=-L/2}\,dt \tag{13.100}$$

Total stored energy is obtained by prescribing $t = T_0$.

13.4.3 Computational Examples

Numerical results shown in Figures 13.30–13.40 are related to the grounding electrode of radius $a = 5$mm, buried at depth $d = 5$m in a lossy medium ($\sigma = 0.1$ ms/m,

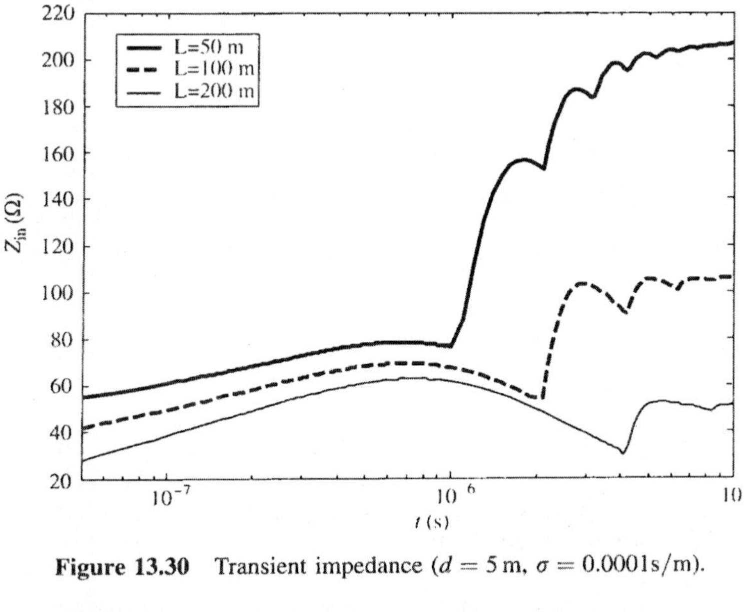

Figure 13.30 Transient impedance ($d = 5$ m, $\sigma = 0.0001$s/m).

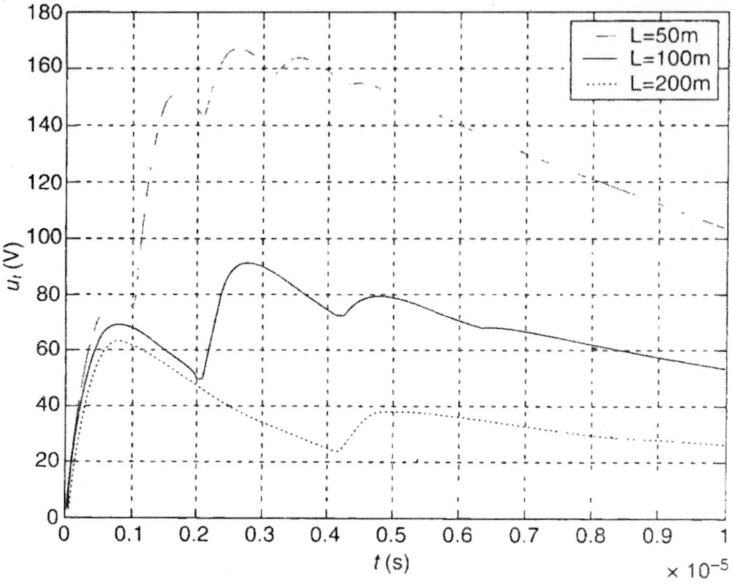

Figure 13.31 Transient voltage for various wire lengths ($d = 5$ m, $\sigma = 0.0001$ s/m).

$\varepsilon_r = 10$). Figures 13.30–13.40 show transient voltage, transient impedance, instantaneous power, and energy within the considered time interval of 10 μs for various set of parameters. It is obvious that the transient response measures are particularly dependent on electrode length.

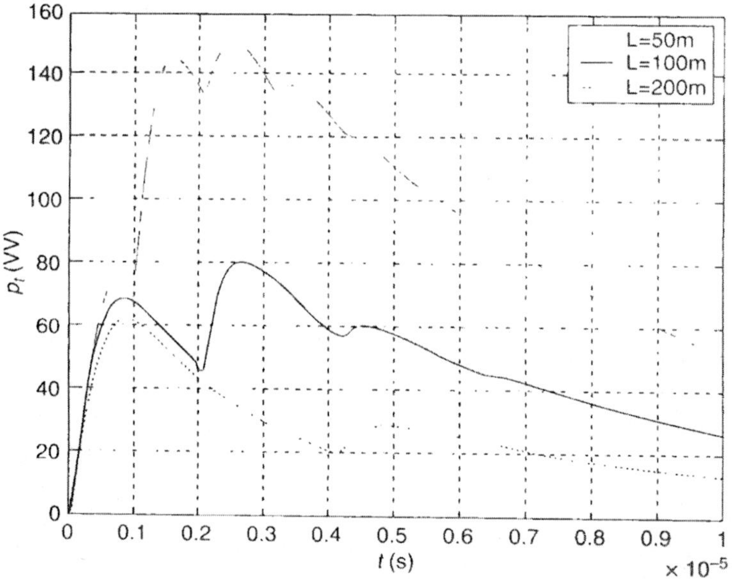

Figure 13.32 Instantaneous power for various wire lengths ($d = 5$ m, $\sigma = 0.0001$ s/m).

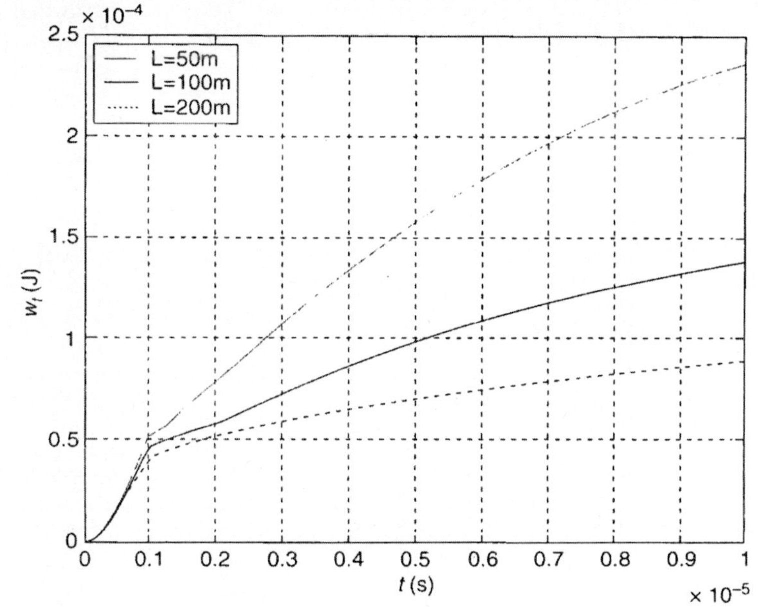

Figure 13.33 Total energy for various wire lengths ($d = 5\,\mathrm{m}$, $\sigma = 0.0001\mathrm{s/m}$).

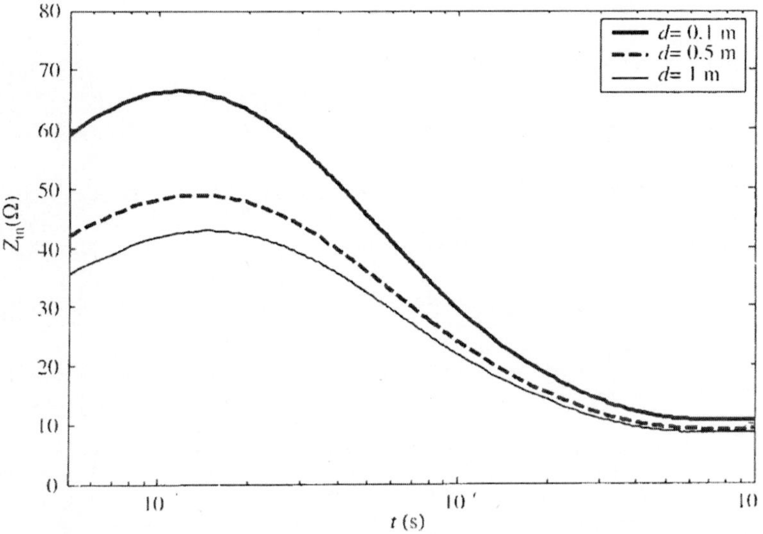

Figure 13.34 Transient impedance ($L = 100\,\mathrm{m}$, $\sigma = 0.001\mathrm{s/m}$).

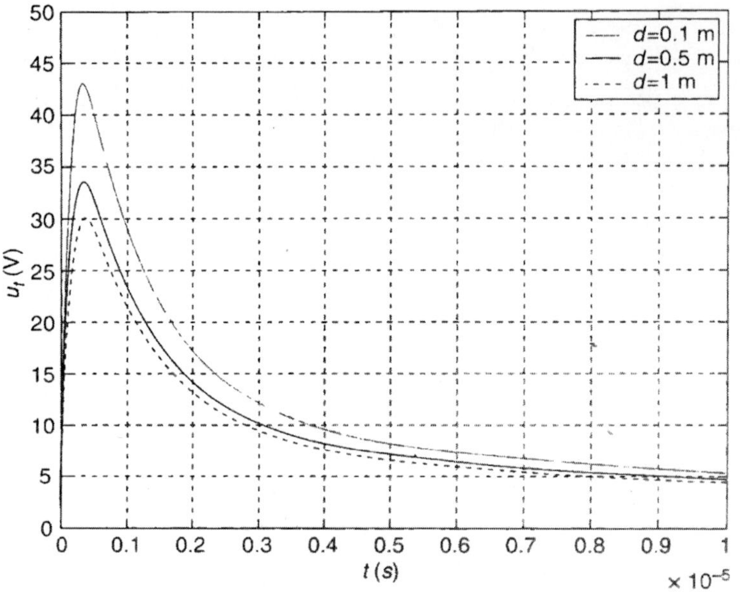

Figure 13.35 Transient voltage for various burial depths ($L = 100\,\mathrm{m}$, $\sigma = 0.001\mathrm{s/m}$).

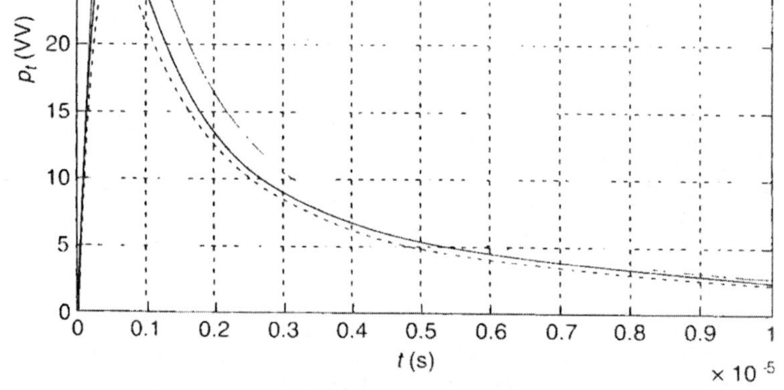

Figure 13.36 Instantaneous power for various burial depths ($L = 100\,\mathrm{m}$, $\sigma = 0.001\mathrm{s/m}$).

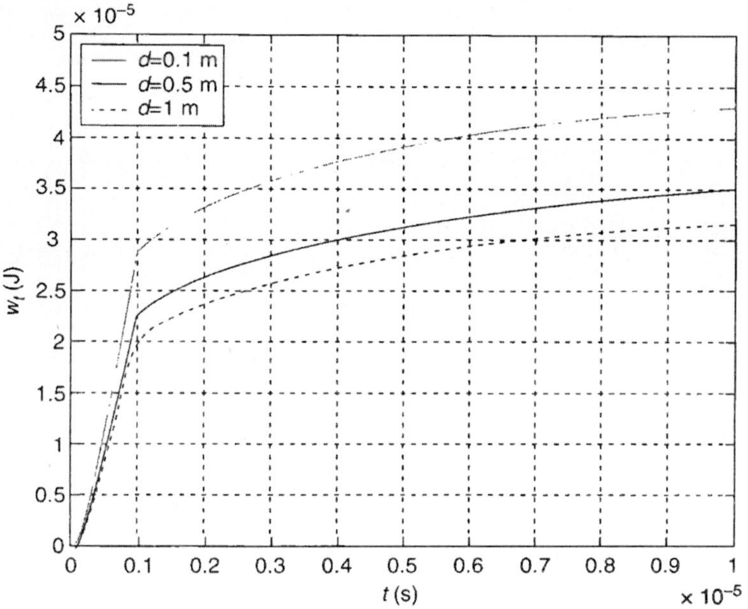

Figure 13.37 Total energy for various burial depths ($L = 100\,\text{m}$, $\sigma = 0.001\,\text{s/m}$).

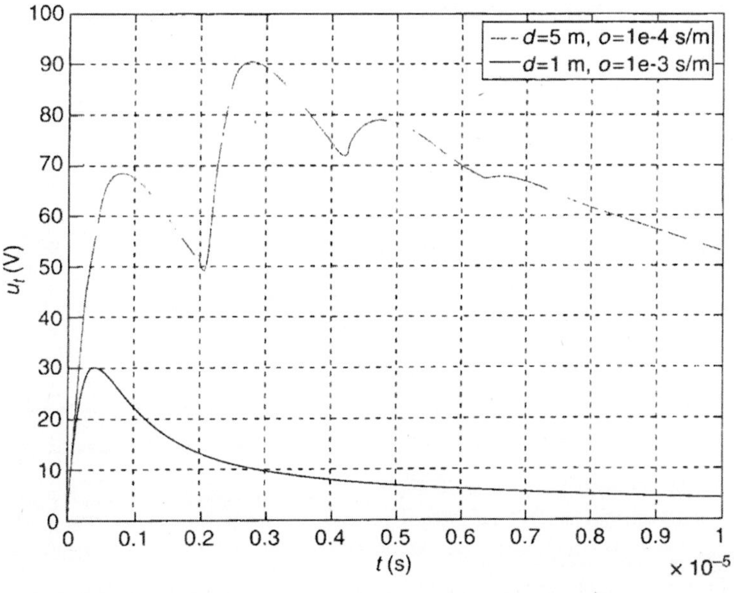

Figure 13.38 Transient voltage for various conductivities and burial depths, $L = 100\,\text{m}$.

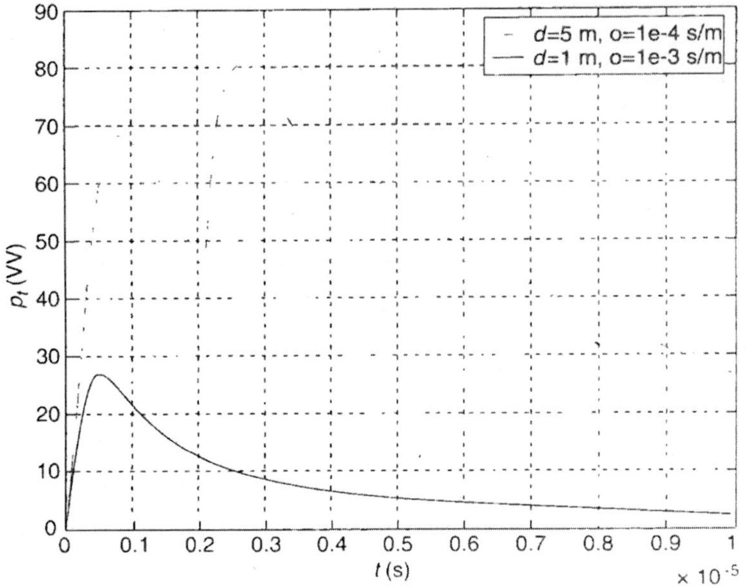

Figure 13.39 Instantaneous power for various conductivities and burial depths, $L = 100\,\mathrm{m}$.

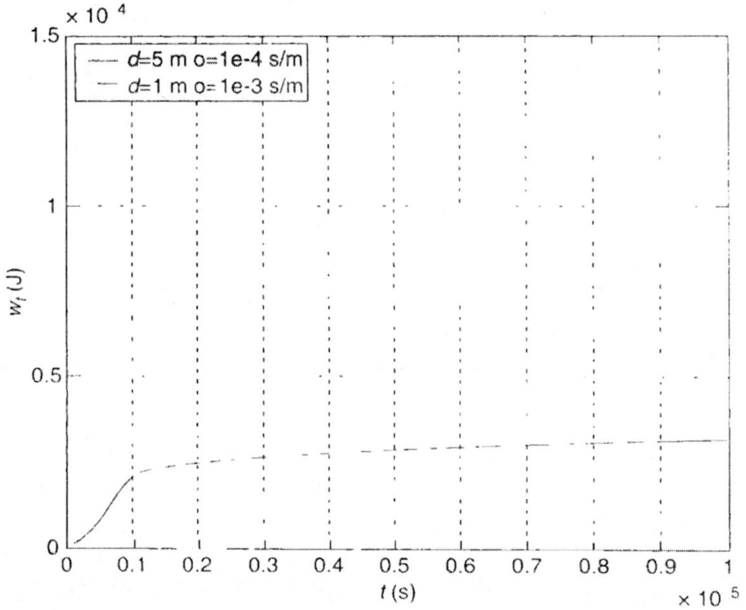

Figure 13.40 Total energy for various conductivities and burial depths, $L = 100\,\mathrm{m}$.

Further extension of the present analysis to the treatment of complex grounding systems including interconnected conductors will be carried out subsequently.

13.5 REFERENCES

[1] R. Velazquez, D. Muhkedo: Analytical modeling of grounding electrodes transient behavior, *IEEE Trans. Power Appar. Sys., 103*, June 1984, pp 1314–1322.

[2] Y. Liu, M. Zitnik, R. Thottappillil: An improved transmission line model of grounding system, *IEEE Trans. EMC, 43 (3)*, 2001, pp. 348–355.

[3] G. Ala, M. L. Di Silvestre: A simulation model for electromagnetic transients in lightning protection systems, *IEEE Trans. EMC 44 (4)*, 2003, pp 539–534.

[4] L. Grcev, F. Dawalibi: An electromagnetic model for transients in grounding systems, *IEEE Trans. Power Delivery (4)*, October 1990, pp 1773–1781.

[5] L. D. Grcev, F. E. Menter: Transient electro-magnetic fields near large earthing systems, *IEEE Trans. Mag., 32*, May 1996, pp 1525–1528.

[6] G. E. Bridges: Transient plane wave coupling to bare and insulated cables buried in a lossy half-space, *IEEE Trans. EMC, 37 (1)*, February 1995, pp 62–70.

[7] D. Poljak, V. Roje: The integral equation method for ground wire impedance, in C. Constanda, J. Saranen, S. Seikkala (Ed): Integral Methods in Science and Engineering, Vol. I, *Longman, UK*, 1997, pp 139–143.

[8] R. G. Olsen, M. C. Willis: A comparison of exact and quasi-static methods for evaluating grounding systems at high frequencies, *IEEE Trans. Power Delivery, 11 (2)*, April 1996, pp 1071–1081.

[9] D. Poljak, I. Gizdic, V. Roje: Plane wave coupling to finite length cables buried in a lossy ground, *Eng. Anal. Boundary Elements, 26 (1)*, January 2002, pp 803–806.

[10] D. Poljak, V. Roje: Boundary element approach to calculation of wire antenna parameters in the presence of dissipative half-space, *IEE Proc. Microw. Antennas Propag., 142 (6)*, December 1995, pp 435–440.

[11] D. Poljak: Electromagnetic Modelling of Wire Antenna Structures, *WIT Press, Southampton, Boston*, 2002.

[12] G. J. Burke, E. K. Miller: Modeling antennas near to and penetrating a lossy interface, *IEEE Trans. AP, 32*, 1984, pp 1040–1049.

[13] D. Poljak, C. A. Brebbia: Boundary Element Methods for Electrical Engineers, *WIT Press, Southampton, Boston*, 2005.

[14] R. E. Ziemer, W. H. Tranter: Principles of Communications, *Houghton Mifflin Company, Boston*, 1995.

[15] D. Poljak: New numerical approach in the analysis of a thin wire radiating over a lossy half-space, *Int. J. Num.Meth. in Eng., 38 (22)*, November 1995, pp 3803–3816.

[16] G. Ala, M. L. Di Silvestre: A simulation model for electromagnetic transients in lightning protection systems, *IEEE Trans. EMC, 44 (4)*, 2003, pp 539–534.

[17] D. Poljak, V. Doric: Time domain modeling of electromagnetic field coupling to finite length wires embedded in a dielectric half-space, *IEEE Trans. EMC*, 2005.

[18] D. Poljak, I. Gizdic, V. Roje: Plane wave coupling to finite length cables buried in a lossy ground, BEM 23, *Lemnos Island, Greece*, 7–9 May 2001, pp 185–193.

[19] L. Grcev: Calculation of the transient impedance of grounding systems, PhD Thesis, *University of Zagreb*, 1986, (in Croatian).

[20] M. I. Lorentzou, N. D. Hatziargyriou, C. Papadias: Time domain analysis of grounding electrodes impulse response, *IEEE Trans. Power Delivery, (2)*, April 2003, pp 517–524.

[21] D. Poljak, R. Lucic, V. Roje: Transient analysis of electromagnetic field coupling to buried cables, ICEAA 01 International Conference on Electromagnetics in Advanced Applications, *Torino, Italy*, 2001, pp 335–338.

[22] D. Poljak, V. Doric, R. Lucic, S. Antonijevic, V. Murko: Comparison of wire antenna and transmission line model for transient analysis of grounding electrodes, *ICLP*, 2006.

[23] E. K. Miller, A. J. Poggio, G. J. Burke, E. S. Selden: Analysis of wire antennas in the presence of a conducting half-space, Part II: The horizontal antenna in free space, *Canad. J. Phys., 50*, 1972, pp 2614–2627.

[24] R. Velazquez, D. Muhkedo: Analytical modeling of grounding electrodes transient behavior, *IEEE Trans. Power Appar. Systems, 103*, June 1984, pp 1314–1322.

[25] D. Poljak, V. Doric: Postprocessing of the grounding wire transient response, *EI*, 2006.

14

HUMAN EXPOSURE TO ELECTROMAGNETIC FIELDS

14.1 ENVIRONMENTAL RISK OF ELECTROMAGNETIC FIELDS: GENERAL CONSIDERATIONS

As electricity plays one of the most important roles in modern society, there has been a tremendous growth in the use of electrical energy for industrial and domestic purposes. Nowadays, it is impossible to imagine modern society without electrical appliances.

Unfortunately, electricity always comes with risks. Namely, there is always a risk of life from contact with ungrounded electrical conductors, and the use of electricity results in the generation of electromagnetic fields as well. Since the very beginning of the twentieth century, one has been surrounded with a high density of electromagnetic energy due to the dramatic growth of power grids, radio and television stations, radars, base stations, cellular phones, numerous domestic appliances, and appliances, at work places.

Nevertheless, the presence of electromagnetic fields in the environment has become an essential part of our society through their various applications; there has been an increasing concern associated with the possible adverse health effects that could exist due to the exposure to these fields.

Public concern regarding possible bioeffects due to exposure to electromagnetic fields and the related research started after World War II. By the middle of seventies, the concern was mostly directed toward possible health hazards due to the exposure to radio frequency (RF) fields. In the years to come, public concern would be directed to extremely low frequency (ELF) fields. During the last decade, public concern regarding RF exposure from cellular base stations and cellular phones has grown rapidly.

14.1.1 Nonionizing and Ionizing Radiation

All types of electromagnetic radiation have the same physical properties of divergence, interference, coherence, and polarization, but differ in terms of energy.

Depending on the frequency, electromagnetic radiation is classified as either nonionizing or ionizing.

Ionizing and nonionizing radiations are separated on the electromagnetic spectrum. The division between them is generally accepted to be at wavelengths around 1nm in the far UV region.

Nonionizing radiation is electromagnetic radiation that does not have sufficient energy to cause ionization in living systems. Natural sources of nonionizing radiation are the sun, distant radio stars, other cosmic sources, and terrestrial sources like lightning. These sources are extremely weak. With the tremendous growth of electricity applications, the density of artificial electromagnetic energy in the environment is much higher than in the natural levels.

Generally, the nonionizing part of the electromagnetic spectrum is often divided into three main ranges: extremely low frequency (ELF) fields, radio frequency (RF) fields, and noncoherent light.

On the contrary, ionizing radiation has sufficient energy to remove electrons from atoms [1]. One source of ionizing radiation is the nuclei of unstable atoms. For these radioactive atoms to become more stable, the nuclei emit subatomic particles and high-energy photons. Included in this part are X-rays, gamma rays, and cosmic rays. The high frequency radiation ($f > 10^{15}$ Hz) is characterized by short wavelengths and high energy.

Atoms of all elements may become ionized; however, only gamma rays, X-rays, alpha particles, and beta particles have enough energy to create ions. Because ions are charged particles, they are chemically more active than their electrically neutral forms. Chemical changes that occur in biological systems may be cumulative, detrimental or even fatal.

The considerations presented in this chapter are strictly related to the nonionizing radiation effects on humans.

14.1.2 Electrosmog or Radiation Pollution at Low and High Frequencies

Electrosmog is a relatively new term for radiation pollution, recently used to describe all artificially generated electromagnetic fields. The nuisance caused by electromagnetic fields can be reduced by paying attention to the quality of electrical and electronic equipment.

Electrosmog occurs wherever voltages and currents are present: in the home, industry, and commerce. Mostly, human senses do not register these fields, but they are nevertheless affected by them.

Electromagnetic fields are split into two main categories: low frequency (up to about 30 kHz) and high frequency (from 30 kHz to 300 GHz). Above this frequency lies the infra-red, visible light, ultra-violet, X-ray, and gamma ray spectra, following ascending order of frequency.

From the beginning of the 1990s, public concern on the presence of electromagnetic fields in the environment has been awakened by a number of articles in the general press. The sources of electromagnetic fields include power lines, mobile phones, and broadcasting antennas. The subject generated much controversy on the possible link between the fields and health risks, particularly in certain forms of cancer in humans. As a result, continuing research has intensified in the biomedical and health aspects. While safety regulations on the exposure to radiation have been well established in the nuclear power industry, it is only recently that new proposals or revisions to the existing ones are being made for limiting the exposure to radio frequency electromagnetic fields.

14.1.2.1 ***Interaction of Humans with Electromagnetic Fields*** The senses of the human body and regulatory systems operate using minute electric currents and voltages. These quantities can be measured, for example, the "brain waves" can be recorded on an electroencephalograph (EEG). Artificially created electromagnetic fields induce much higher currents and voltages in the human body than are naturally present. Although the human body can compensate for a great many things, it is not yet fully known what effect electromagnetic fields have on human health.

14.1.3 The Effects of Low Frequency Fields

Low frequency (LF) fields induce currents in the human body generating surface effects (like making our hair stand on end). These fields can also affect heart pacemakers and other active implants. On the contrary, electric shock, caused by touching electrically charged objects, is in direct contrast with these effects. Some of the LF field effects are listed in Table 14.1.

It is well known that LF fields cause irritation of sensory, nerve, and muscle cells. The higher is the field strength, the stronger are the associated effects. The

TABLE 14.1 The effects of the induced current density in the body

Current density [mA/m^2]	Effects
1000 and higher	Slight to severe heart dysfunction, acute danger to health.
100 to 1000	The excitability of the central nervous system can change that is, the irritation of muscle tissue has been seen in this range.
10 to 100	Clearly demonstrated effects, such as changes in protein and DNA synthesis, changes in enzyme activity, evident visual, and possible nervous effects.
1 to 10	Subtle biological effects such as changes in calcium metabolism or suppression of melatonin production (Melatonin controls the day/night rhythm). The background current densities of the heart and brain lie in this range.
Below 1	No known effects. The background current densities in most body organs are in this range.

human body can often compensate for weak interactions [1]. Higher intensity fields cause stresses that can lead to irreversible damage to health under certain circumstances. Nobody knows for sure yet, if LF electromagnetic fields lead to an increase in the number of cancer cases.

14.1.4 The Effects of High Frequency Fields

Humans are particularly sensitive to high frequency (HF) electromagnetic fields. Basically, the human body absorbs the radiated energy, and the related heating effects are more important in this case [1]. Some research has been undertaken with respect to other effects, such as irritation of the nervous system and long-term consequences. However, the importance of these in connection with health and safety is not yet well established.

HF fields generate heat, either locally or throughout the body. Since the heat rise is generated internally, it is either not noticed or noticed too late because the heat is basically sensed by our skin. Eventually, the temperature control mechanisms may be affected. These effects depend strongly on the frequency.

In addition, there is an increasing evidence that pulsed radiation (such as radar or digital mobile radio telephones) has a greater effect on biosystems than nonpulsed signals [2]. Effects on cell membrane are also noticed. The human body absorbs a great deal of energy at certain frequencies, since the body acts as an antenna. When the body size is half the wavelength, the resonant frequency is reached. Humans can absorb large amounts of energy from the field at resonant frequencies between 30 and 300MHz. It is also important to emphasize that children have a higher resonant frequency than adults.

A measure of radiation power converted to heat, a so called specific absorption rate (*SAR*), can be evaluated for each body.

The *SAR* is used as the basis for setting limit values among other things. The limit values are split into three sections as determined by the absorption curve low, medium, and high absorption sections.

The degree at which parts of the body absorb heat from HF field depends on the circulation and on the ability of heat conduction. As it can be seen in Table 14.2,

TABLE 14.2 Exposure to the HF homogeneous HF field (unmodulated signal)

Body section	Absorption level
Brain	++
Lens of eye	+++
Lungs	+
Heart	+
Skin	+
Internal organs	+
Knee cap	+++
Metallic implants	+++

the lens of the eye and the kneecap are the most susceptible parts. The heart, lungs, and skin on the contrary are relatively insensitive because of the high blood circulation.

For the case of the modulated field, such as GSM, cell irritation may occur in addition to the thermal effects.

14.1.5 Remarks on Electromagnetic Fields and Related Possible Hazard to Humans

The presence of electromagnetic fields in the environment and their hazards to humans are controversial scientific, technical, and often public, issue. The electromagnetic fields are a product of technology that must be used in everyday life despite the unknown risks.

In everyday life, people are assaulted by the amount of information, often conflicting, by the media on hazards from power utilities and communication antennas. The widespread use of electrical energy implies that in all residences and workplaces, there are levels of electromagnetic fields that would be considered normal.

There is a controversy if field emissions from such sources may cause cancer or other diseases. There are people who believe in adverse health effects due to exposure to electromagnetic fields, and consequently they are usually cautious about their health and they keep protesting. They are often grouped as conservationists. On the contrary, the developers and people from industry do not believe in electromagnetic bioeffects.

The only groups in the society staying away from the area of argument are the researchers working in good faith to find scientific facts. However, the reliable results are still not forthcoming.

The public concern is constantly swinging between ELF and HF frequency range, mostly depending on widely publicized claims of these issues in the media.

In the last few decades, public anxiety was directed to the safety of radar equipment at the workplace and microwave ovens in homes. However, it is now accepted that microwave ovens are harmless, (at least when used properly), while a number of investigations regarding radar radiation resulted in certain safety precautions that could minimize some established thermal effects.

Today, base stations, cellular phones, and other pieces of wireless communication equipment are the main sources of concern. In particular, the idea of health effects from cellular phones is rapidly becoming the focus of the research in this area.

The starting point in the analysis of possible health risk is the incident field dosimetry, (including the evaluation of incident fields from various electromagnetic sources in the near- and far-field region), and internal field dosimetry (including the various techniques for determination of internal electromagnetic field), [1], [3].

Evaluation of any health risk due to the exposure to electromagnetic fields relies on the results of a well-established body of research based on experimental data from biological systems, epidemiological and human studies, as well as on understanding the various mechanisms of interaction.

Unfortunately, information, which provides the public with a satisfactory understanding of exposure to electromagnetic fields and effect-related issues, is still rarely available.

The investigation of electromagnetic bioeffects integrates several aspects of electromagnetic fields such as the biological, medical, biochemical, epidemiological, environmental, risk assessment, and health policy.

Only a very small number of scientists can fully grasp all these subjects.

14.2 ASSESSMENT OF HUMAN EXPOSURE TO ELECTROMAGNETIC FIELDS: FREQUENCY AND TIME DOMAIN APPROACH

The presence of electromagnetic fields in the environment have become an essential part of modern society and have caused increasing public concern associated with the possible adverse health effects that exposure to these fields may generate.

The sources of electromagnetic fields include power lines, mobile phones, broadcasting antennas, and many others. The subject has generated much controversy on the possible links between the fields and radiation hazard, particularly on certain forms of cancer. As a result, research has intensified in the fields of biomedical and health aspects.

The starting point in the analysis of possible health risk is the calculation or the measurement of the external fields from different electromagnetic sources to which humans are exposed [4–6]. This provides the assessment of the induced currents and of the fields inside the human body [8–15].

The procedure is based on the boundary element method (BEM).

The simplified approach is based on the equivalent parasitic antenna representation of the body (as it will be shown in the following section) [11–14] and the related Pocklington integral equation formulation. The electrical properties of the body related to the LF and HF exposures are taken into account via the load term in the Pocklington equation for a thick cylinder.

Obtaining the current distribution along the body, as the indirect BEM solution of Pocklington equation, one can readily calculate the induced current density and the induced electric field. This provides simple and efficient procedures for a rapid estimation of these electromagnetic phenomena.

This Chapter focuses on the analysis to LF and transient exposures. Details regarding the HF exposure assessment can be found elsewhere, for example in [3]. Assessment of human exposure to LF fields is carried out using both the simplified and realistic body models. The exposure of humans to transient radiation is analyzed using the simplified body representation only.

14.2.1 Frequency Domain Cylindrical Antenna Model

When exposed to electromagnetic radiation, the human body acts as an imperfectly conducting straight wire antenna.

The currents and fields induced in human organs may give rise to thermal and nonthermal effects. When the body is exposed to LF electromagnetic fields, the thermal effects are found to be negligible and possible nonthermal effects are often related to the behavior at the cell level [7–8].

14.2.2 Realistic Models of the Human Body for ELF Exposures

In addition to cylindrical model of the human body exposed to ELF electric fields, this chapter deals with a multidomain body of revolution and an anatomically based representation of the human body. The formulation of the problem is based on the quasistatic approximation and the related Laplace equation variant of the continuity equation. The Laplace equation is solved by the BEM, thus providing the distribution of scalar potential and the induced current density inside the body.

14.2.3 Human Exposure to Transient Electromagnetic Fields

The mathematical model of human beings exposed to transient electromagnetic fields is based on the cylindrical representation of the human body and the corresponding space-time Hallen integral equation. The Hallen integral equation is solved via the Galerkin-Bubnov scheme of the indirect boundary element method (GB-IBEM) and the equivalent space-time current distribution along the cylindrical body model is obtained. Besides the well-established concept of specific absorption (SA), the transient current flowing through the human body is postprocessed in terms of certain measures of quantifying the transient response. These measures arise from circuit theory and they are *average* and *root-mean square* value of time-varying current, *instantaneous power* dissipated in the body, and *total absorbed energy* in the body.

14.3 HUMAN EXPOSURE TO EXTREMELY LOW FREQUENCY (ELF) ELECTROMAGNETIC FIELDS

The presence of ELF fields in the environment has become an essential part of modern society, which is also associated with public concern regarding possible adverse effects [1]-[4].

At low frequencies, the displacement currents can be neglected providing the electric and magnetic fields to be analyzed separately. The human body can be exposed to low voltage/high intensity systems, and in which the principal radiation is generated from the magnetic field or to high voltage/low intensity systems, and in which the principal radiation is generated by the electric field. In the case of magnetic field exposure the induced currents in the body form close loops, while in the electric field exposure the related currents induced in the body have the axial character.

This section is devoted to the assessment of human exposure to ELF electric fields used for power utilities. The current density is the main parameter for the evaluation of low frequency exposure hazard proposed by ICNIRP basic restrictions [3] and is a key point in any further analysis of possible adverse biological effects of ELF fields.

This section deals with two different approaches to the electromagnetic modeling of the human body exposed to extremely low frequency (ELF) fields. Analysis presented in subsection 14.3.1 is based on the wire antenna theory and the simplified cylindrical representation of the human body, while the formulation presented in subsection 14.3.2 deals with a quasistatic approximation and continuity equation applied to realistic body models.

14.3.1 Parasitic Antenna Representation of the Human Body

The human body consists of various tissues and organs with different electrical properties such as conductivity σ and relative permittivity ε_r. At low frequencies near the power frequency $f = 50$ Hz or 60 Hz, the body is a good conductor with an average conductivity of approximately $\sigma \approx 0.5$ S/m.

Although the body cross-section changes in size and shape, this does not have a significant effect on the total current as long as the tissue material is predominantly conducting and the body can be represented as an equivalent cylindrical antenna model with a uniform cross section and conductivity.

Maximum current flows from the body into the ground when it is assumed that the person is standing bare footed on a good conducting ground.

The thick wire model can be applied not only to ELF and VLF but also to the case of RF and GSM exposures [14]. At the GSM frequencies of nearly 900 MHz the average conductivity of the body is approximately $\sigma = 1.4$ S/m, while the corresponding average permittivity is around $\varepsilon_r = 55$. The behavior of tissues at higher frequencies is taken into account by the load term in the Pocklington equation.

A person standing vertically on the earth and exposed to the electromagnetic field from ELF to GSM frequencies can be represented by an imperfectly conducting cylinder of length L and radius a, as it is shown in Figure 14.1.

The current distribution along the human body is obtained as the solution of the Pocklington integro-differential equation for thick loaded straight wire [11–14], contrary to the Pocklington equation type for thin wire discussed in Chapter 8.

This integro-differential equation can be derived starting from the time harmonic electric field E given by the magnetic potential A and the electric scalar potential φ, that is:

$$\vec{E} = -\nabla\varphi - j\omega\vec{A} \tag{14.1}$$

The vector and magnetic potential are coupled through the previously defined Lorentz Gauge:

$$\nabla\vec{A} = -j\omega\mu\varepsilon\varphi \tag{14.2}$$

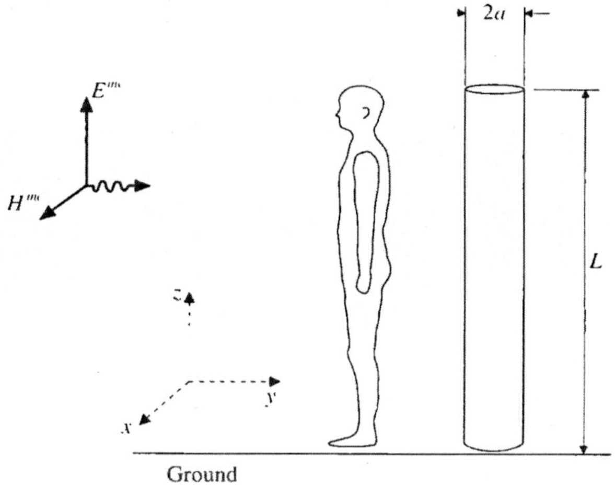

Figure 14.1 The equivalent antenna model of the human body. (Poljak, D.. Tham, C.Y., Gandhi, O., Sarolic, A., Human Equivalent Antenna Model for Transient Electromagnetic Exposure, *IEEE Trans. EMC*, Feb. 2003. 45 (2003), 1; 141–145).

In addition, since only axial component of the magnetic potential A along the cylinder exists, the combination of eqns (14.1) and (14.2) results in:

$$E_z = \frac{1}{j\omega\mu\varepsilon}\left[\frac{\partial^2 A_z}{\partial z^2} + k^2 A_z\right] \tag{14.3}$$

Applying the image theory, the magnetic vector potential A_z can be represented by the integral of the axial current $I(z)$ flowing along the cylinder, that is:

$$A_z = -\frac{\mu}{4\pi}\int_{-L}^{L}\frac{1}{2\pi}\int_{0}^{2\pi}\frac{e^{-jkR}}{R}I(z')\mathrm{d}z'\mathrm{d}\phi \tag{14.4}$$

where R is the distance from the source point to the observation point and k denotes the wave number of the free space.

The continuity conditions for the tangential electric field components can be written as following:

$$E_z^{inc} + E_z^{sct} = I(z)Z_L(z) \tag{14.5}$$

where E_z^{inc} is the excitation function in the form of incident field, E_z^{sct} is the scattered field due to the presence of the imperfectly conducting cylinder, $I(z)$ is the axial current distribution, and $Z_L(z)$ is the impedance per unit length of the cylinder.

Consequently, the tangential component of the electric field on the antenna surface is given by:

$$E_z^{sct}(z,a) = \frac{1}{j4\pi\omega\varepsilon_0} \int_{L}^{L} \left(\frac{\partial^2}{\partial z^2} + k^2 \right) g_E(z,z') I(z') dz' \tag{14.6}$$

where ε_0 is the permittivity of the free space, ω is the applied frequency.

Combining relations (14.3)-(14.6), one obtains the Pocklington integro-differential equation which determines the current along a thick wire:

$$E_z^{inc} = -\frac{1}{4j\pi\omega\varepsilon_0} \int_{-L}^{L} \left(\frac{\partial^2}{\partial z^2} + k^2 \right) g_E(z,z') I(z') dz' + Z_L(z) I(z) \tag{14.7}$$

where $g_E(z, z')$ is the so-called "exact" kernel of the integral equation and is given by [3]:

$$g_E(z,z') = \frac{1}{2\pi} \int_0^{2\pi} \frac{e^{-jkR}}{R} d\phi \tag{14.8}$$

in which R denotes the distance from the source point z to the observation point z', both located on the thick wire antenna surface.

The distance R is given by:

$$R = \sqrt{(z - z')^2 + 4a^2 \sin^2 \frac{\phi}{2}} \tag{14.9}$$

It is worth emphasizing that both the conducting and dielectric properties of the body are taken into account by the properties of the impedance Z_L.

14.3.1.1 Boundary Element Solution of the Pocklington Equation The boundary element solution of the Pocklington integral equations for thin wires has already been discussed in Chapter 8. A similar procedure can be applied to the solution of thick wire Pocklington eqn (14.7). Eqn (14.7) can be, for convenience, written in an operator form:

$$KI = E \tag{14.10}$$

where K is a linear operator, I is the unknown function to be found for a given excitation E.

The unknown current is then expanded into a finite sum of linearly independent basis functions $\{f_i\}$ with unknown complex coefficients α_i, that is:

$$I \cong I_n = \sum_{i=1}^{n} \alpha_i f_i \tag{14.11}$$

Substituting (14.11) into (14.10) yields:

$$KI \cong KI_n = \sum_{i=1}^{n} \alpha_i K f_i \tag{14.12}$$

Now a residual R_n can be defined as:

$$R_n = KI_n - E \tag{14.13}$$

Finally, according to the definition of the scalar product of functions in Hilbert space, the error R_n can be weighted with respect to certain weighting functions $\{W_j\}$ as already discussed:

$$\int_{\Omega} R_n W_j^* \mathrm{d}\Omega = 0; \quad j = 1, 2, \ldots, n \tag{14.14}$$

where $W_j{}^*$ is the complex conjugate value of weighting function W_j.

The expression

$$\langle R_n, W_j \rangle = \int_{\Omega} R_n W_j{}^* \mathrm{d}\Omega \tag{14.15}$$

denotes the scalar product of functions R and W, and Ω is the domain of interest.

Since the operator K is linear, one obtains a system of algebraic equations and by choosing $W_j = f_j$, the Galerkin-Bubnov procedure yields:

$$\sum_{i=1}^{n} \alpha_i \int_{\Omega} K(f_i) f_j \mathrm{d}\Omega = \int_{\Omega} E f_j \mathrm{d}\Omega; \quad j = 1, 2, \ldots, n \tag{14.16}$$

Eqn (14.16) is the Galerkin formulation of the Pocklington integral eqn (14.7). Using the integral equation kernel symmetry, taking into account the boundary conditions for current at the free ends of the thin wires, and after integration by parts eqn (14.16) it can be written:

$$\sum_{i=1}^{n} \alpha_i \frac{1}{j4\pi\omega\varepsilon} \left\{ -\int_{-L}^{L} \frac{\mathrm{d}f_j(z)}{\mathrm{d}z} \int_{-L}^{L} \frac{\mathrm{d}f_i(z)}{\mathrm{d}z'} g_E(z, z')\mathrm{d}z'\mathrm{d}z \right.$$
$$\left. + k_1^2 \int_{L}^{L} f_j(z) \int_{L}^{L} f_i(z') g_E(z, z)\mathrm{d}x'\mathrm{d}x + \int_{-L}^{L} Z_s(z) f_j(z) f_i(z)\mathrm{d}z \right\} \tag{14.17}$$
$$- \int_{L}^{L} E_z^{inc}(z) f_j(z)\mathrm{d}z; \qquad j = 1, 2, \ldots, n$$

Eqn (14.17) represents the weak Galerkin formulation of the integral eqn (14.10).

The resulting matrix equation system that follows from discretization of (14.17) is given by:

$$\sum_{i=1}^{M} [Z]_{ji}\{I\}_i = \{V\}_j, \quad \text{and} \quad j = 1, 2, \ldots, M \tag{14.18}$$

where $[Z]_{ji}$ is the submatrix representing the interaction of the ith source boundary element to the jth observation element, that is:

$$[Z]_{ji} = -\frac{1}{4j\pi\omega\varepsilon}\left(\int_{\Delta l_j} \{D\}_j \int_{\Delta l_i} \{D'\}_i^T g_E(z, z')\mathrm{d}z'\mathrm{d}z + k^2 \int_{\Delta l_j} \{f\}_j \int_{\Delta l_i} \{f\}_i^T g_E(z, z')\mathrm{d}z'\mathrm{d}z\right) + \int_{\Delta l_j} Z_L(z)\{f\}_j\{f\}_i^T \mathrm{d}z \tag{14.19}$$

The vector $\{I\}$ contains the unknown coefficients of the solution, and it represents the local voltage vector. Matrices $\{f\}$ and $\{f'\}$ contain shape functions while $\{D\}$ and $\{D'\}$ contain their derivatives, where M is the total number of segments, and Δl_i, Δl_j are the lengths of ith and jth segment, respectively.

The functions fk(x) are required to be differentiable at least once. A convenient choice for the shape functions is the family of Lagrange's polynomials given by:

$$L_i(z) = \prod_{j=1}^{m} \frac{z - z_j}{z_i - z_j}, j \neq i \tag{14.20}$$

Lagrange's polynomials provide accurate results and are easily integrable in closed form. The local right-side vector $\{V\}_j$ for jth observation boundary element can be written as

$$\{V\}_j = \int_{\Delta l_j} E_z^{inc}\{f\}_j \mathrm{d}z \tag{14.21}$$

and represents the local voltage vector.

The linear approximation is used since it was shown that this choice provides accurate and stable results [3]. The evaluation of the right-hand side vector can be performed in closed form if a constant incident electric field is assumed along the wire, that is:

$$E_z^{inc} = E_0 \tag{14.22}$$

Thus, using the linear basis functions it follows that:

$$V_{ij} = \int_{-\Delta l/2}^{\Delta l/2} E_0 \frac{z_{j+1} - z}{\Delta l} dz = E_0 \frac{\Delta l}{2} \tag{14.23}$$

$$V_{2j} = \int_{\Delta l/2}^{\Delta l/2} E_0 \frac{z - z_j}{\Delta l_g} dz = E_0 \frac{\Delta l}{2} \tag{14.24}$$

It is worth mentioning that the excitation function in the form of an incident field E_0 is usually obtained from the measurement.

14.3.1.2 Evaluation of Internal Currents and Fields In the *ELF* region ($f - 50/60$ Hz), the cylinder representing the human body can be considered as a conducting medium whose impedance per unit length is [3]:

$$Z_L(z) = \frac{1}{a^2 \pi \sigma} + Z_c \tag{14.25}$$

where $Z_c = 1/j\omega C$, and C is the capacitance between the soles of the feet and their image on the earth. If the foot-soles are bare and in direct contact with the moist earth (well-grounded body) then the capacity can be neglected, that is: $Z_c = 0$.

Once the axial current is computed from the numerical solution, it is possible to calculate the induced current density, that is:

$$J_z(z) = \frac{I_z(z)}{a^2 \pi} \tag{14.26}$$

as well as the induced electric field:

$$E_z(z) = \frac{J_z(z)}{\sigma} \tag{14.27}$$

The macroscopic average electric field discussed so far could be subsequently used for the calculation of the corresponding local electric fields induced in different organs [7], [8].

14.3.1.3 Numerical Results for Cylindrical Body Model The first example deals with a well-grounded human being exposed to the 60Hz overhead power line electric field. The configuration of a power line used in many parts of the world including USA, is shown in Figure 14.2.

The conductors are spaced horizontally at a distance $s = 3$ m from each other located at a height $d = 15$ m from the ground [16]. The person is standing at a certain point y from the reference origin facing the positive y direction.

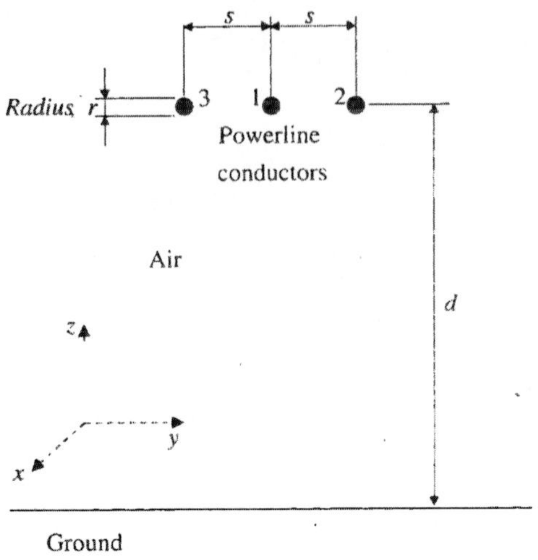

Figure 14.2 Geometry of a three-wire power line.

Although the body is exposed to all six components of the field; it has been shown [3] that $E_{0z}(0, y, z)$, the component parallel to the upright cylindrical body axis, is the largest within the practical ranges of body height z and distance y from the origin. In this case the component of the electric field tangential to the body is assumed to be $E = 10\,\text{kV/m}$.

The foot current results obtained via BEM are compared with other methods in Table 14.3.

The current density induced by the external field in individual organs inside the body may be estimated from the axial current distribution at various heights. At a particular body cross section with the skin effect negligible as shown earlier, the total current results from the sum of all currents flowing in the tissues, which are proportional to their conductivities, for example, $\sigma = 0.01\,\text{S/m}$ for bone and $0.6\,\text{S/m}$ for muscle.

The axial current distribution along the body for this case is shown in Figure 14.3. The results obtained using boundary elements are compared with results obtained by solving 2D surface electric field integral equation using the Method of Moments MoM [16].

At the low power frequency of 60 Hz, the physical dimensions of the body are electrically very small, resulting in the MoM solver computation error [16]. The computation has to be performed at double precision at a higher frequency of 30 kHz and scaled to 60 Hz. Similarly, the finite difference time domain (FDTD)

TABLE 14.3 Foot current on antenna model of the human body $E_z^{inc} = 10\,\text{kV/m}$, 60 Hz

Foot Current	BEM	FDTD [9]	Analytical [1,7]
$I_z(0)$, (μA)	191	179	183

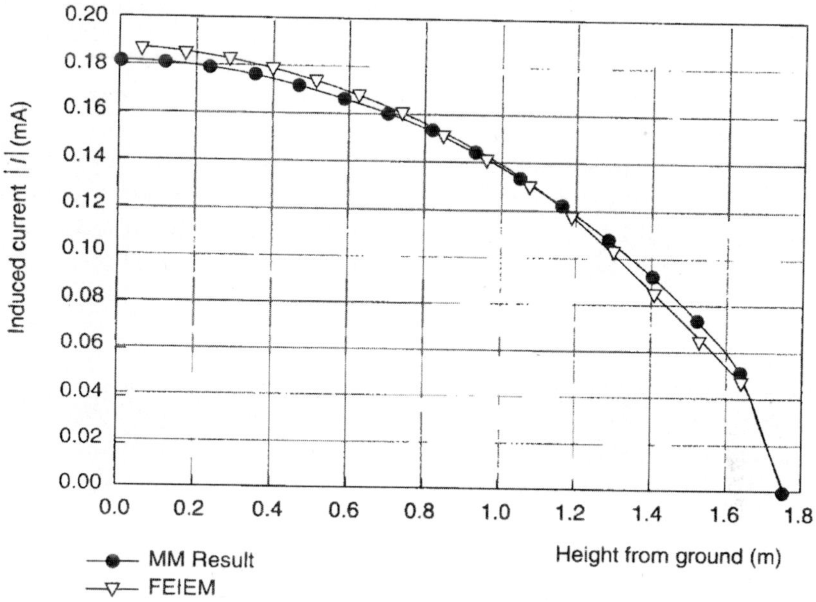

Figure 14.3 Current induced by incident field of 10 kV/m, 60 Hz.

method result used for the comparison above is also performed at several MHz and scaled. Moreover, the FDTD uses 45,024 number of 1.37 cm cubical cells for modeling which is computationally very expensive.

This clearly demonstrates the efficiency of BEM (FEIEM) which uses only 31 linear elements for the entire cylinder discretization. Knowing the current distribution, one can easily determine the current density and the electric field induced in the body. The results are presented in Table 14.4.

The basic restrictions for ELF exposure according to the *International Commission on NonIonizing Radiation Protection* (ICNIRP Guidelines) [2] are given in terms of permissible values for the induced current density inside the body.

It has to be underlined that the obtained value of 3.1 mA/m^2 is less than the limit of 10 mA/m^2 for occupational exposure, but exceeds the limit of 2 mA/m^2 for general public exposure.

14.3.2 Realistic Modeling of the Human Body

Unlike the simplified cylindrical body representation [13] and the computationally very expensive domain methods such as the finite element method (FEM) [17] and

TABLE 14.4 ELF exposure parameters

$I_z(0)$, (μA)	$J_z(0)$ (mA/m^2)	E_z (mV/m)
191	3.1	6.2

finite difference time domain (FDTD) method [9], [18] the boundary element method (BEM) variant has been recently applied to different body models for ELF exposures [19]-[22]. This efficient BEM scheme is more sophisticated approximation than FDTD and at the same time computationally less expensive than FEM, as only the domain boundary has to be discretized.

The formulation is based on the quasistatic approximation of the ELF electric field and the related equation of continuity. The general continuity equation is simplified to the Laplace equation form for the electric scalar potential which is numerically handled via BEM with the domain decomposition concept [23].

Knowing the scalar potential along the body, the induced current density inside the body is determined.

A multidomain inhomogeneous body of revolution and an anatomically based representation of the human body are presented in this work. The mathematical model is based on the quasistatic approximation and the related Laplace equation version of the continuity equation. The quasistatic approximation can be used since the dimension of the body model is electrically short, that is, it is rather small compared to the wavelength of the impressed field.

14.3.2.1 Theoretical Background The continuity equation is usually written as follows:

$$\nabla \vec{J} + \frac{\partial \rho}{\partial t} = 0 \tag{14.28}$$

where $\vec{J}$ is the current density and ρ represents the volume charge density.

The induced current density is expressed in terms of the scalar potential, in accordance to the differential form of Ohm's Law:

$$\vec{J} = -\sigma \nabla \varphi \tag{14.29}$$

where σ is the conductivity of the medium.

Furthermore, the volume charge density ρ and scalar potential φ are related through the Poisson equation:

$$\nabla(\varepsilon \nabla \varphi) = -\rho \tag{14.30}$$

where ε is the corresponding permittivity of the medium.

Combining eqns (14.28)–(14.30), for the case of time-harmonic exposures at extremely low frequencies the equation of continuity transforms into the Laplace equation:

$$\nabla[(\sigma + j\omega\varepsilon)\nabla \varphi] = 0 \tag{14.31}$$

where $\omega = 2\pi f$ is the operating frequency.

In the ELF range all organs behave as good conductors, while the surrounding air is a lossless dielectric medium. Solving the Laplace eqn (14.31) in the body region, the induced current density can be obtained from Ohm's Law (14.29).

The condition for the tangential component of the electric field in the vicinity of the two-media interface is given by:

$$\vec{n} \times (\vec{E}_b - \vec{E}_a) = 0 \tag{14.32}$$

where $\vec{n}$ denotes the unit normal to the interface, while quantities $\vec{E}_a$ and $\vec{E}_b$ represent the fields in the air and in the human body, respectively.

Expressing the electric field in terms of scalar potential it follows:

$$\vec{n} \times (\nabla\varphi_b - \nabla\varphi_a) = 0 \tag{14.33}$$

The interface condition for the normal component of the induced current density near the body-air surface is given by:

$$\vec{n} \cdot \vec{J} = -j\omega\rho_s \tag{14.34}$$

where ρ_s is the surface charge density.

Substituting eqn (14.29) into (14.30) yields:

$$\sigma_b \vec{n} \nabla\varphi_b = -j\omega\rho_s \tag{14.35}$$

where σ_b is the corresponding tissue conductivity, and φ_b is the scalar potential at the body surface.

The interface condition for the normal component of the electric flux density at the air-body surface is:

$$\vec{n} \cdot \vec{D} = \rho_s \tag{14.36}$$

Representing the electric flux density in terms of scalar potential yields:

$$\varepsilon_0 \vec{n} \nabla\varphi_a = \rho_s \tag{14.37}$$

where φ_a denotes the potential in the air in close proximity of the body.

At extremely low frequencies dielectric properties can be neglected that is $\sigma \gg \omega\varepsilon$, that is all organs behave as good conductors. The Laplace equation in the body simplifies into:

$$\nabla(\sigma\nabla\varphi) = 0 \tag{14.38}$$

The corresponding electrical conductivities for various tissue types are given in Table 14.5.

A simplified cylindrical body model is already shown in Figure 14.2 within the framework of the wire antenna theory, and the same geometry is used within the quasistatic approximation. The multidomain body of revolution representation of human being promoted in [17] and also used in [20] consists of nine portions in total, as shown in Figure 14.4. The dimensions are given in centimeters.

TABLE 14.5 Tissue conductivities

Tissue	σ [S/m]
Brain	0.12
Eye	0.11
Heart	0.11
Liver	0.13
Kidney	0.16
Intestine	0.16
Muscle	0.5

The body is located between the parallel plate electrodes and in the middle of the lower one. A calculation domain with the related boundary conditions is shown in Figure 14.5.

The lower plate electrode is assumed to be at zero potential while the upper plate electrode is assumed to be at a given potential of a high voltage power line.

The anatomically based representations of the human body are shown in Figure 14.6 from which the main features and dimensions are visible.

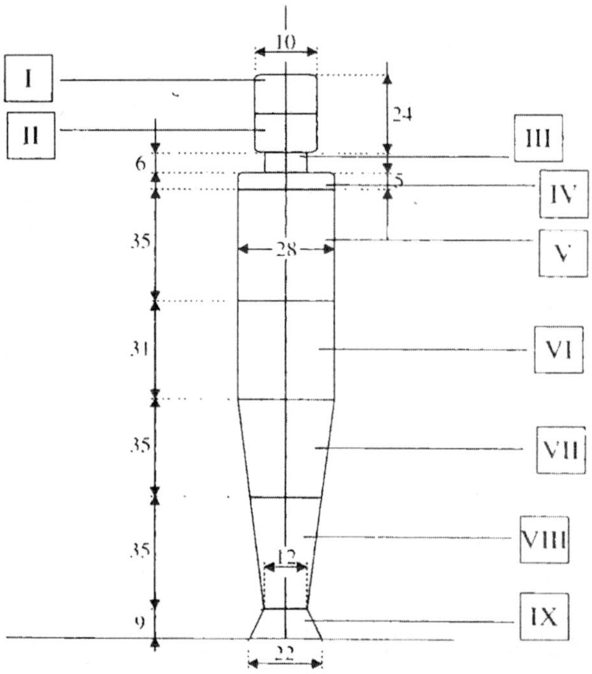

Figure 14.4 Body of revolution representation of human being. (D.Poljak, C.Gonzales, A.Peratta, Assessment of Human Exposure to Extremely Low Frequency Electric Fields using Different Body Models and the Boundary Element Analysis, // *ICECom 2005, Conference Proceedings, 18th International Conference on Applied Electromagnetics and Communications.* 2005. 109–112).

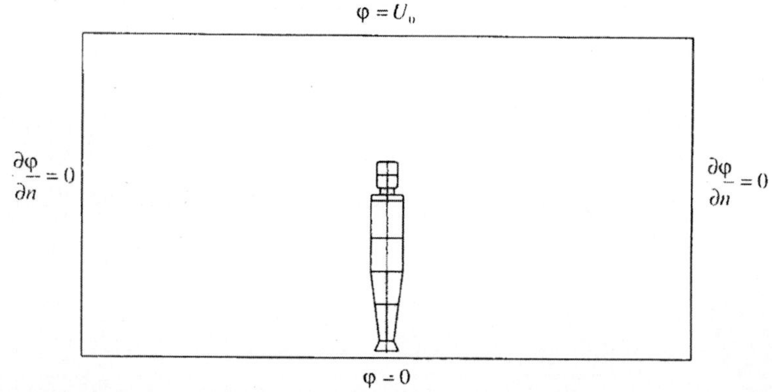

Figure 14.5 Calculation domain with the specified boundary conditions. (D.Poljak, C. Gonzales, A.Peratta, Assessment of Human Exposure to Extremely Low Frequency Electric Fields using Different Body Models and the Boundary Element Analysis. // *ICECom 2005, Conference Proceedings, 18th International Conference on Applied Electromagnetics and Communications*. 2005. 109–112).

Both models, one without arms (Fig. 14.6a) and the other with the arms up (Fig. 14.6b and 14.6c), are analyzed. The prescribed boundary conditions are identical to the ones used in the case of multidomain body of revolution representation of human being. These body models have been analyzed in [20, 21].

14.3.2.2 ***Boundary Element Method of Solution*** The Laplace equation is solved by means of the Boundary Element Method (BEM) with domain decomposition [23] which is outlined in this section. The fundamentals of the method are presented in Chapter 3.

Using Green's theorem for scalar functions, the following integral representation of eqn (14.38) for a subdomain is obtained:

$$c(\xi)\varphi(\xi) + \int_{\Gamma_k} \varphi \frac{\partial \varphi^*}{\partial n} \mathrm{d}\Gamma = \int_{\Gamma_k} \frac{\partial \varphi}{\partial n} \varphi^* \mathrm{d}\Gamma \tag{14.39}$$

where φ^* is the three dimensional fundamental solution of Laplace equation, $\partial\varphi^*/\partial n$ is the derivative in normal direction to the boundary, and $c(\xi)$ is the geometrically dependent term by which the Cauchy type singularity is taken into account. Discretization of eqn (14.39) with N_k elements results in the following expression:

$$c_i\varphi_i + \sum_{j=1}^{N_k} \int_{\Gamma_{kj}} \varphi \frac{\partial \varphi^*}{\partial n} \mathrm{d}\Gamma = \sum_{j=1}^{N_k} \int_{\Gamma_{kj}} \frac{\partial \varphi}{\partial n} \varphi^* \mathrm{d}\Gamma \tag{14.40}$$

where i stands for the source point and Γ_{kj}, represents the jth boundary element of Ω_k.

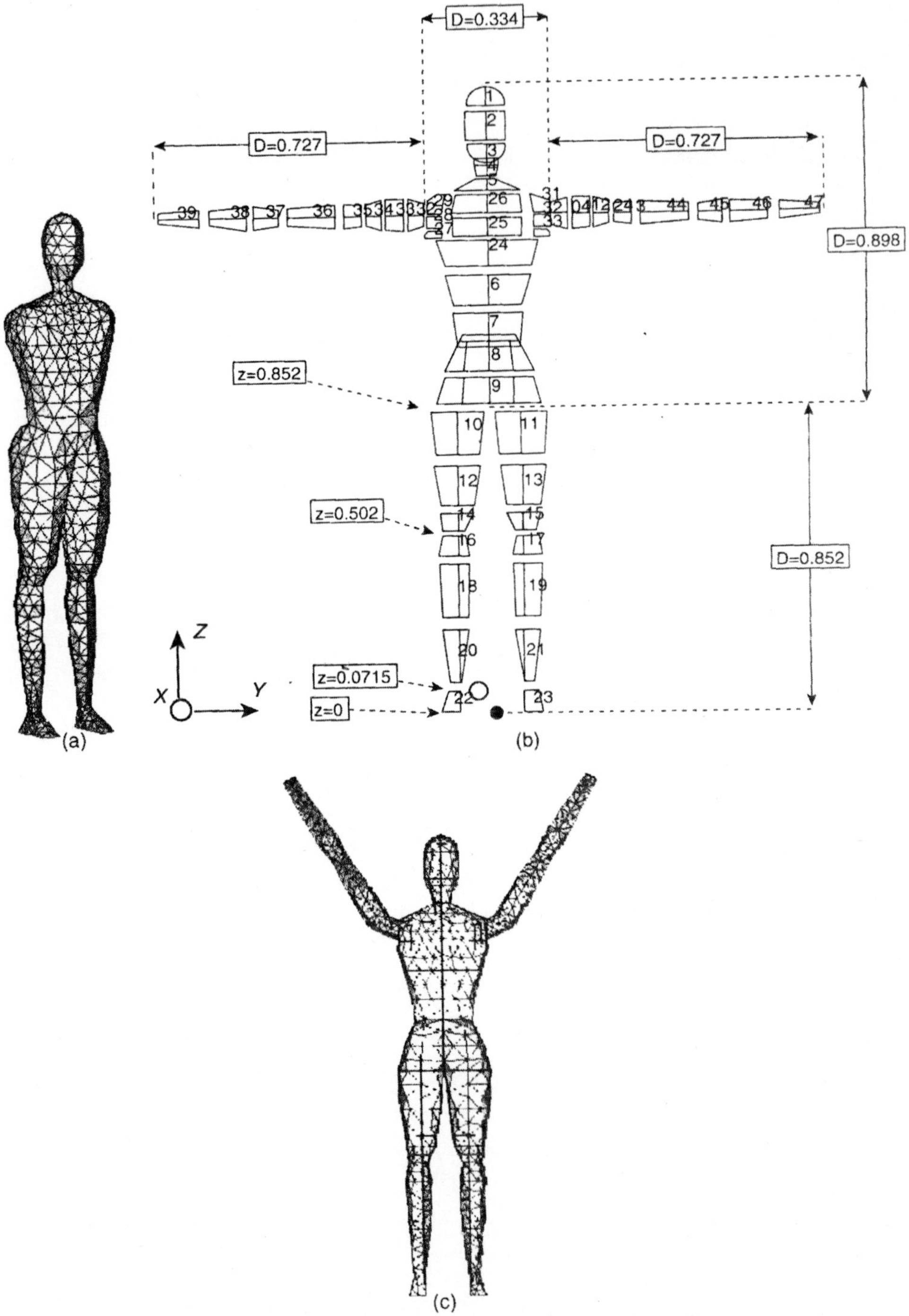

Figure 14.6 Meshes and Dimensions for the realistic human body models.

The present implementation of BEM is based on the isoparametric approach with quadratic interpolation functions defined for triangular elements.

The potential, or its normal derivative, at any point of the jth boundary element can be written as a linear combination of their corresponding values at the collocation nodes n and the interpolation functions f_n:

$$\varphi(\xi) = \sum_{a-1}^{6} f_n(\xi)\phi_n \quad \text{and} \quad \frac{\partial\varphi(\xi)}{\partial n} = \sum_{a=1}^{6} f_n(\xi)\phi_n \qquad (14.41)$$

The ξ is the dimensionless coordinate spanning from the computational square domain to the triangular element.

Combining eqns (14.40) and (14.41), the following system of equations for each subdomain is obtained:

$$H\varphi - G\frac{\partial\varphi}{\partial n} = 0 \qquad (14.42)$$

where H and G are matrices defined by:

$$H_{ij}^{n} = \int_{\Gamma_{k,j}} f_n \left(\frac{\partial\varphi^*}{\partial n}\right)_j \mathrm{d}\Gamma \qquad (14.43)$$

$$G_{ij}^{n} = \int_{\Gamma_{k,j}} f_n \varphi^* \mathrm{d}\Gamma \qquad (14.44)$$

while n stands for the collocation nodes inside the jth observation element.

The individual system of equations arising from each subdomain is to be assembled.

14.3.2.3 Numerical Results for Realistic Models The first numerical example deals with the 175 cm tall body of revolution model of human being exposed to the 10 kVm^{-1}/60 Hz overhead power line electric field. The height of the power line is 10 m above the ground.

The distribution of the induced current densities inside the body is shown in Figure 14.7. It can be clearly observed that the induced current density values increase at narrow sections such as ankle and neck.

Table 14.6 shows the comparison of current densities computed via BEM with FEM results and with experimental results [17]. BEM results agree satisfactorily with the both.

Further examples deal with the anatomically based body models. The calculation domain for the case of such a realistic body representation is shown in Figure 14.8. Distances are given in meters.

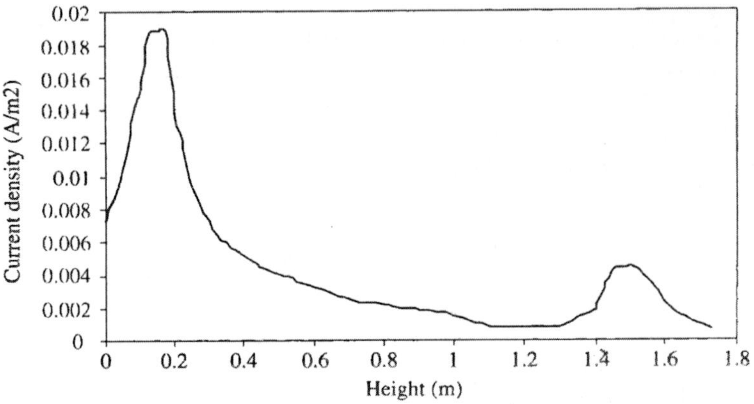

Figure 14.7 The current density distribution. (D.Poljak, C.Gonzales, A.Peratta, Assessment of Human Exposure to Extremely Low Frequency Electric Fields using Different Body Models and the Boundary Element Analysis, // *ICECom 2005, Conference Proceedings, 18th International Conference on Applied Electromagnetics and Communications*, 2005, 109–112).

TABLE 14.6 Comparison between the BEM, FEM and experimental results expressed in [mA/m^2]

Part of the body [17]	BEM	FEM [17]	Experimental
Neck	4.52	4.62	4.66
Pelvis	2.32	2.27	2.25
Ankle	18.91	19.16	18.66

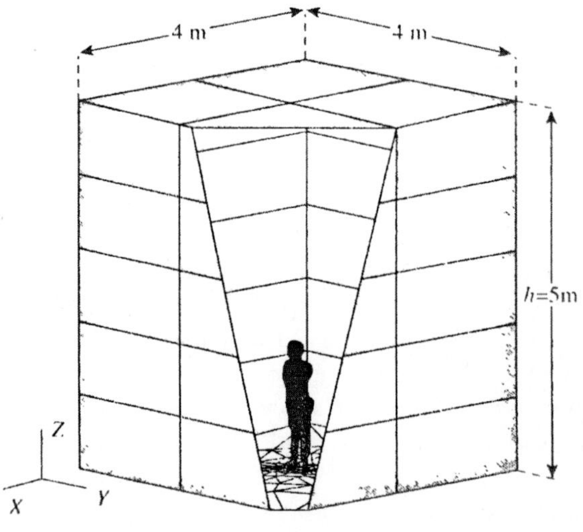

Figure 14.8 A plan view of the integration domain.

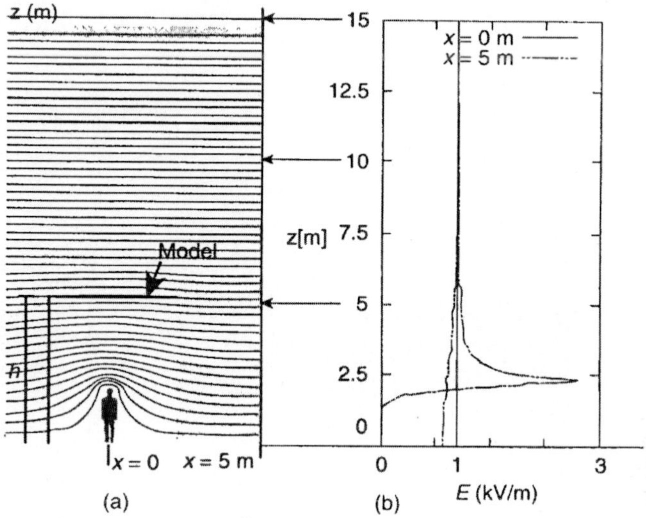

Figure 14.9 Electric field in the air near the human body.

Figure 14.9 shows the scalar potential and electric field distribution in the air around the human body. The computational domain is reduced to the parallelepiped (5 m × 4 m).

The corresponding current density distribution for the realistic human body model without arms is shown in Figure 14.10. The presence of peaks in current density values again, as in the body of revolution model, corresponds to the position of the ankle and the neck.

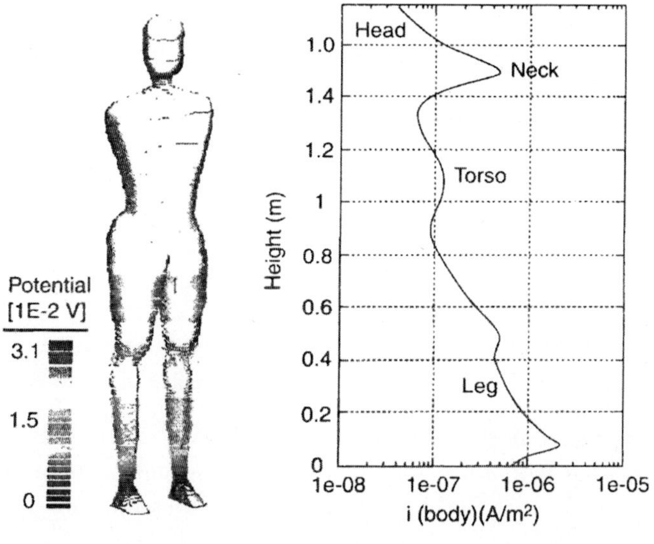

Figure 14.10 Distribution of the internal current density.

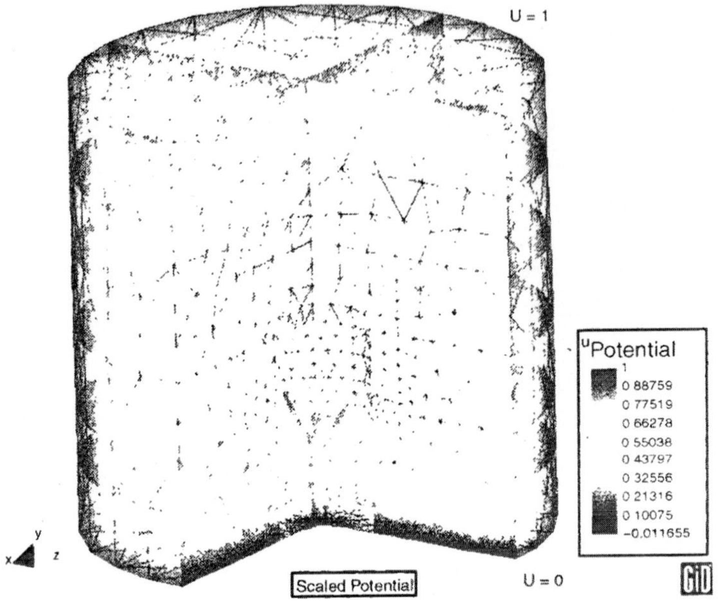

Figure 14.11 Meshing for the realistic model of the body with arms up.

The current density obtained by the realistic model is compared with the one calculated with the simplified cylindrical representation of the human body. It is evident that oversimplified cylindrical body model is unable to capture the current density peaks in the regions with narrow cross section. Yet, the cylindrical approximation is at least useful for determining the order of average value of current density in the torso.

The meshing of the realistic body model with the arms up is shown in Figure 14.11, while the corresponding scalar potential distribution around the body is presented in Figure 14.12. The number of elements used for the discretization is nearly 20,000.

A comparison of the current density values obtained via different body models is shown in Figure 14.13.

It is visible that when the arms are extended upwards the peak decreases, thus causing the screening effect.

Table 14.7 shows the peak value of the current density induced in the neck for some typical values of the electric field under power lines near the ground level.

Analyzing the obtained numerical results, the following conclusions can be drawn:

Wherever a reduction of the cross section of the human body exists, there is a significant increase of the current density, that is the peaks occur in neck and pelvis.

The arms extended upward cause a screening of the electric field from the top, thus reducing the peak of current density in the neck.

Oversimplified cylindrical representation of the human body suffers from the inability to capture the effect of high current density values in regions of reduced cross section.

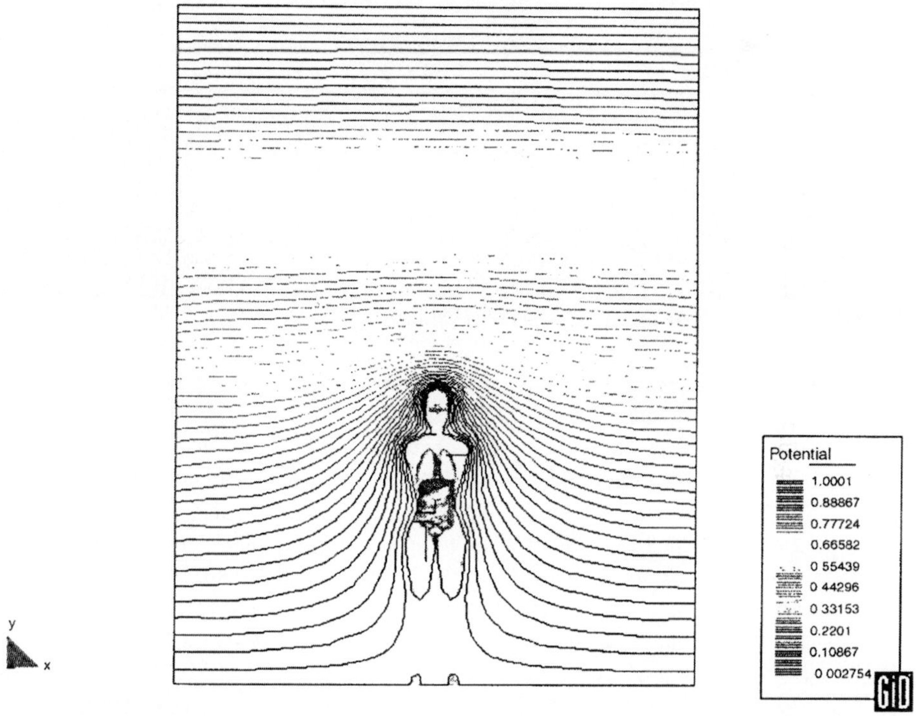

Figure 14.12 Scalar potential distribution in the vicinity of the human body.

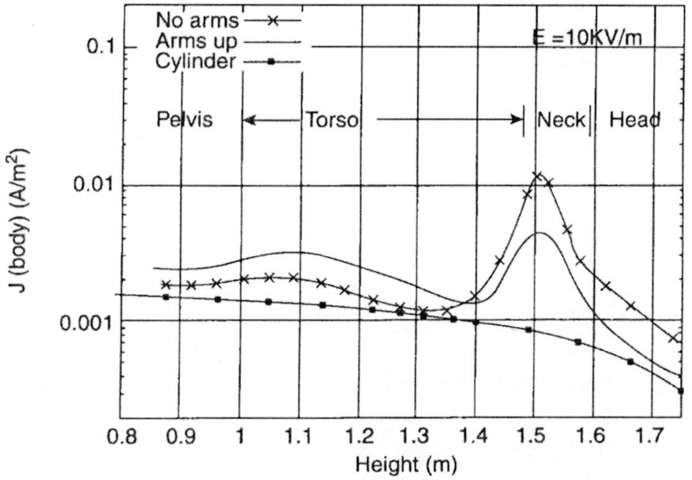

Figure 14.13 Induced current density distribution for the various body models.

TABLE 14.7 Peak values of the current density in the neck versus electric fields

E [kV/m]	j_z[mA/m2]
1	2
5	10
10	19

14.4 EXPOSURE OF HUMANS TO TRANSIENT RADIATION: CYLINDRICAL MODEL OF THE HUMAN BODY

Consequences of human exposure to transient electromagnetic fields are induced transient currents and fields inside the body.

The analysis of the human body transient response plays the key role in understanding the interaction of humans with transient electromagnetic radiation [24–27].

A direct time-domain simulation of the human body exposed to a transient electromagnetic field can be performed either by using realistic, anatomically based models [24–26] or by using the simplified human equivalent antenna model [27]. The coupling of transient electromagnetic fields to anatomically based body models has been usually analyzed via the finite-difference time-domain (FDTD) method [24].

The human equivalent antenna model has been designed for experimental dosimetry, and is valid within the frequency range from 50 to 110 MHz [28]. It is worth noting that the dimensions of the human equivalent antenna are within the thin wire approximation and the effective frequency bandwidth of the pulsed electromagnetic waveforms corresponds to the frequency range of the human equivalent antenna model.

The time domain human equivalent antenna model has been based on the Hallen integral equation and on dimensions available from [28]. A solution of this integral equation using the time domain Galerkin-Bubnov scheme of the boundary element method (GB-BEM) has been presented in [27].

As an extension of the work reported in [27], some convenient measures for postprocessing the human body transient response have been proposed in [29] and [30]. An extension of the work is reported in [27].

Further to the specific absorption (SA) concept for quantifying transient exposures, commonly used within the bioelectromagnetic community [24], [30] additional measures of the body transient response promoted in [29] are given in terms of *average* and *root-mean square* value of a time-varying current, *instantaneous power* dissipated in the body, and *total absorbed energy* by the body. It is worth noting that in [24] and [26], the total energy absorbed by the corresponding body model has been expressed in terms of fields, while the analysis presented in [29] uses the circuit theory representations.

After determining the transient response of the human body by solving the Hallen equation [27], one can readily calculate the distribution of the average and root mean square values of the space-time varying current flowing through the body as a measure of the transient behavior of the induced current.

Electromagnetic energy absorbed by the body can be quantified using the circuit-theory quantities in terms of instantaneous power and total absorbed energy.

It is worth noting that the main feature of the proposed formulation, when compared to more complex realistic models, is its simplicity and efficiency in getting the rapid estimation of the physical phenomena.

14.4.1 Time Domain Model of the Human Body

The time-domain analysis of the transient electromagnetic field that illuminates the well-grounded human body standing vertically on the perfectly conducting ground, as shown in Figure 14.1., is based on a cylindrical representation of the body, in accordance with the human equivalent antenna concept [27, 28]. Technical details of the human equivalent antenna are available in [28].

The dimensions of the human equivalent antenna ($L = 1.8\,\mathrm{m}$, $a = 5\,\mathrm{cm}$) are within the thin wire approximation and the effective bandwidth of the EMP frequency spectrum is 5MHz. This bandwidth is also within the frequency range of the human equivalent antenna.

The space-time dependent current along the equivalent cylinder representing the human body is governed by the time domain Hallen integral equation for loaded wires:

$$\int_0^L \frac{I(z', t - R/c)}{4\pi R} \mathrm{d}z' = F_0\left(t - \frac{z}{c}\right) + F_L\left(t - \frac{L - z}{c}\right)$$
$$+ \frac{1}{2Z_0} \int_0^L E_z^{inc}\left(z', t - \frac{|z - z'|}{c}\right) \mathrm{d}z' - \frac{1}{2Z_0} \int_0^L Z_L(z') I\left(z', t - \frac{|z - z'|}{c}\right) \mathrm{d}z' \tag{14.45}$$

where $I(z', t - R/c)$ is the unknown current to be determined, E_z^{inc} is the incident field, c is the velocity of light, R is the distance from the source point in the cylinder axis to the observation point at the cylinder surface, Z_0 is the wave impedance of a free space, and the unknown signals F0(t) and FL(t) are related with the multiple reflections of the current wave from the wire ends.

The resistance per unit length of the antenna length R_L derived in [27] is given by:

$$R_L = \frac{1}{a^2 \pi \sigma} \tag{14.46}$$

The average conductivity σ of the human body is assumed to be 0.5 S/m (conductivity at 60 Hz) since the predominant component of the induced transient current are at ELF frequencies.

Full details on the derivation of this equation have been carried out in Chapter 9.

Transient current induced in the human body because of any transient field can be obtained by solving the time domain Hallen integral eqn (14.45) via the space-time domain Galerkin-Bubnov scheme of the boundary element method GB-BEM, as it is presented in Chapter 9.

14.4.2 Measures of the Transient Response

Once the transient current flowing through the human body is obtained, it is possible to calculate certain parameters providing a measure of this body transient response.

The convenient parameters evaluating the human body transient response (arising from the basic theory of electric circuits), in addition to specific absorption concept are the spatial distribution of *average* and rms values of a space-time varying current induced in the body, *instantaneous power* dissipated in the body, and *total absorbed energy* in the body.

14.4.2.1 Average Value of the Transient Current The average value of the transient current is associated with a DC component in the spectrum of a particular transient waveform. The average value of a time varying current $i(t)$ is defined as follows:

$$I_{av} = \frac{1}{T_0}\int_0^{T_0} i(t)\mathrm{d}t \tag{14.47}$$

where T_0 is the time interval of interest.

The distribution of average values of current is simply given by:

$$I_{av}(x) = \frac{1}{T_0}\int_0^{T_0} i(x,t)\mathrm{d}t \tag{14.48}$$

Obviously, for any waveform having approximately equal area above and below an abscissa the average value tends to be equal to zero. Therefore, from this parameter one can obtain a quick estimation of the character of the given transient waveform properties.

When the space-time varying current along the cylinder at each space node z_i and time instant t_k is determined by solving the integral eqn (14.45), the *average value* of this current can be computed from the following relation:

$$I_{av} = \frac{1}{T_0}\sum_{k=1}^{N_t}\int_{t_k}^{t_{k+1}} (\{T\}^T\{I\})_i\mathrm{d}t \tag{14.49}$$

Performing of a straight-forward integration yields the following:

$$I_{av}\big|_{z_i} = \frac{\Delta t}{2T_0}\sum_{k=1}^{N_t}[(I_i^k) + (I_i^{k+1})] \tag{14.50}$$

where $\{T\}$ is the vector containing the time-domain linear shape functions, and $\{I\}$ denotes the vector containing time-dependent values of current.

14.4.2.2 ***Root-Mean-Square Value of the Transient Current*** A time-varying current delivers an average power to the resistive load. The amount of delivered power strongly depends on the particular waveform.

A measure of comparing the power delivered by different waveforms is the root-mean-square (rms) or effective value of a transient current. The rms value of a time-varying current is a constant that is equal to the direct current value that would deliver the same average power to a given resistance R_L and that would produce the same heating effect on R_L.

Instantaneous power delivered to a resistance R_L by a transient current $i(t)$ is:

$$p(t) = R_L i^2(t) \tag{14.51}$$

while the corresponding average power P_{av} is determined by the following integral relation:

$$P_{av} = \frac{1}{T_0}\int_0^{T_0} p(t)\mathrm{d}t = \frac{1}{T_0}\int_0^{T_0} R_L i^2(t)\mathrm{d}t = R_L I_{rms}^2 \tag{14.52}$$

from which the rms current is as follows:

$$I_{rms} = \sqrt{\frac{1}{T_0}\int_0^{T_0} i^2(t)\mathrm{d}t} \tag{14.53}$$

Consequently, the spatial distribution of the rms values of the space time current along the cylinder is given by:

$$I_{rms}(x) = \sqrt{\frac{1}{T_0}\int_0^{T_0} i^2(x,t)\mathrm{d}t} \tag{14.54}$$

When the current along the wire at each node and time instant is determined, the rms value of the wire current can be computed from the following relation:

$$I_{rms} = \sqrt{\frac{1}{T_0}\sum_{k=1}^{N_t}\int_{t_k}^{t_{k+1}}(\{T\}^T\{I\})_i^2 \mathrm{d}t} \tag{14.55}$$

and by performing a straightforward integration it follows:

$$I_{rms}|_{z_i} = \sqrt{\frac{\Delta t}{3T_0}\sum_{k=1}^{N_t}[(I_i^k)^2 + I_i^k I_i^{k+1} + (I_i^{k+1})^2]} \tag{14.56}$$

Therefore, the rms value of the transient current flowing through the body appears to be a more interesting parameter from the bioelectromagnetics point of view as it is, by its definition, directly associated with the thermal effect of a time-varying current flowing through a lossy material.

14.4.2.3 Instantaneous Power Instantaneous power delivered to a certain resistance R_L or to some resistive medium having equivalent resistance R_l by a transient current is defined by the relation (14.51). On the contrary, the absorbed power in the human body expressed by the field quantities is equivalent to the concept of instantaneous power arising from the circuit theory and it is usually defined as a volume integral over power density, that is:

$$p_{rad}(t) = \int_V \sigma|\vec{E}(\vec{r},t)|^2 \mathrm{d}V = \int_V \frac{|\vec{J}(\vec{r},t)|^2}{\sigma}\mathrm{d}V \tag{14.57}$$

Assuming the transient current to be approximately constant over the cylinder cross-section yields:

$$i(z,t) = J(z,t)\cdot S \tag{14.58}$$

where $J(z,t)$ denotes the current density along the cylinder, and $S = a^2\pi$ is the cylinder cross-section.

The instantaneous power can be obtained by spatially integrating the squared space-time varying current:

$$p_{rad}(t) = \frac{1}{\sigma\cdot S}\int_0^L i^2(z,t)\mathrm{d}z \tag{14.59}$$

Knowing the transient induced current the instantaneous power can be represented by the following relation:

$$P_{rad}(t) = \frac{1}{\sigma \cdot S} \sum_{i=1}^{M} \int_{\Delta l_i}^{L} (\{f\}_i^T \{I\}_i)^2 dz \tag{14.60}$$

where $\{f\}$ denotes the vector containing linear interpolation functions and $\{I\}$ stands for the induced current coefficients.

Performing of straightforward integration yields:

$$P_{rad}|_{t_k} = \frac{1}{\sigma \cdot S} \frac{\Delta z}{3} \sum_{i=1}^{M} [(I_i^k)^2 + I_i^k I_{i+1}^k + (I_{i+1}^k)^2] \tag{14.61}$$

where M denotes the total number of spatial segments along the cylinder.

14.4.2.4 Total Absorbed Energy In accordance to the circuit theory, the total absorbed energy in the resistance or resistive material can be obtained by temporally integrating the instantaneous power:

$$W_{tot}(t) = \int_0^t P_{rad}(t) dt \tag{14.62}$$

Using the numerical representation of the instantaneous power it follows:

$$W_{tot}(t) = \sum_{k=1}^{N_t} \int_{t_k}^{t_k+\Delta t} \{T\}_k^T \{P\}_k dt \tag{14.63}$$

and performing the straightforward integration gives:

$$W_{tot}|_{t_k} = \frac{\Delta t}{2} \sum_{k=1}^{N_t} [(P^k + P^{k+1})] \tag{14.64}$$

where P_k and P^{k+1} stand for a discrete value of instantaneous power dissipated in the resistive medium at a time instant t_k and t_{k+1}, respectively.

14.4.2.5 The Specific Absorption Once the transient current flowing through the human body is calculated, the specific absorption, a principal measure of this human body transient response, can also be determined.

According to the general definition, the already mentioned total energy absorbed by the resistive material can be obtained by temporally integrating the radiation power $p_{rad}(t)$:

$$W_{tot} = \int_0^{T_0} p_{rad}(t)\mathrm{d}t \tag{14.65}$$

where T_0 is the interval of interest.

The power dissipated in the human body is usually represented as a volume integral over power density $\bar{P}_d$:

$$p_{rad}(t) = \int_V \bar{P}_d \mathrm{d}V \tag{14.66}$$

where $\bar{P}_d$ is given by:

$$\bar{P}_d = \sigma|\vec{E}(\vec{r},t)|^2 = \frac{|\vec{J}(\vec{r},t)|^2}{\sigma} \tag{14.67}$$

while $\vec{E}(\vec{r},t)$ and $\vec{J}(\vec{r},t)$ represent the electric field and current density induced inside the body, respectively.

Combining the eqns (14.65) to (14.67) yields:

$$W_{tot} = \int_0^{T_0}\int_V \sigma|\vec{E}(\vec{r},t)|^2 \mathrm{d}V\mathrm{d}t = \int_0^{T_0}\int_V \frac{|\vec{J}(\vec{r},t)|^2}{\sigma}\mathrm{d}V\mathrm{d}t \tag{14.68}$$

Specific absorption (*SA*) is defined as a quotient of the incremental energy (d*W*) absorbed by an incremental mass (d*m*) contained in the volume (d*V*) of a given density (ρ):

$$SA = \frac{\mathrm{d}W}{\mathrm{d}m} = \frac{\mathrm{d}W}{\rho \mathrm{d}V} \tag{14.69}$$

Substituting expression (14.68) into (14.69) results in:

$$SA = \int_0^{T_0} \frac{1}{\rho}\frac{d}{\mathrm{d}V}\left(\int_V \frac{|\vec{J}(\vec{r},t)|^2}{\sigma}\mathrm{d}V\right)\mathrm{d}t \tag{14.70}$$

rearranging the eqn (14.70) it follows:

$$SA = \int_0^{T_0} \frac{1}{\rho}\frac{|\vec{J}(\vec{r},t)|^2}{\sigma}\mathrm{d}t \tag{14.71}$$

Assuming the transient current distribution to be approximately constant over the cylinder cross-section gives:

$$i(z,t) = J(z,t) \cdot S = J(z,t) \cdot a^2\pi \tag{14.72}$$

where $J(z,t)$ denotes the space-time varying current density along the cylindrical body.

Finally, the SA can be expressed as:

$$SA(z) = \frac{1}{\rho\sigma S^2} \int_0^{T_0} i^2(z,t)\mathrm{d}t = \frac{1}{\rho\sigma(a^2\pi)^2} \int_0^{T_0} i^2(z,t)\mathrm{d}t \tag{14.73}$$

It is worth pointing out that the specific absorption defined with relation (14.73) and the spatial distribution of root mean square values of transient current is given by:

$$I_{rms}^2(z) = \frac{1}{T_0} \int_0^{T_0} i^2(z,t)\mathrm{d}t \tag{14.74}$$

and are simply related as follows:

$$SA(z) = \frac{T_0}{\rho\sigma S^2} I_{rms}^2(z) \tag{14.75}$$

Knowing the solution for induced current using the boundary element formalism the SA can be represented by the following relation:

$$SA = \frac{1}{\rho\sigma(a^2\pi)^2} \frac{\Delta t}{3} \sum_{k=1}^{N_t} \int_{t_k}^{t_k+\Delta t} \{T\}_k^T \{I\}_k \mathrm{d}t \tag{14.76}$$

where vector $\{T\}$ contains the linear temporal interpolation functions while vector $\{I\}$ represents the values of transient current distribution.

Performing the straightforward integration leads to a simple formula:

$$SA = \frac{1}{\rho\sigma(a^2\pi)^2} \frac{\Delta t}{3} \sum_{k=1}^{N_t} [(I_i^k)^2 + I_i^k I_{i+1}^k + (I_{i+1}^k)^2] \tag{14.77}$$

where N_t denotes the total number of time increments. and I_i^k denotes the ith node current at kth time instant.

14.4.2.6 Numerical Results Incident field is assumed to be vertically polarized in order to analyze the maximum coupling to the body. Figure 14.14 shows the transient current induced in the feet due to the Gaussian pulse exposure:

$$E_z^{inc}(t) = E_0 e^{-g^2(t-t_0)^2} \tag{14.78}$$

with following parameters: $E_0 = 1\,\text{v/m}$, $g = 2 \times 10^9\,\text{s}^{-1}$, and $t_0 = 2\,\text{ns}$.
As this pulse is a numerical equivalent of Dirac pulse, the obtained transient exposure can be regarded as an impulse response of the human body. Transient response to any other incident waveform can then be obtained by performing a simple convolution [25], [26].

While the spatial distribution of the average values of the space-time varying current flowing through the cylindrical body model is in the case of Gaussian pulse nearly zero, more information regarding the heating effect due to the Gaussian pulse exposure can be obtained from the spatial distribution of the rms values of the transient current, shown in Figure 14.15. What can be observed from Figure 14.15 is that Gaussian pulse that induces the peak value of current around 1.5 mA in the feet corresponds to the equivalent DC current that is equal to approximately 0.4 mA. In other words, the transient current induced in the feet shown in Figure 14.14 would produce the same heating effect as the constant DC current of 0.4 mA.

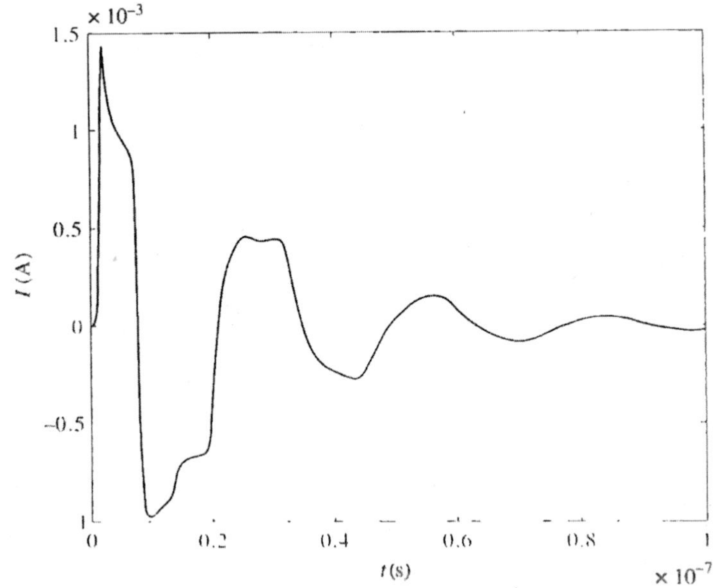

Figure 14.14 Transient current induced in the feet due to the Gaussian pulse exposure. (Poljak, D., Postprocessing of the Human Body Response to Transient Electromagnetic Fields, in J.A.Kong (Ed) Electromagnetic Waves, PIER 49, Chapter 12, pp. 219–238, 2004).

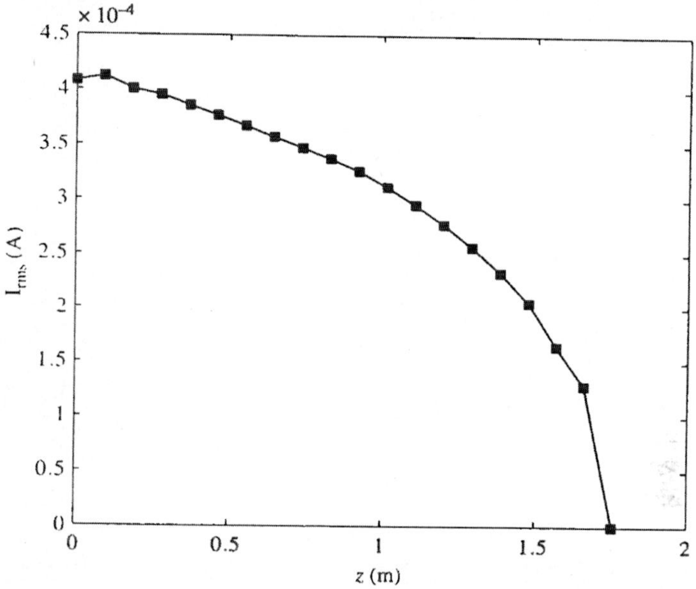

Figure 14.15 Spatial distribution of rms values of the transient current due to the Gaussian pulse exposure. (Poljak, D., Postprocessing of the Human Body Response to Transient Electromagnetic Fields, in J.A.Kong (Ed) Electromagnetic Waves, PIER 49, Chapter 12, pp. 219–238, 2004).

Furthermore, the transient behavior of the instantaneous power dissipated in the body is shown in Figure 14.16. It can be noticed that the power dissipation, with the peak value slightly above 2.5 mW occurs in the early time within the first 50 ns.

The same conclusion can be drawn from the Figure 14.17 representing the total energy absorbed in the body vs time. It is clearly visible that body does not absorb any significant amount of energy after first 50 ns.

Therefore, a very little energy is absorbed by the human body exposed to pulsed field. This occurs because of a very limited time duration (up to 100 ns) of the incident pulse for which significant values of current can be induced in the body. It is worth noting that very little energy is absorbed by the human body due to an exposure to transient fields of rather limited time duration.

$$E_z^{inc}(t) = u(t) \tag{14.79}$$

where $u(t)$ denotes the unit step, and standard double-exponential electromagnetic pulse (EMP) waveform:

$$E_z^{inc}(t) = E_0(e^{-at} - e^{-bt}) \tag{14.80}$$

where $E_0 = 1.05\,\mathrm{Vm}^{-1}$, $a = 4 \times 10^6\,\mathrm{s}^{-1}$, $b = 4.76 \times 10^8\,\mathrm{s}^{-1}$.

It is visible that the highest value of absorption is achieved in the case of EMP exposure, while the lowest value is reached in the case of Gaussian pulse exposure.

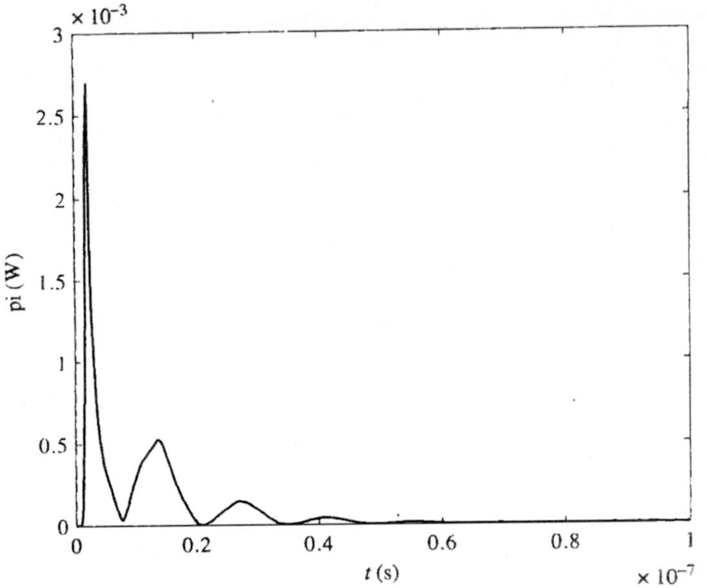

Figure 14.16 Instantaneous power dissipated within the body due to the Gaussian pulse exposure. (Poljak, D., Postprocessing of the Human Body Response to Transient Electromagnetic Fields, in J.A.Kong (Ed) Electromagnetic Waves, PIER 49, Chapter 12, pp. 219–238, 2004).

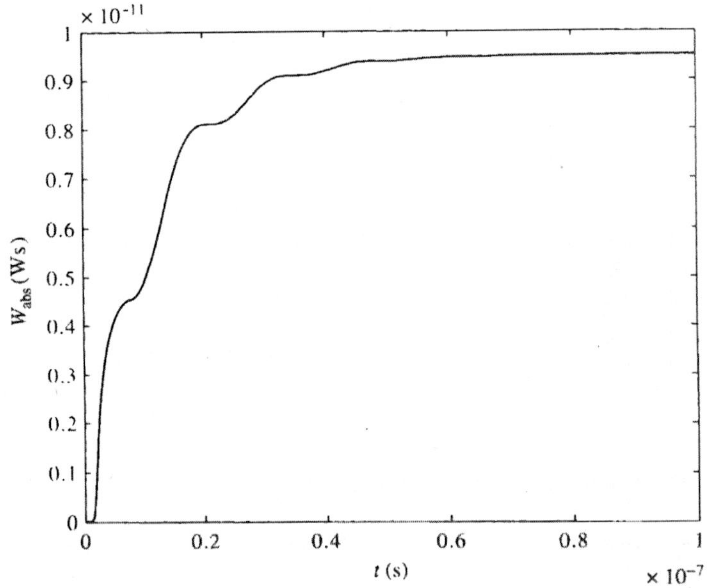

Figure 14.17 Total energy absorbed in the body due to the Gaussian pulse exposure. (Poljak, D., Postprocessing of the Human Body Response to Transient Electromagnetic Fields, in J.A.Kong (Ed) Electromagnetic Waves, PIER 49, Chapter 12, pp. 219–238, 2004).

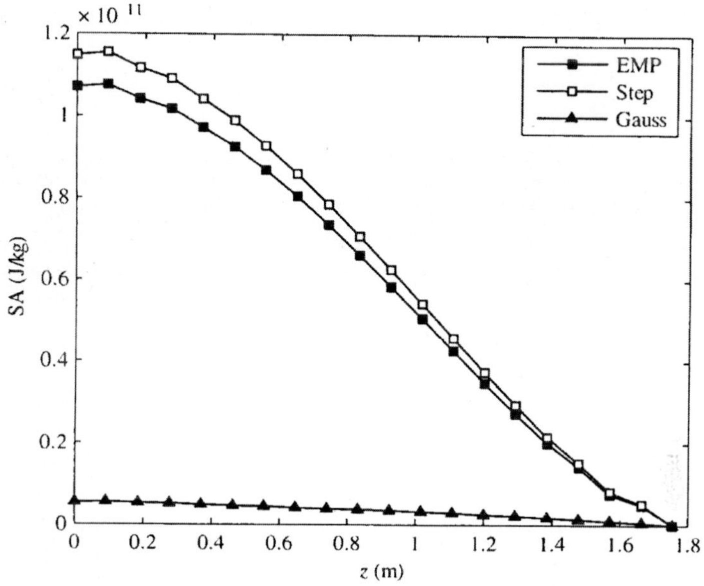

Figure 14.18 Specific absorption for various transient exposure waveforms.

The maximal values of SA (below 12 pJ/kg), compared to the limits of 28.8 J/kg for whole body average, and 576 J/kg for 1-g spatial peak per pulse proposed by an IEEE Standard, stay far below the given threshold. It is worth mentioning that these SA results are obtained for the case of unit pulse amplitudes.

However, even by scaling the results with realistic values of pulse amplitudes corresponding to the order of magnitude from 50 kV/m to 100 kV/m, and also by taking into account very thin radius of antenna body model, the computed SAs, though considered rather underestimated, still remain negligible.

14.5 REFERENCES

[1] R. W. Y. Habash: Electromagnetic Fields and Radiation, *Marcel Dekker, New York*, 2002.

[2] International Nonionizing Radiation Committee of the International Radiation Protection Association: Guidelines for Limiting Exposure to Time-Varying Electric, Magnetic, and Electromagnetic Fields (up to 300 GHz), *Health Phys., 74 (4)*, 1998, pp 494–522.

[3] D. Poljak: Human Exposure to electromagnetic Fields, *WIT Press, Southampton-Boston*, 2003.

[4] Q. Balzano, A. Faraone: Human Exposure to Cellular Base Station Antennas, IEEE International Symposium on Electromagnetic Compatibility, Seattle, Washington, USA, August 2–6 1999, pp 924–927.

[5] P. Bernardi, M. Cavagnaro, S. Pisa, E. Piuzzi: Human Exposure in the Vicinity of Radio Base Station Antennas, 4th European Symposium on Electromagnetic Compatibility, *EMC Europe 2000*, Brugge, Belgium, September 11–15 2000, pp 187–192.

[6] P. Bernardi, M. Cavagnaro, S. Pisa, E. Piuzzi: Human Exposure to Radio Base Station Antennas in Urban Environment, *IEEE Trans. MTT, 48 (11)*, November 2000, pp 1996–2002.

[7] R. W. P. King, S.S. Sandler: Electric Fields and Currents Induced in Organs of the Human Body When Exposed to ELF and VLF Electromagnetic Fields, *Radio Sci., 31*, September–October 1996, pp 1153–1167.

[8] R. W. P King: Fields and Currents in the Organs of the Human Body When Exposed to Power Lines and VLF Transmitters, *IEEE Trans. Biomedical Eng., 45 (4)*, April 1998, pp 520–530.

[9] O. P. Gandhi, J. Y. Chen: Numerical Dosimetry at Power Line Frequencies Using Anatomically Based Models, *Bioelectromagn. Suppl., 1*, 1992, pp. 43–60.

[10] O. P. Gandhi: Some Numerical Methods for Dosimetry: Extremely Low Frequencies to Microwave Frequencies, *Radio Sci. 30 (1)*, January–February 1995, pp 161–177.

[11] D. Poljak, V., Roje: Currents Induced in Human Body Exposed to the Power Line Electromagnetic Field, Proceedings of the 20th Annual Conference of the IEEE Engineering in Medicine and Biology Soc., Hong-Kong, 29. October-1. November 1998, pp 3281–3284.

[12] D. Poljak, C. Y. Tham, V. Roje, T. Zemunik: Parasitic Cylindrical Antenna Representation of the Human Body Exposed to the Low Frequency (LF) Electromagnetic Radiation, Millennium Conference on Antennas & Propagation, Davos, Switzerland, 9–14 April 2000.

[13] D. Poljak, Y. Rashed: The Boundary Element Modeling of the Human Body exposed to the ELF Electromagnetic Fields, *Engineering Analysis with Boundary Elements, 26*, 2002, pp 871–875.

[14] D. Poljak, A. Sarolic, V. Roje: Human Interaction with the Electromagnetic Field radiated from a Cellular Base Station Antennas, Proceedings of the IEEE International Symposium on EMC, EMCEUROPE 2002, Sorrento, Italy, 2002, pp 965–968.

[15] R. W. P. King: Electric Current and Electric Field Induced in the Human Body When Exposed to an Incident Electric Field Near the Resonant Frequency, *IEEE Trans. Microwave Theory and Tech., 48* (9), September 2000, pp 1537–1543.

[16] D. Poljak, C. Y. Tham: Integral Equation Techniques in Electromagnetics, *WIT Press, Southampton-Boston*, 2003.

[17] A. Chiba, K. Isaka, Y. Yokoi, M. Nagata, M. Kitagav, T. Matsuo: Application of Finite Element Method to Analysis of Induced Current Densities Inside Human Model Exposed to 60 Hz Electric Field, *IEEE Trans. Power Apparatus Syst. 103* (7), July 1984, pp 1895–1901.

[18] T. W. Dawson, K. Caputa, M. A. Stuchly: High Resolution Organ Dosimetry for Human Exposure to Low Frequency Electric Fields, *IEEE Trans. Power Delivery, 13* (2), April 1998, pp 366–373.

[19] D. Poljak, C. A. Brebbia: Boundary Elements for Electrical Engineers, *WIT Press, Southampton-Boston*, 2005.

[20] D. Poljak, A. Peratta, C.A. Brebbia: A 3D BEM Modeling of Human Exposure to Extremely Low Frequency (ELF) Electric Fields, *Boundary Elements XXVII*, Orlando, USA, March 2005, pp 441–451.

[21] C. Gonzals, A. Peratta, D. Poljak: Current Density Induced in the Human Body due to Low Frequency Electric Fields using the Boundary Element Method, *SoftCOM 2005*, Split, September 2005.

[22] D. Poljak, C. Gonzales, A. Peratta: Assessment of Human Exposure to Extremely Low Frequency Electric Fields using Different Body Models and the Boundary Element Analysis, *ICECOM 2005*, Dubrovnik, October 2005.

[23] C. A. Brebbia, J. C. F. Telles, L. C. Wrobel: Boundary Elements Techniques, Theory and Applications in Engineering. *Springer-Verlag, Berlin, Heidelberg, New York and Tokyo*, 1984.

[24] J. Y. Chen, O. P. Gandhi: Currents Induced in an Anatomically Based Model of a Human for Exposure to Vertically Polarized Electromagnetic Pulses, *IEEE Trans. Micro. Theo. Tech., 39* (*1*), Jan. 1991., pp 31–39.

[25] J. Y. Chen, C. M. Furse, O. P. Gandhi: A Simple Convolution Procedure for Calculating Currents Induced in the Human Body for Exposure to Electromagnetic Pulses, *IEEE Trans. Micro. Theo. Tech., 42* (*7*), July 1994, pp 1172–1175.

[26] O. P. Gandhi, C. M. Furse: Currents Induced in the Human Body for Exposure to Ultrawideband Electromagnetic Pulses, *IEEE Trans. Electromag. Compat., 39* (*7*), July 1994, pp 1172–1175.

[27] D. Poljak, C. Y. Tham, O. P. Gandhi, A. Sarolic: Human Equivalent Antenna Model for Transient Electromagnetic Exposure, *IEEE Trans. EMC*, February 2003.

[28] O. P. Gandhi, E. Aslan: Human Equivalent Antenna for Electromagnetic Fields, *US Patent No. 5,394* (*164*), February 28, 1995.

[29] D. Poljak: Postprocessing of the Human Body Response to Transient Electromagnetic Fields, in *J. A. Kong (Ed): Electromagnetic Waves, PIER 49, Chapter 12*, 2004, pp 219–238.

[30] D. Poljak, A. Sarolic: Specific Absorption in the Human Body due to an Exposure to Transient Radiation, *Proc SoftCOM 2005*, Split, September 2005.

INDEX